Increased industrial and agricultural activity this century has led to vast quantities of the earth's soil and groundwater resources becoming contaminated with hazardous chemicals. Bioremediation provides a technology based on the use of living organisms, usually bacteria and fungi, to remove pollutants from soil and water, preferably *in situ*. This approach, which is potentially more cost-effective than traditional techniques such as incineration of soils and carbon filtration of water, requires an understanding of how organisms transform chemicals, how they survive in polluted environments and how they should be employed in the field. This book examines these issues for many of the most serious and common environmental contaminants, resulting in a volume which presents the most recent position on the application of bioremediation to the cleanup of polluted soil and water.

Biotechnology Research Series: 6
Series Editor: James Lynch

Bioremediation: Principles and Applications

Series Editor
James Lynch, School of Biological Sciences, University of Surrey, UK

Advisory Editors
Donald Crawford, Department of Microbiology, Molecular Biology and Biochemistry, University of Idaho, USA
Michael Daniels, Sainsbury Laboratory, John Innes Centre, Norwich, UK
Jonathan Knowles, Glaxo Wellcome Research and Development S.A., Geneva, Switzerland

Biotechnology Research
This series of contributed volumes provides authoritative reviews of selected topics in biotechnology aimed at advanced undergraduates, graduate students, university researchers and teachers, as well as those in industry and government research. The scope of the series is as broad-ranging as the applications of the research itself, encompassing those topics in applied science and engineering, ranging from agriculture to the design of pharmacueticals, which are not well served by the existing literature. Each volume therefore represents a significant and unique contribution to the literature, comprising chapters by the leading workers in the field who review the current status of the specialisms and point the way to future research and potential applications.

Titles in the series

1. Plant Protein Engineering
   Edited by P. R. Shewry and S. Gutteridge

2. Release of Genetically Engineered and Other Microorganisms
   Edited by J. C. Fry and M. J. Day

3. Transformation of Plants and Soil Microorganisms
   Edited by K. Wang, A. Herrera-Estrella and M. Van Montagu

4. Biological Control: Benefits and Risks
   Edited by H. M. T. Hokkanen and J. M. Lynch

5. Microbial Biofilms
   Edited by H. Lappin-Scott and J. Costerton

(Volumes 1–5 were originally published under the title Plant and Microbial Biotechnology Research Series.)

Bioremediation: Principles and Applications

Ronald L. Crawford and Don L. Crawford

University of Idaho, Moscow, Idaho, USA

CAMBRIDGE
UNIVERSITY PRESS

CAMBRIDGE UNIVERSITY PRESS
Cambridge, New York, Melbourne, Madrid, Cape Town, Singapore, São Paulo

Cambridge University Press
The Edinburgh Building, Cambridge CB2 2RU, UK

Published in the United States of America by Cambridge University Press, New York

www.cambridge.org
Information on this title: www.cambridge.org/9780521470414

First published 1996
Reprinted 1998
This digitally printed first paperback version 2005

A catalogue record for this publication is available from the British Library

Library of Congress Cataloguing in Publication data

Bioremediation: principles and applications / R.L. Crawford and D.L. Crawford.
   p.   cm. – (Plant and microbial biotechnology research series; 5)
Includes index.
ISBN 0-521-47041-2 (hardcover)
1. Bioremediation.   I. Crawford, Ronald L., 1947–   .
II. Crawford, Don L.   III. Series.
TD192.5.B5573   1996
628.5′2–dc20   95-51193 CIP

ISBN-13 978-0-521-47041-4 hardback
ISBN-10 0-521-47041-2 hardback

ISBN-13 978-0-521-01915-6 paperback
ISBN-10 0-521-01915-X paperback

Contents

Contents

Contributors

W. Admassu   Department of Chemical Engineering, University of Idaho, Moscow, ID 83844-1021, USA

M. E. Caldwell   Department of Botany and Microbiology, The University of Oklahoma, Norman, OK 73019, USA

C. E. Cerniglia   National Center for Toxicological Research, Division of Microbiology, 3900 NCTR Road, HFT-250, Jefferson, AR 72079, USA

D. L. Crawford   Department of Microbiology, Molecular Biology, and Biochemistry, University of Idaho, Moscow, ID 83844–3052, USA

R. L. Crawford   Center for Hazardous Waste Remediation Research, University of Idaho, Moscow, ID 83844-1052, USA

S. C. Francesconi   US Environmental Protection Agency, Gulf Ecology Division, Gulf Breeze, FL 32561, USA

S. B. Funk   Department of Microbiology, Molecular Biology and Biochemistry, University of Idaho, Moscow, ID 83844-3052, USA

R. A. Korus   Department of Chemical Engineering, University of Idaho, Moscow, ID 83844-1021, USA

L. Krumholz   Department of Botany and Microbiology, The University of Oklahoma, Norman, OK 73019, USA

E. S. Melin   Nordic Envicon Oy, Kanslerinkatu 8, FIN-33720 Tampere, Finland

R. M. Miller   Soil and Water Science Department, The University of Arizona, Tucson, AZ 85721, USA

Contributors

M. J. Morra   Department of Plant Soil and Entomological Sciences, University of Idaho, Moscow, ID 83844-2339, USA

J. G. Mueller   U.S. EPA Environmental Research Laboratory, 1 Sabine Island Drive, Gulf Breeze, FL 32561, USA

L. I. Pepper   Soil and Water Science Department, The University of Arizona, Tucson, AZ 85721, USA

P. H. Pritchard   U.S. EPA Environmental Research Laboratory, 1 Sabine Island Drive, Gulf Breeze, FL 32561, USA

J. A. Puhakka   Institute of Water and Environmental Engineering, Tampere University of Technology, FIN-33101 Tampere, Finland

T. M. Roane   Soil and Water Science Department, The University of Arizona, Tucson, AZ 85721, USA

E. Z. Ron   Department of Molecular Microbiology and Biotechnology, Tel Aviv University, Ramat Aviv, Israel

E. Rosenberg   Department of Molecular Microbiology and Biotechnology, Tel Aviv University, Ramat Aviv, Israel

M. S. Shields   The Center for Environmental Diagnostics and Bioremediation, The University of West Florida, Pensacola, FL 32514-5751, USA

J. M. Suflita   Department of Botany and Microbiology, The University of Oklahoma, Norman, OK 73019, USA

R. Unterman   Envirogen, Inc., 4100 Quakerbridge Road, Lawrenceville, NJ 08648, USA

L. P. Wackett   Department of Biochemistry & Institute for Advanced Studies in Biological Process Technology, University of Minnesota, St Paul, MN 55108, USA

Preface

Bioremediation is not a new concept in the field of applied microbiology. Microorganisms have been used to remove organic matter and toxic chemicals from domestic and manufacturing waste effluents for many years. What is new, over the past few decades, is the emergence and expansion of bioremediation as an industry, and its acceptance as an effective, economically viable alternative for cleaning soils, surface water, and groundwater contaminated with a wide range of toxic, often recalcitrant, chemicals. Bioremediation is becoming the technology of choice for the remediation of many contaminated environments, particularly sites contaminated with petroleum hydrocarbons.

Bioremediation has also become an intensive area for research and development in academia, government, and industry. Partly because of new laws requiring stricter protection of the environment and mandating the cleanup of contaminated sites, funding for both basic and applied research on bioremediation by government agencies, as well as by private industry, has increased dramatically over the past decade. As a result, rapid progress has been made in developing effective, economical microbial bioremediation processes. In an even broader sense, this increased activity has led to a surge of interest in 'environmental microbiology', a field covering a spectrum of disciplines, including microbial physiology and ecology, molecular genetics, organic chemistry, biochemistry, soil and water chemistry, geology, hydrology, and engineering. In the academic research environment, bioremediation has, in effect, become so scientifically broad and complex in both its basic and applied aspects that it has of necessity evolved into a multidisciplinary field that requires a 'research center' approach. At the University of Idaho, as just one example, we have built our multidisciplinary environmental remediation research program within the University of Idaho Center for Hazardous Waste Remediation Research. While difficult to quantify, the total amount of research funding now being devoted to bioremediation is quite substantial. A recent estimate suggests that in 1997, industry will spend over $4 billion on designing, engineering, constructing, and equipping wastewater systems (*HazTech News*, May 11, 1995, p. 72).

It is a challenge for researchers to stay abreast of this rapidly advancing field, particularly in light of the diversity of environments and contaminants, as well as the varied approaches to bioremediation that have emerged in recent years. It is even more challenging to predict where the field is headed in the future, an important consideration for staying on the leading edge technologically. Indeed, perhaps even more than in other areas of scientific endeavor, it is easier to become a follower than it is to be a leader in research in bioremediation. Many scientists jump on a specific bandwagon and embrace their favorite approach to bioremediation – the 'intrinsic', the genetic engineering, the *in situ* approach. As a result, much repetitive research takes place, at considerable cost. The bioremediation field needs more scientists who understand the broader implications of various approaches, and who can work with colleagues in a multidisciplinary environment to advance both the basic and applied science of bioremediation.

In this book, we have attempted in several ways to broaden the perspectives of students and scientists working on bioremediation. First, since we wanted to review the principal topics relevant to bioremediation of contaminated soils, sediments, surface waters, and aquifers, we asked each author to provide, in effect, a status report on current research and development in his or her topic area. However, in such a rapidly changing field, even the best reviews become historic within a few years. Therefore, we also asked our authors to offer some thoughts on what knowledge is lacking within their specific research areas, and to comment on where future research emphasis should be placed. Finally, we asked them to be forward looking and offer some conjectures about the potential utility of bioremediation in their specific area of expertise. We think that our authors have accomplished these goals, producing a book that will be particularly useful to teachers and thesis advisors, as well as offering a predictive, paradigm-challenging resource for bioremediation researchers. We hope that the words of our authors will help generate new ideas and approaches to research and development in the field of bioremediation.

We appreciate the hard and thorough work of the authors of the chapters in this book. Thanks for a job well done. Also, we thank Ms. Connie Bollinger at the University of Idaho for the considerable effort she put into preparing the manuscripts as they arrived. She helped us greatly by taking care of the necessary details before the manuscripts were submitted to the publisher. We thank the editors at Cambridge University Press who have been involved in this book – Dr. Robert M. Harington, who got the project underway; Dr. Alan Crowden, who gave us valuable advice on preparing the manuscripts, and Dr. Maria Murphy, who saw the book to completion. Finally, we would like to thank Professor James M. Lynch of the University of Surrey, who first convinced us that this book would be a worthwhile project. His encouragement is deeply appreciated.

Don L. Crawford   Ronald L. Crawford   June 1995

Introduction

Ronald L. Crawford

Most organic chemicals and many inorganic ones are subject to enzymatic attack through the activities of living organisms. Most of modern society's environmental pollutants are included among these chemicals, and the actions of enzymes on them are usually lumped under the term *biodegradation*. However, biodegradation can encompass many processes with drastically differing outcomes and consequences. For example, a xenobiotic pollutant might be mineralized, that is, converted to completely oxidized products like carbon dioxide, transformed to another compound that may be toxic or nontoxic, accumulated within an organism, or polymerized or otherwise bound to natural materials in soils, sediments, or waters. More than one of these processes may occur for a single pollutant at the same time. The chapters in this book will discuss these phenomena in relation to specific groups of xenobiotic pollutants, since these processes ultimately determine the success or failure of bioremediation technologies.

Bioremediation refers to the productive use of biodegradative processes to remove or detoxify pollutants that have found their way into the environment and threaten public health, usually as contaminants of soil, water, or sediments. Though biodegradation of wastes is a centuries-old technology, it is only in recent decades that serious attempts have been made to harness nature's biodegradative capabilities with the goal of large-scale technological applications for effective and affordable environmental restoration. This development has required a combination of basic laboratory research to identify and characterize promising biological processes, pilot-scale development and testing of new bioremediation technologies, their acceptance by regulators and the public, and, ultimately, field application of these processes to confirm that they are effective, safe, and predictable. Examples of these research and development strategies, both successful and unsuccessful, can be found in chapters dealing with specific types or classes of pollutants.

Levin and Gealt (1993, p. 4) estimated the costs of biotreatment of biodegradable contaminants in soils to range between $40 and $100 per cubic yard, as compared with costs as high as \$250–\$800 per cubic yard for incineration and \$150–\$250 per yard for landfilling. Considering that billions of dollars in cleanup costs may be saved when these rates are projected over the whole of the industrialized world, and also that incineration and landfilling are no longer an alternative for many wastes, it is no surprise that bioremediation is receiving so much attention from the scientific and regulatory communities.

Bioremediation can be applied to an environmental problem in a variety of ways. Litchfield (1991) listed five general approaches to bioremediation: aboveground bioreactors, solid phase treatment, composting, landfarming, and *in situ* treatment. These five types of bioprocess largely cover the variations among bioremediation procedures, though there is considerable diversity in technologies within any one area.

Aboveground bioreactors are used to treat liquids (e.g., industrial process streams, pumped groundwater), vapors (e.g., solvents vented from contaminated subsurface environments, factory air), or solids in a slurry phase (e.g., excavated soils, sludges, or sediments; plant materials). They may use suspended microorganisms or adsorbed biofilms, singly or in combination; native microbial populations indigenous to the material being treated; pure microbial cultures isolated from appropriate environments; or genetically engineered microorganisms (GEMs) designed specifically for the problem at hand. They may operate with or without additions of oxygen or other electron acceptors (nitrate, carbon dioxide, sulfate, oxidized metals) and nutrient feeds (nitrogen, phosphorus, trace minerals, co-substrates). Other variables that may be controlled or that must otherwise be accounted for may include ratio of solids to water, biodegradation rates for specific pollutants or mixtures of pollutants, adsorption/desorption of pollutants to matrices such as soil or carriers used to provide surfaces for biofilms, pH, temperature, and redox potential. Bioreactors, as compared with other bioremediation techniques, can be controlled and their processes modeled mathematically with great precision. There are many physical designs of bioreactors, some of them quite novel and designed specifically for bioremediation. Bioreactors may be used in treatment trains (e.g., sequencing batch reactors), with multiple designs being used concurrently. Chapter 1 will discuss some of these designs and their application to environmental cleanup. Other bioreactor technologies will be discussed in chapters dealing with specific pollutants.

Soils are often treated by solid-phase technologies. This usually means placing excavated soils within some type of containment system, e.g., a lined pit with leachate collection and/or volatile compound entrapment equipment, and then percolating water and nutrients through the pile. Oxygen may or may not be supplied, depending on the bioremedial process being encouraged or supported. Inocula may or may not be added, depending on whether or not an indigenous microbial population can be stimulated to remove the target pollutant(s). Solid-phase treatments are particularly useful for petroleum-contaminated soils, but they are constantly finding new uses. For example, fungal mycelia carried on materials such as wood chips have been incorporated into contaminated soils to promote biodegradation of xenobiotic contaminants, a process receiving considerable interest among bioremediation scientists (Lamar, 1990). A typical aerated static pile system for the bioremediation of petroleum-contaminated soil is shown in Figure 1.

Composting is a variation of solid-phase treatment that involves adding large amounts of readily degradable organic matter to a contaminated material, followed by incubations, usually aerobic, lasting several weeks or months. Adjustments of carbon: nitrogen ratios in composting systems require particular attention, and the

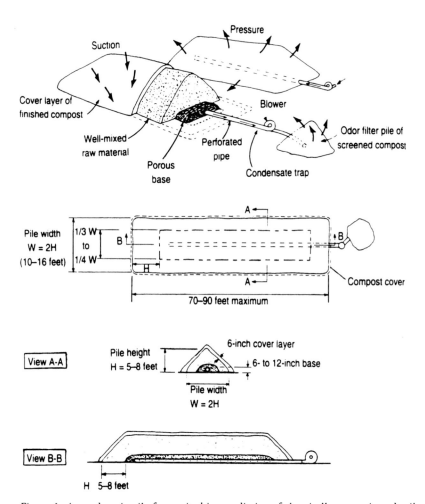

Figure 1. Aerated static pile for *ex situ* bioremediation of chemically contaminated soil. Adapted from Rynk (1992). Reprinted with permission from Northeast Regional Agricultural Engineering Service (NRAES), Cooperative Extension, Ithaca, New York.

need for frequent turning of the piles makes this a labor-intensive treatment technology. Special equipment may be needed for this task as well. Though this technology is well known as a means to convert waste organic matter (leaves, agricultural wastes, manures) to useful soil amendments, it only recently has been applied to the problem of bioremediation of hazardous compounds. Composting can be carried out at mesophilic (20–30 °C) or thermophilic (50–60 °C) temperatures. Compost piles typically contain aerobic, microaerophilic, and anaerobic microhabitats, promoting simultaneous growth of fungi, actinomycetes, and eubacteria. Thus, biodegradative processes in these systems can be very complex. For example, some xenobiotic compounds are polymerized into the organic compartments of composted soils (Williams, Ziegenfuss & Sisk, 1992). Composting is a technology whose true limits and effectiveness have not yet been established, but the technology shows considerable promise for some specific applications such as treatment of petroleum-contaminated soils.

In *landfarming*, contaminated soils, sludges, or sediments are spread on fields and cultivated in much the same manner as a farmer might plow and fertilize agricultural land. It has been used most commonly as an inexpensive and effective process for the treatment of petroleum-contaminated materials. Though simple in design, landfarming must be performed carefully to avoid creating a second, possibly larger, hazardous waste contamination problem, should the process fail. Clearly, many toxic materials should not be landfarmed. It is a technique to be used only for readily biodegradable chemicals. Even with easily degraded compounds there may be potential for leaching of contaminants to groundwater, so most regulatory agencies require that groundwater be deep below a landfarm site, or that there be some type of confining layer (natural clays) or barrier (reinforced liner) between the cultivated material and the subsurface water. A landfarm site must be managed for its moisture content, and fertilization with nitrogen and/or phosphorus may often be required. Landfarming and its variations, as applied to petroleum contaminants, will be discussed in detail in Chapter 4. A photograph of a typical landfarming operation is shown in Figure 2.

Since so many instances of environmental contamination involve soil, and since soil is a highly complex biological, chemical, and physical matrix, we have provided a full chapter (Chapter 2) on the influences of soil properties on bioremediation processes. These influences must be understood if bioremediation is ever to become a dependable technology.

In situ bioremediation is currently receiving a great deal of attention among bioremediation researchers. This attention is warranted because *in situ* processes, which would not require impossibly expensive excavations of vast amounts of contaminated vadose-zone soils, or unending pump-and-treat schemes for large, deep aquifers, might save vast amounts of money and potentially solve problems that are not approachable by off-the-shelf technologies. Most *in situ* processes involve the stimulation of indigenous microbial populations so that they become metabolically active and degrade the contaminant(s) of concern. The best examples,

Figure 2. Landfarming of petroleum-contaminated soil. A 10-acre site for treatment of about 20 000 cubic yards of excavated soil.

some from as early as the mid-1970s (Raymond, 1974), involve treatment of aquifers contaminated by petroleum hydrocarbons. Free product floating on the aquifer surface was first removed by pumping appropriately located wells. Nutrients (primarily nitrogen and phosphorus) were then injected and electron acceptors (usually oxygen, but sometimes nitrate) were supplied either by sparging wells (air) or through the addition of solutions of hydrogen peroxide or nitrate salts. Problems encountered during *in situ* stimulation of microbial populations include the plugging of wells and subsurface formations by the tremendous amounts of biomass that may be generated through microbial growth on hydrocarbons, difficulties in supplying sufficient oxygen to the subsurface, and the inability to move nutrients and electron acceptors to all regions of heterogeneous subsurface environments. Also, it is rarely possible to remove all free product, so reservoirs of slowly released contamination may be present for many years in some situations.

A process known as 'bioventing' is becoming an attractive option for promoting *in situ* biodegradation of readily biodegradable pollutants like petroleum hydrocarbons. Bioventing involves the forced movement of air (oxygen) through the vadose zone of contaminated sites (Hinchee, 1994). When oxygen is provided as a terminal electron acceptor, indigenous microorganisms multiply at the expense of the carbon present in the contaminating material. The goal is to provide sufficient oxygen to allow degradation of the pollutants to proceed to completion, without vaporizing contaminants to the surface. Air may be forced through the vadose zone either by injection through sparging wells, or by vacuum extraction through appropriately located infiltration and withdrawal wells (Figure 3). In soils that are impermeable (e.g., clay-rich soils) bioventing may not be possible since air cannot be moved through these soils at sufficient rates to supply microbial populations the electron acceptors required. Sometimes such soils can be 'opened' by fracturing with pressurized air or water, improving their permeability for air transport. In theory, bioventing should work for both volatile and non-volatile contaminants, and petroleum or non-petroleum compounds, as long as the contaminants are inherently biodegradable and an indigenous degrader population of microbes exists in the zone of contamination. If nitrogen is also limiting microbial growth in a contaminated vadose zone, it might be supplied by gaseous ammonia vapors along with the infiltrated air. Expect to see bioventing become a major technique in future bioremediation efforts worldwide.

Since petroleum and petroleum-derived products are the single most pervasive environmental contamination problem, we have provided two chapters directly related to hydrocarbon treatment. Chapter 4 specifically addresses petroleum, while Chapter 5 discusses polycyclic aromatic hydrocarbons, a particularly problematic component of petroleum and other fossil residues.

In situ bioremediation has also been proposed for the cleanup of aquifers contaminated by solvents such as trichloroethylene (TCE) and dichloroethylene (DCE). Compounds like TCE are usually not degraded as sole sources of carbon and

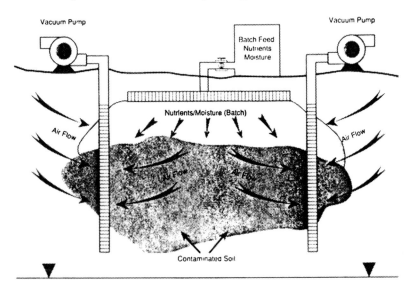

Figure 3. Common bioventing system for treatment of vadose-zone contaminants using oxygen as a terminal electron acceptor. Reprinted from *In Situ Bioremediation: When Does It Work?* Copyright 1993 by the National Academy of Sciences. Courtesy of the National Academy Press, Washington, DC.

energy for microbial growth. They are, however, degraded by processes generally called *cometabolism*. During cometabolic processes contaminants are degraded fortuitously by enzymes that microbes normally employ to degrade substrates that do provide carbon and energy. The primary example studied for its application to *in situ* bioremediation is cometabolism of halogenated solvents during microbial growth on methane. Methane-oxidizing bacteria produce a methane monooxygenase (MMO) that oxidizes methane to methanol using reducing power derived from cellular-reduced pyridine nucleotides (NADH). Certain forms of this enzyme show extraordinarily broad substrate specificities, oxidizing perhaps hundreds of substrates in the place of methane (Henry & Grbić-Galić, 1991), including compounds like TCE and DCE. Once oxidized by MMO, TCE, for example, decomposes to largely innocuous products. Through sparging a TCE-contaminated aquifer with an appropriate mixture of methane and oxygen, the growth of methane-oxidizing bacteria and the concomitant cometabolism to TCE can be stimulated. This practice has worked well in some model systems and in small, well-controlled aquifers (Semprini *et al.*, 1990) but has not yet shown great success in real-world situations. The outlook for this technology, however, is improving as more work is done to perfect the *in situ* techniques. Biodegradation of chlorinated aliphatic compounds will be discussed in depth in Chapter 9.

The phenomenon of cometabolic degradation of pollutants will undoubtedly be harnessed in the future for many compounds, including classes of chemicals beyond the chlorinated solvents. Chapter 7 will specifically discuss the poly-chlorinated biphenyls (PCBs), where cometabolic processes play a very important part in the biodegradation of multiple pollutant isomers. Work at the University of Idaho has shown that munitions compounds such as 2,4,6-trinitrotoluene (TNT) and hexahydro-1,3,5-trinitro-1,3,5-triazine (RDX), and herbicides such as 2-*sec*-butyl-4,6-dinitrophenol (dinoseb) are degraded to innocuous products, i.e., volatile organic acids, by consortia of microorganisms fermenting carbohydrates (Kaake *et al.*, 1992; Funk *et al.*, 1993; U.S. EPA Fact Sheet, March 1994). *Clostridium* spp. appear to be prime players in these processes (Regan & Crawford, 1994), in which the anaerobic bacteria get most of their energy from fermentable carbohydrates, simultaneously reducing and degrading the nitrated contaminants. It should be possible to reproduce this process *in situ* by injecting the appropriate nutrients (soluble starch or molasses) into nitro-compound-contaminated aquifers to stimulate growth and reductive activities of endogenous clostridia. If such clostridia are not present in sufficient numbers, they could be introduced in the form of spores. Because nitroaromatic compounds are one of the world's major environmental problems, Chapter 6 is provided as an entire chapter on their bioremediation.

Anaerobic processes are now known to be much more diverse in biodegradation of pollutants than was thought even a few years ago. Anaerobic bioremediation of nitro-substituted compounds, halogenated molecules, and even hydrocarbons now appears possible, employing electron acceptors such as nitrate, halogenated compounds themselves, carbon dioxide, sulfate, and oxidized metals such as iron. This rapidly moving field is summarized in Chapter 3. An important subclass of pollutants treatable both anaerobically and aerobically, the chlorinated phenols, is covered in Chapter 8 to illustrate one of the better model systems for bioremediation.

In another approach to *in situ* bioremediation, investigators have developed methods to encapsulate pure microbial cultures in small beads (5–10 μm diameter) that might be used as transport and survival vehicles for introducing unique microorganisms into aquifers (Stormo & Crawford, 1992, 1993). Preliminary work at the University of Idaho in near-surface, heterogeneous aquifers has shown that bacteria-loaded microspheres can be introduced into and transported within the subsurface. Problems still to be overcome include the cost of preparing large quantities of beads in the smallest size ranges, and developing dependable ways to assure good distribution of the beads throughout the aquifer or in the path of a contaminant plume. The U.S. EPA has developed a method to use hydrofracturing to introduce lenses of porous sands into subsurface soils (Vesper *et al.*, 1994). The addition of encapsulated bacteria and nutrients to such sand lenses to intercept pollutant plumes is a possible variation of the encapsulated microorganism theme. A photograph of a microencapsulated *Flavobacterium* that degrades pentachlorophenol (PCP) is shown in Figure 4.

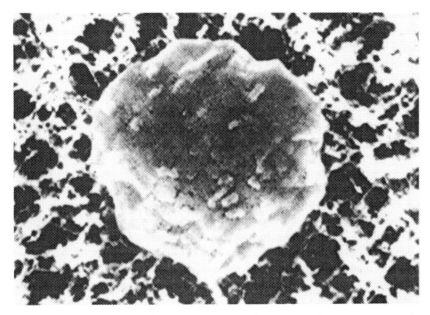

Figure 4. Microencapsulated *Flavobacterium* cells. The bead (10 μm diameter) is composed of alginate. Similar beads can be prepared for agar, polyurethane, and other polymeric materials. From Stormo & Crawford (1992). Reprinted with permission of the American Society for Microbiology.

Chapter 11 will discuss the rapidly developing potential for use of genetically engineered microorganisms (GEMs) by the bioremediation industry. Some very important environmental pollutants are not readily biodegradable by known biological processes; that is, they are not subject to attack by existing enzymes, and microorganisms apparently have never evolved the capability to degrade these structures. These recalcitrant chemicals may not provide sufficient energy for growth of microorganisms, may be too toxic to allow for growth and mutation over long periods of time, may not be inducers of appropriate enzymes, or may be so new to nature that evolution has not proceeded to the point of modifying existing pathways sufficiently to allow for enzymic attack on the novel structures. The modern tools of molecular biology allow for human intervention in this process of evolution. Genes from different organisms can now be cloned and reassembled under proper regulatory control into new pathways for biodegradation of previously nondegradable or highly recalcitrant compounds. These novel pathways can be placed in new hosts, from *Escherichia coli* and *Pseudomonas* to a variety of other microbial strains, depending on their intended uses. For example, several of the genes encoding enzymes for the complete biodegradation of pentachlorophenol (PCP) have been cloned, sequenced, and moved from a *Flavobacterium* into *Escherichia coli* and placed under the control of the *lac* operon promoter (Xun & Orser, 1991; Orser *et al.*, 1993a,b; Lange, 1994). The recombinant *E. coli* was shown to detoxify PCP faster than the original host of the pathway. In the original *Flavobacterium*, PCP is first oxidatively dechlorinated by a PCP-4-monooxygenase, encoded by the *pcpB* gene. Two subsequent dechlorinations are catalyzed by a glutathione-dependent reductive dehalogenase, encoded by the *pcpC* gene. The product of the *pcpA* gene may be a ring-fission oxygenase or an oxygenase component, but this remains to be established; the gene has been cloned and sequenced. The pathway also contains a *lysR* type regulatory gene and a gene encoding a reductase that probably functions with the hydroxylase. The recombinant *E. coli* strain that converts PCP to dichlorohydroquinone has been constructed, as shown in Figure 5 (Lange, 1994) by cloning *pcpB* and *pcpC* into the recombinant host. Efforts are underway to move the remaining PCP pathway genes into the recombinant. Such recombinant microorganisms will probably first see use in bioreactors, where it should be easier to contain their novel genotypes than it would be after direct release to soil or water. However, we should expect to see releases of GEMs directly to the environment for bioremediation of specific pollutants in the near future, as regulatory questions and concerns about environmental risks are addressed. This approach to bioremediation is one of the most exciting areas of the new discipline known as environmental biotechnology.

Toxic metals are a special class of environmental pollutants. Metals cannot be degraded, but only changed from one form (oxidation state) to another. Thus, bioremediation processes for metal-contaminated environments aim at sequestering the metals to make them unavailable to biological components of the ecosystem, or mobilizing them in a manner that allows their 'flushing' from the system for

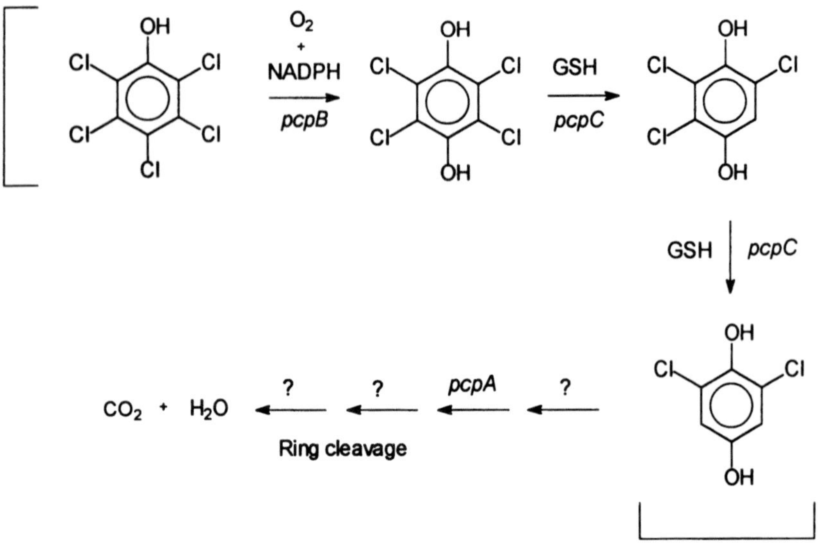

Figure 5. *Flavobacterium* ATCC 39723 pentachlorophenol pathway genes expressed in *Escherichia coli*. Brackets indicate genes cloned into and functional in a recombinant *E. coli*.

collection and disposal. Chapter 10 covers this uniquely difficult class of toxins.

Throughout the world, the problems of environmental contamination by toxic chemicals are enormous. The projected costs of cleaning up just the worst instances by means of available technologies run into the hundreds of billions of dollars. In some cases, no appropriate technologies are available at any cost. Bioremediation offers a partial solution to this dilemma. As compared with incineration or landfilling, biodegradation of pollutants can be inexpensive. It can be a permanent solution when pollutants are mineralized, and it can be combined with other procedures in treatment trains to deal with the complex problems associated with many sites. Bioremediation may be the only possible approach for cleaning some environments, such as deep aquifers. Yet, bioremediation is still an unpredictable technology that may be simple in concept, but sometimes hard to apply in practice. The bioremediation business has suffered from some overselling of the technology, which has not always worked as advertised. In some cases bioremediation simply is not the technology of choice. Research, however, is changing the status quo. Some of the world's best scientists are using their skills to design experiments that lead to a better understanding and tighter control of biodegradative processes. As those processes are being applied in well-engineered systems to the treatment of contaminated environments, the role for bioremediation in environmental restoration is steadily increasing. We hope the discussions in this book will convey to our readers our excitement about the progress being made in the field of bioremediation.

References

Funk, S. B., Roberts, D. J., Crawford, D. L. & Crawford, R. L. (1993). Initial-phase optimization for bioremediation of munition compound-contaminated soils. *Applied and Environmental Microbiology*, 59, 2171–7.

Henry, S. & Grbić-Galić, D. (1991). Influence of endogenous and exogenous electron donors and trichloroethylene toxicity on trichloroethylene oxidation by methanotrophic cultures from a groundwater aquifer. *Applied and Environmental Microbiology*, 57, 236–44.

Hinchee, R., ed. (1994). *Air Sparging for Site Remediation*. Boca Raton, FL: Lewis Publishers.

Kaake, R. H., Roberts, D. J., Stevens, T. O., Crawford, R. L. & Crawford, D. L. (1992). Bioremediation of soils contaminated with the herbicide 2-*sec*-butyl-4,6-dinitrophenol (dinoseb). *Applied and Environmental Microbiology*, 58, 1683–9.

Lamar, R. T. (1990). In situ depletion of pentachlorophenol from contaminated soil by *Phanerochaete* spp. *Applied and Environmental Microbiology*, 56, 3093–100.

Lange, C. (1994). Molecular analysis of pentachlorophenol degradation by *Flavobacterium* sp. strain ATCC 39723. Ph.D. Dissertation, University of Idaho.

Levin, M. A. & Gealt, M. A. (1993). *Biotreatment of Industrial and Hazardous Waste*. New York: McGraw-Hill.

Litchfield, C. D. (1991). Practices, potential, and pitfalls in the application of biotechnology to environmental problems. In *Environmental Biotechnology for Waste Treatment*, ed. G. Sayler *et al.*, pp. 147–57. New York: Plenum Press.

National Research Council, Water Science and Technology Board (1993). *In Situ Bioremediation: When Does It Work?* Washington, DC: National Academy Press.

Orser, C. S., Dutton, J., Lange, C., Jablonski, P., Xun, L. & Hargis, M. (1993a). Characterization of a *Flavobacterium* glutathion S-transferase gene involved in reductive dechlorination. *Journal of Bacteriology*, 175, 2640–4.

Orser, C. S., Lange, C. C., Xun, L., Zahrt, T. C. & Schneider, B. J. (1993b). Cloning, sequence analysis, and expression of the *Flavobacterium* pentachlorophenol-4-hydroxylase gene in *Escherichia coli*. *Journal of Bacteriology*, 175, 411–16.

Raymond, R. (1974). *Reclamation of Hydrocarbon Contaminated Waters*. U.S. patent 3 846 290.

Regan, K. M. & Crawford, R. L. (1994). Characterizatin of *Clostridium bifermentans* and its biotransformation of 2,4,6-trinitrotoluene (TNT) and 1,3,5-triaza-1,3,5-trinitrocyclohexane (RDX). *Biotechnology Letters*, Oct., 1081–6.

Rynk, R., ed. (1992). *On-Farm Composting Handbook*. Ithaca, NY: Northeast Regional Agricultural Engineering Service.

Semprini, L., Roberts, P., Hopkins, G. & McCarty, P. (1990). A field evaluation of *in situ* biodegradation of chlorinated ethenes: Part 2, Results of biostimulation and biotransformation experiments. *Ground Water*, 28, 715–27.

Stormo, K. E. & Crawford, R. L. (1992). Preparation of encapsulated microbial cells for environmental applications. *Applied and Environmental Microbiology*, 58, 727–30.

Stormo, K. E. & Crawford, R. L. (1993). Pentachlorophenol degradation by microencapsulated Flavobacteria and their enhanced survival for *in situ* aquifer bioremediation. In *Applied Biotechnology for Site Remediation*, ed. R. Hinchee *et al.*, pp. 422–7. Boca Raton, FL: Lewis Publishers.

U.S. EPA (1994). *Fact Sheet, March 1994, Superfund Innovative Technology Evaluation (SITE). Demonstration of the J. R. Simplot ex Situ Bioremediation Technology for Treatment of*

Nitroaromatic Contaminants at the Weldon Spring Ordnance Works Site in Weldon Spring, Missouri. TNT. Washington, DC: U.S. Government Printing Office.

Vesper, S. J., Narayanaswamy, M., Murdoch, L. C. & Davis-Hoover, W. J. (1994). Hydraulic fracturing to enhance in situ bioreclamation of subsurface soils. In *Applied Biotechnology for Site Remediation*, ed. R. Hinchee *et al.*, pp. 36–48. Boca Raton, FL: Lewis Publishers.

Williams, R. T., Ziegenfuss, P. S. & Sisk, W. E. (1992). Composting of explosives and propellant contaminated soils under thermophilic and mesophilic conditions. *Journal of Industrial Microbiology*, 9, 137–44.

Xun, L. & Orser, C. S. (1991). Purification of a *Flavobacterium* pentachlorophenol-induced periplasmic protein [pcpA] and nucleotide sequence of the corresponding gene. *Journal of Bacteriology*, 173, 2920–6.

1

Engineering of bioremediation processes: needs and limitations

Wudneh Admassu and Roger A. Korus

1.1 Introduction

The selection of microbiological processes for treating soils and groundwater contaminated with organic pollutants requires characterization of the waste, selection of an appropriate microorganism or consortium, and information about degradation pathway and rates. This chapter will focus on *ex situ* processes and the selection of reactors for the treatment of soils and groundwater.

Bioremediation must compete economically and functionally with alternate remediation technologies, which are often incineration and chemical treatments. Bioremediation usually competes well on a cost basis, especially with petroleum products and many solvents. A drawback, however, is the large amount of preliminary information necessary to support process design. When information on waste characteristics, microbial physiology, and the complex options for process design and operation is lacking, bioremediation can be more difficult to apply than alternate technologies. The variable end results of bioremediation are due in large part to these complexities of process design.

The engineering of bioremediation processes relies on information about the site and about candidate microorganisms. Process analysis usually begins with fixed waste characteristics but with options for microbial cultures, reactor types, waste pretreatment, and process operating conditions. Laboratory measurements are necessary to explore these options and to design an efficient process. Laboratory tests that examine degradation rates as functions of critical operating parameters such as pH, oxygen and nutrient concentrations, microbial composition, soil particle size, temperature, and redox potential shape the design of a bench-scale process. At a small scale mass transfer effects such as agitation and aeration are also explored. These tests form the basis for scale-up to the field scale and for the implementation of process control.

Many aspects of this design strategy have been presented. A screening protocol for evaluating and implementing bioremediation should be based on treatability studies performed to determine the effectiveness of bioremediation for specific contaminants and media (Rogers, Tedaldi & Kavanaugh, 1993). In phase I of this protocol, chemical properties of the contaminant, as well as physical, chemical, and microbiological properties of the site, are considered with regard to selecting the bioremediation method and determining the metabolic pathway. In phase II, the kinetics of bioremediation and the feasibility of attaining the desired end point are determined, along with the parameters and costs for full-scale implementation as design criteria. A rapid screening protocol could involve site study, regulatory analysis, biological screening, and treatability testing (Block, Stroo & Swett, 1993).

A main objective of biological remediation design is to remove the limiting factors in the growth of bacteria (Nyer, 1992). Options for reactor design are discussed in several texts (Jackman & Powell, 1991; Nyer, 1992; Armenante, 1993; Alexander, 1994). Applications to soil and to groundwater are often discussed separately. Common options for soil treatment are land treatment or 'landfarming' (Eckenfelder & Norris, 1993) and slurry reactors (LaGrega, Buckingham & Evans, 1994), while supported biofilm reactors are often used to treat groundwater (Nyer, 1992; Sayles & Suidan, 1993).

Bioremediation has been most successful with petroleum products. Of the 132 bioremediation activities with clearly identified target compounds reported to the U.S. Environmental Protection Agency (EPA), 75 dealt with petroleum and related materials (Devine, 1992), almost always in soil or groundwater. Oils and fuels often enter the environment from leaking storage tanks or accidental spills. Steel tanks have an average life expectancy of 25 years, and failure rate increases with age (Robison, 1987). It is estimated that approximately 25% of underground petroleum storage tanks are leaking (U.S. EPA, 1988). Technology for landfarming has been well developed for the treatment of surface spills (Riser-Roberts, 1992), but subsurface contamination at depths greater than approximately 3 m is complex and difficult to treat since soil can tightly bind petroleum contaminants.

1.2 Process analysis

Before process options can be selected, site characterization and feasibility studies must determine key chemical, physical, and microbiological properties of the site. The objectives of these initial studies are to identify rate-limiting factors for later field applications and to obtain kinetic and equilibrium data for process design.

1.2.1 Site characterization

The main objective of site characterization is to identify the contaminants, their concentration, and the extent of contamination. The distribution of contaminants between soil and groundwater will largely determine whether soil or groundwater treatment is applicable, while the extent of contamination will largely determine the applicability of soil excavation and treatment.

Physical properties of organic contaminants that are important to bioremediation processes have been tabulated (Eckenfelder & Norris, 1993). Water solubilities, octanol/water partition coefficients, vapor pressures, and Henry's Law constants are available for most contaminants. The octanol/water partition coefficient is defined as the ratio of a compound's concentration in the octanol phase to its concentration in the aqueous phase of a two-phase system. Measured values for organic compounds range from 10^{-3} to 10^{7}. Compounds with low values (<10) are hydrophilic, with high water solubilities, while compounds with high values ($>10^{4}$) are very hydrophobic. Compounds with low water solubilities and high octanol/water coefficients will be adsorbed more strongly to solids and are generally less biodegradable. Highly soluble compounds tend to have low adsorption coefficients for soils and tend to be more readily biodegradable.

Vapor pressures and Henry's Law constant, H_A, measure liquid–air partitioning. Henry's Law states that the equilibrium partial pressure of a compound in the air above the air/water interface, P_A, is proportional to the concentration of that compound in the water, usually expressed as the mole fraction, X_A.

$$P_A = H_A X_A \qquad (1.1)$$

Aeration is often employed to strip volatile organic compounds from water and is favored by large H_A values. Conversely, volatilization must be controlled or contained in many bioremediation processes. Henry's Law constant is highly temperature sensitive, and temperature changes of $10\,°C$ can give threefold increases in H_A.

In soil bioremediation the rate-limiting step is often the desorption of contaminants, since sorption to soil particles and organic matter in soils can determine the bioavailability of organic pollutants. Bioavailability is also an important toxicity characteristic, as determined by the toxicity characteristic leaching procedure (TCLP) (Johnson & James, 1989) established by the U.S. Environmental Protection Agency. Determining the feasibility of *in situ* bioremediation requires extensive characterization of hydrological and soil properties. Rates of *in situ* soil bioremediation are governed by mass transfer of contaminants (desorption and diffusion), the convective–dispersive flux of oxygen and nutrients, and the microbiological content of the soil. On-site testing to determine the rate and extent of biodegradation can be done immediately in consideration of *ex situ* process feasibility (Autry & Ellis, 1992).

Abiotic soil desorption measurements are probably the most important tests to precede measurement of microbial degradation. Desorption tests measure the site-specific soil/water partition coefficients for the contaminants of interest. Several experimental protocols are available for measuring partition coefficients (Wu & Gschwend, 1986; Rogers, Tedaldi & Kavanaugh, 1993). At the two extremes of bioavailability and biodegradability, contaminants can either be detected near their solubility limit or can be undetectable in the aqueous phase. Measurements of aqueous phase and soil phase concentrations in equilibrium may be sufficient to indicate potential problems with soil sorption. Many organic contaminants are hydrophobic, have a low water solubility, and are tightly held to the soil phase. Desorption of such contaminants is likely to be rate-limiting, especially for *ex situ* bioremediation where addition of nutrients and microorganisms and careful control of environmental parameters can minimize these potential rate-limiting effects. Failure to bioremediate polyaromatic hydrocarbon (PAH) compounds has been attributed to strong sorption to soil at former manufactured gas plant sites where gaseous fuels were produced from soft coal (Rogers *et al.*, 1993).

The use of surfactants and cosolvents has been investigated in attempts to increase bioavailability of contaminants that are strongly bound to soil. However, high surfactant concentrations can be required to achieve small increases in solubility. Typically, 2% surfactant solutions are needed to remove a high percentage of compounds such as higher-ringed PAHs, polychlorinated biphenyls (PCBs), and higher molecular weight hydrocarbons from soils. Similar enhanced removal can be obtained with biosurfactants (Scheibenbogen *et al.*, 1994).

1.2.2 Microbiological characterization

The measurement of biodegradation rates by indigenous microorganisms is the first step in microbiological characterization. These measurements can be complicated by low microbial populations or by the absence of species capable of degrading contaminants. Also, optimum conditions of temperature, oxygen nutrient supply, and contaminant availability due to low solubility and sorption can limit degradation rates, especially in early tests where these limiting factors are not well defined.

The main objective of microbial degradation tests is to determine whether the indigenous microorganisms are capable of bioremediation when conditions are optimized, or if inoculation by nonindigenous microorganisms will be required. For example, in the bioremediation sites contaminated by 2-sec-butyl-4,6-dinitrophenol (dinoseb), the first site was remediated with indigenous microorganisms, but a second site required inoculation with microorganisms from the first site (Kaake *et al.*, 1992). While pure cultures or consortia of bacteria do not in general persist when introduced into a natural environment, soil and groundwater reactors can be operated to maintain a population of introduced microorganisms.

1.2.3 Environmental factors

Along with biodegradation tests, some attention must be given to the environmental factors affecting biodegradation rate measurements. Chemical analyses that support process design include measurements of pH, COD, TOC, nitrogen, phosphorus, and iron; and inhibitory, toxic, or essential metals. Soil type, clay and organic matter content, and particle size distribution analyses are used, and, for water, total suspended solids. Microbiological analyses supporting process design include BOD, plate counts, and shake flask and/or column degradation studies with indigenous microbes or introduced cultures. Bioremediation is usually carried out near neutral pH, although fungi often require an acidic environment. Most microorganisms are mesophilic, requiring temperatures in the 25 to 37 °C range. More difficult to optimize are the oxygen and nutrient supply. Most bacteria capable of degrading organic compounds are heterotrophic and require an organic compound as a source of carbon and energy. It may be necessary to add a readily metabolizable carbon source such as glucose to maintain cell viability or to increase cell growth and degradation rates. Many xenobiotic compounds can be transformed by cometabolism, in which the transformation does not serve as an energy source for synthesis of cell biomass, which therefore requires a separate carbon source.

Although many compounds appear to be cometabolized in soil, water, and sewage (Alexander, 1994), the concept of cometabolism is of limited use in bioremediation process analysis and design. A chemical that is cometabolized at one concentration or in one environment may be mineralized under other conditions (Alexander, 1994). Cometabolism can often be accelerated by the addition of a mineralizable compound with a structure analogous to the target compound. This method of analog enrichment has been used to enhance the cometabolism of PCBs by addition of biphenyl, in which the unchlorinated biphenyl serves as a carbon source for microorganisms that cometabolize PCBs with the enzymes induced by biphenyl (Brunner, Sutherland & Focht, 1985). In another example of analog enrichment, the addition of 2-hydroxybenzoate was shown to increase naphthalene degradation in soil (Ogunseitan & Olson, 1993). Usually, a more empirical approach is taken to the selection of a carbon source and determination of the optimum concentration. Degradation is often coupled to growth and microbial mass, so the carbon source that best supports growth also gives the highest rate of degradation.

Since nitrogen and phosphorus supplies in soil systems are inadequate to support microbial growth and degradation of organic compounds, most bioremediation processes supply these two compounds. The availability of nitrogen and phosphorus usually limits the degradation of petroleum hydrocarbons in aqueous and terrestrial environments, but comparatively little attention has been given to the effects of nitrogen and phosphorus on the degradation of other compounds. Even less work has been done to examine other nutrients that might affect biodegradation, but fortunately these are required only in small or trace amounts, which are generally supplied in a natural soil.

Biodegradation can occur aerobically or anaerobically. The most effective biodegradation of hydrocarbon compounds is mediated by aerobic bacteria. Oxygen supply can be a rate-limiting factor, and large-scale aerobic bioremediation of hydrocarbons must include an aeration system as a critical design component. Early bioremediation work focused almost exclusively on aerobic systems, but recent improvements in anaerobic technologies have shown that a wide range of nitroaromatic compounds, chlorinated phenols, PAHs, and PCBs can be degraded as well or better by anaerobic processes (Singleton, 1994).

In the selection of a microbial system and bioremediation method, some examination of the degradation pathway is necessary. At a minimum, the final degradation products must be tested for toxicity and other regulatory demands for closure. Recent advances in the study of microbial metabolism of xenobiotics have identified potentially toxic intermediate products (Singleton, 1994). A regulatory agency sets treatment objectives for site remediation, and process analysis must determine whether bioremediation can meet these site objectives. Specific treatment objectives for some individual compounds have been established. In other cases total petroleum hydrocarbons; total benzene, toluene, ethyl benzene, and xylene (BTEX); or total polynuclear aromatics objectives are set, while in yet others, a toxicology risk assessment must be performed.

1.2.4 Predicting degradation rate

Degradation rates and equilibrium properties (Table 1.1) are useful for process design. During wastewater or soil treatment, target pollutants can be degraded or mineralized, volatilized, adsorbed onto effluent solids, or discharged in the liquid effluent. Volatilization will occur for compounds with relatively high Henry's Law constants, and octanol–water partition constants give a measure of the adsorption potential to hydrophobic matrices such as sludge and organic materials. With soils, the clay content and cation exchange capacity will give measures of adsorption for hydrophilic compounds. Volatilization and sorption must be minimized so that the principal fate of contaminants in a bioremediation process is biodegradation.

Table 1.1 is a portion of the 196-compound data base used in a fate model to estimate volatile emissions, concentrations of toxic compounds in sludges, and

Table 1.1. *Biodegradation kinetics in wastewater treatment plants and equilibrium constants*

Compound	Biodegradation rate constant (l/g·h)	Henry's Law constant (atm·m^3/mole)	Octanol–water partition constant
Acetone	0.20	2.5×10^{-5}	0.57
Anthracene	3	8.6×10^{-5}	2.8×10^4
Chlorobenzene	3	3.6×10^{-3}	690
2,4-Dichlorophenol	100	9.4×10^{-6}	790
Fluorene	10	6.4×10^{-5}	1.5×10^4
Hexachlorobenzene	3×10^{-3}	6.8×10^{-4}	2.6×10^6
Naphthalene	100	4.6×10^{-4}	2000
Nitrobenzene	15	1.3×10^{-5}	72
Pyrene	0.1	5.1×10^{-6}	8.0×10^6
Tetrachloroethane	1.0×10^{-3}	0.011	1100.00

From Sayles & Suidan (1993); Govind *et al.* (1991).

removal of toxic compounds during activated sludge wastewater treatment (Govind, Lai & Dobbs, 1991). The major error in predicting the fate of toxic compounds in wastewater treatment plants was in estimating the value of the biodegradation rate constant. The biodegradation rate constants in Table 1.1 are second-order overall, and first-order in both cell mass in the activated sludge process and in concentration of the target compound. For a compound such as tetrachloroethylene, with a low biodegradation rate constant and a high Henry's Law constant, the principal fate will be volatilization. An alternative process would be air stripping of the influent and air treatment with a biofilter.

Published biodegradation rate constants do not provide a good comparative basis for process design since experimental conditions vary greatly. Many unreported factors can influence biodegradation rates, including transport effects, acclimation of microorganisms to toxic chemicals, inhibition, and cometabolism. Also, zero-order kinetics have been frequently reported for biodegradation (Alexander, 1994) rather than the first-order kinetics with respect to substrate as given in Table 1.1. Zero-order kinetics may indicate rate limitation by another factor such as oxygen supply, another nutrient, or the limiting solubility of the target compound. Also, fewer than 1% of the more than 2000 new compounds submitted to the U.S. Environmental Protection Agency each year for regulatory review contain data on biodegradability (Boethling & Sabjlic, 1989). Therefore, degradation rate constants are used mainly for order-of-magnitude estimates, and other approaches to predicting biodegradability have been developed that are based on physical or chemical properties of compounds.

A measure of hydrophobicity can be used to estimate relative biodegradation rates among homologous compounds. For example, polycyclic aromatic hydrocarbons (PAHs) become more recalcitrant as water solubilities decrease (Cerniglia, 1993). Degradation rates can also be correlated with the octanol–water partition constant (Table 1.1). Rates of anaerobic degradation of halogenated aromatic compounds are correlated with the strength of the carbon halide bond that is cleaved in the rate-determining step (Peijnenburg et al., 1992).

A promising method for predicting biodegradation rates is the group contribution approach, which estimates biodegradability from the type, location, and interactions of the substituent groups that make up a compound. Experimental data are insufficient for assessing the feasibility of biodegradation for most compounds. This problem is especially acute in EPA programs involved in chemical scoring, which are often mandated by legislation. An experimental data base that assesses mixed culture biodegradation data for 800 organic compounds is available (Howard, Hueber & Boethling, 1987). A training set with 200 chemicals and expert scoring from 17 authorities was used to determine the best parameter values in models for predicting primary and ultimate biodegradability (Boethling et al., 1994). Chemical compounds were accurately classified as rapidly or slowly biodegradable on the basis of the biodegradability parameters of the fragments that compose a compound, and on the molecular weight.

Group contribution models are available for predicting biodegradation rate constants (Govind *et al.*, 1991; Sayles & Suidan, 1993). Good predictions of degradation rates can be obtained for mixed-culture aerobic degradation in processes such as activated sludge treatment. The activated sludge reactor can be optimized with respect to aeration using a model to calculate biodegradation and volatilization rates of organics as functions of aeration rate. The fates of organic compounds can be simulated in a conventional wastewater treatment plan, showing that many of the hazardous chemicals are not readily biodegradable by conventional treatment processes (Sayles & Suidan, 1993).

A nonlinear group contribution method has been developed using neural networks trained with published data on first-order biodegradation rate constants for priority pollutants (Tabak & Govind, 1993). The training set contained first-order biodegradation rate data for acids, ketones, amines, single-ring aromatics, and hydroxy compounds. The resulting group contribution values were able to predict the natural log of first-order biodegradation constants for the 18 training compounds and 10 test compounds with errors of 11% or less, relative to experimentally determined rate constants. More data are required to predict the unique contributions due to positional effects of groups and to include more functional groups in structure–activity relationships. As the data base grows and as more sophisticated models give a better description of the interactions among group fragments within complex molecules, quantitative estimates of biodegradation rates will be more reliable.

The data base and correlation methods are more limited for anaerobic biodegradation rates. Anaerobic kinetic data are scarce, often not well defined, and difficult to interpret and classify because of the complex interactions and compositional variables in anaerobic environments. Whereas aerobic conditions can be well defined by good aeration, anaerobic conditions are dependent on redox potential and the presence of alternate electron acceptors. Also, the variability of culture acclimation and strong inhibition effects greatly complicate anaerobic data interpretation. Essentially no kinetic parameters are available for biodegradation of toxic organic chemicals under methanogenic conditions. Therefore, an experimental protocol has been developed for obtaining and analyzing kinetic data for biotransformations of toxic organic chemicals in anaerobic treatment processes (Sathish, Young & Tabak, 1993). This protocol requires development of an acclimated culture and determination of inhibition parameters that define biodegradation rates.

1.3 Reactor options

Reactor options are determined primarily by the physical properties of the waste and the chemical and biochemical properties of the contaminants. System characteristics can favor a particular reactor option. If the waste is found in groundwater, then a continuous supported reactor is desirable, while a suspended batch reactor is preferable with contaminated soil. A polar target compound favors an aerobic reactor, while nonpolar compounds favor anaerobic reactors. Groundwater is easily treated in a continuous process in which the microbial biomass must be retained in the reactor by adhering to a support. The usual reactor configuration is a packed bed column with an attached biofilm. Soils are difficult to transport, so batch reactors are favored. Agitation is a critical design parameter required for aeration, cell suspension, and transport of contaminants and nutrients during soil remediation.

1.3.1 Groundwater remediation

The technology for the removal of suspended and dissolved contaminants from municipal and industrial wastewater has been well established for wastes that are readily biodegradable under aerobic conditions. After preliminary treatment such as screening, grit removal, and sedimentation, a biological reactor degrades the organic matter in the wastewater during secondary treatment. Typically, the reactor is an aerobic fixed film process or a suspended-growth, activated sludge process. The theory, design, and operational characteristics of these wastewater treatment systems are well described in engineering texts. Industry relies heavily on treatment of their chemical wastes by publicly owned wastewater treatment facilities. In the U.S., over 70 million kg of chemical wastes were sent to domestic treatment plants in 1990 alone (Thayer, 1992). However, removal rates of xenobiotic compounds are not well described. Inhibitory organic compounds are common constituents of industrial wastewaters, and these wastewaters are usually treated by activated sludge processes. Very few of the 114 priority organic pollutants listed by the U.S. EPA pass through an activated sludge plant without some form of attenuation (Allsop, Moo-Young & Sullivan, 1990).

Activated sludge reactors

Petrasek *et al.* (1983) spiked the influent of an activated sludge plant with 50 μg/liter of 22 organic compounds from the EPA list of priority pollutants. Polychlorinated phenols and biphenyls, phenols, phthalates, and PAHs were tested, with average removal rates of 97%. However, higher concentrations can destabilize an activated sludge system. The presence of cyanide, pentachlorophenol, 1,2-dichloropropane, acrylonitrile, phenolics, and ammonia can cause instability in the operation of activated sludge plants (Allsop *et al.*, 1990).

Mobile units are often required for the treatment of groundwater pumped out of a contaminated aquifer, or wash water used to remove chemicals from a contaminated soil. Such processes can often be designed for the continuous treatment of a low-organic-content wastewater. The advantages of an activated sludge design for treating high-organic-content wastewaters that produce a flocculating sludge are lost in these cases. Low microbial growth, poor flocculation, and instability problems can make an activated sludge process very difficult to apply to groundwater treatment (Nyer, 1992). Fixed film processes, similar to the trickling filter designs for wastewater treatment, can generally be operated at lower oxygen supply costs, with retention of bacteria in the reactor and a more stable operation.

Fixed film reactors

The BioTrol aqueous treatment system is a good example of a fixed film, continuous reactor used successfully for the treatment of groundwater (Pflug & Burton, 1988). This reactor is packed with corrugated polyvinyl chloride media on which a bacterial biofilm is grown (U.S. EPA, 1991). The combination of groundwater flow, air sparging, and design of the support media facilitates the upward and lateral distribution of water and air in the reactor. Removal of more than 95% pentachlorophenol (PCP) is achieved at a 1.8 h residence time. Other organic contaminants such as PAHs have been mineralized at a total capital and operating cost as low as $0.78/1000 liters.

The most widely accepted bioreactor is the fixed film, plug flow reactor. Fixed film or attached growth systems require degradative microorganisms with the ability to attach to the surfaces of an inert packing. Fixed film reactors can treat low concentrations of organics in wastewater because of biomass retention, and they can also treat concentrations as high as 1000 ppm. The high biomass loadings of fixed film reactors render them insensitive to shock loadings, that is, to high fluctuations in organic loadings. The outer layers of the biofilm protect the inner cells from the toxicity of high loadings, and the adsorption of contaminants within the biofilm reduces the soluble concentrations of contaminants (Hu *et al.*, 1994).

The fixed film reactor used for municipal sewage is the trickling filter reactor, with large packing, typically 50 mm diameter stones or synthetic plastic, and downward liquid flow, producing a liquid film over the biofilm and leaving air voidage within the packed bed. With mineral media the large weight severely limits the depth of the beds used, and their low voidage (40 to 55% with 40–50 mm diameter stones) permits the bed to be more easily blocked by microbial growth. Plastic packings have up to 90% voidage with about one-tenth the weight of mineral packings. Liquid hold-up is low and retention times are short, so that liquid recirculation is necessary to adequately reduce contaminant concentrations. Recirculated effluent is usually taken from a secondary clarifier output, rather than directly from the trickling filter effluent, to minimize the risk of clogging the trickling filter with biomass released from the filter.

Wastewater contaminated with xenobiotics is usually treated in a fixed film reactor with upflow through a submerged plastic packing. Air sparging is used since air voidage is not produced, as in the trickling filter. Nitrogen and phosphorus may be added to the inflow so that the C : N : P ratio is approximately 100 : 5 : 1. With contaminated water there is usually a low level of organics, so the production and sloughing of biomass is low. Reactor cells may be staged, and typically a 2-hour residence time is sufficient to give a high level of contaminant degradation. The major advantage of the upflow fixed film reactor versus the trickling bed reactor is the longer and better control of residence time. Increased residence time is possible, since air voidage is largely eliminated in the upward flow operation of a fixed film reactor.

A good example of a successful application of a fixed film bioreactor is the treatment of a lagoon contaminated with 36 ppm pentachlorophenol (PCP), 37 ppm polynuclear aromatics (PNAs), 52 ppm solvents, and a total chemical oxygen demand (COD) of 6700 ppm (King, 1992). Biological treatment was the method of choice because equipment could rapidly be brought to the site and remediation could be completed within several months. A total volume of 10.2 million liters was treated, with PCP, PNAs, and phenols each reduced to 0.5 ppm or less.

Anaerobic processes

While anaerobic processes for bioremediation of groundwater are uncommon, anaerobic treatment is used for municipal sludge processing and for agricultural and food-processing wastewaters with a high content of biodegradable matter. The resultant methane production provides a valuable energy source. Since energy production is not an option in the treatment of groundwater, anaerobic processes are considered only for compounds recalcitrant to aerobic degradation. Anaerobic microorganisms have great potential for the reductive dehalogenation of multihalogenated aromatic compounds. With increasing levels of halogenation, bioremediation may be feasible only with an initial anaerobic dehalogenation. Once lightly halogenated or nonhalogenated compounds are produced, subsequent degradations are more rapid in an aerobic environment. This suggests an anaerobic–aerobic process for the degradation of compounds that are highly chlorinated or nitrated, or for PAHs.

Anaerobic bioreactors have been used since the 1880s to treat wastewaters with large amounts of suspended solids. However, anaerobic reactors are sensitive to toxic pollutants and vulnerable to process upsets, and have been used mainly for municipal sludge digestion. For methane production the sequential metabolism of the anaerobic consortia must be balanced, and the methanogens in particular are vulnerable to process upsets. Recently, anaerobic–aerobic processes (Figure 1.1) have been developed for the mineralization of xenobiotics. These processes take advantage of an anaerobic reactor for the initial reductive dechlorination of polychlorinated compounds or the reduction of nitro substituents to amino substituents. If the reduced compounds are more readily mineralized in an aerobic reactor, an anaerobic–aerobic process is feasible.

The anaerobic bioreactor is typically an upflow packed-bed reactor. The reactor is completely filled with liquid except for gas formed during the process. Coarse packing (2 to 6 cm) is used since anaerobic organisms can form large flocs instead of thin attached films, causing clogging. Anaerobic reactors have a number of advantages, including high efficiency at low organic loading, high loading capacity, stability with toxic substances, and low energy requirements. COD

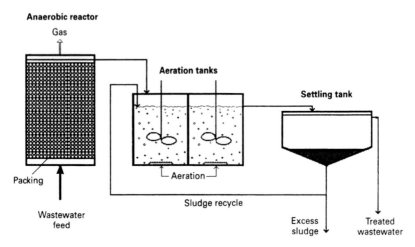

Figure 1.1. Anaerobic–aerobic process for wastewater treatment.

reductions of 4 to $10 \, kg/m^3$/day can be obtained with residence times from 4 to 18 h (Schuegerl, 1987).

A two-stage anaerobic–aerobic biofilm reactor process has been used to metabolize hexachlorobenzene (HCB), tetrachloroethylene (PCE), and chloroform (CF) (Fathepure & Vogel, 1991). Reductive chlorination is relatively rapid for these compounds and other highly chlorinated compounds such as polychlorinated biphenyls, trichloroethylene, carbon tetrachloride and 1,1,1-trichloroethane. These are some of the most pervasive groundwater contaminants. An anaerobic biofilm developed by seeding from a primary anaerobic digester reduced HCB to dichlorobenzene, PCE to dichloroethylene, and CF to dichloromethane, with acetate used as a primary carbon source. The second-stage aerobic biofilm reactor further metabolized degradation intermediates (Fathepure & Vogel, 1991). A pure anaerobic culture was capable of similar reductions but required 3-chlorobenzoate as an inducing substrate (Fathepure & Tiedje, 1994). Reductive dehalogenation may yield vinyl chloride (VC) or CF, which may be more dangerous than the parent compounds, but VC was not detected during dechlorination of PCE (Fathepure & Tiedje, 1994).

In a process similar to that shown in Figure 1.1, a packed bed anaerobic reactor followed by an air-sparged reactor has been used to mineralize 2,4,6-trichlorophenol (Armenante et al., 1992). 2,4,6-Trichlorophenol was degraded to 4-chlorophenol in the anaerobic reactor, and the 4-chlorophenol was mineralized in the subsequent aerobic operation. Also, a sequential anaerobic–aerobic process similar to the process shown in Figure 1.1 has been used to degrade nitrobenzene (Dickel, Haug & Knackmuss, 1993). Under anaerobic conditions nitrobenzene was converted to aniline, a reaction accelerated by the addition of glucose. Complete mineralization of aniline was accomplished in the aerobic stage. Other nitroaromatic compounds

Table 1.2. *Methods of soil bioremediation*

Method	Advantages	Disadvantages
Landfarming	Simple and inexpensive	Long incubation times
	Currently accepted method for petroleum contaminants	Residual contamination
		Requires lining, monitoring, and prevention of leakage
Slurry reactor	Good control of reactor conditions	Expensive
	Aerobic or anaerobic	Limited by reactor size
	Enhances desorption from soil	Requires soil pretreatment
	Short incubation periods	

are serious contaminants, especially 2,4,6-trinitrotoluene (TNT) and other munitions compounds. Oxidative biodegradation of nonpolar nitroaromatics is slow, due to the electrophilic character of the nitro groups, and under aerobic conditions dimerization and polymerization of TNT biotransformation products occurs. However, nitroaromatics are readily converted to aromatic amines, which are readily oxidized by dioxygenases. A two-stage anaerobic–aerobic process can reduce TNT to the amino derivatives in the first stage, with the second aerobic stage producing aromatic ring cleavage (Funk *et al.*, 1994).

Anaerobic–aerobic processes have a high potential for the treatment of pulp mill wastewater containing xenobiotic compounds. The pulp and paper industry is under great pressure to remove chlorophenols, chlorinated aliphatic hydrocarbons, and chlorinated dioxins and furans from wastewater. The Canadian government requires that pulp mill effluent contain no measurable level of dioxins or furans (Murray & Richardson, 1993). Such regulations require novel wastewater treatment technologies for the complete removal of target compounds.

1.3.2 Soil remediation

With *ex situ* treatment of contaminated soils, a controlled environment for soil treatment can be maintained. With mixing, nutrient addition, aeration, and other environmental controls, mass transfer rates that typically limit *in situ* bioremediation can be greatly increased. Of course, the disadvantages of *ex situ* bioremediation are the costs of soil excavation and reactor operation. Thus, *ex situ* bioremediation is favored by localized, shallow soil contamination.

Landfarming

Biological soil treatment by landfarming is a relatively simple and inexpensive method for treating soil contaminated by compounds that are readily degraded aerobically (Table 1.2). Contaminated soil is evacuated and usually treated in pits lined with a high-density synthetic or clay liner (Figure 1.2). Perforated pipes can be placed in a layer of sand between the liner and contaminated soil to collect

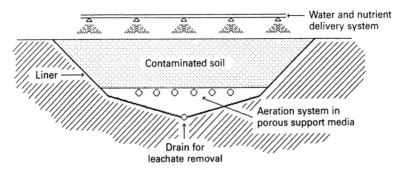

Figure 1.2. Landfarming of contaminated soil.

drainage that can be separately treated or recycled. Alternatively, the treatment area can be graded to a sump where runoff is collected. Aeration can be accomplished by tilling the soil or with forced aeration. With tillage, soil is usually spread to a depth of 15 to 50 cm. For forced aeration, soil is placed above slotted PVC pipes that are manifolded to a blower (Figure 1.2). Nutrients may be added and pH adjusted. Phosphorus is typically added as a salt of phosphoric acid and nitrogen as an ammonium salt, a nitrate salt, or urea. Nutrient requirements are estimated from contamination concentrations or laboratory treatability tests, and water is added or sprayed onto the soil to maintain optimum moisture.

Landfarming has been widely implemented at petroleum production and storage sites, and at sites contaminated with polynuclear aromatic residues (PNAs) or pentachlorophenol (PCP). Soil contaminated with 300–1750 mg/kg gasoline and diesel fuels was treated by placing excavated soil on clay-lined pads at depths of 0.15–1.2 m, biweekly tilling, and addition of nutrients and water to maintain moisture above 10% by weight (Hogg et al., 1994). The pads covered 6.6 hectares and treated $45\,875\,m^3$ of petroleum-contaminated soils. Similar technology was applied to cleanup of a field inundated with approximately 1.9 million liters of kerosene (Dibble & Bartha, 1979). The oil content of the soil was 0.87% in the top 30 cm and about 0.7% at 30–45 cm. After treatment the oil content in soil that received 200 kg N, 20 kg P, and lime was reduced to <0.1% in the upper 30 cm, but only to 0.3% at lower depth.

The efficacy and design of land treatment for petroleum-contaminated soil has been studied in controlled laboratory experiments. The effects of soil type, fuel type, contamination level, and temperature on the kinetics of fuel disappearance were determined for a land treatment process using lime to raise the pH to 7.5–7.6, addition of 60 μmol of N as NH_4NO_3 and 5 μmol of P as K_2HPO_4, and tilling (Song, Wang & Bartha, 1990). Disappearance of hydrocarbons was maximal at 27 °C. The C_6 to C_9 components of gasoline were removed primarily by evaporation. The C_{10} to C_{11} components were removed by biodegradation. The medium distillates responded well to bioremediation and increased in persistence in this order: jet fuel, heating oil, and diesel oil.

Biodegradation can be accelerated in a prepared bed reactor with forced aeration. These reactors (Figure 1.2) are used at many Superfund sites for bioremediation of PAHs and BTEX (benzene, toluene, ethylbenzene, and xylene) (Alexander, 1994). This method, with recirculating leachate, was used to reduce the average total petroleum hydrocarbon concentration in a diesel-contaminated soil from 6200 mg/kg dry soil to 280 mg/kg in approximately 7 weeks (Reynolds *et al.*, 1994). A bed reactor with forced aeration was also used to treat 115 000 m^3 of soil contaminated with bunker C fuel oil (Compeau, Mahaffey & Patras, 1991) and 23 000 m^3 of soil contaminated with gasoline and fuel oil (Block, Clark & Bishop, 1990).

Land treatment has been successfully used to treat soil contaminated with creosote and pentachlorophenol. With weekly tilling, periodic irrigation, and addition of nitrogen and phosphorus when needed, 4347 m^3 of soil was successfully treated so that pyrene was reduced from ≈ 100 mg/kg to ≈ 5 mg/kg, PCP from ≈ 150 mg/kg to ≈ 10 mg/kg, and carcinogenic PAH from ≈ 300 mg/kg to ≈ 20 mg/kg (Piotrowski *et al.*, 1994). Treatment times varied from 32 to 163 days, depending on climate conditions and initial concentrations of contaminants.

A potential problem in soil treatment is the residual contaminant concentration that is slowly or not noticeably degraded by soil microorganisms. This nonbioavailable fraction is recognized by its slow transport out of soil micropores. Slow desorption can result from entrapment in intraparticle micropores, especially in the presence of organic matter, which can tightly bind nonionic organic contaminants. Soil leaching experiments can demonstrate whether contaminants are slowly released from the soil matrix, and the addition of surfactants can increase the desorption rate. However, the mixing and surfactant addition that may be necessary to release contaminants increase the cost of landfarming and favor the use of more intensive bioremediation methods such as slurry reactors.

Slurry reactors

If increased agitation, aeration, or soil pretreatment are necessary for an acceptable biodegradation rate, a slurry reactor is often the preferred method. Slurry reactors are closed reactors with high-intensity agitators for soil–water mixing. They may be simple lined lagoons into which contaminated soil, sludge, or sediment is fed, or bioreactors typically 3 to 15 m in diameter and 4.5 to 8 m high, holding 60 000 to 1 million liters. Agitation can vary from intermittent mixing to the intense mixing required to continuously suspend soil particles.

Slurry reactors were initially developed in the 1950s for use in the chemical process industries. These reactors typically operate at 5–10% solids with 100–200 μm particles. Soil slurry reactors differ in having larger particles with strongly adsorbed pollutants and in operating at higher solids loading. Slurry phase treatment is more efficient and requires less land area than landfarming (Table 1.2). Slurry reactors are selected when the simpler landfarming option is severely limited by contaminant desorption. Slurry processes overcome this limitation by soil pretreatment and by reactor agitation and aeration. Some specific reactor designs, such as the EIMCO soil slurry reactor, require fine milling and classification of the soil prior to introduction into the reactor (Figure 1.3). Pretreatment methods include operations that enhance desorption by the reduction of soil particle size and addition of surfactant, and operations that concentrate the waste to be treated through concentrating the smaller particle sizes. Fractionation eliminates the heavier particles that are difficult to suspend in the slurry reactor. An analysis of soil contamination versus particle size will indicate the degree of concentration possible through fractionation (Compeau *et al.*, 1991). Contaminants often adsorb preferentially to finer soil particles.

A crucial aspect of process analysis is the determination of the rate-controlling step or steps. In soil bioremediation, three considerations for the rate-controlling step have been outlined (Li, Annamalai & Hopper, 1993) as follows:

1. *Contaminant mass transfer rate.* This rate at which the contaminant is desorbed from the soil, R_{des} (mg/s), can be expressed as:

$$R_{des} = k_l a (C^* - C) \tag{1.2}$$

where $k_l a$ is the mass transfer coefficient multiplied by the solid/liquid interfacial area and is determined experimentally as a single mass transfer parameter (cm^3/s) and C^* and C are the solubility and aqueous concentration of the contaminant, respectively (mg/cm^3). This desorption rate is strongly influenced by agitation, surfactant concentration, soil composition and structure, soil particle or aggregate

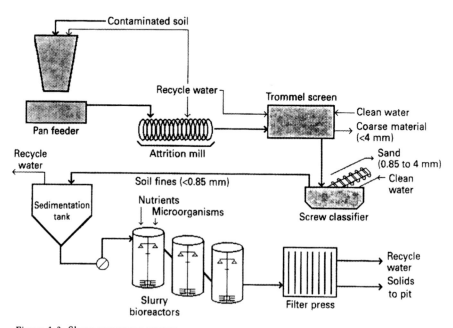

Figure 1.3. Slurry treatment system.

size, contaminant partition coefficient, and age of contamination. Desorption experiments can determine $k_l a$, but care must be taken to control properly the variables that affect desorption. Adsorption isotherms and desorption rates can be determined by sequential batch washing (Linz, Neuhauser & Middleton, 1991) or column desorption tests (Luthy *et al.*, 1994). Desorption rates can be determined as a function of particle size (Compeau *et al.*, 1991).

An accurate evaluation of $k_l a$ is complicated by the heterogeneous nature and poor definition of contaminant/soil systems. Some success has been achieved in modeling mass transfer from a separate contaminant phase. During degradation these nonaqueous phase liquids (NAPLs) often dissolve under conditions where phase equilibrium is not achieved and dissolution is proportional to $k_l a$. Experimental determinations and correlations for $k_l a$ depend on interfacial area of the NAPL and liquid velocity at the interface (Geller & Hunt, 1993). For adsorbed contaminants, $k_l a$ varies with soil composition and structure, concentration and age of contamination, and therefore with time. For example, slurry reactor tests indicate that the rate of naphthalene mass transfer decreases with time, with media size, and with aging of the tar prior to testing (Luthy *et al.*, 1994).

2. *Oxygen supply.* The mass transfer rate of oxygen, R_o (mg/s), can be defined between the oxygen in air bubbles and the dissolved oxygen in aqueous solution or between dissolved oxygen in the bulk aqueous phase and a lower concentration of oxygen at the site where oxygen is consumed by microorganisms. The first case will give the higher, more favorable rate and can be expressed as:

$$R_o = k_o a_o (C^*_s - C_o) \tag{1.3}$$

where $k_o a_o$ is the transfer coefficient for oxygen (cm^3/s) and is analogous to the transfer coefficient for contaminant desorption with a_o representing the gas/liquid interfacial area. Concentrations are defined analogously for oxygen solubility (C^*_s) and aqueous oxygen concentration (C_o). The value of $k_o a_o$ is strongly dependent on solids loading, agitation, and aeration rate. Oxygen supply can be estimated from oxygen transfer correlations and the physical properties of the slurry (Shah, 1979), but $k_o a_o$ is difficult to predict for landfarming tillage, and no published correlations exist.

In aerobic soil slurry reactors it is difficult to maintain high oxygen concentrations due to the tendency for gas bubbles to coalesce (Andrews, 1990). Also, since the reactors are usually low in profile, there is a very short liquid–gas contact time and a small surface area to volume ratio for the bubbles. Mechanical agitation is required to disperse gas bubbles and give smaller gas bubbles, but as the concentration of solids increases this agitation effect decreases (Andrews, 1990). Operational problems such as diffuser clogging, solids settling, and materials corrosion must be avoided. The design of agitators and aeration diffusers is critical to the performance of soil slurry reactors, especially at high solids loading. Solids loading is usually in the 20 to 40% range.

3. *Biodegradation rate.* (Tabulations and predictive methods were discussed

earlier.) In soil remediation, biodegradation rate depends on an optimum nutrient supply, especially nitrogen and phosphorus, as well as on temperature, pH, aeration, and water content.

A protocol for process analysis would require characterization of soils to determine soil size distribution, settling measurements, microbial population, background degradation rate in well-mixed flasks, and chemical analysis for carbon, nitrogen, and phosphorus. Because of the complexity of soil bioremediation, it is difficult to predict individual rates for contaminant desorption, oxygen supply, and biodegradation. However, an understanding of rate limitations is crucial to good process design. The three competing rates above should be estimated, if only qualitatively. The slowest rate with oxygen adjusted for the stoichiometric amount required will identify the limiting rate. Then attention can be given to increasing this rate by optimizing the key factors that affect this limiting rate.

The rate-controlling step in slurry reactor operation is often desorption of contaminants from soil particles (equation 1.2). Assuming that biodegradation occurs in the aqueous phase, the rate of disappearance of a contaminant in the aqueous phase can be expressed as (Luthy et al., 1994):

$$dC(t)/dt = k_l a[C^*(t) - C(t)] - k_{deg}C(t) \qquad (1.4)$$

where k_{deg} is the first-order rate constant for degradation. Other kinetic expressions may be used for the degradation rate.

Mixing design and intensity is critical for particle suspension, aeration, contaminant desorption, and high degradation rates. Maintaining a suspension of dense soil particles requires high-energy mixers, and the mixer/reactor design is the most critical factor in slurry design. Slurry reactors may incorporate high-intensity impeller blades, rake arms at the base of the reactor to disperse settled particles, baffles to increase turbulence, and fine bubble diffusers to minimize air bubble size and enhance aeration. A major problem in aerated slurry reactors is foaming, and a mechanical foam breaker may be necessary.

Recently, many bench-scale tests and field applications of slurry reactors have been reported. There may be as many as 30 applications within the wood-treating industry alone, primarily to close inactive waste lagoons. Several applications have been reported for chlorinated aromatics, PAHs, and petroleum sludges (LaGrega et al., 1994). Slurry reactors are more effective than landfarming in degrading these more recalcitrant wastes. Bench-scale tests using EIMCO slurry reactors showed 90% PAH removal from creosote-contaminated soil in two weeks (Lauch et al., 1992). Two- and three-ring PAHs were 96% removed, while higher-ring compounds were approximately 83% removed in two weeks. Creosote-contaminated soil classified to < 80 μm was successfully treated in 680 000 liter slurry reactors with 20 to 25% solids by weight (Jerger, Cady & Exner, 1994).

Anaerobic slurry reactors have recently been developed to treat soil contaminated with compounds such as trinitrotoluene (TNT) and other munitions compounds that are more readily degraded anaerobically than aerobically. The time frame for the degradation of these recalcitrant compounds and their intermediate degradation products is relatively long, 2 or 3 weeks or longer (Funk *et al.*, 1993). This process has been scaled up to a 60 000 liter reactor to treat soil contaminated with munitions waste and with 2-*sec*-butyl-4,6 dinitrophenol (dinoseb). The mixing intensity for anaerobic slurry reactors can be much less than for aerobic slurry reactors. Off-bottom suspension or on-bottom motion is sufficient, in contrast to the complete uniform suspension required for good aeration. The power requirements, which are a major operational cost, can be reduced 5 to 25 times with these lower mixing intensities.

References

Alexander, M. (1994). *Biodegradation and Bioremediation*. New York: Academic Press.

Allsop, P. J., Moo-Young, M. & Sullivan, G. R. (1990). The dynamics and control of substrate inhibition in activated sludge. *Critical Reviews in Environmental Control*, 20, 115–67.

Andrews, G. (1990). Large-scale bioprocessing of solids. *Biotechnology Progress*, 6, 225–30.

Armenante, P. M. (1993). Bioreactors. In *Biotreatment of Industrial and Hazardous Waste*, ed. M. A. Levin & M. A. Gealt, pp. 65–112. New York: McGraw-Hill.

Armenante, P. M., Kafkewitz, D., Lewandowski, G. & Kung, C-M. (1992). Integrated anaerobic–aerobic process for the biodegradation of chlorinated aromatic compounds. *Environmental Progress*, 11, 113–22.

Autry, A. R. & Ellis, G. M. (1992). Bioremediation: an effective remedial alternative for petroleum hydrocarbon-contaminated soil. *Environmental Progress*, 11, 318–23.

Block, R., Stroo, H. & Swett, G. H. (1993). Bioremediation – why doesn't it work sometimes? *Chemical Engineering Progress*, 89 (8), 44–50.

Block, R. N., Clark, T. P. & Bishop, M. (1990). Biological treatment of soils contaminated by petroleum products. In *Petroleum Contaminated Soils*, ed. P. T. Kostecki & E. J. Calabrese, Vol. 3, pp. 167–75. Chelsea, MI: Lewis Publishers.

Boethling, R. S. & Sabjlic, A. (1989). Screening-level model for aerobic biodegradability based on a survey of expert knowledge. *Environmental Science & Technology*, 23, 672–9.

Boethling, R. S., Howard, P. H., Meylan, W., Stiteler, W., Beauman, J. & Tirado, N. (1994). Group contribution method for predicting probability and rate of aerobic biodegradation. *Environmental Science & Technology*, 28, 459–65.

Brunner, W., Sutherland, F. H. & Focht, D. D. (1985). Enhanced biodegradation of polychlorinated biphenyls in soil by analog enrichment and bacterial inoculation. *Journal of Environmental Quality*, 14, 324–8.

Cerniglia, C. E. (1993). Biodegradation of polycyclic aromatic hydrocarbons. *Current Opinion in Biotechnology*, 4, 331–8.

Compeau, G. C., Mahaffey, W. D. & Patras, L. (1991). Full-scale bioremediation of contaminated soil and water. In *Environmental Biotechnology for Waste Treatment*, ed. G. S. Sayler, R. Fox & J. W. Blackburn, pp. 91–109. New York: Plenum.

Devine, K. (1992). *Bioremediation Case Studies: An Analysis of Vendor Supplied Data*. Publ. EPA/600/R-92/043. Office of Engineering and Technology Demonstration, Washington, DC: U.S. Environmental Protection Agency.

Dibble, J. T. & Bartha, R. (1979). Rehabilitation of oil-inundated agricultural land: a case history. *Soil Science*, 128, 56–60.

Dickel, O., Haug, W. & Knackmuss, H-J. (1993). Biodegradation of nitrobenzene by a sequential anaerobic–aerobic process. *Biodegradation*, 4, 187–94.

Eckenfelder, W. W. Jr & Norris, R. D. (1993). Applicability of biological processes for treatment of soils. In *Emerging Technologies in Hazardous Waste Management III*, ACS Symposium Series 518, ed. D. W. Tedder & F. G. Pohland, pp. 138–58. Washington, DC: American Chemical Society.

Fathepure, B. Z. & Tiedje, J. M. (1994). Reductive dechlorination of tetrachloroethylene by a chlorobenzoate-enriched biofilm reactor. *Environmental Science & Technology*, 28, 746–52.

Fathepure, B. Z. & Vogel, T. M. (1991). Complete degradation of polychlorinated hydrocarbons by a two-stage biofilm reactor. *Applied and Environmental Microbiology*, 57, 3418–22.

Funk, S. B., Crawford, D. L., Roberts, D. J. & Crawford, R. L. (1994). Two-stage bioremediation of TNT contaminated soils. In *Bioremediation of Pollutants in Soil and Water*, ed. B. S. Schepart, pp. 177–89. ASTM STP 1235, Philadelphia: American Society for Testing and Materials.

Funk, S. B., Roberts, D. J., Crawford, D. L. & Crawford, R. L. (1993). Initial-phase optimization for bioremediation of munition compound-contaminated soils. *Applied and Environmental Microbiology*, 59, 2171–7.

Geller, J. T. & Hunt, J. R. (1993). Mass transfer from nonaqueous phase organic liquids in water-saturated porous media. *Water Resources Research*, 29, 833–45.

Govind, R., Lai, L. & Dobbs, R. (1991). Integrated model for predicting the fate of organics in wastewater treatment plants. *Environmental Progress*, 10, 13–23.

Hogg, D. S., Piotrowski, M. R., Masterson, R. P., Jorgensen, M. R. & Frey, C. (1994). Bioremediation of hydrocarbon-contaminated soil. In *Hydrocarbon Bioremediation*, ed. R. E. Hinchee, B. C. Alleman, R. E. Hoeppel & R. N. Miller, pp. 398–404. Boca Raton, FL: Lewis Publishers, CRC Press.

Howard, P. H., Hueber, A. E. & Boethling, R. S. (1987). Biodegradation data evaluation for structure/biodegradability relations. *Environmental Toxicology and Chemistry*, 6, 1–10.

Hu, Z-C., Korus, R. A., Levinson, W. E. & Crawford, R. L. (1994). Adsorption and biodegradation of pentachlorophenol by polyurethane-immobilized *Flavobacterium*. *Environmental Science & Technology*, 28, 491–6.

Jackman, A. P. & Powell, R. L. (1991). *Hazardous Waste Treatment Technologies*. Park Ridge, NJ: Noyes Publications.

Jerger, D. E., Cady, D. J. & Exner, J. H. (1994). Full-scale slurry-phase biological treatment of wood-preserving wastes. In *Bioremediation of Chlorinated and Polycyclic Aromatic Hydrocarbon Compounds*, ed. R. E. Hinchee, A. Leeson, L. Semprini & S. K. Ong, pp. 480–3. Boca Raton, FL: Lewis Publishers, CRC Press.

Johnson, L. D. & James, R. H. (1989). Sampling and analysis of hazardous wastes. In *Standard Handbook of Hazardous Waste Treatment and Disposal*, ed. H. M. Freeman, pp. 13.3–13.44. New York: McGraw-Hill.

Kaake, R. H., Roberts, D. J., Stevens, T. O., Crawford, R. L. & Crawford, D. L. (1992). Bioremediation of soils contaminated with the herbicide 2-*sec*-butyl-4,6-dinitrophenol (dinoseb). *Applied and Environmental Microbiology*, 58, 1683–9.

King, R. B. (1992). *Practical Environmental Bioremediation*. Boca Raton, FL: CRC Press.

LaGrega, M. D., Buckingham, P. L. & Evans, J. C. (1994). *Hazardous Waste Management*. New York: McGraw-Hill.

Lauch, R. P., Herrmann, J. G., Mahaffey, W. R., Jones, A. B., Dosani, M. & Hessling, J. (1992). Removal of creosote from soil by bioslurry reactors. *Environmental Progress*, 11, 265–71.

Li, K. Y., Annamalai, S. N. & Hopper, J. R. (1993). Rate controlling model for bioremediation of oil contaminated soil. *Environmental Progress*, 12, 257–61.

Linz, D. G., Neuhauser, E. F. & Middleton, A. C. (1991). Perspectives on bioremediation in the gas industry. In *Environmental Biotechnology for Waste Treatment*, ed. G. S. Sayler, R. Fox & J. W. Blackburn, pp. 25–36. New York: Plenum Press.

Luthy, R. G., Dzombak, D. A., Peters, C. A., Roy, S. B., Ramaswami, A., Nakles, D. V. & Nott, B. R. (1994). Remediating tar-contaminated soils at manufactured gas plant sites. *Environmental Science & Technology*, 28, 266A–76A.

Murray, W. D. & Richardson, M. (1993). Development of biological and process technologies for the reduction and degradation of pulp mill wastes that pose a threat to human health. *Critical Reviews in Environmental Science & Technology*, 23, 157–94.

Nyer, E. K. (1992). *Groundwater Treatment Technology*. New York: Van Nostrand Reinhold.

Ogunseitan, O. A. & Olson, B. H. (1993). Effect of 2-hydroxybenzoate on the rate of naphthalene mineralization in soil. *Applied Microbiology and Biotechnology*, 38, 799–807.

Peijnenburg, W. J. M., Hart, M. J. T., den Hollander, H. A., van de Meent, D., Verboom, H. H. & Wolfe, N. L. (1992). QSAR's for predicting reductive transformation rate constants of halogenated aromatic hydrocarbons in anoxic sediment systems. *Environmental Toxicology and Chemistry*, 11, 301–14.

Petrasek, A. C., Kugelmann, I. J., Austern, B. M., Pressley, T. A., Winslow, L. A. & Wise, R. H. (1983). Fate of toxic organic compounds in wastewater treatment plants. *Journal of the Water Pollution Control Federation*, 55, 1286–96.

Pflug, A. D. & Burton, M. B. (1988). Remediation of multimedia contamination from the wood-preserving industry. In *Environmental Biotechnology*, ed. G. S. Omenn, pp. 193–201. New York: Plenum Press.

Piotrowski, M. R., Doyle, J. R., Cosgriff, D. & Parsons, M. C. (1994). Bioremedial progress at the Libby, Montana, Superfund site. In *Applied Biotechnology for Site Remediation*, ed. R. E. Hinchee, D. B. Anderson, F. B. Metting, Jr & G. D. Sayles, pp. 240–55. Boca Raton, FL: Lewis Publishers.

Reynolds, C. M., Travis, M. D., Braley, W. A. & Scholze, R. J. (1994). Applying field-expedient bioreactors and landfarming in Alaskan climates. In *Hydrocarbon Bioremediation*, ed. R. E. Hinchee, B. C. Alleman, R. E. Hoeppel & R. N. Miller, pp. 100–6. Boca Raton, FL: Lewis Publishers.

Riser-Roberts, E. (1992). *Bioremediation of Petroleum Contaminated Sites*. Boca Raton, FL: C. K. Smoley.

Robison, R. (1987). Regulations target underground tanks. *Civil Engineering*, February, 72–4.

Rogers, J. A., Tedaldi, D. J. & Kavanaugh, M. C. (1993). A screening protocol for bioremediation of contaminated soil. *Environmental Progress*, 12, 146–56.

Sathish, N., Young, J. C. & Tabak, H. H. (1993). Protocol for determining the rate of biodegradation of toxic organic chemicals in anaerobic processes. In *Emerging Technologies in Hazardous Waste Management III*, ed. D. W. Tedder & F. G. Pohland, ACS Symposium Series 518, pp. 203–18. Washington, DC: American Chemical Society.

Sayles, G. D. & Suidan, M. T. (1993). Biological treatment of industrial and hazardous wastewater. In *Biotreatment of Industrial and Hazardous Waste*, ed. M. A. Levin & M. A. Gealt, pp. 245–67. New York: McGraw-Hill.

Scheibenbogen, K., Zytner, R. G., Lee, H. & Trevors, J. T. (1994). Enhanced removal of selected hydrocarbons from soil by *Pseudomonas aeruginosa* UG2 biosurfactants and some chemical surfactants. *Journal of Chemical Technology and Biotechnology*, 59, 53–9.

Schuegerl, K. (1987). *Bioreaction Engineering*, Vol. 2. New York: John Wiley & Sons.

Shah, Y. T. (1979). *Gas–Liquid–Solid Reactor Design*. New York: McGraw-Hill.

Singleton, I. (1994). Microbial metabolism of xenobiotics: fundamental and applied research. *Journal of Chemical Technology and Biotechnology*, 59, 9–23.

Song, H-G., Wang, X. & Bartha, R. (1990). Bioremediation potential of terrestrial fuel spills. *Applied and Environmental Microbiology*, 56, 652–6.

Tabak, H. H. & Govind, R. (1993). Development of nonlinear group contribution method for prediction of biodegradation kinetics from respirometrically derived kinetic data. In *Emerging Technologies in Hazardous Waste Management III*, ed. D. W. Tedder & F. G. Pohland, ACS Symposium Series 518, pp. 159–90. Washington, DC: American Chemical Society.

Thayer, A. M. (1992). Pollution reduction. *Chemical Engineering News*, 70 (46), 22–52.

U.S. Environmental Protection Agency (1988). *Rules and Regulations on Underground Storage Tanks*. Washington, DC: USEPA (December).

U.S. Environmental Protection Agency (1991). *Biological Treatment of Wood Preserving Site Groundwater by BioTrol, Inc.* EPA/540/A5-91/001. Washington, DC: U.S. EPA.

Wu, S. & Gschwend, P. M. (1986). Sorption kinetics of hydrophobic compounds to natural sediments and soils. *Environmental Science & Technology*, 20, 717–25.

2

Bioremediation in soil: influence of soil properties on organic contaminants and bacteria

Matthew J. Morra

2.1 Introduction to the soil system

One of the major obstacles in bioremediation of soils contaminated with synthetic organic compounds is the failure of laboratory remediation schemes to simulate the impact of field soil conditions on both the contaminant and the microorganism (Rao *et al.*, 1993). The purpose of this chapter is to introduce those topics which must be considered in order to develop an effective bioremediation strategy for soils contaminated with organic pollutants. My emphasis is on providing a comprehensive overview of the complexity of the soil system as it relates to bioremediation.

The soil environment is a dynamic one which includes gas, liquid, and solid phases. It is imperative in any soil bioremediation process to have a clear understanding of these phases and how they interact. This includes not only the chemical characteristics of soil colloids, but the physical arrangement of components. The first section is meant as a brief introduction to soil chemical and physical properties and is intended primarily for those unfamiliar with the area.

2.1.1 The inorganic solid phase in soil

The solid phase consists of both inorganic and organic components, which in many cases form a heterogeneous mixture defined as organo-mineral complexes. In-depth treatments of soil chemistry, humus chemistry, and mineralogy are available (Aiken *et al.*, 1982; Bohn *et al.*, 1985; Dixon and Weed, 1989; Sposito, 1989; McBride, 1994; Stevenson, 1994). Soil inorganic components include both crystalline materials in the form of layer silicates as well as more poorly crystalline oxides, hydroxides, and oxyhydroxides (collectively termed hydrous oxides). Although these minerals are present in various soil size fractions, those minerals in the clay-sized fraction ($< 2.0 \, \mu$m diameter) generally exert the greatest influence on chemical reactions because of their large surface areas.

Layer silicates or phyllosilicates play a key role in soil reactions because they typically exist in the clay-sized fraction and participate as cation and, in some cases, anion exchangers. The tetrahedral layer or sheet contains SiO_4 tetrahedra in which the centrally located Si is surrounded by three basal and one apical O^{2-} ion (Figure 2.1). The SiO_4 tetrahedra are linked in the phyllosilicates by corner-sharing of the basal oxygens. In these minerals, the apical O^{2-} links with an octahedral sheet containing either divalent or trivalent cations surrounded in octahedral coordination with four OH groups and $2 O^{2-}$. The interaction of the tetrahedral and octahedral sheets is such that, under ideal conditions, charge neutrality is maintained.

The number and exact composition of the sheets is used to classify the phyllosilicates. The most important classification for our purposes is the distinction between 1:1 and 2:1-type minerals (Figure 2.1). In 1:1 minerals such as kaolinite, the basal oxygens of the tetrahedral sheet are free to interact with octahedral OH groups forming hydrogen bonds. In contrast, 2:1 minerals such as pyrophyllite or talc contain two tetrahedral sheets sandwiched around an octahedral sheet. These minerals have only basal oxygens exposed on the faces of the tetrahedral sheets and are linked by weak van der Waals forces. The weaker interaction of one 2:1 layer with a second 2:1 layer results in interlayer spaces which, depending on the particular mineral, may be available for contaminant intercalation.

The cation exchange capacity of phyllosilicates results from deviation of the ideal structure such that charge neutrality is no longer maintained. The net negative charge may be pH dependent or permanent in origin. The 1:1-type minerals have a relatively low cation exchange capacity because the structure closely resembles that of the ideal lattice and charge results mainly from unsatisfied bonds on broken mineral edges. In 2:1-type minerals, isomorphous substitution of a similar-sized cation in either the tetrahedral or octahedral sheet may occur. If the substituted cation is of lower valence, then a permanent negative charge results. As a consequence, 2:1 minerals typically have much higher cation exchange capacities than 1:1 minerals.

Also contributing to larger cation exchange capacities for 2:1 minerals is the

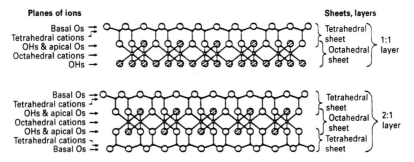

Figure 2.1. Basic structures of 1:1 and 2:1 phyllosilicates. Adapted from Schulze (1989). Reprinted with permission from the Soil Society of America.

possible availability of large interlayer surface areas. The negative charge deficit on 2 : 1 minerals is satisfied by interlayer cations which link adjacent sheets via mutual attraction. If the layer charge is weak enough, molecules such as water can enter this interlayer space. The effective surface area of these minerals is increased much as the total outer surface area of a book is increased by summing the surface areas of the individual pages.

Aluminum, iron, manganese, and titanium oxides, hydroxides, or oxyhydroxides also exist in soils. These minerals occur in the very small particle size fraction and thus exert a great influence on soil reactions because of their large surface areas. A characteristic of aluminum and iron oxides that is particularly relevant to bioremediation is the pH-dependent charge resulting from the protona-tion/deprotonation of surface functional groups (Figure 2.2). Consequently, at pH values below their zero point of charge (ZPC), aluminum and iron oxides have a net positive charge and thus exhibit an anion exchange capacity. Manganese oxides likewise exhibit a pH-dependent charge, but in most cases the ZPC is so far below typical soil pH values that they primarily participate only in cation exchange. Manganese and iron oxides may also participate in redox reactions resulting from oxidation state changes, reactions which are enhanced by the fact that these oxides possess such large surface areas.

2.1.2 Natural organic materials in soils

Natural organic materials in soils exist in two distinct forms: those that are recognizable as originating from a particular organism and those that have been chemically altered such that their origin is not easily determined. Recognizable compounds include large polymeric materials (e.g., cellulose, protein, lignin) as well as relatively simple molecules (e.g., sugars, organic acids, amino acids). Large polymeric materials without a recognizable, highly organized structure are called humic materials (Figure 2.3). Humic materials are operationally divided into humic and fulvic acids based on solubility differences in acid and base. Humic substances not extracted by base are termed humin. Both humic and fulvic acids can be extracted from soil using NaOH, with acidification of the extract resulting in humic acid precipitation. This differential behavior is interpreted to reflect the

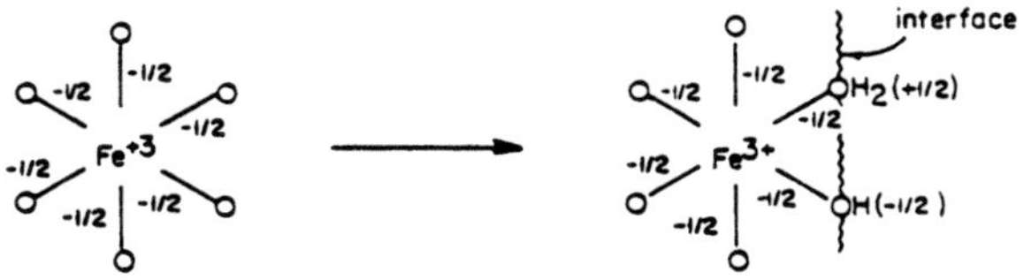

Figure 2.2. Charges of ligands of Fe^{3+} in the interior of hematite and at the surfaces. Adapted from Bohn *et al.* (1985). Reprinted by permission of John Wiley & Sons, Inc.

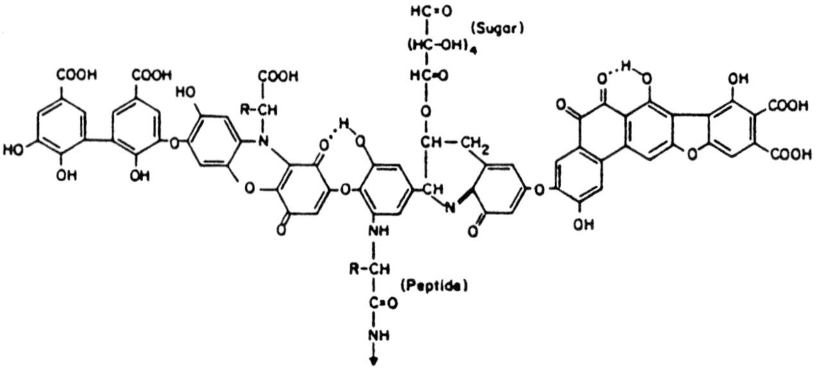

Figure 2.3. Hypothetical structure for humic acid. Adapted from Stevenson (1994). Reprinted by permission of John Wiley & Sons, Inc.

smaller molecular size of fulvic acid and its lower content of carboxyl and phenolic hydroxyl groups. Basic conditions result in deprotonation of these functional groups, thus rendering both humic and fulvic acids soluble. Subsequent acidification of the extract causes protonation, resulting in humic acid precipitation as a consequence of its larger molecular size and greater acidic functional group content. The underlying mechanism is charge neutralization and increased intra- and intermolecular bonding.

Humic materials are, therefore, largely polar materials having a net negative charge, and like most layer lattice clays, a cation exchange capacity. In soil, humic and non-humic materials largely exist in association with mineral surfaces to form organo-mineral complexes. It is these organo-mineral complexes that control contaminant binding, in turn controlling accessibility for microbial attack and bioremediation. Likewise, the interaction of microorganisms with organo-mineral complexes is a controlling factor in their ability to survive and move in the soil environment. Not only are the bonding mechanisms or type of interaction important, but also the physical orientation and organization of the organo-mineral complexes within soil.

2.1.3 Physical characteristics of the soil system

In addition to the chemical characteristics of soil it is also necessary to consider the implications of physical parameters on bioremediation. The organization and orientation of soil particles results in the development of soil structure. Soil particles, irrespective of their chemical composition, are defined on a size basis as sand (2.0–0.05 mm), silt (0.5–0.002 mm), and clay (<0.002 mm). These primary soil particles are cemented together by organic and inorganic materials to form microscopic aggregates. At a larger scale of millimeters to centimeters, the microaggregates are organized into macroaggregates or peds. Aggregate formation results in structures containing various size pores which in turn can be filled by either the soil solution or by various soil gases (Hillel, 1982). Approximately 50%

of a typical surface soil is pore space, with a gas to liquid ratio which varies tremendously both temporally and spatially. Diffusion of gaseous and liquid components into and out of the aggregate has important implications on bioremediation efficacy and degradation kinetics. A graphic example of diffusional constraints is the development of anaerobic microsites within the central portion of soil aggregates as small as 4 mm in diameter (Figure 2.4). Aggregate characteristics controlling the development of microsites have been summarized in the form of a model (Renault & Sierra, 1994a, b).

In many instances a contaminant is distributed vertically within soil to depths in which soil characteristics vary tremendously (Buol *et al.*, 1989) (Figure 2.5). A vertical section of the soil exposing these layers or horizons is termed a soil profile.

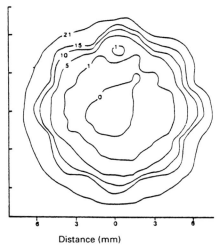

Distance (mm)

Figure 2.4. Anaerobic microsite in the interior of a soil aggregate. Adapted from Sexstone *et al.* (1985). Contour designations represent percentage oxygen concentrations. Reprinted with permission from the Soil Science Society of America.

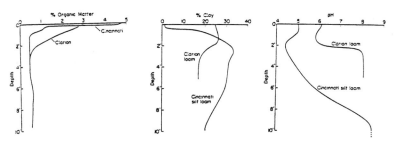

Figure 2.5. Changes in soil characteristics with depth in the soil profile. Depth measured in feet. Adapted from Ahlrichs (1972) by courtesy of Marcel Dekker Inc.

The horizons lie approximately parallel to the soil surface and are the product of weathering processes. The physical and chemical differences among the horizons alter reactions between the soil and contaminant, and the soil and microorganism. As a result, the efficiency of bioremediation at one location in the soil profile may differ radically from that at another point in the same profile.

2.2 Influence of soil on microorganisms

The interaction of bacteria with solid surfaces including soil may have a variety of indirect and direct impacts on the cell (van Loosdrecht et al., 1990). Direct impacts result from changes in microbial membranes (e.g., permeability to various substrates) resulting from a surface interaction. Indirect impacts related to microbial activity are a result of modification of the immediate environment of the cell (e.g., alteration of substrate availability) (Harms & Zehnder, 1994). The influences of soil colloids on general microbial processes (Stotzky, 1986) and biodegradation kinetics of organic contaminants (Scow, 1993) have been summarized. However, two areas specifically pertinent to bioremediation will be described.

2.2.1 Microbial survival

Bioremediation requires either the inoculation of contaminated soil with a specific strain or consortium, or the stimulation of indigenous organisms (Rao et al., 1993). Successful inoculation requires sufficient microbial numbers within the inoculum itself and adequate survival of these organisms once placed in the soil system. The most difficult aspect is maintaining numbers of the introduced bacterial strain or strains. Numerous investigations demonstrate a decline in bacterial numbers with increased time after the original inoculation. A number of mechanisms have been suggested to account for these observations, including both biotic (e.g., predation) (Acea & Alexander, 1988) and abiotic mechanisms (e.g., inappropriate soil pH) (Young & Burns, 1993). Within this context, it is appropriate to consider the impact of minerals on microbial survival.

Minerals differentially affect microbial activity and survival (Marshall, 1975; Stotzky, 1986; Young & Burns, 1993), most often with enhanced microbial survival in the presence of clay (Marshall, 1964, 1975; Bitton et al., 1976; Heijnen et al., 1988, 1993). One of the proposed mechanisms for increased survival is the ability of clays to protect the organism from desiccation (Robert & Chenu, 1992). The most thoroughly investigated relationship between bacterial desiccation and soil mineralogy has been performed with rhizobia. Evidence indicates that Rhizobium survival in soils undergoing desiccation positively correlates with clay content (Al-Rashidi et al., 1982; Osa-Afiana & Alexander, 1982; Hartel & Alexander, 1986). However, Chao & Alexander (1982) found a negative relationship between the clay content of twelve different soils and survival of several rhizobia species.

Soil mineralogy, not just total clay content, must be considered in interpreting the results of these apparently conflicting results. Amendment of soils or sand with montmorillonite increases the desiccation resistance of some rhizobia species (Marshall, 1964; Bushby & Marshall, 1977a; Osa-Afiana & Alexander, 1982), whereas kaolinite amendment results in decreased survival (Osa-Afiana & Alexander, 1982). In addition, montmorillonite- but not kaolinite-amended soil has been shown to protect *R. trifolii* from exposure to high temperatures (Marshall, 1964). Illite-containing or illite-amended soils result in variable survival rates (Chao & Alexander, 1982; Osa-Afiana & Alexander, 1982).

Bacterial desiccation and survival of other species in dried soils and mineral powders has been reported (Bitton *et al.*, 1976; Labeda *et al.*, 1976; Dupler & Baker, 1984; Moll & Vestal, 1992). In many instances the authors did not report soil mineralogical characteristics; however, Bitton *et al.* (1976) showed greater survival of *Klebsiella aerogenes* under desiccation stress when in soils dominated by montmorillonite as compared to kaolinite. Amendment of montmorillonite to a sandy soil also increased the survival of *K. aerogenes* and thus produced the greatest increase in survival in these studies as well as those conducted with rhizobia.

In addition to ecological considerations concerning bacterial survival in soil, the relationship between desiccation sensitivity and mineral characteristics has relevance to carriers selected for the preparation of bacterial inocula (Kloepper & Schroth, 1981; van Elsas & Heijnen, 1990; Caesar & Burr, 1991). Chao & Alexander (1984) explored the potential use of soil-based inoculants for rhizobia. Pesenti-Barili *et al.* (1991) conducted a comprehensive survey of nine potential carriers for *Agrobacterium radiobacter* K84, including kaolinite and vermiculite, and concluded that vermiculite was most suitable.

Increased desiccation resistance and survival in the presence of clays may result for a number of reasons. Particle size as related to the type or amount of clay is the central focus in many of the suggested mechanisms. The survival of *Pseudomonas* spp. was greater in air-dried mineral powders with smaller size particles and higher surface areas (Dandurand *et al.*, 1994). Increased survival of the bacteria was achieved in lower surface area minerals by inclusion of lactose. There was no correlation of survival with any one particular type of lattice structure. A number of investigators argue that increased surface area slows the rate of drying and that a slower desiccation rate results in increased bacterial survival (Bitton *et al.*, 1976; Zechman & Casida, 1982; Chao & Alexander, 1984; Dupler & Baker, 1984; Hartel & Alexander, 1984, 1986). The addition of clay to increase soil surface area, particularly a higher surface area clay such as montmorillonite, would thus be expected to increase desiccation resistance. Others suggest that higher surface area minerals increase bacterial desiccation resistance by removing more water from the bacterial cell, reducing enzymatic activity (Bushby & Marshall, 1977b; Osa-Afiana & Alexander, 1982). Slower metabolic activity would create a resting bacterial stage capable of surviving longer periods of desiccation stress. It is also possible that colloidal materials may coat bacterial cells (Bashan & Levanony, 1988) and in some unknown fashion increase desiccation resistance of the organism (Marshall, 1964). Smaller particles would be expected to produce more complete coverage and a larger protective effect. Although the exact mechanisms by which minerals affect microbial survival remain unknown, it is apparent that investigations to enhance the survival of organisms released in bioremediation strategies should consider mineral–cell interactions.

Increased survival of introduced organisms may also be achieved by amending soil with a substrate utilizable by the introduced organism, but not the indigenous population. Lajoie et al. (1992, 1993, 1994) have explored the possibility of using surfactants as selective substrates to enhance the survival of a genetically modified organism capable of contaminant degradation. In addition to providing a growth substrate for the organism, the surfactant may also increase contaminant desorption from soil constituents, thereby increasing its bioavailability (see Section 2.3).

The second relevant area concerns the participation of soil colloids in genetic exchange processes, a topic of considerable interest in those cases in which genetically altered microorganisms are utilized in remediation strategies, as described in Chapter 10. Conjugal transfer of both chromosomal and plasmid DNA is possible between introduced organisms as well as between an introduced microbe and the indigenous bacterial population (Stotzky, 1986; Walter et al., 1989; Henschke & Schmidt, 1990). Any differences in the soil environment which promote or inhibit donor or recipient populations would be expected to alter the extent of conjugal transfer. Such conditions include soil moisture content, pH, and mineralogy, as well as competition and predation by other soil microflora.

More important for the purposes of this review is the possible participation of soil in promoting transformation involving extracellular DNA stabilized in some manner by its interaction with soil colloids. Extracellular DNA originates from soil bacteria (Lorenz et al., 1991) and genetically modified bacteria placed in aquatic environments (Paul & David, 1989), and exists in environmental matrices (Lorenz et al., 1981; Ogram et al., 1987; Paul et al., 1990).

Once released in the soil system, DNA interaction with organic and inorganic colloids is likely. Binding of DNA to sand, clays, and soils has been reported under laboratory conditions (Greaves & Wilson, 1969, 1970; Lorenz et al., 1981; Aardema et al., 1983; Lorenz & Wackernagel, 1987; Ogram et al., 1988; Romanowski et al., 1991; Khanna & Stotzky, 1992). Given that the isoelectric point of DNA is at approximately pH 5.0, a positive charge develops only at lower pH values. Once DNA becomes positively charged, electrostatic interaction with negatively charged clays and humic materials is possible. Lower pH values do indeed promote DNA binding to montmorillonite (Greaves & Wilson, 1969; Khanna & Stotzky, 1992). Some dispute exists as to the location of the binding and whether or not intercalation into interlayer spaces of 2:1 minerals occurs. Greaves & Wilson (1969, 1970) present convincing X-ray diffraction evidence to illustrate expansion of montmorillonite interlayer spacing, apparently as a result of DNA intercalation. In contrast, Khanna & Stotzky (1992) also used X-ray diffraction and observed no DNA-mediated expansion.

The mechanism of DNA binding has important implications with respect to its enzymatic availability and potential to participate in microbial transformations. Association of DNA with solid surfaces reduces, but does not eliminate, degradation by various DNAses (Greaves & Wilson, 1969, 1970; Lorenz et al., 1981; Lorenz & Wackernagel, 1987; Romanowski et al., 1991; Khanna & Stotzky, 1992). However, significant evidence exists to demonstrate that colloids protect extracellular DNA and that at least a portion of the sorbed DNA is available for transformation (Lorenz et al., 1988, 1992; Lorenz & Wackernagel, 1990; Stewart & Sinigalliano, 1990; Paul et al., 1991; Romanowski et al., 1992; Khanna & Stotzky, 1992; Recorbet et al., 1993). In addition to a direct impact on transforming DNA, the soil environment may indirectly affect transformation as it controls immobilization and competence of the recipient cell (Lorenz & Wackernagel, 1991).

2.2.2 Microbial mobility and transport

In situ soil bioremediation requires transport of the microorganism from the site of introduction to the location of the contaminant within the soil (Young & Burns, 1993). In many cases the contaminant is located some distance from the soil surface and significant vertical transport within the soil profile is necessary. Models are available to predict the transport of bacteria in soils or groundwater matrices (Corapcioglu & Haridas, 1984, 1985; Tan et al., 1992); however, our underlying understanding of the processes which control microbial mobility is incomplete. A recent summary of the variables controlling microbial movement in soil is available (Gammack et al., 1992).

The main processes which govern bacterial transport, excluding survival, are sorption to soil components and physical filtering in which cell size inhibits transport through small soil pores. It is important to remember that inorganic and organic colloids in many instances have a net negative charge, as do most bacterial cells. The mechanisms by which microorganisms may interact and overcome electrostatic repulsion with solid surfaces in soils include anion exchange, H-bonding, cation bridging, ligand exchange, and hydrophobic interactions (Stotzky, 1986). Most investigations indicate that hydrophobic mechanisms control bacterial sorption onto soil constituents and that electrostatic repulsion is insignificant (Wan et al., 1994; Huysman & Verstraete, 1993a, b; Stenström, 1989; DeFlaun et al., 1990). The more hydrophobic the cell membrane, the greater the sorption on solid soil surfaces.

The role of sorption in bacterial transport must be discussed within the context of the soil physical environment. Laboratory studies have focused on the use of leaching columns containing soils or sand and irrigated with controlled water flow rates (Bitton et al., 1974; Wollum & Cassel, 1978; Breitenbeck et al., 1988; Gannon et al., 1991b; Huysman & Verstraete, 1993a, b). Greater cell hydrophobicities appear to decrease vertical transport (Huysman & Verstraete, 1993a, b; Wan et al., 1994), but increased cell size may also contribute to decreased microbial movement (Gannon et al., 1991a). As a result, columns packed with larger sand-sized particles result in greater bacterial transport than columns packed with soils or sand to which clay is added (Breitenbeck *et al.*, 1988; Huysman & Verstraete, 1993a). Although some debate exists as to the relative importance of filtering versus sorption phenomena in the observed decrease in movement (Huysman & Verstraete, 1993b), the overall consequence of smaller-sized soil particles is less vertical transport. It is also generally agreed that increased flow rates result in increased cell transport (Trevors *et al.*, 1990; Huysman & Verstraete, 1993b).

The recent work of Wan *et al.* (1994) may help in interpreting the controversial aspects presented above. This work is especially relevant to the soil system and *in situ* bioremediation because the focus is on microbial transport under unsaturated conditions. Previously, microbial retention on solid surfaces was considered of primary importance and little consideration was given to the gas–water interface. The authors appropriately consider unsaturated soil a three-phase system and, in contrast to previous investigations, placed special emphasis on the role of the gas phase. They conclude that bacterial sorption to the gas–water interface is a hydrophobic process, thus explaining the correlation with cell hydrophobicity previously observed by various investigators. At low flow rates, the gas–water interface facilitates bacterial retention and cell transport is inhibited. With increased water flow rates, this stabilized interface with its associated bacteria is mobilized, resulting in increased cell transport. A re-evaluation of the results of previous investigators within the context of these findings seems warranted.

Although the above studies conducted with packed columns are important from a fundamental standpoint as they relate to the mechanisms of cell sorption to solid surfaces, *in situ* remediation of contaminants in subsoils requires microbial transport in well-structured soils. The presence of soil macropores that facilitate preferential water flow is well appreciated (Thomas & Phillips, 1979). Sorption phenomena are less important when bacterial transport occurs through structured soils in which cells pass unimpeded through relatively large conduits (Smith *et al.*, 1985).

2.3 Interaction of synthetic organics with soil constituents

The fate of organic contaminants in soil, including the potential for bioremediation, is controlled by binding affinities and reactions with soil constituents. Numerous reviews discussing the mechanisms and forces by which organic contaminants bind to soil constituents are available (Green, 1974; Weed & Weber, 1974; Hassett & Banwart, 1989; Koskinen & Harper, 1990; Weber *et al.*, 1993) and a comprehensive discussion of this topic will not be repeated. However, it is highly recommended that those involved in soil bioremediation carefully consider the chemical characteristics of the organic contaminant and the type of reactions possible with the site-specific soil.

The first step in making such predictions is to categorize the pesticide and, in a relative sense, characterize its binding affinity (Figure 2.6). Cationic compounds will bind tightly to soil because of electrostatic interaction with negatively charged minerals and organic matter. Weak bases (e.g., triazine herbicides) will also bind

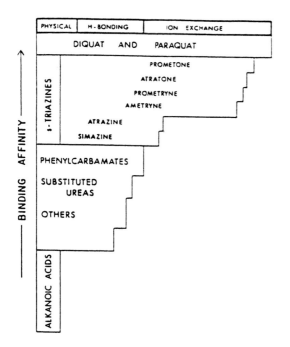

Figure 2.6. Binding affinities of herbicides to soil organic matter. Adapted from Stevenson (1972). Reprinted with permission from the Soil Science Society of America.

electrostatically, but pH will control whether the compound is positively charged or neutral. In a similar but opposite fashion, acidic compounds containing carboxylic or phenolic hydroxyl functional groups will be neutral at low pH, but electrostatically repulsed as deprotonation occurs at higher pH values. Nonionic compounds are perhaps the most difficult and controversial groups of contaminants on which to make predictions. In soils containing water, it is generally assumed that soil organic carbon controls the extent and type of interaction and that binding is a result of hydrophobic sorption reactions. Substantial evidence exists supporting a phase-partitioning mechanism in which the nonionic compound is solvated by soil organic material, thus becoming homogeneously distributed throughout the organic matrix (Chiou, 1989; Chiou & Kile, 1994). However, recent work with organo-mineral complexes suggests that hydrophobic organic compounds interact in a specific fashion such that sorption is actually a surface reaction and not solvation (Murphy et al., 1990, 1994). Others have suggested that specific reactions, in contrast to nonspecific partitioning, of synthetic organics with soil humic materials are important especially at low contaminant concentrations (Spurlock, 1995; Spurlock & Biggar, 1994), but the topic remains controversial (Chiou, 1995). The type of interaction has important implications with respect to the microbial bioavailability of nonionic organic contaminants.

Beyond these generalizations, it is important to consider possible reactions specific to the contaminant and soil combination of interest. For example, if the soil contains iron oxides, the pH-dependent binding of weak acids is possible. Below the ZPC, electrostatic interaction of the positively charged oxide is possible with negatively charged compounds such as 2,4-D, 2,4,5,-T, and MCPA if present in their deprotonated form (Schwertmann *et al.*, 1986).

2.3.1 Impact on contaminant bioavailability

Substrate availability is of tremendous interest, not only for predicting the successful application of soil bioremediation, but as it relates to microbial ecology in general. It is often assumed that bound substrates are not microbially available until desorption occurs (Miller and Alexander, 1991). Investigations specifically designed to determine microbial bioavailability of organic contaminants in soils have resulted in varied conclusions, creating a controversial topic without clear generalizations. However, the contrasting conclusions are perhaps better interpreted not as inconsistencies, but as true bioavailability differences resulting from varied binding affinities of the contaminants of interest. Reviews of sorption impacts on biodegradation of nonionic (Scow, 1993) and ionizable (Ainsworth *et al.*, 1993) compounds are available.

The relatively strong interaction of cationic contaminants with negatively charged soil constituents, for example, is expected to decrease bioavailability. This has been shown to be the case for diquat, in which intercalation into internal clay surfaces eliminates microbial degradation of the compound (Weber and Coble, 1968). Decreased bioavailabilities for benzylamine in association with montmorillonite (Miller & Alexander, 1991), quinoline bound to hectorite or montmorillonite (Smith *et al.*, 1992), and cationic surfactants with humic materials or montmorillonite (Knaebel *et al.*, 1994) have also been reported.

Bioavailability of nonionic and weakly acidic compounds when in association with soil constituents is more difficult to predict. Confusion exists because bonding mechanisms and associated affinities are correspondingly more complex and exhibit more soil-to-soil variability. Ogram *et al.* (1985) present convincing kinetic evidence showing that sorbed 2,4-D is unavailable to a 2,4-D-degrading bacterial culture. Guerin & Boyd (1992) also use kinetic analysis of naphthalene degradation in soil to demonstrate that the sorbed compound is differentially available to two bacterial species. One bacterial strain, *Pseudomonas putida* ATCC 17484, was capable of degrading naphthalene at a faster rate than would be predicted based only on the pool of naphthalene in the aqueous phase as determined from equilibrium calculations. The phenomenon was explained using the rationale that both a labile and nonlabile pool of naphthalene exist in soil and that the bacteria use the labile pool. A concentration gradient is thereby created for transfer of naphthalene from the nonlabile to the labile pool.

When selecting an organism or organisms for bioremediation it is therefore important to consider the possibility that overall degradation rates of the contaminant may be controlled by substrate availability, which in turn may vary among the organisms. The production of microbial biosurfactants is well established (Lin *et al.*, 1994; Georgiou *et al.*, 1992) and should be considered as a possible means to increase contaminant bioavailability when designing bioremediation strategies. Supplementation with exogenous surfactants may also be used to increase contaminant availability and biodegradation rates (Churchill *et al.*, 1995), although increased off-site transport must be considered.

The issue of bioavailability is further clouded by the physical characteristics of soil and the role of a possible mass transfer limitation. Soil constituents are not simply flat surfaces with free and equal access to all bacterial species. The formation of aggregates from sand-, silt-, and clay-sized particles results in stable structures which control microbial contact with the substrate (Figure 2.7). Discussion of sorption mechanisms and binding affinities must include the possible impact of intra-aggregate transport of the substrate. If the substrate is physically inaccessible to the microorganism then both desorption from soil constituents and diffusion to an accessible site are necessary. The impact of intra-aggregate diffusion on degradation kinetics has been modeled for γ-hexachlorocyclohexane (Rijnaarts *et al.*, 1990) and naphthalene (Mihelcic & Luthy, 1991).

Relatively long residence times of synthetic organics in some soils offer additional evidence for the physical isolation and protection of these potential

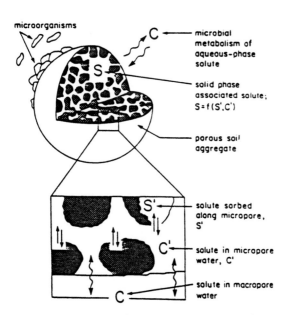

Figure 2.7. Depiction of the physical relationship of microorganisms to the structure of a soil aggregate containing an organic contaminant. Adapted with permission from Mihelcic & Luthy (1991). Copyright 1991, American Chemical Society.

substrates from microbial degradation. Steinberg *et al.* (1987) showed that EDB residues remaining in soils 19 years after they were last treated must result from containment of the compound in micropores inaccessible to degrading microorganisms. This supposition was supported indirectly by amendment experiments in which the freshly added compound was mineralized in several weeks by indigenous microorganisms. In addition, direct evidence for EDB entrapment in micropores was demonstrated by the accelerated release of EDB from the soil caused by soil pulverization. The persistence of parathion and DBCP in soils far beyond the anticipated residence times has also been reported (Stewart *et al.*, 1971).

More recent evidence for atrazine and metolachlor also indicates 'aged residues' of these compounds exist in soil in a form or location affording protection from microbial attack (Pignatello *et al.*, 1993). However, the mechanism was not definitively identified as entrapment in micropores. Desorption took place over long time periods, was positively correlated with soil organic carbon content, and was independent of aggregate size. Both sorption and desorption of the compounds into and out of recalcitrant pools were controlled by diffusion kinetics. Although the possibility of micropore participation in the process was not eliminated, the data are also consistent with a mechanism involving contaminant partitioning and diffusion within soil organic matter (Brusseau & Rao, 1989; Brusseau *et al.*, 1991).

These studies are important from a theoretical standpoint as they relate to the partitioning mechanism proposed to explain nonionic compound interactions with soil organic matter (Chiou, 1989). The slow kinetics related to diffusion through soil organic matter implies contaminant mobility through a three-dimensional network, consistent with the solvation or phase partitioning hypothesis. More important in bioremediation are the consequences with respect to substrate bioavailability. Synthetic organic contaminants existing within this trapped phase are more resistant or perhaps completely unavailable to bioremediation. Experiments showing bioremediation potential of a particular microorganism or consortium using soils recently amended with the contaminant of interest may not simulate contaminant bioavailability actually existing in the field. Scribner *et al.* (1992) clearly demonstrated the lack of bioavailability of aged simazine residues in soil. Similarly, Hatzinger & Alexander (1995) showed decreased biodegradation of both phenanthrene and 4-nitrophenol with longer periods of soil contact. The kinetics of both sorption and desorption would be difficult to simulate under laboratory conditions and if not considered, would greatly alter the time necessary to bioremediate a contaminated field soil.

2.3.2 Transport phenomena

The interaction of synthetic organic chemicals with mobile colloids present in soil solution has been demonstrated (Ballard, 1971). These interactions may alter the environmental behavior and fate of synthetic organic compounds, including toxicity and bioaccumulation, volatility, photolysis, and nonbiological alteration. It has also been proposed that mobile colloids may act as carriers for organic chemicals, increasing their rate of transport in the soil profile (Jury *et al.*, 1986; Enfield, 1985; Enfield *et al.*, 1989) and the subsurface environment (McCarthy & Zachara, 1989).

Although such reactions and the consequences with respect to contaminant fate have primarily focused on soluble humic materials (Carter & Suffet, 1982; Madhun *et al.*, 1986; Traina *et al.*, 1989; Morra *et al.*, 1990; Puchalski *et al.*, 1992; Engebretson & von Wandruszka, 1994), the participation of microbial products in similar reactions is possible. Dohse and Lion (1994) showed that extracellular bacterial polymers enhanced the transport of phenanthrene in sand columns. The mobilization of contaminants might be beneficial to bioremediation if degradation reactions are not inhibited and substrate bioavailability is increased. Conversely, increased contaminant transport may increase the potential for contaminant movement and likewise the extent of environmental contamination.

2.3.3 Abiotic catalysis

One aspect of soil bioremediation frequently overlooked is the possible catalytic participation of inorganic and organic colloids in the alteration of the synthetic organic contaminant. Abiotic alteration does not usually result in mineralization of the compound, but alters the structure such that bioremediation efforts must consider the possibility that the contaminant may no longer exist in its original form. Most authors consider abiotic catalysis to include the participation of recognizable biomolecules (e.g., enzymes and coenzymes) as long as the catalytic activity of these molecules does not require the participation of living organisms. Abiotic reactions including hydrolysis (Torrents & Stone, 1994), elimination, substitution, redox, and polymerization are influenced by minerals, oxides, and natural organic materials. Detailed reviews of reaction mechanisms and kinetic considerations are available (Schwarzenbach *et al.*, 1993; Wolfe *et al.*, 1990; Huang, 1990; Wang *et al.*, 1986; Hayes & Mingelgrin, 1991; Zielke *et al.*, 1989; Voudrias & Reinhard, 1986). No attempt will be made to reiterate information contained in these reviews, but several areas merit special attention.

Abiotic reduction reactions occur in natural matrices, possibly a consequence of the presence of reduced iron or sulfur species (Kriegman-King & Reinhard, 1994; Roberts *et al.*, 1992). However, it has been suggested that direct reduction of an organic contaminant by a reduced sulfur or iron component is too slow to account for observed kinetics. Kriegman-King & Reinhard (1992) showed that the solid surfaces of biotite and vermiculite increase the rate of CCl_4 disappearance when included with H_2S. Alternatively, it has been proposed that electron carriers mediate electron transfer such as nitroaromatic reduction (Schwarzenbach *et al.*, 1990). Tetrapyrroles containing reduced iron, cobalt, or nickel have been proposed as potential electron transfer mediators in natural environments (Tratnyek & Macalady, 1989; Schwarzenbach *et al.*, 1990). Tetrapyrroles containing reduced cobalt (Krone *et al.*, 1989a, b, 1991; Marks *et al.*, 1989; Gantzer & Wackett, 1991; Assaf-Anid *et al.*, 1992, 1994; Schanke & Wackett, 1992) or iron (Wade & Castro,

1973; Khalifa *et al.*, 1976; Klecka & Gonsior, 1984; Baxter, 1990) have been shown to participate in reductive dehalogenation of a wide variety of substrates exclusive of any living organism. Dehalogenation of chlorinated aliphatic hydrocarbons also occurs with participation of coenzyme F_{430}, a nickel(II) porphinoid present in anaerobic bacteria (Krone *et al.*, 1989a, b, 1991). It is likely that extracellular tetrapyrroles potentially available for participation in reductive dehalogenation are associated with solid surfaces (i.e., minerals and organic matter). Zoro *et al.* (1974) demonstrated dehalogenation of DDT by a heme in sewage sludge. More recently Ukrainczyk *et al.* (1995) have demonstrated CCl_4 dechlorination in the presence of a cationic porphyrin, Co tetrakis(N-methyl-4-pyridiniumyl)porphyrin, exchanged on a variety of mineral surfaces. The binding of the tetrapyrrole to the mineral alters tetrapyrrole redox behavior and thus its catalytic activity with respect to dehalogenation (Ukrainczyk *et al.*, 1994, 1995).

These examples illustrate that biomolecules may act as catalysts in soils to alter the structure of organic contaminants. The exact nature of the reaction may be modified by interaction of the biocatalyst with soil colloids. It is also possible that the catalytic reaction requires a specific mineral–biomolecule combination. Mortland (1984) demonstrated that pyridoxal-5'-phosphate (PLP) catalyzes glutamic acid deamination at 20 °C in the presence of copper-substituted smectite. The proposed pathway for deamination involved formation of a Schiff base between PLP and glutamic acid, followed by complexation with Cu^{2+} on the clay surface. Substituted Cu^{2+} stabilized the Schiff base by chelation of the carboxylate, imine nitrogen, and the phenolic oxygen. In this case, catalysis required combination of the biomolecule with a specific metal-substituted clay.

Finally, mention must be made of possible conjugation reactions in which a covalent bond is formed between a contaminant molecule and a second contaminant molecule or soil organic matter. Oxidative coupling reactions of phenolics and aromatic amines are catalyzed by extracellular enzymes, clays, and oxides (Wang *et al.*, 1986; Liu *et al.*, 1987; Huang, 1990). The bioavailability of the synthetic organic within the product is reduced or possibly eliminated (Dec *et al.*, 1990; Allard *et al.*, 1994).

In summary, extrapolation of reactions determined in homogeneous solutions to the soil system may not result in accurate predictions of possible products. Unanticipated abiotic reactions which occur during bioremediation may influence the products, causing altered degradative pathways of the contaminants. These pathways may be site specific because of differences in abiotic catalysts. Abiotic reactions may occur to both the parent compound or to intermediates formed during biotic alteration of the compound. Bioremediation efforts should include the possibility of site specific abiotic reactions.

2.4 Conclusion

During the development of bioremediation strategies it is imperative to consider soil complexity and variability. Frequently, it seems that soil physical and chemical characteristics are considered only after a laboratory-developed remediation strategy fails to ameliorate a pollution problem under field conditions. Instead, the potential impacts of soil on both the contaminant and microorganism should be integrated into the strategy from its inception. The researcher should be familiar with what soil minerals are present and how such minerals might bind the organic contaminant or participate in its abiotic alteration. Likewise, it is necessary to predict binding affinity of the compound to soil organic matter and the effect binding might have on contaminant bioavailability. In those situations in which soils are inoculated with a specific microorganism, it is also necessary to consider what impact the soil will have on microbial survival and activity. Although our knowledge base is far from complete, soil bioremediation would have a much higher potential of success if the remediation strategy adequately anticipated the influence of soil properties.

References

Aardema, B. W., Lorenz, M. G. & Krumbein, W. E. (1983). Protection of sediment-adsorbed transforming DNA against enzymatic inactivation. *Applied and Environmental Microbiology*, 46, 417–20.

Acea, M. J. & Alexander, M. (1988). Growth and survival of bacteria introduced into carbon-amended soil. *Soil Biology & Biochemistry*, 5, 703–9.

Ahlrichs, J. L. (1972). The soil environment. In *Organic Chemicals in the Soil Environment*, Vol. 1, ed. C. A. I. Goring & J. W. Hamaker, pp. 3–46. New York: Marcel Dekker.

Aiken, G. R., McKnight, D. M., Wershaw, R. L. & McCarthy, P., eds. (1982). *Humic Substances in Soil, Sediment, and Water*. New York: John Wiley.

Ainsworth, C. C., Frederickson, J. K. & Smith, S. C. (1993). Effect of sorption on the degradation of aromatic acids and bases. In *Sorption and Degradation of Pesticides and Organic Chemicals in Soil*, ed. D. M. Linn, T. H. Carski, M. L. Brusseau & F-H. Chang, pp. 125–44. Madison, WI: Soil Science Society of America, American Society of Agronomy.

Al-Rashidi, R. K., Loynachan, T. E. & Frederick, L. R. (1982). Desiccation tolerance of four strains of *Rhizobium japonicum*. *Soil Biology & Biochemistry*, 14, 489–93.

Allard, A-S., Hynning, P-A., Remberger, M. & Neilson, A. H. (1994). Bioavailability of chlorocatechols in naturally contaminated sediment samples and of chloroguaiacols covalently bound to C2-guaiacyl residues. *Applied and Environmental Microbiology*, 60, 777–84.

Assaf-Anid, N., Nies, L. & Vogel, T. M. (1992). Reductive dechlorination of a polychlorinated biphenyl congener and hexachlorobenzene by vitamin B_{12}. *Applied and Environmental Microbiology*, 58, 1057–60.

Assaf-Anid, N., Hayes, K. F. & Vogel, T. M. (1994). Reductive dechlorination of carbon tetrachloride by cobalamin(II) in the presence of dithiothreitol: mechanistic study, effect of redox potential and pH. *Environmental Science & Technology*, 28, 246–52.

Ballard, T. M. (1971). Role of humic carrier substance in DDT movement through forest soil. *Soil Science Society of America Proceedings*, 35, 145–7.

Bashan, Y. & Levanony, H. (1988). Adsorption of the rhizosphere bacterium *Azospirillum brasilense* Cd to soil, sand and peat particles. *Journal of General Microbiology*, 134, 1811–20.

Baxter, R. M. (1990). Reductive dechlorination of certain chlorinated organic compounds by reduced hematin compared with their behaviour in the environment. *Chemosphere*, 21, 451–8.

Bitton, G., Lahav, N. & Henis, Y. (1974). Movement and retention of *Klebsiella aerogenes* in soil columns. *Plant and Soil*, 40, 373–80.

Bitton, G., Henis, Y. & Lahav, N. (1976). Influence of clay minerals, humic acid and bacterial capsular polysaccharide on the survival of *Klebsiella aerogenes* exposed to drying and heating in soils. *Plant and Soil*, 45, 65–74.

Bohn, H., McNeal, B. & O'Connor, G. (1985). *Soil Chemistry*, 2nd edn. New York: John Wiley.

Breitenbeck, G. A., Yang, H. & Dunigan, E. P. (1988). Water-facilitated dispersal of inoculant *Bradyrhizobium japonicum* in soils. *Biology and Fertility of Soils*, 7, 58–62.

Brusseau, M. L. & Rao, P. S. C. (1989). Sorption nonideality during organic contaminant transport in porous media. *Critical Reviews of Environmental Control*, 19, 33–99.

Brusseau, M. L., Jessup, R. E. & Rao, P. S. C. (1991). Nonequilibrium sorption of organic chemicals: elucidation of rate-limiting processes. *Environmental Science & Technology*, 25, 134–42.

Buol, S. W., Hole, F. D. & McCracken, R. J. (1989). *Soil Genesis and Classification*, 3rd edn. Ames, IA: Iowa State University Press.

Bushby, H. V. A. & Marshall, K. C. (1977a). Some factors affecting the survival of root-nodule bacteria on desiccation. *Soil Biology & Biochemistry*, 9, 143–7.

Bushby, H. V. A. & Marshall, K. C. (1977b). Water status of *Rhizobia* in relation to their susceptibility to desiccation and to their protection by montmorillonite. *Journal of General Microbiology*, 99, 19–27.

Caesar, A. J. & Burr, T. J. (1991). Effect of conditioning, betaine, and sucrose on survival of rhizobacteria in powder formulations. *Applied and Environmental Microbiology*, 57, 168–72.

Carter, C. W. & Suffet, I. H. (1982). Binding of DDT to dissolved humic materials. *Environmental Science & Technology*, 16, 735–40.

Chao, W-L. & Alexander, M. (1982). Influence of soil characteristics on the survival of *Rhizobium* in soils undergoing drying. *Soil Science Society of America Journal*, 46, 949–52.

Chao, W-L. & Alexander, M. (1984). Mineral soils as carriers for *Rhizobium* inoculants. *Applied and Environmental Microbiology*, 47, 94–7.

Chiou, C. T. (1989). Theoretical considerations of the partition uptake of nonionic organic compounds by soil organic matter. In *Reactions and Movement of Organic Chemicals in Soils*, ed. B. L. Sawhney & K. Brown, pp. 1–29. Madison, WI: Soil Science Society of America and American Society of Agronomy.

Chiou, C. T. (1995). Comment on 'Thermodynamics of organic chemical partition in soils'. *Environmental Science & Technology*, 29, 1421–2.

Chiou, C. T. & Kile, D. E. (1994). Effects of polar and nonpolar groups on the solubility of organic compounds in soil organic matter. *Environmental Science & Technology*, 28, 1139–44.

Churchill, S. A., Griffin, R. A., Jones, L. P. & Churchill, P. F. (1995). Biodegradation rate enhancement of hydrocarbons by an oleophilic fertilizer and a rhamnolipid biosurfactant. *Journal of Environmental Quality*, 24, 19–28.

Corapcioglu, M. Y. & Haridas, A. (1984). Transport and fate of microorganisms in porous media: a theoretical investigation. *Journal of Hydrology*, 72, 149–69.

Corapcioglu, M. Y. & Haridas, A. (1985). Microbial transport in soils and groundwater: a numerical model. *Advances in Water Resources*, 8, 188–200.

Dandurand, L-M., Morra, M. J., Chaverra, M. H. & Orser, C. S. (1994). Survival of

Pseudomonas spp in air-dried mineral powders. *Soil Biology & Biochemistry*, 26, 1423–30.

Dec, J., Shuttleworth, K. L. & Bollag, J-M. (1990). Microbial release of 2,4-dichlorophenol bound to humic acid or incorporated during humification. *Journal of Environmental Quality*, 19, 546–51.

DeFlaun, M. F., Tanzer, A. S., McAteer, A. L., Marshall, B. & Levy, S. B. (1990). Development of an adhesion assay and characterization of an adhesion-deficient mutant of *Pseudomonas fluorescens*. *Applied and Environmental Microbiology*, 56, 112–19.

Dixon, J. B. & Weed, S. B., eds. (1989). *Minerals in Soil Environments*, 2nd edn. Madison, WI: Soil Science Society of America.

Dohse, D. M. & Lion, W. (1994). Effect of microbial polymers on the sorption and transport of phenanthrene in a low-carbon sand. *Environmental Science & Technology*, 28, 541–8.

Dupler, M. & Baker, R. (1984). Survival of *Pseudomonas putida*, a biological control agent, in soil. *Phytopathology*, 74, 195–200.

Enfield, C. G. (1985). Chemical transport facilitated by multiphase flow systems. *Water Science Technology*, 17, 1–12.

Enfield, C. G., Bengtsson, G. & Lindqvist, R. (1989). Influence of macromolecules on chemical transport. *Environmental Science & Technology*, 23, 1278–86.

Engebretson, R. R. & von Wandruszka, R. (1994). Microorganization in dissolved humic acids. *Environmental Science & Technology*, 28, 1934–41.

Gammack, S. M., Paterson, E., Kemp, J. S., Cresser, M. S. & Killham, K. (1992). Factors affecting the movement of microorganisms in soils. In *Soil Biochemistry*, Vol. 7, ed. G. Stotzky & J-M. Bollag, pp. 263–305. New York: Marcel Dekker.

Gannon, J. T., Manilal, V. B. & Alexander, M. (1991a). Relationship between cell surface properties and transport of bacteria through soil. *Applied and Environmental Microbiology*, 57, 190–3.

Gannon, J. T., Mingelgrin, U., Alexander, M. & Wagenet, R. J. (1991b). Bacterial transport through homogenous soil. *Soil Biology & Biochemistry*, 23, 1155–66.

Gantzer, C. J. & Wackett, L. P. (1991). Reductive dechlorination catalyzed by bacterial transition-metal coenzymes. *Environmental Science & Technology*, 25, 715–22.

Georgiou, G., Lin, S. C. & Sharma, M. M. (1992). Surface-active compounds from microorganisms. *Bio/Technology*, 10, 60–5.

Greaves, M. P. & Wilson, M. J. (1969). The adsorption of nucleic acids by montmorillonite. *Soil Biology & Biochemistry*, 1, 317–23.

Greaves, M. P. & Wilson, M. J. (1970). The degradation of nucleic acids and montmorillonite–nucleic-acid complexes by soil microorganisms. *Soil Biology & Biochemistry*, 2, 257–68.

Green, R. E. (1974). Pesticide–clay–water interactions. In *Pesticides in Soil and Water*, ed. W. D. Guenzi, pp. 3–37. Madison, WI: Soil Science Society of America.

Guerin, W. F. & Boyd, S. A. (1992). Differential bioavailability of soil-sorbed naphthalene to two bacterial species. *Applied and Environmental Microbiology*, 58, 1142–52.

Harms, H. & Zehnder, A. J. B. (1994). Influence of substrate diffusion on degradation of dibenzofuran and 3-chlorodibenzofuran by attached and suspended bacteria. *Applied and Environmental Microbiology*, 60, 2736–45.

Hartel, P. G. & Alexander, M. (1984). Temperature and desiccation tolerance of cowpea rhizobia. *Canadian Journal of Microbiology*, 30, 820–3.

Hartel, P. G. & Alexander, M. (1986). Role of extracellular polysaccharide production and clays in the desiccation tolerance of cowpea *Bradyrhizobia*. *Soil Science Society of America Journal*, 50, 1193–8.

Hassett, J. J. & Banwart, W. L. (1989). The sorption of nonpolar organics by soils and sediments. In *Reactions and Movement of Organic Chemicals in Soils*, ed. B. L. Sawhney & K. Brown, pp. 31–44. Madison, WI: Soil Science Society of America and American Society of Agronomy.

Hatzinger, P. B. & Alexander, M. (1995). Effect of aging of chemicals in soil on their biodegradability and extractability. *Environmental Science & Technology*, 29, 537–45.

Hayes, M. H. B. & Mingelgrin, U. (1991). Interactions between small organic chemicals and soil colloidal constituents. In *Interactions at the Soil Colloid – Soil Solution Interface*, ed. G. H. Bolt, M. F. DeBoodt, M. H. B. Hayes, & E. B. A. De Strooper, pp. 323–407. Dordrecht: Kluwer Academic Publishers.

Heijnen, C. E., van Elsas, J. D., Kuikman, P. J. & van Veen, J. A. (1988). Dynamics of *Rhizobium leguminosarum* biovar *trifolii* introduced into soil; the effect of bentonite clay on predation by protozoa. *Soil Biology & Biochemistry*, 20, 483–8.

Heijnen, C. E., Burgers, S. L. G. E. & van Veen, J. A. (1993). Metabolic activity and population dynamics of rhizobia introduced into unamended and bentonite-amended loamy sand. *Applied and Environmental Microbiology*, 59, 743–7.

Henschke, R. B. & Schmidt, R. J. (1990). Plasmid mobilization from genetically engineered bacteria to members of the indigenous soil microflora in situ. *Current Microbiology*, 20, 105–10.

Hillel, D. (1982). *Introduction to Soil Physics*. Orlando: Academic Press.

Huang, P. M. (1990). Role of soil minerals in transformations of natural organics and xenobiotics. In *Soil Biochemistry*, Vol. 6, ed. J-M. Bollag & G. Stotzky, pp. 29–115. New York: Marcel Dekker.

Huysman, F. & Verstraete, W. (1993a). Water-facilitated transport of bacteria in unsaturated soil columns: influence of cell surface hydrophobicity and soil properties. *Soil Biology & Biochemistry*, 25, 83–90.

Huysman, F. & Verstraete, W. (1993b). Water-facilitated transport of bacteria in unsaturated soil columns: influence of inoculation and irrigation methods. *Soil Biology & Biochemistry*, 25, 91–7.

Jury, W. A., Elabd, H. & Resketo, M. (1986). Field study of napropamide movement through unsaturated soil. *Water Resources Research*, 22, 749–55.

Khalifa, S., Holmstead, R. L. & Casida, J. E. (1976). Toxaphene degradation of iron(II) protoporphyrin systems. *Journal of Agricultural and Food Chemistry*, 24, 277–82.

Khanna, M. & Stotzky, G. (1992). Transformation of *Bacillus subtilis* by DNA bound on montmorillonite and effect of DNase on the transforming ability of bound DNA. *Applied and Environmental Microbiology*, 58, 1930–9.

Klecka, G. M. & Gonsior, S. J. (1984). Reductive dechlorination of chlorinated methanes and ethanes by reduced iron(II) porphyrins. *Chemosphere*, 13, 391–402.

Kloepper, J. W. & Schroth, M. N. (1981). Development of a powder formulation of rhizobacteria for inoculation of potato seed pieces. *Phytopathology*, 71, 590–2.

Knaebel, D. B., Federle, T. W., McAvoy, D. C. & Vestal, J. R. (1994). The effect of mineral and organic soil constituents and the microbial mineralization of organic compounds in natural soil. *Applied and Environmental Microbiology*, 60, 4500–8.

Koskinen, W. C. & Harper, S. S. (1990). The retention process: mechanisms. In *Pesticides in the Soil Environment: Processes, Impacts, and Modeling*, ed. H. H. Cheng, pp. 51–77. Madison, WI: Soil Science Society of America.

Kriegman-King, M. R. & Reinhard, M. (1992). Transformation of carbon tetrachloride in the presence of sulfide, biotite, and vermiculite. *Environmental Science & Technology*, 26, 2198–206.

Kriegman-King, M. R. & Reinhard, M. (1994). Transformation of carbon tetrachloride by pyrite in aqueous solution. *Environmental Science & Technology*, 28, 692–700.

Krone, U. E., Laufer, K., Thauer, R. K. & Hogenkamp, H. P. C. (1989a). Coenzyme F_{430} as a possible catalyst for the reductive dehalogenation of chlorinated C_1 hydrocarbons in methanogenic bacteria. *Biochemistry*, 28, 10 061–5.

Krone, U. E., Laufer, K., Thauer, R. K. & Hogenkamp, H. P. C. (1989b). Reductive dehalogenation of chlorinated C_1-hydrocarbons mediated by corrinoids. *Biochemistry*, 28, 4908–14.

Krone, U. E., Thauer, R. K., Hogenkamp, H. P. C. & Steinbach, K. (1991). Reductive formation of carbon monoxide from CCl_4 and FREONs 11, 12, and 13 catalyzed by corrinoids. *Biochemistry*, 30, 2713–9.

Labeda, D. P., Liu, K-C. & Casida, L. E. Jr (1976). Colonization of soil by *Arthrobacter* and *Pseudomonas* under varying conditions of water and nutrient availability as studied by plate counts and transmission electron microscopy. *Applied and Environmental Microbiology*, 31, 551–61.

Lajoie, C. A., Chen, S-Y., Oh, K. C. & Strom, P. F. (1992). Development and use of field application vectors to express nonadaptive foreign genes in competitive environments. *Applied and Environmental Microbiology*, 58, 655–63.

Lajoie, C. A., Zylstra, G. J., DeFlaun, M. F. & Strom, P. F. (1993). Development of field application vectors for bioremediation of soils contaminated with polychlorinated biphenyls. *Applied and Environmental Microbiology*, 59, 1735–41.

Lajoie, C. A., Layton, A. C. & Sayler, G. S. (1994). Cometabolic oxidation of polychlorinated biphenyls in soil with a surfactant-based field application vector. *Applied and Environmental Microbiology*, 60, 2826–33.

Lin, S-C., Minton, M. A., Sharma, M. M. & Georgiou, G. (1994). Structural and immunological characterization of a biosurfactant produced by *Bacillus licheniformis* JF-2. *Applied and Environmental Microbiology*, 60, 31–8.

Liu, S-Y., Minard, R. D. & Bollag, J-M. (1987). Soil-catalyzed complexation of the pollutant 2,6-diethylaniline with syringic acid. *Journal of Environmental Quality*, 16, 48–53.

Lorenz, M. G. & Wackernagel, W. (1987). Adsorption of DNA to sand and variable degradation rates of adsorbed DNA. *Applied and Environmental Microbiology*, 53, 2948–52.

Lorenz, M. G. & Wackernagel, W. (1990). Natural genetic transformation of *Pseudomonas stutzeri* by sand-adsorbed DNA. *Archives of Microbiology*, 154, 380–5.

Lorenz, M. G. & Wackernagel, W. (1991). High frequency of natural genetic transformation of *Pseudomonas stutzeri* in soil extract supplemented with a carbon/energy and phosphorus source. *Applied and Environmental Microbiology*, 57, 1246–51.

Lorenz, M. G., Aardema, B. W. & Krumbein, W. E. (1981). Interaction of marine sediment with DNA and DNA availability to nucleases. *Marine Biology*, 64, 225–30.

Lorenz, M. G., Aardema, B. W. & Wackernagel, W. (1988). Highly efficient genetic transformation of *Bacillus subtilis* attached to sand grains. *Journal of General Microbiology*, 134, 107–12.

Lorenz, M. G., Gerjets, D. & Wackernagel, W. (1991). Release of transforming plasmid and chromosomal DNA from two cultured soil bacteria. *Archives of Microbiology*, 156, 319–26.

Lorenz, M. G., Reipschlager, K. & Wackernagel, W. (1992). Plasmid transformation of naturally competent *Acinetobacter calcoaceticus* in non-sterile soil extract and groundwater. *Archives of Microbiology*, 157, 355–60.

Madhun, Y. A., Young, J. L. & Freed, V. H. (1986). Binding of herbicides by water-soluble organic materials from soil. *Journal of Environmental Quality*, 15, 64–8.

Marks, T. S., Allpress, J. D. & Maule, A. (1989). Dehalogenation of lindane by a variety of porphyrins and corrins. *Applied Environmental Microbiology*, 55, 1258–61.

Marshall, K. C. (1964). Survival of root-nodule bacteria in dry soils exposed to high temperatures. *Australian Journal of Agricultural Research*, 15, 273–81.

Marshall, K. C. (1975). Clay mineralogy in relation to survival of soil bacteria. *Annual Review of Phytopathology*, 13, 357–73.

McBride, M. B. (1994). *Environmental Chemistry of Soils*. New York: Oxford University Press.

McCarthy, J. F. & Zachara, J. M. (1989). Subsurface transport of contaminants. *Environmental Science & Technology*, 23, 496–502.

Mihelcic, J. R. & Luthy, R. G. (1991). Sorption and microbial degradation of naphthalene in soil–water suspensions under denitrification conditions. *Environmental Science & Technology*, 25, 169–77.

Miller, M. E. & Alexander, M. (1991). Kinetics of bacterial degradation of benzylamine in a montmorillonite suspension. *Environmental Science & Technology*, 25, 240–5.

Moll, D. M. & Vestal, J. R. (1992). Survival of microorganisms in smectite clays: implication for Martian exobiology. *Icarus*, 98, 233–9.

Morra, M. J., Corapcioglu, M. O., von Wandruszka, R. M. A., Marshall, D. B. & Topper, K. (1990). Fluorescence quenching and polarization studies of naphthalene and 1-naphthol interaction with humic acid. *Soil Science Society of America Journal*, 54, 1283–9.

Mortland, M. M. (1984). Deamination of glutamic acid by pyridoxal phosphate-Cu^{2+}-smectite catalysts. *Journal of Molecular Catalysis*, 27, 143–55.

Murphy, E. M., Zachara, J. M. & Smith, S. C. (1990). Influence of mineral-bound humic substances on the sorption of hydrophobic organic compounds. *Environmental Science & Technology*, 24, 1507–16.

Murphy, E. M., Zachara, J. M., Smith, S. C., Phillips, J. L. & Wietsma, T. W. (1994). Interaction of hydrophobic organic compounds with mineral-bound humic substances. *Environmental Science & Technology*, 28, 1291–9.

Ogram, A. V., Jessup, R. E., Ou, L. T. & Rao, P. S. C. (1985). Effects of sorption on biological degradation rates of (2,4-dichlorophenoxy)acetic acid in soils. *Applied and Environmental Microbiology*, 49, 582–7.

Ogram, A., Sayler, G. S. & Barkay, T. (1987). The extraction and purification of microbial DNA from sediments. *Journal of Microbiological Methods*, 7, 57–66.

Ogram, A., Sayler, G. S., Gustin, D. & Lewis, R. J. (1988). DNA adsorption to soils and sediments. *Environmental Science & Technology*, 22, 982–4.

Osa-Afiana, L. O. & Alexander, M. (1982). Clays and the survival of *Rhizobium* in soil during desiccation. *Soil Science Society of America Journal*, 46, 285–8.

Paul, J. H. & David, A. W. (1989). Production of extracellular nucleic acids by genetically altered bacteria in aquatic-environment microcosms. *Applied and Environmental Microbiology*, 55, 1865–9.

Paul, J. H., Cazares, L. & Thurmond, J. (1990). Amplification of the rbcL gene from dissolved and particulate DNA from aquatic environments. *Applied and Environmental Microbiology*, 56, 1963–6.

Paul, J. H., Frischer, M. E. & Thurmond, J. M. (1991). Gene transfer in marine water column and sediment microcosms by natural plasmid transformation. *Applied and Environmental Microbiology*, 57, 1509–15.

Pesenti-Barili, B., Ferdani, E., Mosti, M. & Degli-Innocenti, F. (1991). Survival of *Agrobacterium radiobacter* K84 on various carriers for crown gall control. *Applied and Environmental Microbiology*, 57, 2047–51.

Pignatello, J. J., Ferrandino, F. J. & Huang, L. Q. (1993). Elution of aged and freshly added herbicides from a soil. *Environmental Science & Technology*, 27, 1563–71.

Puchalski, M. M., Morra, M. J. & von Wandruszka, R. (1992). Fluorescence quenching of synthetic organic compounds by humic materials. *Environmental Science & Technology*, 26, 1787–92.

Rao, P. S. C., Bellin, C. A. & Brusseau, M. L. (1993). Coupling biodegradation of organic chemicals to sorption and transport in soils and aquifers: paradigms and paradoxes. In *Sorption and Degradation of Pesticides and Organic Chemicals in Soil*, ed. D. M. Linn, T. H. Carski, M. L. Brusseau, & F-H. Chang, pp. 1–26. Madison, WI: Soil Science Society of America, American Society of Agronomy.

Recorbet, G., Picard, C., Normand, P. & Simonet, P. (1993). Kinetics of the persistence of chromosomal DNA from genetically engineered *Escherichia coli* introduced to soil. *Applied and Environmental Microbiology*, 59, 4289–94.

Renault, P. & Sierra, J. (1994a). Modeling oxygen diffusion in aggregated soils: I. Anaerobiosis inside the aggregates. *Soil Science Society of America Journal*, 58, 1017–23.

Renault, P. & Sierra, J. (1994b). Modeling oxygen diffusion in aggregated soils: II. Anaerobiosis in topsoil layers. *Soil Science Society of America Journal*, 58, 1023–30.

Rijnaarts, H. H. M., Bachmann, A., Jumelet, J. C. & Zehnder, A. J. B. (1990). Effect of desorption and intraparticle mass transfer on the aerobic biomineralization of α-hexachlorocyclohexane in a contaminated calcareous soil. *Environmental Science & Technology*, 24, 1349–54.

Robert, M. & Chenu, C. (1992). Interactions between soil minerals and microorganisms. In *Soil Biochemistry*, Vol. 7, ed. G. Stotzky & J-M. Bollag, pp. 307–404. New York: Marcel Dekker.

Roberts, A. L., Sanborn, P. N. & Gschwend, P. M. (1992). Nucleophilic substitution reactions of dihalomethanes with hydrogen sulfide species. *Environmental Science & Technology*, 26, 2263–74.

Romanowski, G., Lorenz, M. G. & Wackernagel, W. (1991). Adsorption of plasmid DNA to mineral surfaces and protection against DNase I. *Applied and Environmental Microbiology*, 57, 1057–61.

Romanowski, G., Lorenz, M. G., Sayler, G. & Wackernagel, W. (1992). Persistence of free plasmid DNA in soil monitored by various methods, including a transformation assay. *Applied and Environmental Microbiology*, 58, 3012–19.

Schanke, C. A. & Wackett, L. P. (1992). Environmental reductive elimination reactions of polychlorinated ethanes mimicked by transition-metal coenzymes. *Environmental Science & Technology*, 26, 830–3.

Schulze, D. G. (1989). An introduction to soil mineralogy. In *Minerals in Soil Environments*, 2nd edn, ed. J. B. Dixon & S. B. Weed, pp. 1–34. Madison, WI: Soil Science Society of America.

Schwarzenbach, R. P., Stierli, R., Lanz, K. & Zeyer, J. (1990). Quinone and iron porphyrin mediated reduction of nitroaromatic compounds in homogeneous aqueous solution. *Environmental Science & Technology*, 24, 1566–74.

Schwarzenbach, R. P., Gschwend, P. M. & Imboden, D. M. (1993). *Environmental Organic Chemistry*. New York: John Wiley & Sons.

Schwertmann, U., Kodama, H. & Fischer, W. R. (1986). Mutual interactions between organics and iron oxides. In *Interactions of Soil Minerals with Natural Organics and Microbes*, ed. P. M. Huang & M. Schnitzer, pp. 223–50. Madison, WI: Soil Science Society of America.

Scow, K. M. (1993). Effect of sorption-desorption and diffusion processes on the kinetics of biodegradation of organic chemicals in soil. In *Sorption and Degradation of Pesticides and Organic Chemicals in Soil*, ed. D. M. Linn, T. H. Carski, M. L. Brusseau & F-H. Chang, pp. 73–114. Madison, WI: Soil Science Society of America, American Society of Agronomy.

Scribner, S. L., Benzing, T. R., Sun, S. & Boyd, S. A. (1992). Desorption and bioavailability of aged simazine residues in soil from a continuous corn field. *Journal of Environmental Quality*, 21, 115–20.

Sexstone, A. J., Revsbech, N. P., Parkin, T. B. & Tiedje, J. M. (1985). Direct measurement of oxygen profiles and denitrification rates in soil aggregates. *Soil Science Society of America Journal*, 49, 645–51.

Smith, M. S., Thomas, G. W., White, R. E. & Ritonga, D. (1985). Transport of *Escherichia coli* through intact and disturbed soil columns. *Journal of Environmental Quality*, 14, 87–91.

Smith, S. C., Ainsworth, C. C., Traina, S. J. & Hicks, R. J. (1992). The effect of sorption on the biodegradation of quinoline. *Soil Science Society of America Journal*, 56, 737–46.

Sposito, G. (1989). *The Chemistry of Soils*. New York: Oxford University Press.

Spurlock, F. C. (1995). Estimation of humic-based sorption enthalpies from nonlinear isotherm temperature dependence: theoretical development and application to substituted phenylureas. *Journal of Environmental Quality*, 24, 42–9.

Spurlock, F. C. & Biggar, J. W. (1994). Thermodynamics of organic chemical partition in soils. 2. Nonlinear partition of substituted phenylureas from aqueous solution. *Environmental Science & Technology*, 28, 996–1002.

Steinberg, S. M., Pignatello, J. J. & Sawhney, B. L. (1987). Persistence of 1,2-dibromoethane in soils: entrapment in intraparticle micropores. *Environmental Science & Technology*, 21, 1201–8.

Stenström, T. A. (1989). Bacterial hydrophobicity, an overall parameter for the measurement of adhesion potential to soil particles. *Applied and Environmental Microbiology*, 55, 142–7.

Stevenson, F. J. (1972). Organic matter reactions involving herbicides in soil. *Journal of Environmental Quality*, 1, 333–43.

Stevenson, F. J. (1994). *Humus Chemistry*, 2nd edn. New York: John Wiley.

Stewart, D. K. R., Chisholm, D. & Ragab, M. T. H. (1971). Long term persistence of parathion in soil. *Nature (London)*, 229, 47.

Stewart, G. J. & Sinigalliano, C. D. (1990). Detection of horizontal gene transfer by natural transformation in native and introduced species of bacteria in marine and synthetic sediments. *Applied and Environmental Microbiology*, 56, 1818–24.

Stotzky, G. (1986). Influence of soil mineral colloids on metabolic processes, growth, adhesion, and ecology of microbes and viruses. In *Interactions of Soil Minerals with Natural Organics and Microbes*, ed. P. M. Huang & M. Schnitzer, pp. 305–428. Madison, WI: Soil Science Society of America.

Tan, Y., Bond, W. J. & Griffin, D. M. (1992). Transport of bacteria during unsteady unsaturated soil water flow. *Soil Science Society of America Journal*, 56, 1331–40.

Thomas, G. W. & Phillips, R. E. (1979). Consequences of water movement in macropores. *Journal of Environmental Quality*, 8, 149–52.

Torrents, A. & Stone, A. T. (1994). Oxide surface-catalyzed hydrolysis of carboxylate esters and phosphorothioate esters. *Soil Science Society of America Journal*, 58, 738–45.

Traina, S. J., Spontak, D. A. & Logan, T. J. (1989). Effects of cations on complexation of naphthalene by water-soluble organic carbon. *Journal of Environmental Quality*, 18, 221–7.

Tratnyek, P. G. & Macalady, D. L. (1989). Abiotic reduction of nitro aromatic pesticides in anaerobic laboratory systems. *Journal of Agricultural and Food Chemistry*, 37, 248–54.

Trevors, J. T., van Elsas, J. D., van Overbeek, L. S. & Starodub, M-E. (1990). Transport of a genetically engineered *Pseudomonas fluorescens* strain through a soil microcosm. *Applied and Environmental Microbiology*, 56, 401–8.

Ukrainczyk, L., Chibwe, M., Pinnavaia, T. J. & Boyd, S. A. (1994). ESR study of cobalt(II) tetrakis(N-methyl-4-pyridiniumyl)porphyrin and cobalt(II) tetrasulfophthalocyanine intercalated in layered aluminosilicates and a layered double hydroxide. *Journal of Physical Chemistry*, 98, 2668–76.

Ukrainczyk, L., Chibwe, M., Pinnavaia, T. J. & Boyd, S. A. (1995). Reductive dechlorination of carbon tetrachloride in water catalyzed by mineral-supported biomimetic cobalt macrocycles. *Environmental Science & Technology*, 29, 439–45.

van Elsas, J. D. & Heijnen, C. E. (1990). Methods for the introduction of bacteria into soil: a review. *Biology and Fertility of Soils*, 10, 127–33.

van Loosdrecht, M. C. M., Lyklema, J., Norde, W. & Zehnder, A. J. B. (1990). Influence of interfaces on microbial activity. *Microbiological Reviews*, 54, 75–87.

Voudrias, E. A. & Reinhard, M. (1986). Abiotic organic reactions at mineral surfaces. In *Geochemical Processes at Mineral Surfaces*, ed. J. A. Davis & K. F. Hayes, pp. 462–86. Washington, DC: American Chemical Society.

Wade, R. S. & Castro, C. E. (1973). Oxidation of iron(II) porphyrins by alkyl halides. *Journal of the American Chemical Society*, 95, 226–30.

Walter, M. V., Porteous, A. & Seidler, R. J. (1989). Evaluation of a method to measure conjugal transfer of recombinant DNA in soil slurries. *Current Microbiology*, 19, 365–70.

Wan, J., Wilson, J. L. & Kieft, T. L. (1994). Influence of the gas–water interface on transport of microorganisms through unsaturated porous media. *Applied and Environmental Microbiology*, 60, 509–16.

Wang, T. S. C., Huang, P. M., Chou, C-H. & Chen, J-H. (1986). The role of soil minerals in the abiotic polymerization of phenolic compounds and formation of humic substances. In *Interactions of Soil Minerals with Natural Organics and Microbes*, ed. P. M. Huang & M. Schnitzer, pp. 251–81. Madison, WI: Soil Science Society of America.

Weber, J. B. & Coble, H. D. (1968). Microbial decomposition of diquat adsorbed on montmorillonite and kaolinite clays. *Journal of Agricultural and Food Chemistry*, 16, 475–8.

Weber, J. B., Best, J. A. & Gonese, J. U. (1993). Bioavailability and bioactivity of sorbed organic chemicals. In *Sorption and Degradation of Pesticides and Organic Chemicals in Soil*, ed. D. M. Linn, T. H. Carski, M. L. Brusseau, & F-H. Chang, pp. 153–96. Madison, WI: Soil Science Society of America, American Society of Agronomy.

Weed, S. B. & Weber, J. B. (1974). Pesticide–organic matter interactions. In *Pesticides in Soil and Water*, ed. W. D. Guenzi, pp. 39–66. Madison, WI: Soil Science Society of America.

Wolfe, N. L., Mingelgrin, U. & Miller, G. C. (1990). Abiotic transformations in water, sediments, and soil. In *Pesticides in the Soil Environment: Processes, Impacts, and Modeling*, ed. H. H. Cheng, pp. 103–68. Madison, WI: Soil Science Society of America.

Wollum, A. G., III & Cassel, D. K. (1978). Transport of microorganisms in sand columns. *Soil Science Society of America Journal*, 42, 72–6.

Young, C. S. & Burns, R. G. (1993). Detection, survival, and activity of bacteria added to soil. In *Soil Biochemistry*, Vol. 8, ed. J-M. Bollag & G. Stotzky, pp. 1–277. New York: Marcel Dekker.

Zechman, J. M. & Casida, L. E. Jr (1982). Death of *Pseudomonas aeruginosa* in soil. *Canadian Journal of Microbiology*, 28, 788–94.

Zielke, R. C., Pinnavaia, T. J. & Mortland, M. M. (1989). Adsorption and reactions of selected organic molecules on clay mineral surfaces. In *Reactions and Movement of Organic Chemicals in Soils*, ed. B. L. Sawhney & K. Brown, pp. 81–110. Madison, WI: Soil Science Society of America and American Society of Agronomy.

Zoro, J. A., Hunter, J. M., Eglinton, G. & Ware, G. C. (1974). Degradation of p,p'-DDT in reducing environments. *Nature (London)*, 247, 235–6.

3

Biodegradation of 'BTEX' hydrocarbons under anaerobic conditions

Lee R. Krumholz, Matthew E. Caldwell and Joseph M. Suflita

3.1 Introduction

Fossil energy reserves are valuable natural resources that underpin most major world economies. The extraction, transport and utilization of these resources inevitably leads to the release of these substances to environmental compartments where they are deemed undesirable. For instance, petroleum or petroleum distillation products often occur as contaminants in soils, aquifers and surface waters via a myriad of mechanisms including leaking underground storage tanks, aboveground spills and release by marine transport vessels. Since environmental matrices (air, water and soil) are completely integrated, all are susceptible to a reduction in quality and often quantity due to the release of pollutant materials. When this occurs, a great deal of concern is associated with the impact of the hydrocarbons on humans and recipient environments. However, this impact cannot be correctly gauged without information on the transport and fate characteristics of the individual contaminants.

A major factor governing the transport and fate of contaminant hydrocarbons is their susceptibility to metabolism by aerobic and anaerobic microorganisms. While a plethora of information is available on the prospects for aerobic biodegradation, comparatively little is understood about anaerobic hydrocarbon biotransformation. However, it is well recognized that anaerobic microbial activities directly or indirectly impact all major environmental compartments. In many environments, most notably the terrestrial subsurface, oxygen concentrations are often initially low. With rapid utilization by hydrocarbonoclastic microorganisms and limited rates of reoxygenation, oxygen becomes depleted. Without the reactive power of molecular oxygen, the biodegradation rate of hydrocarbons slows down and contamination problems are exacerbated. Nevertheless, some of the most toxicologically important components of petroleum can be metabolized, even in the absence of oxygen.

This review will focus on the anaerobic metabolism of the aromatic hydrocarbons including benzene, toluene, ethylbenzene and the xylene isomers. Collectively, these are referred to as the BTEX compounds. BTEXs have been the focus of a large number and variety of biodegradation and bioremediation studies. The reviewed evidence includes both field and laboratory investigations for which anaerobic degradation of the BTEXs by microbial action provides a reasonable explanation for the observations. The reason for the strong emphasis on this group of compounds is twofold. First, there is often concern with organisms residing in or in contact with water resources tainted with petroleum hydrocarbons. This concern is dictated by the much greater water solubility of BTEXs relative to the aliphatic, alicyclic, and polycyclic hydrocarbons. There is obviously a much greater risk of exposure to higher concentrations of BTEX in comparison with the other major groups of hydrocarbon contaminants in water. Secondly, this exposure risk is amplified by the fact that benzene is considered a human carcinogen (Hughes, Meek & Bartlett, 1994). TEX compounds are generally not considered carcinogens but must be kept to low levels in drinking and recreational waters due to their toxicity (Meek & Chan, 1994a, b).

Bacteria oxidize BTEXs and other organic pollutants to obtain energy and to meet their nutritional needs. In the process, electrons or reducing equivalents from susceptible contaminants are transferred to and ultimately reduce an organic or inorganic electron acceptor. During the electron transfer process, usable energy is recovered via a series of redox reactions through the formation of energy storage compounds or electrochemical gradients. Under anaerobic conditions, and in the presence of suitable electron donors, microbial electron transfer and respiration can proceed if suitable exogenous electron acceptors are available. The potential energy available from the oxidation of a particular substrate coupled with the reduction of various electron acceptors varies considerably. In an ecosystem, higher energy yielding processes tend to predominate (but do not exclude other processes) until a favored electron acceptor is consumed. Nitrate is energetically the best electron acceptor under anoxic conditions and is generally used preferentially by microbial communities containing appropriate denitrifying microorganisms. Oxidized nitrogen utilization will be followed by ferric iron and then manganese respiration. Once sufficiently low redox potential levels have been reached, sulfate reduction and methanogenesis will occur. Often, such processes tend to be spatially distinct. For example, Baedecker et al. (1993) observed that Fe(III) reduction was the major electron accepting process close to a groundwater hydrocarbon spill, while methanogenesis predominated downgradient from the spill.

The understanding of the anaerobic biodegradation of BTEXs has increased substantially since several research groups began to investigate earnestly the process in the early 1980s. The picture that emerges is that anaerobic biodegradation of some aromatic hydrocarbons can occur at rates that rival those observed in aerobic cultures, while other BTEX compounds degrade slowly under anoxic conditions, if at all. Some of the factors governing the susceptibility of these compounds to biodegradation are considered in this review. Moreover, the diversity of proposed metabolic routes for the anaerobic bioconversion of some BTEX compounds is truly impressive and more pathways will likely be discovered in the near future. Our current understanding of this metabolic diversity is a far cry from the historical perception of anaerobes as nutritionally limited and of little environmental consequence.

A number of reviews have been published over the last decade which have reviewed the subject of the anaerobic degradation of BTEX compounds. Christensen *et al.* (1994) have focused on the fate, the chemistry and the microbiology of contaminants in landfill leachate. Grbić-Galić (1990), in an excellent review, focused on the microbiology and biochemistry of the methanogenic decomposition of aromatic hydrocarbons and phenols. In an earlier review, Evans & Fuchs (1988) focused on the pathways for the biodegradation of monoaromatic compounds. More recently, two more field-oriented reviews have been published (Salanitro, 1993; McAllister & Chiang, 1994). These latter reviews are directed at the remediation professional, with McAllister & Chiang (1994) including aspects of monitoring the fate of BTEX compounds. Salanitro (1993) has taken more of a microbiological perspective, including the relative biodegradation rates for specific compounds, but has focused mainly on aerobic studies. In this review we have attempted to summarize the literature on the anaerobic biodegradation of BTEX compounds with a thrust towards developing an understanding of the relative rates of biodegradation of each of the components. This includes both laboratory and field investigations with a discussion of the relative influences of environmental factors on these rates. In addition, we consider the requisite organisms when known and the predominant metabolic pathways.

3.2 Benzene

Benzene is the simplest aromatic hydrocarbon, perhaps the most recalcitrant of the BTEX compounds, and of greatest regulatory concern because of its associated health impacts. Until recently, benzene was believed to be completely resistant to attack under anaerobic conditions, a view supported by the majority of both field and laboratory investigations (Table 3.1).

Almost all field studies indicate negative results for anaerobic benzene biodegradation regardless of the nature of the terminal electron acceptor (Table 3.1). In fact, only a single defined and one undefined example stand out as exceptions. In a subsurface field study of the Leiduin dune site in Amsterdam, Piet & Smeenk (1985) reported that a variety of low-boiling aromatic compounds, including benzene, were reduced in concentration relative to the infiltration water. Under so-called 'deep anaerobic' conditions, benzene was reduced by 95% in 3 weeks. However, only slight experimental detail is given and ill-defined anaerobic conditions are referred to. In this case, the significance of anaerobic metabolism versus other potential removal mechanisms is impossible to evaluate. Field

Table 3.1. *Summary of results from reports on the anaerobic biodegradation of benzene*

Electron acceptor	Inoculum source[a]	Degradation	Concentration (μM)	Rate	Reference
Nitrate	Borden, Ontario	y	38	1.1 μM/d	Major et al. (1988)
	Various enrichments	n	100		Evans et al. (1991b)
	Traverse City, MI	n	49–70		Hutchins (1991a)
					Hutchins (1992)
	Traverse City, MI	n (ind)	154–254		Hutchins et al. (1991b)
					Hutchins (1993)
	North Bay, Ontario	n	1.2		Acton & Barker (1992)
	Borden, Ontario	n	37		Barbaro et al. (1992)
	Northern Michigan	n	256		Anid et al. (1993)
	Empire, MI	n	40		Barlaz et al. (1993)
	Danish groundwater	n	11.5		Flyvbjerg et al. (1993)
	Dutch groundwater	y	256	1.0 μM/d	Morgan et al. (1993)
	Perth, Australia	n	12.8		Patterson et al. (1993)
	BTEX polluted aquifers	n	14.1		Alvarez & Vogel (1994)
	Eastern United States	n (ind)	11.5		Davis et al. (1994)
	Toluene enrichment	n	2.5–6		Jensen & Arvin (1994)
	Seal Beach, CA	n	36		Ball & Reinhard (1996)
Sulfate	Seal Beach, CA	n	79 μmol/kg		Haag et al. (1991)
	Borden, Ontario	n	36		Acton & Barker (1992)
	Seal Beach, CA	y (ind)	40–200	0.4–3.7 μM/d	Edwards & Grbić-Galić (1992)
	Patuxent, MD	n	50–100		Beller et al. (1992a)
	Seal Beach, CA	n	64		Edwards et al. (1992)
	Perth, Australia	n	64		Thierrin et al. (1993)
	Eastern United States	y (ind)	11.8 μmol/kg	0.35 μmol/kg/d	Davis et al. (1994)
	North Sea Oil Tank	n	NR		Rueter et al. (1994)
	Seal Beach, CA	n	14.1–32		Ball & Reinhard (1996)
	San Diego Bay, CA	y	1.5	0.75 μM/d	Lovley et al. (1995)

	Location		Concentration	Rate	Reference
Methanogenic	Mud and sewage	n	12 000		Schink (1985)
	Norman, OK	y	7.8	28 nM/d	Wilson et al. (1986)
	Ferulate enrichment	y	1500–3000	1.4–2.6 μM/d	Vogel & Grbić-Galić (1986) Grbić-Galić & Vogel (1987)
	Gloucester, Ontario	n	7.4		Berwanger & Barker (1988)
	Anoxic sewage sludge	n	641		Battersby & Wilson (1989)
	Rhine River, Germany	y	0.013 + 1.3	0.05/d	Van Beelen & Van Keulen (1990)
	Traverse City, MI	y	4.1		Wilson et al. (1990)
	Various soils	y (ind)	1.2 μmol/kg	0.76–6.7 nmol/kg/d	Watwood et al. (1991)
	Acidogenic bioreactor	n	179	12.8 μM/d	Ghosh & Sun (1992)
	Seal Beach, CA		14.1		Ball & Reinhard (1996)
	Eastern United States	y (ind)	11.8 μmol/kg	0.2 μmol/kg/d	Davis et al. (1994)
Iron(III)	Bemidji, MN	y (ind)	10	NR	Baedecker et al. (1993)
	Hanahan, SC	y (ind)	10	>3.3 μM/d	Lovley (1994)
Field					
Nitrate	Ontario, Canada	n	5.4		Gillham et al. (1990)
	North Bay, Ontario	n	1.9		Acton & Barker (1992)
	Borden, Ontario	n	61		Barbaro et al. (1992)
	Empire, MI	n	40		Barlaz et al. (1993)
	Seal Beach, CA	n	3.2		Ball et al. (1994)
Sulfate	North Bay, Ontario	n	0.025–0.60		Reinhard et al. (1984)
	Leiduin, The Netherlands	y	NR		Piet & Smeenk (1985)
	North Bay, Ontario	n	1.9		Acton & Barker (1992)
	Borden, Ontario	n	60		Barbaro et al. (1992)
	Seal Beach, CA	n	3.2		Ball et al. (1994)
	Seal Beach, CA	n	2–3		Beller et al. (1995)
Methanogenic	Vegen City, Denmark	n	1.5		Nielsen et al. (1992)
	Traverse City, MI	y	0–1.6	0.05–0.17/wk	Wilson et al. (1990)
Iron(III)	Bemidji, MN	y	NR		Cozzarelli et al. (1990)
	Arvida, NC	n	538–16 700		Borden et al. (1994)

[a] All studies use aquifer sediments and/or groundwater unless otherwise indicated. (ind) – tested individually, rather than as a group of BTEX compounds. NR – not reported.

evidence for benzene decay has been presented for the Fe(III)-reducing regions of an aquifer in Bemidji, Minnesota (Cozzarelli, Eganhouse & Baedecker, 1990). In this study the presence of phenol, a possible intermediate of benzene oxidation, is believed to indicate anaerobic benzene biodegradation. This is possibly true, but phenol may also have originated from the metabolism of other compounds.

Similarly, only a few reports have alluded to the loss of benzene in nitrate-reducing incubations (Table 3.1). Morgan, Lewis & Watkinson (1993) reported losses of benzene in groundwater samples held under denitrifying conditions. In that study, no attempt was made to remove oxygen from the groundwater prior to initiating microcosm incubations, so the possibility of oxygen contamination in the experiments cannot be dismissed. In the aquifer study of Major, Mayfield & Barker (1988), most of the added benzene was removed from experimental incubations within 30 days. As stringent measures were taken to preclude oxygen contamination, these experiments would suggest the removal of benzene with nitrate serving as a terminal electron acceptor. However, when these workers attempted a similar experiment using the same aquifer as the source of their inoculum, the earlier results were not reproduced (Barbaro *et al.*, 1992).

The picture of benzene biodegradation under sulfate-reducing conditions is only marginally different. Only three studies report positive results under these conditions. Davis, Klier & Carpenter (1994) suggested that benzene loss was evident in microcosms incubated with 2 mM sulfate, 1 mM sulfide and 1 mg/kg benzene. Degradation began after a 60 day lag and continued for about 20 days until 2–3% of the initial benzene remained. Although it seems clear that benzene was removed relative to control incubations, it is not certain whether its consumption was linked to the reduction of sulfate, since methanogenic incubations showed similar kinetics of benzene loss. The relatively small initial amount of sulfate may have been depleted due to the consumption of other electron donors by the time benzene removal was evident.

A more definitive study was done by Edwards & Grbić-Galić (1992) with Seal Beach, California, sediments incubated with benzene. Their results showed conversion of ^{14}C-benzene to $^{14}CO_2$, (90% recovery). A small amount of methane was also observed equivalent to about 5% of the benzene carbon, but it was not determined whether any of this methane was radiolabeled. Lag periods before the onset of degradation were generally 30–60 days with these sediments but increased to 70–100 days as the benzene concentration was increased to 200 μM. Our laboratory has successfully repeated these findings and similarly demonstrated the conversion of ^{14}C-benzene to $^{14}CO_2$ using inoculum from the Seal Beach sediments (Caldwell & Suflita, 1995). However, in our studies it took approximately 125 days for benzene (100 μM) biodegradation to begin. Similar recoveries of radiolabeled carbon dioxide were also measured. Like Edwards & Grbić-Galić (1992), we also had trouble obtaining evidence indicating that the oxidation of benzene was directly linked to the consumption of sulfate. This difficulty stemmed from the inability to measure a significant amount of sulfate depletion or sulfide formation against the large background when such small but environmentally realistic concentrations of benzene were employed. Conclusive evidence to this effect will remain elusive until stable enrichment cultures are obtained. However, when we amended formerly methanogenic sediments from a gasoline-contaminated aquifer in Michigan with sulfate, we observed an increase in sulfate depletion and sulfide formation along with the conversion of ^{14}C-benzene to ^{14}CO$_2$ (Caldwell & Suflita, 1995).

In another definitive study, Lovely *et al.* (1995) demonstrated the complete mineralization of ^{14}C-benzene by enrichment cultures from San Diego Bay sediments. Their cultures clearly coupled the reduction of sulfate to the oxidation of benzene with no detectable extracellular intermediates.

The success of these research groups can, in part, be attributed to the fact that no other BTEX compounds were included in the incubation systems. It may be that the consumption of preferred substrates limits anaerobic benzene biodegradation. This phenomenon has been previously observed where alternative substrates were shown to inhibit the degradation of *o*-xylene and toluene under methanogenic conditions (Edwards & Grbić-Galić, 1994). These organisms may not be able to metabolize benzene until the TEX compounds are essentially depleted. Alternately, the presence of TEXs may inactivate the benzene-degrading microorganisms in some way. In consistent fashion, the published field studies of BTEX hydrocarbon decay linked with sulfate as the terminal electron acceptor did not observe the degradation of benzene. However, such extrapolations must be conducted with a great deal of caution as many complicating factors could interplay to limit the anaerobic metabolism of benzene.

Under methanogenic conditions, a number of studies have reported the biodegradation of benzene when the latter was either included as a sole substrate of interest or when mixed with the other TEX hydrocarbons. Wilson, Smith & Rees (1986) report that after a 20–40 week lag and 120 weeks of incubation, >99% of the original amount of benzene, added in a BTEX mixture, disappeared in microbiologically active aquifer incubations, while only 30% was removed from autoclaved controls, the difference being attributed to biodegradation. Such studies are difficult to interpret since no benzene metabolites were detected in these studies and no indicators of anaerobiosis were included in the experimental protocol. A similar investigation in our laboratory using inoculum from the same aquifer (Norman Landfill, Oklahoma) was unsuccessful in showing evidence for anaerobic benzene metabolism within a 4 year incubation period (Caldwell & Suflita, 1995).

Other studies reporting benzene degradation under anaerobic conditions are similarly difficult to interpret. For instance, Ghosh & Sun (1992) observed a loss of 16 mg/l benzene over a 44 day period in a bioreactor that did not exhibit a detectable lag period prior to the onset of benzene removal. We can only presume that the bioreactor was not methanogenic as the ambient pH was 5.2–5.5, but no exogenously supplied electron acceptor was available. In this study, the abiotic loss observed in a single control bottle accounted for 50% of the total mass of benzene.

In another study (Watwood, White & Dahm, 1991), benzene mineralization occurred in soils incubated under an inert gas for 4 weeks. No attempt was made to remove residual oxygen from these soils and the possibility exists that benzene mineralization may have been linked to the consumption of oxygen. Alternately, it may have been that benzene was partially oxidized by microorganisms and the resulting product was amenable to anaerobic decay. An earlier study (Van Beelen & Van Keulen, 1990) showed an extremely rapid rate of benzene mineralization; 2% mineralized in 1 h and 5% in 7 days. No samples were taken between 1 and 7 days and further benzene mineralization was not observed.

Insight on anaerobic benzene biodegradation is contributed by Grbić-Galić and Vogel (Vogel & Grbić-Galić, 1986; Grbić-Galić & Vogel, 1987). Using a mixed methanogenic culture enriched from sewage sludge for its ability to mineralize ferulic acid, they found that ^{14}C-labeled benzene could be converted to $^{14}CO_2$. Intermediates formed during benzene metabolism were detected by gas chromatography/mass spectroscopy (GC/MS) and indicated an initial oxidative attack of the ring to form phenol (Grbić-Galić & Vogel, 1987). Cyclohexanone was tentatively identified as a ring saturation product of the phenol intermediate and methane was formed as an end-product. Using ^{18}O-labeling studies, the same authors showed that the source of the oxygen on the phenolic intermediate produced during benzene metabolism originated from water (Vogel & Grbić-Galić, 1986). In similar fashion, Davis *et al.* (1994) showed the disappearance of 1 mg/kg benzene over a period of 20 days in a methanogenic laboratory incubation system following a 60 day lag period. Unfortunately, methane production levels were not reported in the latter study.

Several studies have reported that benzene will not degrade under methanogenic conditions. Ball & Reinhard (1996) waited 50 days and did not observe the degradation of any BTEX compounds. Battersby & Wilson (1989) tested higher levels of benzene (641 μM), a level which may have been inhibitory to unacclimated cells (Edwards & Grbić-Galić, 1992), and again only waited a relatively short 80 days. Schink (1985) also tested levels of benzene which may have been inhibitory to the organisms present, and negative results were obtained. Berwanger & Barker (1988) incubated aquifer material for 160 days but observed no degradation of any BTEX compounds. This was likely due to the availability of many other preferred organic substrates in the leachate-impacted water. As noted above, we have also failed to observe the biodegradation of benzene in landfill-leachate-impacted anaerobic groundwater, even after extended incubation periods. However, we have observed the methanogenic biodegradation of benzene in an aquifer chronically contaminated with fuel hydrocarbons (Caldwell & Suflita, 1995). This degradation ensued only after a lag period of at least 250 days. Such results may reflect the adaptation and selection of benzene-degrading anaerobes in the chronically fuel-contaminated aquifer compared with the leachate-impacted one. One would predict on this basis that benzene degradation would be difficult to detect in most field and laboratory investigations.

Not surprisingly, the methanogenic field study reported a lack of anaerobic benzene biodegradation. *In situ* columns were used with a variety of chemicals over a 90 day period (Nielsen, Holm & Christensen, 1992). In this study, no BTEX degradation was observed, most likely due to the relatively short time of the study and the high level of landfill leachate components in the water (666 mg/l dissolved organic matter).

A notable exception to the general recalcitrance of benzene under anaerobic conditions is found in the studies of Lovley, Woodward & Chapelle (1994). These investigators found that benzene degradation could be linked to the reduction of ferric iron. Low concentrations of benzene (10 μM) were degraded at a rate of 2–5 μmol/kg sediment/day with concomitant reduction of the near stoichiometrically expected amounts of ferric iron. Interestingly, these authors found that under conditions where iron oxides are present in sediments, the addition of a metal chelator was absolutely required to observe the biodegradation of benzene. While it seems likely that the chelator may increase the availability of iron to the requisite microorganisms, it seems equally unlikely that all endogenous ferric iron reserves in the sediment were initially unavailable to these organisms. The dependency of a chelator for anaerobic benzene decay under iron-reducing conditions requires additional investigation. In this respect, it is interesting to note that Baedecker *et al.* (1993) have also shown the degradation of benzene in iron-reducing incubations with 98% of 10 μM benzene removed after 120 days. An increase in the level of dissolved Fe(II) and Mn(II) was also observed.

3.3 Toluene

Toluene, or methylbenzene, is the second most water-soluble hydrocarbon found in refined gasoline and is on the EPA's regulatory list of pollutant chemicals. Contrary to belief of only a very few years ago, toluene is now known to be biodegradable under a wide variety of anaerobic conditions (Table 3.2). Research on this topic has progressed rapidly and reached the point where several isolates have been obtained and different metabolic pathways have been proposed.

Early field evidence of Schwarzenbach *et al.* (1983) argued that toluene could be transformed anaerobically in a riverwater to groundwater infiltration area in Switzerland. This suggestion was confirmed in subsequent laboratory investigations which revealed that the metabolism was linked with nitrate as the terminal electron acceptor and that ^{14}C-toluene could be converted to ^{14}CO$_2$ (Kuhn *et al.*, 1985; Zeyer, Kuhn & Schwarzenbach, 1986; Kuhn *et al.*, 1988). An aviation-fuel-contaminated aquifer in Traverse City, Michigan, has also become an instructive site for understanding the fate and transport of hydrocarbons in the terrestrial subsurface (Wilson *et al.*, 1990). Aquifer slurries from this site exhibited degradation of toluene as well as other fuel hydrocarbons. In a series of papers by Hutchins and colleagues (Hutchins, 1991a, b; Hutchins *et al.*, 1991a, b; Hutchins, 1993), accelerated toluene degradation could be optimized and linked to the reduction of nitrate or nitrous oxide.

Table 3.2. *Summary of results from reports on the anaerobic biodegradation of toluene*

Electron acceptor	Inoculum source[a]	Degradation	Concentration (μM)	Rate	Reference
Nitrate	Leiduin, The Netherlands	y	NR	$t_{1/2}$ = < 7 d	Piet & Smeenk (1985)
	Swiss River Sediment	y	250	~36 μM/d	Zeyer et al. (1986)
	Borden, Ontario	y	33	NR	Major et al. (1988)
	Swiss River Sediment	y	380	~54 μM/d	Kuhn et al. (1988)
	Traverse City, MI	y	33–240	2.2 μM/d	Hutchins (1991a)
				0.16–0.38/d	Hutchins et al. (1991b)
					Hutchins (1991b)
					Hutchins (1992)
					Hutchins (1993)
	Various inocula	y	100	2.15 μM/d	Evans et al. (1991b)
	North Bay, Ontario	n	1		Acton & Barker (1992)
	Borden, Ontario	y	17	NR	Barbaro et al. (1992)
	Danish groundwater	y	1.6	NR	Flyvbjerg et al. (1993)
	Dutch groundwater	y	16	0.05 μM/d	Morgan et al. (1993)
	Northern Michigan	y	217	NR	Anid et al. (1993)
	Perth, Australia	y	11	$(4 \pm 2)10^{-5}$/s	Patterson et al. (1993)
	Toluene enrichment	y	217–870	3 mmol/g cells/d	Alvarez et al. (1994)
	BTEX polluted aquifer	y	106–209	11 μM/d	Alvarez & Vogel (1994)
	Column biofilm	y	0–1000	167 mm/m^2/d	Arcangeli & Arvin (1994)
	Seal Beach, CA	y	30	7.5 μM/d	Ball & Reinhard (1996)
	Toluene enrichment	y	2.1–5.4	0.05 μM/d	Jensen & Arvin (1994)
Sulfate	Patuxent River, MD	y	80	3.9 μM/d	Beller et al. (1991)
	Seal Beach, CA	y	0.6 μmol/kg	0.1–0.5 nmol/g/d	Haag et al. (1991)
	Patuxent River, MD	y	75–250	3.2 μM/d	Beller et al. (1992a,b)
	Seal Beach, CA	y	50	$dt = 22 \pm 4$ d	Edwards et al. (1992)

	Danish groundwater	y	1.6	NR	Flyvbjerg et al. (1993)
	Perth, Australia	y	11	$(4 \pm 3)10^{-5}$/s	Patterson et al. (1993)
	Perth, Australia	y	11	$(1.2)10^{-7}$/s	Thierrin et al. (1993)
	North Sea Oil Tank	y	NR	NR	Rueter et al. (1994)
	Seal Beach, CA	y	30	1.7 μM/d	Ball & Reinhard (1996)
Methanogenic	Various inocula	n	9413	NR	Schink (1985)
	Sewage sludge	y	50–100		Grbić-Galić (1986)
	Norman, OK	y	6	0.02 μM/d	Wilson et al. (1986)
	Sewage sludge	y	1500–3000		Vogel & Grbić-Galić (1986)
					Grbić-Galić & Vogel (1987)
	Gloucester, Ontario	n	3–7		Berwanger & Barker (1988)
	Traverse City, MI	y	4.6	0.3/wk	Wilson et al. (1990)
	Pensacola, FL	y	100		Beller et al. (1991)
	Seal Beach, CA	n	11	4 μM/d	Ball & Reinhard (1996)
	Various aquifers	y	7–108	NR	Liang & Grbić-Galić (1993)
	Activated carbon reactor	y	217	NR	Narayanan et al. (1993)
	Pensacola, FL	y	<1800	$\mu_{max} = 0.11$/d	Edwards & Grbić-Galić (1994)
Iron (III)	Bemidji, MN	y	600	~13 μM/d	Lovley et al. (1989)
	Bemidji, MN	y	10	NR	Baedecker et al. (1993)
	Bemidji, MN	y	10–40	NR	Cozzarelli et al. (1994)
	Denmark	y	2.1	0.037 μM/d	Albrechtsen (1994)
	Hanahan, SC	y	10	1.8/d	Lovley et al. (1994)
Field					
Nitrate	Leiduin, The Netherlands	y	NR		Piet & Smeenk (1985)
	Traverse City, MI	y	0.14	1.3/wk	Wilson et al. (1990)
	North Bay, Ontario	n	2.2		Acton & Barker (1992)
	Borden, Ontario	y	28	NR	Barbaro et al. (1992)
	Seal Beach, CA	y	2.7	1.6 μM/d	Ball et al. (1994)

Table 3.2. *continued*

Electron acceptor	Inoculum source[a]	Degradation	Concentration (µM)	Rate	Reference
Sulfate	Estuarine sediment	y	NR		Ward et al. (1980)
	Leiduin, The Netherlands	y	NR		Piet & Smeenk (1985)
	Perth, Australia	y	2 g	$t_{1/2} = 100$ d	Thierrin et al. (1993)
	Seal Beach, CA	y	2.7	NR	Ball et al. (1994)
	Arvida, NC	y	98–108 000	0.002 1/d	Borden et al. (1994)
	Seal Beach, CA	y	2–3	0.11 µM/d	Beller et al. (1995)
Methanogenic	Woolwich, Ontario	n	<1.5		Reinhard et al. (1984)
	North Bay, Ontario	y	<0.65	NR	Reinhard et al. (1984)
	Traverse City, MI	y	0–0.1	0.42–1.3/wk	Wilson et al. (1990)
	North Bay, Ontario	y	2.2	NR	Acton & Barker (1992)
	Vegen City, Denmark	n	1.3		Nielsen et al. (1992)
Iron (III)	Bemidji, MN	y	NR		Cozzarelli et al. (1990)

[a] All studies use aquifer sediments and/or groundwater unless otherwise indicated.
NR – not reported.

Similar observations of toluene decay under nitrate-reducing conditions were made by many other investigators (Table 3.2) and nitrate has been used in a variety of attempts to stimulate toluene removal from contaminated areas which normally did not contain appreciable concentrations of nitrate (Major et al., 1988; Aelion & Bradley, 1991; Gersberg, Dawsey & Bradley, 1991; Barbaro et al., 1992; Flyvbjerg et al., 1993; Ball & Reinhard, 1996; Morgan et al., 1993). The picture that emerges from such studies is that toluene-degrading, nitrate-reducing microorganisms are widely distributed in soils, sediments, and groundwaters. In a consistent fashion, Evans, Mang & Young (1991b) reported seven different enrichment cultures from a variety of contaminated sources capable of coupling nitrate reduction with toluene as an electron donor and Fries et al. (1994) isolated ten organisms from a variety of locations around the world capable of this transformation (see below). However, Acton & Barker (1992) saw no toluene degradation in microcosms made from North Bay, Ontario, aquifer sediments amended with nitrate. This is not surprising since the authors had added acetylene to all the bottles. The latter is a known inhibitor of the nitrous oxide to N_2 step of denitrification. Hutchins (1992) and Barbaro et al. (1992) showed that alkylbenzene degradation can be inhibited by acetylene under nitrate-reducing conditions.

Toluene degradation has also been frequently observed under sulfate-reducing conditions (Table 3.2). One of the earliest observations to this effect was that of Berwanger & Barker (1988) using landfill leachate impacted sediments from Gloucester, Ontario. The authors observed that toluene degradation ceased after 159 days, presumably due to sulfate depletion or the accumulation of inhibitory levels of hydrogen sulfide. Similarly, Beller et al. (1991; Beller, Grbić-Galić & Reinhard, 1992a; Beller, Reinhard & Grbić-Galić, 1992b) observed toluene degradation in laboratory microcosms under sulfate- and iron-reducing conditions in an aviation-fuel-contaminated site near the Patuxent River, Maryland. $Fe(OH)_3$ enhanced the onset and rate of toluene biodegradation in these experiments. Presumably this occurred because the iron was reduced and abiotically reacted with the hydrogen sulfide formed during the course of sulfate reduction. Thus, free sulfide was less available to potentially inhibit the toluene-degrading microorganisms. Studies indicating sulfate reduction linked to toluene biodegradation were also reported by Haag, Reinhard & McCarty (1991), Edwards et al. (1992), and Ball & Reinhard (1996) in both column and batch experiments with sediment from the contaminated gasoline aquifer at Seal Beach, California, and with creosote-contaminated aquifer sediments from Denmark (Flyvbjerg et al., 1993). Thierrin et al. (1993) calculated a half-life of 0.3 ± 0.1 day for toluene decay in a sulfate-reducing column study using sediments from the Western Australian coastal plain, while comparable rates were estimated by Patterson et al. (1993) using the same sediments and a one-dimensional model.

Many investigators have also shown that toluene can be degraded under methanogenic conditions. The first report of this was that of Ward et al. (1980) who measured a 5% $^{14}CO_2$ recovery and some methane from ^{14}C-toluene after a 233-

day incubation period in anoxic sediments. Toluene disappearance was also reported in methanogenic aquifer slurries from Norman, Oklahoma (Wilson *et al.*, 1986), Traverse City, Michigan (Wilson *et al.*, 1990), Pensacola, Florida (Beller *et al.*, 1991) and the Sleeping Bear Dunes National Lakeshore Park, Michigan (Barlaz *et al.*, 1993). Lag periods prior to the onset of toluene decay in these studies was often over 100 days and relatively slow rates of toluene decay were measured (Table 3.2).

Toluene decay can also be coupled with the consumption of other electron acceptors. In this regard, an enrichment from the Traverse City, Michigan, site could couple the oxidation of toluene to the reductive dehalogenation of tetrachloroethylene (Liang & Grbić-Galić, 1993). Lovley *et al.* (1989) also observed toluene degradation in aquifer sediments coupled to the reduction of ferric iron. An iron-reducing bacterium was isolated (*Geobacter metallireducens*) that used toluene as an electron donor and produced magnetite as the primary reduced iron end-product (Lovley & Lonergan, 1990). Subsequent investigations suggested that the bioavailability of ferric iron to toluene-degrading microbial communities could be enhanced by the addition of chelating agents (Lovley *et al.*, 1994). Toluene decay was also observed in microcosm experiments using inoculum from the Bemidji, Minnesota, site (Baedecker *et al.*, 1993). Toluene loss was concomitant with an increase in Fe^{2+} and Mn^{2+} under field conditions, lending presumptive evidence for the oxidation of the hydrocarbon coupled to iron or manganese reduction.

Generally, less consistent results for toluene degradation have been obtained with field studies (Table 3.2). Nielsen *et al.* (1992) observed no degradation of toluene in methanogenic field experiments using *in situ* microcosms in a landfill-leachate-impacted aquifer in Denmark. The 100 day incubation period may have been too short to overcome the lag period if the studies noted above are reliable indicators. Even under the most energetically favorable redox conditions, field treatments can be relatively slow. For instance, Hutchins *et al.* (1991a, b) found that nitrate amended and recirculated groundwater helped decrease levels of toluene to below drinking water standards after > 100 days of treatment. While Acton & Barker (1992) saw no degradation of toluene under nitrate-reducing conditions in laboratory experiments, they did observe field evidence for toluene metabolism under both nitrate- and sulfate-reducing conditions, as did Barbaro *et al.* (1992) and Ball *et al.* (1994). Thierrin *et al.* (1993) reported a half-life for toluene decay under sulfate-reducing conditions in the laboratory that was 200 to 500 times faster than the 100 ± 40 days observed in the field. Presumptive field evidence for the anaerobic conversion of toluene to benzoic acid in the anaerobic zone of a crude oil-contaminated aquifer was obtained by Cozzarelli *et al.* (1990, 1994). A similar gasoline-impacted site in Galloway, New Jersey, was described by Cozzarelli *et al.* (1994; Cozzarelli, Baedecker & Hopple, 1991) in which nitrate, iron, and sulfate reduction zones were spatially distinct and presumably linked to the consumption of the spilled hydrocarbons. Lastly, field studies of the methanogenic aquifer sediments from Sleeping Bear Dunes National Lakeshore Park near Empire,

Michigan, produced a first-order toluene decay rate that ranged from 0.023–0.067 per day depending on distance downgradient from the contaminant source (Barlaz *et al.*, 1993).

There are 18 documented pure cultures of denitrifying, sulfate-reducing or iron-reducing bacteria that are capable of anaerobic toluene degradation (Table 3.3). The first isolate with this activity was the iron-reducing bacterium *Geobacter metallireducens* (Lovley *et al.*, 1989; Lovley & Lonergan, 1990). This was also the first pure culture capable of anaerobic oxidation of an aromatic hydrocarbon, although it was not originally isolated on toluene as the sole carbon source. Dolfing *et al.* (1990) isolated a toluene-degrading bacterium, strain T, that could use nitrate or nitrous oxide as an electron acceptor. A toluene-degrading denitrifying bacterium strain T1, not to be confused with strain T, was isolated by Evans *et al.* (1991a). Strain T1 was used by Evans *et al.* (1992a) and Frazer, Ling & Young (1993) to probe the toluene substituent addition reactions described below. Further work by Coschigano, Häggblom & Young (1994) showed that a recombinant strain of T1 was able to hydrolytically dechlorinate 4-chlorobenzoate to 4-hydroxybenzoate with the latter being used as a sole source of carbon for growth. In this same study, the transconjugate strain T1-pUK45-10C was also capable of the simultaneous degradation of toluene and 4-chlorobenzoate under nitrate-reducing conditions.

Seyfried, Tschech & Fuchs (1991) were one of the earliest to observe anaerobic degradation of toluene under nitrate-reducing conditions with pure cultures of microorganisms. Two different isolates were obtained and tentatively identified as *Pseudomonas* sp., strains SP and S2. Schocher *et al.* (1991) documented several strains of denitrifying bacteria capable of anaerobic toluene metabolism. Four *Pseudomonas* sp. were examined, including strains T, S2, S100, and K172. Of these, only strain T was originally isolated for its ability to utilize toluene. Strain S2 was isolated on salicylate and the latter two bacteria were isolated on phenol. The widespread distribution of similar organisms was demonstrated by Fries *et al.* (1994). They were able to isolate ten toluene-degrading nitrate-reducing microorganisms from a variety of locations around the world; four from noncontaminated sources and six from contaminated areas. Of these, eight had 16S rRNA sequences indicating they were closely related and belonged to the genus *Azoarcus*. (Table 3.3).

Rabus *et al.* (1993) were the first to isolate a toluene-degrading organism that utilized sulfate as its terminal electron acceptor. *Desulfobacula toluolica* was isolated from an anoxic marine sediment in a biphasic cultivation system in which the toluene was supplied in an immiscible layer of heptamethylnonane or mineral oil. This may circumvent some of the problems associated with the cultivation of cells on hydrocarbon substrates that are potentially toxic to the organisms at high substrate concentrations. Similarly, in a very recent report, Beller, Ding & Reinhard (1995) obtained a toluene-degrading isolate, strain PRTOL1, that is a sulfate-reducing bacterium obtained from a fuel-contaminated subsurface site near the Patuxent River, Maryland. This isolate is noteworthy because it was used to document the production of several different dead-end metabolites of toluene and

Table 3.3. *Isolates which have been documented to degrade toluene under anaerobic conditions*

Organism (strain)	Electron acceptor	Concentration max.	Degradation rate	Reference
Geobacter metallireducens	Fe(III)	10 mM	dt = 18 d[a]	Lovley *et al.* (1989)
				Lovley & Lonergan (1990)
Pseudomonas sp. strain T	NO_3, N_2O	0.3 mM	5–12 μmol/min./g	Dolfing *et al.* (1990)
Strain T1	NO_3, N_2O	3.0 mM	1.8 μmol/min./l	Evans *et al.* (1991a)
	NO_2		gr = 0.14 h^{-1}	Evans *et al.* (1992b)
			117 μmol/l/h	Coschigano *et al.* (1994)
Strain T1-pUK45-10C	NO_3	0.4 mM	95 μmol/l/h	Coschigano *et al.* (1994)
Pseudomonas sp. strain SP	NO_3	0.5 mM	nd	Seyfried *et al.* (1991)
Pseudomonas sp. strain K172	NO_3, N_2O	2.0 mM	dt = 24 h	Altenschmidt & Fuchs (1991)
			20–50 μmol/min./g	Schocher *et al.* (1991)
Pseudomonas sp. strain S100	NO_3, N_2O	2.0 mM	nd	Schocher *et al.* (1991)
Pseudomonas sp. strain S2	NO_3, N_2O	2.0 mM	nd	Schocher *et al.* (1991)
	NO_3	0.5 mM	nd	Seyfried *et al.* (1991)
Desulfobacula toluolica	SO_4	0.5 mM	dt = 27 h	Rabus *et al.* (1993)
Azoarcus sp. strain Tol-4	NO_3	0.54 mM	dt = 8–13 h	Fries *et al.* (1994)
Azoarcus sp. strain Td-1	NO_3	0.54 mM	dt = 6–7 h	Fries *et al.* (1994)
Azoarcus sp. strain Td-2	NO_3	0.54 mM	dt = 5–7 h	Fries *et al.* (1994)
Azoarcus sp. strain Td-3	NO_3	0.54 mM	dt = 7–8 h	Fries *et al.* (1994)
Azoarcus sp. strain Td-15	NO_3	0.54 mM	dt = 6–7 h	Fries *et al.* (1994)
Azoarcus sp. strain Td-17	NO_3	0.54 mM	dt = 5–7 h	Fries *et al.* (1994)
Azoarcus sp. strain Td-19	NO_3	0.54 mM	dt = 5–7 h	Fries *et al.* (1994)
Azoarcus sp. strain Td-21	NO_3	0.54 mM	dt = 5–7 h	Fries *et al.* (1994)
strain PRTOL1	SO_4	nd	nd	Beller *et al.* (1995)

dt = doubling time.
[a] = approximate doubling time.
nd = not determined.
gr = growth rate.

xylene isomers that were products of substituent addition reactions (see below).

Toluene-degrading methanogenic enrichments have played pivotal roles in helping establish the initial biotransformation reactions associated with the anaerobic decay of this hydrocarbon. Proof for the anaerobic biodegradation of toluene was first obtained using a mixed methanogenic enrichment cultivated for its ability to mineralize ferulic acid. Using this enrichment, Vogel & Grbić-Galić (1986), showed that the heavy isotope of ^{18}O-labeled water was incorporated into p-cresol when toluene served as the parent substrate. Thus, the initial transformation was oxidative with the oxygen functionality derived from water. Later, Grbić-Galić & Vogel (1987) established a convincing mass balance relationship between the toluene and mineralized end products. The authors further suggested that p-cresol, o-cresol, benzyl alcohol, and methylcyclohexane were early toluene transformation intermediates. Such findings suggested that at least three routes of anaerobic toluene decay occurred simultaneously in this mixed culture. These routes included, (i) separate ring oxidations, (ii) the hydroxylation of the aryl methyl group, and (iii) ring reduction (Figure 3.1).

Ring and aryl methyl group oxidation were the initial toluene-degradation routes speculated on for the nitrate-reducing enrichment obtained by Kuhn *et al.* (1988) and the metabolically diverse iron-reducing bacterium *Geobacter metallireducens* (Lovley & Lonergan, 1990). The speculation was consistent with the fact that both of these cultures could metabolize the appropriate suite of putative intermediates. However, conclusive evidence as to which pathway was actually involved was not obtained.

Aryl methyl group hydroxylation to benzyl alcohol was also suggested by coadaptation studies where toluene-degrading, sulfate-reducing aquifer enrichments dosed with benzyl alcohol degraded it without a lag (Haag *et al.*, 1991). Similarly, Schocher *et al.* (1991) suggested that the initial attack on the toluene molecule by two nitrate-reducing isolates, *Pseudomonas* sp. strain K172 and strain T, was via hydroxylation of the methyl group. This view was supported by the use of [*methyl*-^{14}C]toluene in the presence of fluoroacetate with the accumulation of [*methyl*-^{14}C]benzoate. In addition, benzaldehyde and benzoate were found to accumulate when cell suspensions of the same two isolates were given toluene and incubated at 5 °C (Seyfried *et al.*, 1994). The oxidation of the aryl methyl group would lead to the formation of benzoic acid (Figure 3.1), which is known to be a metabolite of anaerobic toluene decomposition (e.g., Kuhn *et al.*, 1988; Schocher *et al.*, 1991; Altenschmidt & Fuchs, 1991; Frazer *et al.*, 1993; Beller *et al.*, 1992b; Seyfried *et al.*, 1994).

Confirmation of this proposed pathway was established by Altenschmidt & Fuchs (1991, 1992) in their study of the biochemistry of toluene decay by the denitrifying *Pseudomonas* sp. strain K172. These investigators confirmed the presence of benzyl alcohol dehydrogenase, benzaldehyde dehydrogenase, and benzoyl-CoA synthetase in cell-free extracts of this isolate. Further, [^{14}C]benzyl

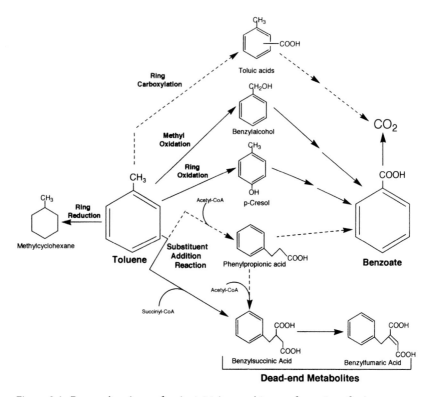

Figure 3.1. Proposed pathways for the initial anaerobic transformation of toluene. Dashed lines indicate pathways for which identification of the intermediate is based on indirect evidence.

alcohol and $[^{14}C]$benzaldehyde were detected in high density cell suspensions of strain K172 grown on $[^{14}C]$toluene. The demonstration of the intermediates and the appropriate enzymes is strong evidence consistent with the methyl hydroxylation mechanism for anaerobic toluene metabolism by this bacterium. Evidence from an isotope-trapping experiment caused Edwards, Edwards & Grbić-Galić (1994) to hypothesize that the aryl methyl group of toluene was similarly hydroxylated by a mixed methanogenic culture, although some labeled *p*-cresol was also detected. This enrichment was derived from creosote-contaminated sediments and was capable of completely mineralizing toluene to stoichiometric amounts of carbon dioxide and methane (Edwards & Grbić-Galić, 1994). Toluene decay followed Monod kinetics with a $K_s = 30 \ \mu M$ and a $\mu_{max} = 0.11/day$.

An interesting initial transformation reaction for the anaerobic metabolism of toluene was reported by Evans *et al.* (1992a) using a denitrifying bacterium. These investigators found benzylsuccinic acid and benzylfumaric acid as metabolic dead-end products of toluene decay. They hypothesized the addition of the four carbon fragment of succinyl-CoA to the methyl group of toluene to form the derivative of these dead-end metabolites. The analogous reaction involving the addition of acetyl-CoA to the aryl methyl group of toluene was speculated to be a more biologically productive route of parent substrate decay. The same dead-end metabolites were also detected in studies of toluene decay in sulfate-reducing incubations (Beller *et al.*, 1992b) as well as other nitrate-reducing incubations (Frazer *et al.*, 1993; Seyfried *et al.*, 1994). Chee-Sanford and Tiedje (1994) speculated that the substituent group addition to toluene was not a single attack by succinyl-CoA, but two separate reactions involving two molecules of acetyl-CoA to give rise to the benzylsuccinic acid end-product (Figure 3.1). In all cases described thus far, varying amounts (up to 10–20%) of toluene carbon are converted to these dead-end products.

Another interesting speculation on the initial anaerobic bioconversion of toluene is that of Tschech & Fuchs (1989). They note that toluene might be carboxylated to form a toluic acid isomer (Figure 3.1). To date, there is no evidence to support this contention.

3.4 Ethylbenzene

There are relatively few studies that argue the case for or against the anaerobic biodegradation of ethylbenzene (Table 3.4). The majority of studies focus on the removal of ethylbenzene at relatively low concentrations (0.9–189 μM) under nitrate-reducing conditions. In several studies employing aquifer material (see Table 3.4), ethylbenzene depletion was rapid relative to the number of sampling opportunities. In several of these studies biodegradation began without a lag with a rate comparable to that observed with toluene. In others, a lag of 1–2 months was encountered with degradation ensuing over the next one to two weeks.

These studies represent an exciting prospect for those interested in the anaerobic metabolism of alkylbenzenes under nitrate-reducing conditions, but rapid biodegradation of ethylbenzene is not a universal phenomenon. Microcosms prepared from the Borden aquifer (Barbaro *et al.*, 1992) revealed that ethylbenzene degradation proceeded slowly with only about 30% of the parent substrate removed over an 18 month period. In experiments using inoculum from a southwestern Australian aquifer (Patterson *et al.*, 1993), a lag period of approximately 100 days was measured before the biodegradation of ethylbenzene began. Several other studies report on the recalcitrance of ethylbenzene under anaerobic conditions. Alvarez & Vogel (1994) indicated that no ethylbenzene decay could be obtained over an 11 week period when inocula from four different sediments were incubated anaerobically in the presence of nitrate as a terminal electron acceptor. Similar results were found over the same time frame when primary sewage effluent was used as inoculum (Bouwer & McCarty, 1983), in the column studies of Anid, Alvarez & Vogel (1993) following a 6 week incubation, and in another column study after only a 2–6 day incubation (Kuhn *et al.*, 1988).

Table 3.4. *Summary of results from reports on the anaerobic biodegradation of ethylbenzene*

Electron acceptor	Inoculum source[a]	Degradation	Concentration (μM)	Rate	Reference
Nitrate	Sewage sludge	n	0.4–1.1		Bouwer & McCarty (1983)
	Swiss River Sediment	n	220		Kuhn et al. (1988)
	Traverse City, MI	y	188	11 μM/d	Hutchins (1991a)
			207	0.016–0.065/d	Hutchins et al. (1991b)
			36	1.4 μM/d	Hutchins (1991b)
	North Bay, Ontario	n	0.9		Acton & Barker (1992)
	Borden, Ontario	y	2.0	NR	Barbaro et al. (1992)
	Traverse City, MI	y	20	3.3 μmol/d	Hutchins (1992)
	Dutch groundwater	y	1.5	0.03 μM/d	Morgan et al. (1993)
	Northern MI	n	28		Anid et al. (1993)
	Perth, Australia	y	9	$(4 \pm 3)10^{-5}$/s	Patterson et al. (1993)
	Toluene enrichment	y	1.8	0.6 μM/d	Jensen & Arvin (1994)
	BTEX polluted aquifers	n	16–35		Alvarez & Vogel (1994)
	Seal Beach, CA	y	10	1.0 μM/d	Ball & Reinhard (1996)
Sulfate	Seal Beach, CA	n	54 μmol/kg		Haag et al. (1991)
	Patuxent River, MD	n	40–100		Beller et al. (1991)
	Perth, Australia	n	9		Thierrin et al. (1993)
	Perth, Australia	n	9		Patterson et al. (1993)
	Woods Hole, MA pond sediment	n	250		Rabus et al. (1993)
	North Sea Oil Tank	n	NR		Rueter et al. (1994)
	Seal Beach, CA	n	10		Ball & Reinhard (1996)
	Seal Beach, CA	n	2–3		Beller et al. (1995)

Methanogenic	Norman, OK	y	80 nM/wk	2.1	Wilson et al. (1986)
	Sewage sludge	y	NR	0.05	Grbić-Galić (1986)
	Gloucester, Ontario	n		3–6	Berwanger & Barker (1988)
	Various aquifers	n		94	Liang & Grbić-Galić (1993)
	Pensacola, FL	n		40–320	Edwards & Grbić-Galić (1994)
Iron (III)	Bemidji, MN	y	NR	32.7	Cozzarelli et al. (1994)
Field					
Nitrate	Traverse City, MI	y	NR	8	Hutchins et al. (1991a)
	North Bay, Ontario	n		1.9	Acton & Barker (1992)
	Borden, Ontario	y		3.5	Barbaro et al. (1992)
	Seal Beach, CA	y	0.38 µM/d	2.4	Ball et al. (1994)
Sulfate	North Bay, Ontario	n		0.13–4.4	Reinhard et al. (1984)
	Leiduin, The Netherlands	y		NR	Piet & Smeenk (1985)
	North Bay, Ontario	y	NR	1.9	Acton & Barker (1992)
	Perth, Australia	y	4.65 kg/y	155	Thierrin et al. (1993)
	Seal Beach, CA	n		9	Ball et al. (1994)
Methanogenic	Woolwich, Ontario	n		1.1	Reinhard et al. (1984)
Iron(III)	Bemidji, MN	y		NR	Cozzarelli et al. (1990)

[a]All studies use aquifer sediments and/or groundwater unless otherwise indicated.
NR – not reported.

There are several possible reasons for the lack of biodegradation in the above cases. Hutchins *et al.* (1991b) showed that the presence of BTEX contamination in sediments may inhibit degradation of some substrates, especially ethylbenzene. This effect may have been important in the experiments of Alvarez & Vogel (1994) who used sediments previously exposed to BTEX contamination as inoculum. Nutrients also seem to influence rates of metabolism with the addition of phosphate having a positive effect on biodegradation (Hutchins, 1991b: Morgan *et al.*, 1993). These factors and conceivably many others may have contributed to the apparent recalcitrance of ethylbenzene in the studies noted above.

In field studies, it is not uncommon for the anaerobic biodegradation of ethylbenzene to be observed (see Table 4.3). When BTEXs were added to a well as a slug and monitored with time at the Seal Beach site, most of the ethylbenzene was removed in 10 days (Ball *et al.*, 1994). In parallel bioreactor experiments the investigators observed that ethylbenzene was removed but to a much lesser extent than observed *in situ*. In an injection experiment at the Borden aquifer site (Barbaro *et al.*, 1992), ethylbenzene was effectively removed and no longer appeared at a sampling well located 5 m downgradient from an injection well, even after 300 days of continuous injection. In a biorestoration experiment at the Traverse City site (Hutchins *et al.*, 1991a), ethylbenzene concentrations were reduced approximately 200-fold, but oxygen was occasionally present in the injected groundwater. The only field study which did not show biodegradation of ethylbenzene was done in an *in situ* column at the North Bay site (Acton & Barker, 1992). In that experiment, acetylene was added to the column, which may have negatively influenced the anaerobic biodegradation of ethylbenzene.

There has been no conclusive laboratory studies linking the anaerobic biodegradation of ethylbenzene to the availability of sulfate as a terminal electron acceptor. As noted above, sulfate-reducing bacteria are known to attack methylbenzenes, possibly by activating the methyl group. The difficulty in similarly activating the ethyl moiety by these cells may be a factor limiting the biodegradation of ethylbenzene in laboratory studies. Ethylbenzene has been observed to degrade in the field under sulfate-reducing conditions with a study by Thierrin *et al.* (1993) reporting degradation within an Australian aquifer at a rate of 4.65 kg/year. *In situ* columns at the North Bay site showed degradation of ethylbenzene (Acton & Barker, 1992) although no biodegradation was detected in the aquifer injection study. Piet & Smeenk (1985) also reported losses. In the field, it is more difficult to determine the exact nature of the electron acceptor and more than one may exist at any one site. We must therefore exercise caution in interpreting the field results to mean that ethylbenzene can be degraded under sulfate-reducing conditions. Further studies will be required to determine the biodegradability of ethylbenzene.

Only a few studies have clearly shown the biodegradation of ethylbenzene under methanogenic conditions (Grbić-Galić, 1986; Wilson *et al.*, 1986). Grbić-Galić (1986) observed the accumulation of ethylbenzene over a period of 1 month by a consortium growing on ferulic acid. Subsequently, the ethylbenzene was degraded by this enrichment culture. In order to more easily detect potential metabolites, the culture was treated with bromoethane sulfonic acid (BESA), an inhibitor of methanogenesis. Mass spectral analysis of metabolites in the culture fluids suggested at least two possible routes for the transformation of the parent substrate. The presence of ethylcyclohexane suggested the involvement of a ring saturation mechanism. In addition, the oxidation of the ring in a manner possibly analogous to that proposed for the oxidation of benzene to phenol (Grbić-Galić & Vogel, 1987) was suggested by the presence of 3-hydroxy-4-ethylphenol which may have been formed from 2-ethylphenol. Since this consortium was growing on ferulate, there is the possibility that the ethylphenols may have been produced directly from ferulate. In our study with an inoculum from a gasoline-contaminated aquifer, we similarly observed the bioconversion of ethylbenzene to the stoichiometrically expected amount of methane following a lag period of at least 125 days (unpublished results). In consistent fashion, Wilson et al. (1986) observed the disappearance of ethylbenzene in landfill-leachate-impacted aquifer microcosms following a 20–40 week lag period. There have also been studies where the anaerobic biodegradation of ethylbenzene was not observed under methanogenic conditions. In one such study, the incubation period of 7 days was undoubtedly too short (Liang & Grbić-Galić, 1993), while the reason negative results were obtained in the other study is not clear (Edwards & Grbić-Galić, 1994). Possibilities include the lack of appropriate organisms, inhibitory concentrations of BTEXs (160 μM total BTEX), or it may have been that the lag was longer than the incubation period (greater than 300 days). Many other potential biotic and abiotic factors could also limit the anaerobic biodegradation of ethylbenzene.

Field studies alluding to the methanogenic biodegradation of ethylbenzene are more difficult to interpret. Indirect evidence for methanogenic degradation of ethylbenzene is provided by Cozzarelli et al. (1990) as the concentration of this compound was observed to decrease in the region of the plume where methanogenesis was occurring. Although methane production occurs in this region, reduction of Fe(III) is also known to occur and thereby confounds interpretation of the results.

3.5 *m*-, *p*-Xylenes

The xylene isomers are the final class of chemicals pertinent to this review. The *meta* and *para* isomers are considered together because they are often measured simultaneously due to the difficulty in resolving them by gas chromatography. As with most of the BTEX compounds, the majority of the work on these two xylenes has been carried out with nitrate available as the terminal electron acceptor. Both compounds tested either individually (Evans et al., 1991b; Hutchins, 1991b, 1993; Hutchins et al., 1991b; Kuhn et al., 1988; Zeyer et al., 1986) or as a group with other BTEX compounds (see Table 3.5) are known to be degradable in acclimated laboratory cultures at the rate of about 1 μM/day; this rate is reported in several of these

Table 3.5. *Summary of results from reports on the anaerobic biodegradation of the xylenes*

Electron acceptor	Isomer	Inoculum source[a]	Degradation	Concentration (µM)	Rate	Reference
Nitrate	p	Swiss River Sediment	y	0.5	NR	Kuhn et al. (1985)
	m		y	0.5		
	o		y	0.5		
	m (ind)	Swiss River Sediment	y	190, 400	0.45/h	Kuhn et al. (1988)
						Zeyer et al. (1986)
	m	Borden, Ontario	y	28	0.30 µM/d	Major et al. (1988)
	o		y	28	0.29 µM/d	
	p	Various enrichments	n	100		Evans et al. (1991b)
	m (ind)		y tol. dep.[1]	100	7.1 µM/d	
	o		y	100	4.5 µM/h	
	p (ind)	Traverse City, MI	y	151	7.5 µM/d	Hutchins (1991b)
	m (ind)		y	151	7.5 µM/d	
	o (ind)		n	141		
	m (ind)	Traverse City, MI	n	148		Hutchins et al. (1991b)
	o (ind)		y	226	1.6 µM/d	
	m,o	North Bay, Ontario	n	1.2		Acton & Barker (1992)
	p	Borden, Ontario	y	2	0.0017 µM/d	Barbaro et al. (1992)
	m		y	5	0.0043 µM/d	
	o		y	3	0.0032 µM/d	
	p	Traverse City, MI	y	20	1.4 µM/d	Hutchins (1992)
	m		y	20	1.4 µM/d	Hutchins (1991a)
	o		y	23	1.1 µM/d	
	(m + p)	Empire, MI	y	NR	0.0006/d	Barlaz et al. (1993)
	o		y		0.0006/d	
	m,o,p	Danish groundwater	n	0.28–0.75		Flyvbjerg et al. (1993)
	m (ind)	Traverse City, MI	y	94	31 µM/d	Hutchins (1993)
	o (ind)		y tol. dep.	113		
	(m + p)	Netherlands groundwater	y	2.7	0.015 µM/d	Morgan et al. (1993)
	o		n	6.6		
	m + o	Perth, Australia	y	9.4		Patterson et al. (1993)
	m		y	45		
	p	BTEX polluted aquifers	y	17	0.5 µM/d	Alvarez & Vogel (1994)
	o		y	17	0.18 µM/d	

Process	Compound	Location		Conc.	Rate	Reference
	(m+p)	Northern Michigan	y	123	NR	Anid et al. (1993)
			y	38		
	m,p,o	Toluene enrichment	y	1.8–4.7	NR	Jensen & Arvin (1994)
	(m+p)	Seal Beach, CA	y	30	1.2 µM/d	Ball & Reinhard (1996)
	o		n	19		
Sulfate	p	Seal Beach, CA	y	0.053 µmol/g	0.2 nmol/g/d	Haag et al. (1991)
	o		n	0.056 µmol/g		
	m,p,o	Borden, Ontario	n	1.9–4.8		Barbaro et al. (1992)
	p	Seal Beach, CA	y	47	0.78 µM/d	Edwards et al. (1992)
	m		y	47	1.3 µM/d	
	m,p,o	Seal Beach enrichment	y	47		
	m	Danish groundwater	n	0.28–0.75		Flyvbjerg et al. (1993)
	p	North Sea Oil Tank	y	NR		Rueter et al. (1994)
	(m+p)	Seal Beach, CA	y	21	1.0 µM/d	Ball & Reinhard (1996)
	o		n	15		
Methanogenic	m,p,o (ind)	Mud + sewage	n	9434		Schink (1985)
	o	Norman, OK	y	2.4	0.0034 µM/d	Wilson et al. (1986)
	m,p,o	Gloucester, Ontario	n	2.8–5.6	NR	Berwanger & Barker (1988)
	(m+p)	Traverse City, MI	y	4.2	0.5/wk	Wilson et al. (1990)
	o		y	4.2	32 µM/d	
	o	Pensacola, FL enrichment	y	40	$\mu_{max} = 0.07$/d	Edwards & Grbić-Galić (1994)
Iron (III)	p	Bemidji, MN	n	40		Cozzarelli et al. (1994)
	o		y	29	0.5 µM/d	
Field Nitrate	o,m	North Bay, Ontario	n	1.8 µM	NR	Acton & Barker (1992)
	m,p,o	Borden, Ontario	y	3.4–8.9 µM	0.002–0.01/d	Barbaro et al. (1992)
	p	Empire, MI	y		0.004–0.014/d	Barlaz et al. (1993)
	m				0.004–0.016/d	
	m	Seal Beach, CA	y	250 µg/l	5.3 µg/l/h	Ball et al. (1994)
	o		y	250 µg/l	0.2 µg/l/h	
	(m+p)	Seal Beach, CA, bioreactor	y	NR		Ball et al. (1994)
	o		n			

Table 3.5. *continued*

Electron acceptor	Isomer	Inoculum source[a]	Degradation	Concentration (μM)	Rate	Reference
Sulfate	$(m+p)$	North Bay, Ontario	y	0.004–7.7	NR	Reinhard et al. (1984)
	o		y	0.006–5.0		
	$(m+p)$	Leiduin, The Netherlands	y	NR		Piet & Smeenk (1985)
	o		y			
	m	North Bay, Ontario	y	1.8	NR	Acton & Barker (1992)
	o		n			
	m,p,o	Borden, Ontario	n	2.5–6.5		Barbaro et al. (1992)
	$(m+p)$	Perth, Australia	y	4.2–170	$t_{1/2}$ = 170 d	Thierrin et al. (1993)
	o		y	0–72	$t_{1/2}$ = 125 d	
	$(m+p)$	Seal Beach, CA	y	2.3	NR	Ball et al. (1994)
	o		y	2.3		
	m	Seal Beach, CA	y	2–3	0.09 μM/d	Beller et al. (1995)
	o		y	2–3	0.09 μM/d	
Methanogenic	o	Denmark	n	1.1		Nielsen et al. (1992)
Iron (III)	m,p,o	Bemidji, MN	y	NR		Cozzarelli et al. (1990)
	$(m+p)$	Arvida, NC	y	1433–75471	0.0013/d	Borden et al. (1994)
	o		y	123–35800	0.0021/d	

[1] tol. dep. = degradation was dependent on the presence of toluene.
[a] All studies use aquifer sediments and/or groundwater unless otherwise indicated.
(ind) – tested individually, rather than as a group of BTEX compounds.
NR – not reported.

papers. However, there are notable exceptions. In a study using Borden aquifer sediments as inoculum (Barbaro *et al.*, 1992), rates of xylene degradation were about three orders of magnitude slower than that indicated above. This finding may be partially explained by the use of lower initial xylene concentrations (one to two orders of magnitude less) and the use of an unacclimated inoculum.

In several studies, no degradation of the xylene isomers was observed at all. In one example (Hutchins *et al.*, 1991b), *m*-xylene did not degrade significantly when incubated in the absence of other co-contaminants. This was an unexpected finding, since in a similar paper published the same year (Hutchins, 1991b) *m*-xylene was indeed transformed under these conditions in incubation systems containing multiple potential substrates. Evans *et al.* (1991b) were unable to measure either *m*- or *p*-xylene degradation in their enrichments containing a mixture of BTX compounds. They did observe significant degradation of toluene and *o*-xylene when the incubations were reamended with toluene. Under conditions where the toluene concentration was relatively high (100 μM), toluene-degrading bacteria may have outcompeted the *p*- and *m*-xylene degraders for available nutrients or preferentially metabolized toluene over the xylenes. Later experiments provided additional evidence for the latter possibility. A subculture of a soil enrichment was incubated without toluene and with only *m*-xylene. With no competition for other substrates the *m*-xylene was rapidly degraded (within 2 weeks). As well, a pure culture isolated with toluene and nitrate was shown also to degrade *m*-xylene (Fries *et al.*, 1994).

In one of the few studies in which only groundwater was used as an inoculum, no xylene degradation was reported (Flyvbjerg *et al.*, 1993). The relatively low abundance of the requisite microorganisms as well as the use of the microbial inhibitor Na_2SO_3 to remove residual oxygen may help explain such findings. Similarly, two other studies allude to the recalcitrance of *m*-xylene under anaerobic conditions. The short incubation time of 17 days may have been a factor in the denitrifying study of Patterson *et al.* (1993), while the inhibitory effect of acetylene may have precluded xylene decomposition in the North Bay aquifer sediments (Acton & Barker, 1992).

When aquifers were amended with nitrate in order to stimulate biodegradation, the results were generally consistent with those obtained in laboratory investigations. In a field injection experiment at Seal Beach, California, Ball *et al.* (1994) demonstrated complete removal of *m*-xylene and the *m*-, *p*-xylene fraction decreased significantly in parallel bioreactor experiments. A Canadian study showed a decrease in *m*- and *p*-xylene of 14% and 15%, respectively, over a 1 m flowpath in the Borden aquifer (Barbaro *et al.*, 1992). Very little degradation was observed beyond that point, presumably due to the availability of preferred electron donors in the landfill leachate impacted aquifer.

As with nitrate, the anaerobic biodegradation of *m*- and *p*-xylene could also be linked with the reduction of sulfate at a rate of about 1 μM/day. Edwards *et al.* (1992) demonstrated the sulfate-dependent degradation of all three xylene isomers. Interestingly, the xylenes were not degraded until toluene was almost depleted.

Free sulfide (≥ 1 mM) and higher BTEX (≥ 300 μM) concentrations inhibited the bioconversion. In another study with Seal Beach sediments (Haag *et al.*, 1991), *p*-xylene was degraded in sulfate-reducing columns at a similar rate (0.2 μmol/kg/day).

Such results contrast with those of Flyvbjerg *et al.* (1993) and Barbaro *et al.* (1992), which failed to demonstrate xylene degradation with sulfate as an electron acceptor. The former study was negative, presumably for reasons analogous to their nitrate-reducing experiments noted above, while the latter may have been undermined by the relatively high level of dissolved organic matter in the groundwater and/or by the absence of an exogenous reductant in the experiments.

Field experience has yielded consistent results with sulfate-reducing studies at Seal Beach (Ball *et al.*, 1994; Beller *et al.*, 1995), North Bay, Ontario (Acton & Barker, 1992; Reinhard, Goodman & Barker, 1984) and in an Australian aquifer (Thierrin *et al.*, 1993) indicating loss of the *m*-, *p*-xylene relative to a conservative tracer. Anaerobic biodegradation of *m*- and *p*-xylene was suggested in a study of the Arvida, North Carolina, site (Borden, Gomez & Becker, 1994). Nitrate, iron and sulfate are potentially available at this site and it is not clear if the consumption of the xylenes is coupled with the reduction of these or any other electron acceptor. While the reduction of *m*-xylene was measured in a field study at the North Bay site (Acton & Barker, 1992), no direct evidence for biodegradation was observed at the Borden site. This may be for the same reasons as those that may have beset their findings under nitrate-reducing conditions.

Until now, only a single study using aquifer material from Traverse City, Michigan (Wilson *et al.*, 1990), suggested the loss of *m*-, *p*-xylene under methanogenic conditions. In this study, the authors reported losses in control bottles of greater than 50% of the losses in live bottles and reported no error values. The relatively few other laboratory methanogenic studies failed to demonstrate biodegradation of *m*- or *p*-xylene. *m*-Xylene was tested in two studies, one of which employed potentially inhibitory substrate levels (Schink, 1985), while the other used leachate-impacted groundwater which, as noted above, may have had attendant problems (Berwanger & Barker, 1988). These workers also did not include a reducing agent in their experiments; this is often needed for methanogenic laboratory cultures.

There are relatively few field hydrocarbon biodegradation studies in which methanogenesis was reported to be the dominant electron-accepting process. Wilson *et al.* (1990) have reported the loss of all three xylene isomers at a rate of 0.03/wk. However, the lack of tracers in this study make it more difficult to interpret. Therefore, the prospects for the methanogenic biodegradation of both *meta* and *para* xylene must be considered with caution.

In a single field study (Cozzarelli *et al.*, 1990) where Fe(III) served as the terminal electron acceptor, oxidized products of *m*- and *p*-xylene were observed in the groundwater at Bemidji, Minnesota. While it is not inconceivable that similar products can be formed by either aerobic or anaerobic metabolic routes, the study provides presumptive evidence for xylene degradation under iron-reducing conditions.

3.6 o-Xylene

The anaerobic biodegradation of *ortho*-xylene has been considered in almost all papers that have investigated the other isomers. These studies indicate that o-xylene can also be degraded under a variety of anaerobic conditions. In the presence of nitrate, o-xylene was degraded at the same or slightly less rapid rate than the other xylenes. Several studies (Evans *et al.*, 1991b; Hutchins, 1993) have shown that the degradation of o-xylene is a cometabolic process in which o-xylene is transformed by toluene-degrading bacteria. Evans *et al.* (1991b) reported that toluene depletion from a culture containing a denitrifying bacterium caused the degradation o-xylene to cease. In most of the other reports, the degradation of o-xylene was observed in the presence of toluene or the influence of toluene was not determined. There is one notable exception (Hutchins *et al.*, 1991b), possibly indicating alternative routes for the anaerobic metabolism of o-xylene (see below).

There are several cases where o-xylene degradation did not occur under nitrate-reducing conditions (see Table 4.5). Possible reasons for these observations were considered above. However, two reports not considered in the previous subsection failed to measure the biodegradation of o-xylene (Morgan *et al.*, 1993; Flyvbjerg *et al.*, 1993) In both cases there was probably very little microbial biomass as only groundwater was used as an inoculum and rates of o-xylene metabolism were likely too slow to measure.

In the field, o-xylene is also observed to degrade with nitrate in all but a few of the reported cases (Table 3.5). No o-xylene degradation was observed in either microcosms (Ball & Reinhard, 1996) or in field-scale bioreactors by Ball *et al.* (1994). However, the removal of this substrate was apparent in a field injection experiment. Since the sediments were initially sulfate reducing, there may have been an adaptation period for substrate depletion with nitrate that was longer than the 40 and 60 day incubation periods associated with the microcosm and bioreactor study, respectively. As noted for the other alkylbenzenes, the second field study not showing anaerobic o-xylene degradation (Acton & Barker, 1992) was compounded by the presence of acetylene in the experiment.

There have been several reports that have examined the degradation of o-xylene under sulfate-reducing conditions. Edwards *et al.* (1992) observed the decay of this substrate to about 20% of the original level by BTEX-adapted enrichment cultures. The critical observation was that the culture did not begin to degrade o-xylene until toluene and p-xylene were almost completely removed. This again points to the possibilities of (a) competition for available nutrients or for available electron acceptors among BTEX-degrading organisms, or (b) that a single organism selectively metabolizes preferred hydrocarbons prior to the attack on others. Such observations may also help explain why other groups were unable to demonstrate the degradation of o-xylene under sulfate-reducing conditions (Ball & Reinhard, 1996; Haag *et al.*, 1991; Barbaro *et al.*, 1992; Flyvbjerg *et al.*, 1993).

Positive results in field investigations of o-xylene metabolism were obtained at

Seal Beach (Ball *et al.*, 1994; Beller *et al.*, 1995) and in an Australian aquifer contaminated with gasoline (Thierrin *et al.*, 1993). *o*-Xylene was degraded in an initial study at the North Bay site (Reinhard *et al.*, 1984), but in a later more comprehensive study (Acton & Barker, 1992) was not observed to degrade. Again, no biodegradation of this substrate was observed at the Borden site (Barbaro *et al.*, 1992; see previous discussion of this site). In each of the above studies showing biodegradation, *o*-xylene was present with other methylbenzenes. The difference between field and laboratory observations may, in part, be a reflection of the level of compounds; lower field concentrations allowing for the simultaneous degradation of different compounds. The laboratory evidence of Edwards *et al.* (1992) is consistent with the field observation that *o*-xylene is the most slowly degraded of the dimethylbenzene isomers.

o-Xylene is sometimes degraded by methanogenic aquifer enrichment cultures (Table 3.5). In the positive studies (Edwards & Grbić-Galić, 1994; Wilson *et al.*, 1986) a much longer lag was encountered with *o*-xylene than with toluene, leading to the conclusion that a similar competition to that noted above also exists in these enrichments. However, when Edwards & Grbić-Galić (1994) reamended their enrichments with toluene and *o*-xylene, degradation began for both without a lag. Furthermore, no lag was observed for any of the BTEXs at the Traverse City site (Wilson *et al.*, 1990). Edwards & Grbić-Galić (1994) measured a K_s of 20 μM and a μ_{max} of 0.07/day for *o*-xylene decay by the enrichment. Concentrations of *o*-xylene > 500 μM were slightly inhibitory to the cells, while no degradation was observed at substrate concentrations above 1.3 mM. These cultures were also inhibited by the presence of other organic compounds, including acetate, acetone and glucose. Therefore, the presence of endogenous organic compounds may be the reason for the inability of investigators to demonstrate anaerobic *o*-xylene biodegradation in microcosms from the Gloucester landfill site (Berwanger & Barker, 1988).

Very little is known about xylene-degrading microorganisms and the pathways they have evolved for the metabolism of the various isomers. The first report of an anaerobe capable of xylene decomposition appeared in 1990 (Dolfing *et al.*, 1990). In addition to utilizing toluene, *Pseudomonas* sp. strain T was capable of metabolizing *m*-xylene with nitrate as an electron acceptor. Xylene could act as a sole carbon source for this organism and a 50% recovery of $^{14}CO_2$ from radiolabeled *m*-xylene could be measured. Subsequent work with strain T by Seyfried *et al.* (1994) showed that when strain T was grown in the presence of toluene, the bacterium could transform *m*- and *p*-xylene to the 3- and 4-methylbenzoic acids, with the transient accumulation of the respective methylbenzaldehydes. Thus, oxidation of the methyl group of xylene was the predominant mode of attack and the information is consistent with the initial steps of toluene decay by this organism (see above).

Other organisms and routes of xylene metabolism are also known. *Azoarcus* sp. strain Td-15 was the only one of eight toluene-degrading, nitrate-reducing bacteria isolated by Fries *et al.* (1994) capable of metabolizing *m*-xylene, but the predominant pathway employed is unknown. Evans *et al.* (1991a) obtained a denitrifying isolate, strain T1, that could transform *o*-xylene in presence of toluene, but was unable to utilize the former as a sole source of carbon. Subsequent work with strain T1 (Evans *et al.*, 1992a) identified the *o*-xylene transformation products as 2-methylbenzylsuccinic and 2-methylbenzylfumaric acids, which are the methylated derivatives of the dead-end metabolites produced by this organism during toluene decay (see above). Similarly, Beller *et al.* (1995) detected the same metabolites of *o*-xylene decay under sulfate-reducing conditions by strain PRTOL1 and found 4-methylbenzylsuccinic acid as a product of *p*-xylene metabolism. In analogous fashion, Frazer *et al.* (1993) reported observing analogous dead-end intermediates for the degradation of *m*-xylene by another isolate under nitrate-reducing conditions. These reports argue strongly for a series of methyl group addition reactions as described above for toluene.

Negative results were obtained in field studies that specifically examined the biodegradation of *o*-xylene under methanogenic conditions. Nielson *et al.* (1992) did not observe BTEX degradation over a 100 day period, indicating that the resident microflora may have been inhibited, possibly by a high level of organic carbon at the site. Laboratory studies (Edwards & Grbić-Galić, 1994; Wilson *et al.*, 1986) have indicated long lag periods (140–200 days) prior to the onset of *o*-xylene biodegradation.

In the single field study (Cozzarelli *et al.*, 1990) where Fe(III) likely functioned as the predominant terminal electron acceptor, *o*-toluic acid was speculated to be an oxidized product of *o*-xylene decomposition at the Bemidji, Minnesota, aquifer site. This study provides presumptive evidence that *o*-xylene can be degraded under iron-reducing conditions.

3.7 Concluding remarks

In light of the information reviewed above, it is now quite apparent that the anaerobic biotransformation of petroleum hydrocarbons cannot be easily dismissed as an environmentally insignificant process. Since oxygen reserves are so readily depleted from many environments, anaerobic microorganisms and their activities may, in fact, turn out to be major determinants governing the transport and fate of many contaminants including the hydrocarbon examples considered herein. The biodegradation of BTEX hydrocarbons has been clearly demonstrated to occur under a variety of anaerobic conditions and catalyzed by diverse cell types. To date, relatively little is known about the mechanisms that anaerobes employ to metabolize these substrates. However, the state of the art is far from complete and must be viewed as a work in progress. In our opinion, it is quite probable that novel pathways for the metabolism of many other hydrocarbons will be discovered in the very near future. Such bioconversions will undoubtedly be of both fundamental and practical significance. For instance, knowledge of the pathways and kinetics of these anaerobes will allow for more accurate assessments of the risks associated with

BTEX contamination and will facilitate the directed search for unique metabolites that attest to the occurrence of these microbial processes in the field. The information may also help us to predict the rates of intrinsic bioremediation processes as well as to promote alternate strategies for the restoration of contaminated environments. Ultimately, the fundamental knowledge base will serve as a solid foundation upon which solutions to biotechnological problems may be built.

Dedication

This work is dedicated to the memory of Dunja Grbić-Galić. Her initial work in this area was critical in demonstrating that the anaerobic transformation of BTEX compounds can occur. Subsequent studies done in her lab, as illustrated in this review, has provided a basis for our general interpretation of the fate of BTEX compounds in the environment.

References

Acton, D. W. & Barker, J. F. (1992). In situ biodegradation potential of aromatic hydrocarbons in anaerobic groundwaters. *Journal of Contaminant Hydrology*, 9, 325–52.

Aelion, C. M. & Bradley, P. M. (1991). Aerobic biodegradation potential of subsurface microorganisms from a jet fuel-contaminated aquifer. *Applied and Environmental Microbiology*, 57(1), 57–63.

Albrechtsen, H-J. (1994). Bacterial degradation under iron-reducing conditions. In *Hydrocarbon Bioremediation*, ed. R. E. Hinchee, B. C. Alleman, R. E. Hoeppel & R. N. Miller, pp. 418–23. Boca Raton, FL: Lewis Publishers.

Altenschmidt, U. & Fuchs, G. (1991). Anaerobic degradation of toluene in denitrifying *Pseudomonas* sp.: indication for toluene methylhydroxylation and benzyl-CoA as central aromatic intermediate. *Archives of Microbiology*, 156, 152–8.

Altenschmidt, U. & Fuchs, G. (1992). Anaerobic toluene oxidation to benzyl alcohol and benzaldehyde in a denitrifying *Pseudomonas* strain. *Journal of Bacteriology*, 174(14), 4860–2.

Alvarez, P. J. J. & Vogel, T. M. (1994). Degradation of BTEX and their aerobic metabolites by indigenous microorganisms under nitrate reducing conditions. *Abstracts of the 94th Annual Meeting, American Society for Microbiology*, Q-339.

Alvarez, P. J. J., Anid, P. J. & Vogel, T. M. (1994). Kinetics of toluene degradation by denitrifying aquifer microorganisms. *Journal of Environmental Engineering*, 120(5), 1327–36.

Anid, P. J., Alvarez, P. J. J. & Vogel, T. M. (1993). Biodegradation of monoaromatie hydrocarbons in aquifer columns amended with hydrogen peroxide and nitrate. *Water Research*, 27(4), 685–91.

Arcangeli, J. P. & Arvin, E. (1994). Biodegradation of BTEX compounds in a biofilm system under nitrate-reducing conditions. In *Hydrocarbon Bioremediation*, ed. R. E. Hinchee, B. C. Alleman, R. E. Hoeppel & R. N. Miller, pp. 374–82. Boca Raton, FL: Lewis Publishers.

Baedecker, M. J., Cozzarelli, I. M., Eganhouse, R. P., Siegel, D. I. & Bennett, P. C. (1993). Crude oil in a shallow sand and gravel aquifer. III. Biogeochemical reactions and mass balance modeling in anoxic groundwater. *Applied Geochemistry*, 8, 569–86.

Ball, H. A. & Reinhard, M. (1996). Monoaromatic hydrocarbon degradation under anaerobic conditions at Seal Beach, California; Laboratory studies. *Environmental Toxicology and Chemistry* (in press).

Ball, H. A., Hopkins, G. D., Orwin, E. & Reinhard, M. (1994). Anaerobic bioremediation of aromatic hydrocarbons at Seal Beach, CA: Laboratory and field investigations. *Environmental Protection Agency Symposium on Intrinsic Bioremediation of Groundwater*, August 30–September 1, 1994, Denver, Colorado.

Barbaro, J. R., Barker, J. F., Lemon, L. A. & Mayfield, C. I. (1992). Biotransformation of BTEX under anaerobic, denitrifying conditions: field and laboratory observations. *Journal of Contaminant Hydrology*, 11, 245–72.

Barlaz, M. A., Shafer, M. B., Borden, R. C. & Wilson, J. T. (1993). Rate and extent of natural anaerobic bioremediation of BTEX compounds in ground water plumes. *Symposium on Bioremediation of Hazardous Wastes: Research, Development, and Field Evaluations*. May 4–6 Dallas, Texas.

Battersby, N. S. & Wilson, V. (1989). Survey of the anaerobic biodegradation potential of organic chemicals in digesting sludge. *Applied and Environmental Microbiology*, 55(2), 433–9.

Beller, H. R., Ding, W-R. & Reinhard, M. (1995). Byproducts of anaerobic alkylbenzene metabolism useful as indicators of in situ bioremedication. *Environmental Science and Technology*, 29(11), 2864–70.

Beller, H. R., Edwards, E. A., Grbić-Galić, D., Hutchins, S. R. & Reinhard, M. (1991). *Microbial Degradation of Alkylbenzenes under Sulfate-reducing and Methanogenic Conditions*. United States Environmental Protection Agency Project Summary, EPA/600/S2-91/027.

Beller, H. R., Grbić-Galić, D. & Reinhard, M. (1992a). Microbial degradation of toluene under sulfate-reducing conditions and the influence of iron on the process. *Applied and Environmental Microbiology*, 58(3), 786–93.

Beller, H. R., Reinhard, M. & Grbić-Galić, D. (1992b). Metabolic by-products of anaerobic toluene degradation by sulfate-reducing enrichment cultures. *Applied and Environmental Microbiology*, 58(9), 3192–5.

Berwanger, D. J. & Barker, J. F. (1988). Aerobic biodegradation of aromatic and chlorinated hydrocarbons commonly detected in landfill leachates. *Water Pollution Research Journal of Canada*, 23(3), 460–75.

Borden, R. C., Gomez, C. A. & Becker, M. T. (1994). Natural bioremediation of a gasoline spill. In *Hydrocarbon Bioremediation*, ed. R. E. Hinchee, B. C. Alleman, R. E. Hoeppel & R. N. Miller, pp. 290–5, Boca Raton, FL: Lewis Publishers.

Bouwer, E. J. & McCarty, P. L. (1983). Transformations of halogenated organic compounds under denitrification conditions. *Applied and Environmental Microbiology*, 45(4), 1295–9.

Caldwell, M. E. & Suflita, J. M. (1995). Anaerobic and microaerophilic degradation of benzene in aquifer sediments. *Abstracts of the Third International Symposium on In Situ and On Site Bioreclamation*, San Diego, California, April 24–27, 1995, E9.6.

Chee-Sanford, J. C. & Tiedje, J. M. (1994). Detection and identification of metabolites produced during anaerobic toluene degradation by *Azoarcus* sp. (TOL-4). *Abstracts of the 94th Annual Meeting, American Society for Microbiology*, Q-335.

Christensen, T. H., Kjeldsen, P., Albrechtsen, H-J., Heron, G., Nielsen, P. H., Bjerg, P. L. & Holm, P. E. (1994). Attenuation of landfill leachate pollutants in aquifers. *Critical Reviews in Environmental Science and Technology*, 24(2), 119–202.

Coschigano, P. W., Häggblom, M. M. & Young, L. Y. (1994). Metabolism of both

4-chlorobenzoate and toluene under denitrifying conditions by a constructed bacterial strain. *Applied and Environmental Microbiology*, 60(3), 989–95.

Cozzarelli, I. M., Eganhouse, R. P. & Baedecker, M. J. (1990). Transformation of monoaromatic hydrocarbons to organic acids in anoxic groundwater environment. *Environmental Geology and Water Science*, 16(2), 135–41.

Cozzarelli, I. M., Baedecker, M. J. & Hopple, J. A. (1991). Geochemical gradients in shallow ground water caused by the microbial degradation of hydrocarbons at Galloway Township, New Jersey. *U. S. Geological Survey Toxic Substances Hydrology Program – Proceedings of the Technical Meeting*, Monterey, California, U. S. Geological Survey Water-Resources Investigations Report 91-4034, pp. 256–62.

Cozzarelli, I. M., Baedecker, M. J., Eganhouse, R. P. & Goerlitz, D. F. (1994). The geochemical evolution of low-molecular-weight organic acids derived from the degradation of petroleum contaminants in groundwater. *Geochimica et Cosmochimica Acta*, 58(2), 863–77.

Davis, J. W., Klier, N. J. & Carpenter, C. L. (1994). Natural biological attenuation of benzene in ground water beneath a manufacturing facility. *Ground Water*, 32(2), 215–26.

Dolfing, J., Zeyer, J., Binder-Eicher, P. & Schwarzenbach, R. P. (1990). Isolation and characterization of a bacterium that mineralizes toluene in the absence of molecular oxygen. *Archives of Microbiology*, 154, 336–41.

Edwards, E. A. & Grbić-Galić, D. (1992). Complete mineralization of benzene by aquifer microorganisms under strictly anaerobic conditions. *Applied and Environmental Microbiology*, 58(8), 2663–6.

Edwards, E. A. & Grbić-Galić, D. (1994). Anaerobic degradation of toluene and *o*-xylene by a methanogenic consortium. *Applied and Environmental Microbiology*, 60(1), 313–22.

Edwards, E. A., Wills, L. E., Reinhard, M. & Grbić-Galić, D. (1992). Anaerobic degradation of toluene and xylene by aquifer microorganisms under sulfate-reducing conditions. *Applied and Environmental Microbiology*, 58(3), 794–800.

Edwards, E. A., Edwards, A. M. & Grbić-Galić, D. (1994). A method for detection of aromatic metabolites at very low concentrations: application of detection of metabolites of anaerobic toluene degradation. *Applied and Environmental Microbiology*, 60(1), 323–7.

Evans, P. J., Mang, D. T., Kim, K. S. & Young, L. Y. (1991a). Anaerobic degradation of toluene by a denitrifying bacterium. *Applied and Environmental Microbiology*, 57(4), 1139–45.

Evans, P. J., Mang, D. T. & Young, L. Y. (1991b). Degradation of toluene and *m*-xylene and transformation of *o*-xylene by denitrifying enrichment cultures. *Applied and Environmental Microbiology*, 57(2), 450–4.

Evans, P. J., Ling, W., Goldschmidt, B., Ritter, E. R. & Young, L. Y. (1992a). Metabolites formed during anaerobic transformation of toluene and *o*-xylene and their proposed relationship to the initial steps of toluene mineralization. *Applied and Environmental Microbiology*, 58(2), 496–501.

Evans, P. J., Ling, W., Palleroni, N. J. & Young, L. Y. (1992b). Quantification of denitrification by strain T1 during anaerobic degradation of toluene. *Applied Microbiology and Biotechnology*, 37, 136–40.

Evans, W. C. & Fuchs, G. (1988). Anaerobic degradation of aromatic compounds. *Annual Reviews of Microbiology*, 42, 289–317.

Flyvbjerg, J., Arvin, E., Jensen, B. K. & Olsen, S. K. (1993). Microbial degradation of phenols and aromatic hydrocarbons in creosote-contaminated groundwater under nitrate-reducing conditions. *Journal of Contaminant Hydrology*, 12, 133–50.

Frazer, A. C., Ling, W. & Young, L. Y. (1993). Substrate induction and metabolite accumulation during anaerobic toluene utilization by the denitrifying strain T1. *Applied and Environmental Microbiology*, 59(9), 3157–60.

Fries, M. R., Zhou, J., Chee-Sanford, J. & Tiedje, J. M. (1994). Isolation, characterization, and distribution of denitrifying toluene degraders from a variety of habitats. *Applied and Environmental Microbiology*, 60(8), 2802–10.

Gersberg, R. M., Dawsey, W. J. & Bradley, M. D. (1991). Biodegradation of monoaromatic hydrocarbons in groundwater under denitrifying conditions. *Bulletin of Environmental Contamination and Toxicology*, 47, 230–7.

Ghosh, S. & Sun, M. L. (1992). Anaerobic biodegradation of benzene under acidogenic fermentation condition. In *Proceedings of the Conference on Hazardous Waste Research*, ed. L. E. Erickson, S. C. Grant & J. P. McDonald, pp. 208–18. Manhattan, KS: Engineering Extension.

Gillham, R. W., Starr, R. C. & Miller, D. J. (1990). A device for *in situ* determination of geochemical transport parameters. 2. Biochemical reactions. *Ground Water*, 28(6), 858–62.

Grbić-Galić, D. (1986). Anaerobic production and transformation of aromatic hydrocarbons and substituted phenols by ferulic acid-degrading BESA-inhibited methanogenic consortia. *FEMS Microbiology Ecology*, 38, 161–9.

Grbić-Galić, D. (1990). Methanogenic transformation of aromatic hydrocarbons and phenols in groundwater aquifers. *Geomicrobiology Journal*, 8, 167–200.

Grbić-Galić, D. & Vogel, T. M. (1987). Transformation of toluene and benzene by mixed methanogenic cultures. *Applied and Environmental Microbiology*, 53(2), 254–60.

Haag, F., Reinhard, M. & McCarty, P. L. (1991). Degradation of toluene and *p*-xylene in anaerobic microcosms: evidence for sulfate as a terminal electron acceptor. *Environmental Toxicology and Chemistry*, 10, 1379–89.

Hughes, K., Meek, M. E. & Bartlett, S. (1994). Benzene: Evaluation of risks to health from environmental exposure in Canada. *Journal of Environmental Science and Health, Part C, Environmental Carcinogenesis and Ecotoxicology Reviews*, C12(2), 161–71.

Hutchins, S. R. (1991a). Biodegradation of monoaromatic hydrocarbons by aquifer microorganisms using oxygen, nitrate, or nitrous oxide as the terminal electron acceptor. *Applied and Environmental Microbiology*, 57(8), 2403–7.

Hutchins, S. R. (1991b). Optimizing BTEX biodegradation under denitrifying conditions. *Environmental Toxicology and Chemistry*, 10, 1437–48.

Hutchins, S. R. (1992). Inhibition of alkylbenzene biodegradation under denitrifying conditions by using the acetylene block technique. *Applied and Environmental Microbiology*, 58(10), 3395–8.

Hutchins, S. R. (1993). Biotransformation and mineralization of alkylbenzenes under denitrifying conditions. *Environmental Toxicology and Chemistry*, 12, 1413–23.

Hutchins, S. R., Downs, W. C., Wilson, J. T., Smith, G. B., Kovacs, D. A., Fine, D. D., Douglass, R. H. & Hendrix, D. J. (1991a). Effect of nitrate addition on biorestoration of fuel-contaminated aquifer: field demonstration. *Ground Water*, 29(4), 571–80.

Hutchins, S. R., Sewell, G. W., Kovacs, D. A. & Smith, G. A. (1991b). Biodegradation of aromatic hydrocarbons by aquifer microorganisms under denitrifying conditions. *Environmental Science and Technology*, 25(1), 68–76.

Jensen, B. K. & Arvin, E. (1994). Aromatic hydrocarbon degradation specificity of an enriched denitrifying mixed culture. In *Hydrocarbon Bioremediation*, ed. R. E. Hinchee, B. C. Alleman, R. E. Hoeppel & R. N. Miller, pp. 411–17. Boca Raton, FL: Lewis Publishers.

Kuhn, E. P., Colberg, P. J., Schnoor, J. L., Wanner, O., Zehnder, A. J. B. & Schwarzenbach, R. P. (1985). Microbial transformations of substituted benzenes during infiltration of river water to groundwater: laboratory column studies. *Environmental Science and Technology*, 19(10), 961–8.

Kuhn, E. P., Zeyer, J., Eicher, P. & Schwarzenbach, R. P. (1988). Anaerobic degradation of alkylated benzenes in denitrifying laboratory aquifer columns. *Applied and Environmental Microbiology*, 54(2), 490–6.

Liang, L-N. & Grbić-Galić, D. (1993). Biotransformation of chlorinated aliphatic solvents in the presence of aromatic compounds under methanogenic conditions. *Environmental Toxicology and Chemistry*, 12, 1377–93.

Lovley, D. R. & Lonergan, D. J. (1990). Anaerobic oxidation of toluene, phenol, and *p*-cresol by the dissimilatory iron-reducing organism, GS-15. *Applied and Environmental Microbiology*, 56(6), 1858–64.

Lovley, D. R., Baedecker, M. J., Lonergan, D. J., Cozzarelli, I. M., Phillips, E. J. P. & Siegel, D. I. (1989). Oxidation of aromatic contaminants coupled to microbial iron reduction. *Nature*, 339, 297–300.

Lovley, D. R., Woodward, J. C. & Chapelle, F. H. (1994). Stimulated anoxic biodegradation of aromatic hydrocarbons using Fe(III) ligands. *Nature*, 370, 128–31.

Lovley, D. R., Coates, J. D., Woodward, J. C. & Phillips, E. J. P. (1995). Benzene oxidation coupled to sulfate reduction. *Applied and Environmental Microbiology*, 61(3), 953–8.

Major, D. W., Mayfield, C. I. & Barker, J. F. (1988). Biotransformation of benzene by denitrification in aquifer sand. *Ground Water*, 26, 8–14.

McAllister, P. M. & Chiang, C. Y. (1994). A practical approach to evaluating natural attenuation of contaminants in ground water. *Ground Water Monitoring and Remediation*, 14, 161–73.

Meek, M. E. & Chan, P. K. L. (1994a). Toluene: Evaluation of risks to human health from environmental exposure in Canada. *Journal of Environmental Science and Health, Part C, Environmental Carcinogenesis and Ecotoxicology Reviews*, C12(2), 507–15.

Meek, M. E. & Chan, P. K. L. (1994b). Xylenes: Evaluation of risks to health from environmental exposure in Canada. *Journal of Environmental Science and Health, Part C, Environmental Carcinogenesis and Ecotoxicology Reviews*, C12(2), 545–56.

Morgan, P., Lewis, S. T. & Watkinson, R. J. (1993). Biodegradation of benzene, toluene, ethylbenzene, and xylenes in gas-condensate-contaminated ground-water. *Environmental Pollution*, 82(2), 181–90.

Narayanan, B., Suidan, M. T., Gelderloos, A. B. & Brenner, R. C. (1993). Treatment of VOCs in high strength wastes using an anaerobic expanded-bed GAC reactor. *Water Research*, 27(1), 181–94.

Nielsen, P. H., Holm, P. E. & Christensen, T. H. (1992). A field method for determination of groundwater and groundwater-sediment associated potentials for degradation of xenobiotic organic compounds. *Chemosphere*, 25(4), 449–62.

Patterson, B. M., Pribac, F., Barber, C., Davis, G. B. & Gibbs, R. (1993). Biodegradation and retardation of PCE and BTEX compounds in aquifer material from Western Australia using large-scale columns. *Journal of Contaminant Hydrology*, 14, 261–78.

Piet, G. J. & Smeenk, J. G. M. M. (1985). Behavior of organic pollutants in pretreated Rhine water during dune infiltration. In *Ground Water Quality*, ed. C. H. Ward, W. Giger & P. L. McCarty, pp. 122–44. New York: John Wiley & Sons.

Rabus, R., Nordhaus, R., Ludwig, W. & Widdel, F. (1993). Complete oxidation of toluene under strictly anoxic conditions by a new sulfate-reducing bacterium. *Applied and Environmental Microbiology*, 59(5), 1444–51.

Reinhard, M., Goodman, N. L. & Barker, J. F. (1984). Occurrence and distribution of organic chemicals in two landfill leachate plumes. *Environmental Science and Technology*, 18(12), 953–61.

Rueter, P., Rabus, R., Wilkes, H., Aeckersberg, F., Rainey, F. A., Jannasch, H. W. & Widdel, F. (1994) Anaerobic oxidation of hydrocarbons in crude oil by new types of sulphate-reducing bacteria. *Nature*, 372, 455–8.

Salanitro, J. P. (1993). The role of bioattenuation in the management of aromatic hydrocarbon plumes in aquifers. *Ground Water Monitoring Review*, 13, 150–61.

Schink, B. (1985). Degradation of unsaturated hydrocarbons by methanogenic enrichment cultures. *FEMS Microbiology Ecology*, 31, 69–77.

Schocher, R. J., Seyfried, B., Vazquez, F. & Zeyer, J. (1991). Anaerobic degradation of toluene by pure cultures of denitrifying bacteria. *Archives of Microbiology*, 157, 7–12.

Schwarzenbach, R. P., Giger, W., Hoehn, E. & Schneider, J. K. (1983). Behavior of organic compounds during infiltration of river water to groundwater: field studies. *Environmental Science and Technology*, 17(8), 472–9.

Seyfried, B., Tschech, A. & Fuchs, G. (1991). Anaerobic degradation of phenylacetate and 4-hydroxyphenylacetate by denitrifying bacteria. *Archives of Microbiology*, 155, 249–55.

Seyfried, B., Glod, G., Schocher, R., Tschech, A. & Zeyer, J. (1994). Initial reactions in the anaerobic oxidation of toluene and *m*-xylene by denitrifying bacteria. *Applied and Environmental Microbiology*, 60(11), 4047–52.

Thierrin, J., Davis, G. B., Barber, C., Patterson, B. M., Pribac, F., Power, T. R. & Lambert, M. (1993). Natural degradation rates of BTEX compounds and naphthalene in a sulphate reducing groundwater environment. *Hydrological Sciences Journal*, 38(4), 309–22.

Tschech, A. & Fuchs, G. (1989). Anaerobic degradation of phenol via carboxylation to 4-hydroxybenzoate: in vitro study of isotope exchange between $^{14}CO_2$ and 4-hydroxybenzoate. *Archives of Microbiology*, 152, 594–9.

Van Beelen, P. & Van Keulen, F. (1990). The kinetics of the degradation of chloroform and benzene in anaerobic sediment from the River Rhine. *Hydrobiological Bulletin*, 24(1), 13–21.

Vogel, T. M. & Grbić-Galić, D. (1986). Incorporation of oxygen from water into toluene and benzene during anaerobic fermentative transformation. *Applied and Environmental Microbiology*, 52(1), 200–2.

Ward, D. M., Atlas, R. M., Boehm, P. D. & Calder, J. A. (1980). Microbial biodegradation and chemical evolution of oil from the Amoco spill. *Ambio*, 9, 277–83.

Watwood, M. E., White, C. S. & Dahm, C. N. (1991). Methodological modifications for accurate and efficient determination of contaminant biodegradation in unsaturated calcareous soils. *Applied and Environmental Microbiology*, 57(3), 717–20.

Wilson, B. H., Smith, G. B. & Rees, J. F. (1986). Biotransformations of selected alkylbenzenes and halogenated aliphatic hydrocarbons in methanogenic aquifer material: a microcosm study. *Environmental Science and Technology*, 20(10), 997–1002.

Wilson, B. H., Wilson, J. T., Kampbell, D. H., Bledsoe, B. E. & Armstrong, J. M. (1990).

Biotransformation of monoaromatic and chlorinated hydrocarbons at an aviation gasoline spill site. *Geomicrobiology Journal*, 8, 235–40.

Zeyer, J., Kuhn, E. P. & Schwarzenbach, R. P. (1986). Rapid microbial mineralization of toluene and 1,3-dimethylbenzene in the absence of molecular oxygen. *Applied and Environmental Microbiology*, 52(4), 944–7.

Note:

After completion of the manuscript, publications appeared on the following topics:

Isolation of denitrifying bacteria capable of toluene, ethylbenzene, and *m*- or *p*-xylene degradation

Rabus, R. & Widdel, F. (1995). Anaerobic degradation of ethylbenzene and other aromatic hydrocarbons by new denitrifying bacteria. *Archives of Microbiology*, 163, 96–103.

Su, J-J. & Kafkewitz, D. (1994). Utilization of toluene and xylenes by a nitrate-reducing strain of *Pseudomonas maltophilia* under low oxygen and anoxic conditions. *FEMS Microbiology Ecology*, 15, 249–58.

Enrichment of denitrifying bacteria capable of toluene, and *m*- or *p*-xylene degradation

Häner, A., Höhener, P. & Zeyer, J. (1995). Degradation of *p*-xylene by a denitrifying enrichment culture. *Applied and Environmental Microbiology*, 61(8), 3185–8.

Classification of denitrifying isolates capable of toluene degradation

Anders, H-J., Kaetzke, A., Kämpfer, P., Ludwig, W. & Fuchs, G. (1995). Taxonomic position of aromatic-degrading denitrifying Pseudomonad strains K172 and KB740 and their description as new members of the genera *Thauera*, as *Thauera aromatica* sp. nov., and *Azoarcus*, as *Azoarcus evansii* sp. nov., respectively, members of the beta subclass of the *Proteobacteria*. *International Journal of Systematic Bacteriology*, 45(2), 327–33.

Zhou, J., Fries, M. R., Chee-Sanford, J. C. & Tiedje, J. M. (1995). Phylogenetic analyses of a new group of denitrifiers capable of anaerobic growth on toluene and description of *Azoarcus tolulyticus* sp. nov. *International Journal of Systematic Bacteriology*, 45(3), 500–6.

Enzyme studies on the pathway for toluene degradation under denitrifying conditions

Biegert, T. & Fuchs, G. (1995). Anaerobic oxidation of toluene (analogues) to benzoate (analogues) by whole cells and by cell extracts of a denitrifying *Thauera* sp. *Archives of Microbiology*, 163, 407–17.

Biegert, T., Altenschmidt, U., Eckerskorn, C. & Fuchs, G. (1995). Purification and properties of benzyl alcohol dehydrogenase from a denitrifying *Thauera* sp. *Archives of Microbiology*, 163, 418–23.

Field studies on the degradation of BTEX in a gasoline-contaminated aquifer

Borden, R. C., Gomez, C. A. & Becker, M. T. (1995). Geochemical indicators of intrinsic bioremediation. *Ground Water*, 33(2), 180–9.

Anaerobic degradation of benzene under methanogenic conditions

Chaudhuri, B. K. & Wiesmann, U. (1995). Enhanced anaerobic degradation of benzene by enrichment of mixed microbial culture and optimization of the culture medium. *Applied Microbiology and Biotechnology*, 43, 178–87.

4

Bioremediation of petroleum contamination

Eugene Rosenberg and Eliora Z. Ron

4.1 Introduction

During this century the demand for petroleum as a source of energy and as primary raw material for the chemical industry has resulted in an increase in world production to about 3500 million metric tons per year (Energy Information Administration, 1992). It has been estimated that approximately 0.1%, 35 million tons, enters the sea per annum (National Research Council, 1985). A great part of the oil pollution problem results from the fact that the major oil-producing countries are not the major oil consumers. It follows that massive movements of petroleum have to be made from areas of high production to those of high consumption. Although the large crude oil spills following tanker accidents (e.g., Table 4.1) receive the most public attention, tanker accidents represent only a small fraction, about one million tons, of the total input. By comparison, oil input to the sea from natural sources, principally seeps, is about 0.5 million tons annually. The major sources of oil pollution are municipal and industrial wastes and runoffs, leaks in pipelines and storage tanks, and discharge of dirty ballast and bilge waters. Since the residence time of hydrocarbons in the sea is decades or centuries (Button *et al.*, 1992), dissolved hydrocarbons circulate in the deep oceans, thereby constituting a global problem.

The toxicity of petroleum hydrocarbons to microorganisms, plants, animals and humans is well established. In fact, many bioassays for petroleum pollution have been developed which depend on low concentrations (5–100 mg/l) of crude oil or petroleum fractions killing or inhibiting the growth of microalgae and juvenile forms of marine animals. The toxic effects of hydrocarbons on terrestrial higher plants and their use as weedkillers have been ascribed to the oil dissolving the lipid portion of the cytoplasmic membrane, thus allowing cell contents to escape (Currier & Peoples, 1954). The polycyclic aromatic hydrocarbon (PAH) fraction of petroleum is particularly toxic and is on the United States Environmental

Table 4.1. *Recent large oil spills*

Source	Place	Date	10^3 tons of oil
Iraq/Kuwait	Persian Gulf	February 1991	1000
Exxon Valdez	Gulf of Alaska	April 1989	33
IXTOC I well	Campeche Bay, Mexico	June 1979	350
Amoco Cadiz	Brittany, France	March 1978	223
Torrey Canyon	Cornwall, England	March 1967	117

Protection Agency's priority pollutant list (Cerniglia, 1992). The environmental persistence and genotoxicity of PAHs increases as the molecular size increases up to four or five fused benzene rings, and toxicological concern shifts to chronic toxicity, primarily carcinogenesis (Miller & Miller, 1981). In short, oil pollution on land and in the sea presents a serious problem to public health, commercial fisheries, land development, and recreational and water resources.

It has been known for 80 years that certain microorganisms are able to degrade petroleum hydrocarbons and use them as a sole source of carbon and energy for growth. The early work was summarized by Davis in 1967. A more recent volume (Atlas, 1984) covers the microbial metabolism of alkanes, cyclic alkanes, aromatic and gaseous hydrocarbons, the genetics of hydrocarbon-utilizing microorganisms, the fate of petroleum in soil and water, and various applied aspects of petroleum microbiology. Recent reviews on various aspects of hydrocarbon bioremediation include Morgan & Watkinson (1989), Leahy & Colwell (1990), Atlas (1991) and Rosenberg (1993). In this chapter we will review some general principles of hydrocarbon microbiology and then present three case studies of petroleum bioremediation.

4.2 Principles of hydrocarbon microbiology

The use of hydrocarbons as substrates for bacterial growth presents special problems to both the microorganisms using them as a source of carbon and energy (Table 4.2) and to investigators in the field of hydrocarbon microbiology. There are two essential characteristics that define hydrocarbon-oxidizing microorganisms:

1. Membrane-bound, group-specific oxygenases, and
2. Mechanisms for optimizing contact between the microorganisms and the water-insoluble hydrocarbon.

4.2.1 Distribution of hydrocarbon-degrading microorganisms

The localization of hydrocarbon-oxidizing bacteria in natural environments has received considerable attention because of the possibility of utilizing their

Table 4.2. *Requirements for biodegradation of petroleum*

A. Microorganisms with:
1. Hydrocarbon-oxidizing enzymes
2. Ability to adhere to hydrocarbons
3. Emulsifier-producing potential
4. Mechanisms for desorption from hydrocarbons
B. Water
C. Oxygen
D. Phosphorus
E. Utilizable nitrogen source

biodegradation potential in the treatment of oil spills. Because of the enormous quantities of crude and refined oils that are transported over long distances and consumed in large amounts, the hydrocarbons have now become a very important class of potential substrates for microbial oxidation. It is not surprising, therefore, that hydrocarbon-oxidizing microorganisms have been isolated in large numbers from a wide variety of natural aquatic and terrestrial environments (Rosenberg, 1991), including hydrothermal vent sites (Bazylinski, Wirsen & Jannasch, 1989). Several investigators have demonstrated an increase in the number of hydrocarbon-oxidizing bacteria in areas that suffer from oil pollution. Walker & Colwell (1976a, b) observed a positive correlation between the percentage of petroleum-degrading bacteria in the total population of heterotrophic microorganisms and the amount of heptane-extractable material in sediments of Colgate Creek, a polluted area of Chesapeake Bay. In contrast, no correlation was found when total numbers of hydrocarbon oxidizers (rather than percentages) were compared with hydrocarbon levels. Horowitz & Atlas (1977a) observed shifts in microbial populations in an Arctic freshwater lake after the accidental spillage of 55 000 gallons of leaded gasoline. The ratio of hydrocarbon-utilizing to total heterotrophic bacteria was reported to be an indicator of the gasoline contamination. These investigators also studied shifts in microbial populations in Arctic coastal water using a continuous flow-through system, following the introduction of an artificial oil slick (Horowitz & Atlas, 1977b). The addition of the oil appeared to cause a shift to a greater percentage of petroleum-degrading bacteria. Atlas & Bartha (1973b) found similar results in an oil-polluted area in Raritan Bay off the coast of New Jersey. Hood *et al.* (1975) compared microbial populations in sediments of a pristine salt marsh with those of an oil-rich marsh in southeastern Louisiana. These investigators also found a high correlation between the percentage of hydrocarbon oxidizers and the level of hydrocarbons in the sediments. Significant increases in the number of hydrocarbon-utilizing microorganisms were found in field soils following the addition of several different oil samples (Raymond *et al.*, 1976). No estimate of the ratio of hydrocarbon oxidizers to the total heterotrophic population was presented.

Table 4.3. *Changes in the number of hydrocarbon degraders with time in jet-fuel-contaminated loam soil*[a]

Treatment	Weeks after treatment	Microorganisms/g soil
None	0	4×10^4
	4	2×10^4
	16	2×10^4
50 mg jet fuel/g soil	4	4×10^8
	16	1×10^6
50 mg jet fuel/g soil plus	4	3×10^{10}
bioremediation	16	1×10^7

[a]Data from Song & Bartha (1990). Reprinted with permission of the American Society for Microbiology.

From the studies discussed above, it is clear that the presence of hydrocarbons in the environment frequently brings about a selective enrichment *in situ* for hydrocarbon-utilizing microorganisms. The data presented in Table 4.3 demonstrate a typical response of microbial populations to oil pollution. The relatively low levels of hydrocarbon degraders in the soil increased by three orders of magnitude in the 4 weeks following addition of the oil, and then slowly declined. Supplementation of nitrogen and phosphorus fertilizers (bioremediation) caused another tenfold increase in the hydrocarbon degraders.

The practical conclusions from these data are: (1) seeding an oil-polluted area with microorganisms is generally not necessary; and (2) genetically engineered microorganisms (GEMs) are not required for petroleum bioremediation. However, under some circumstances, seeding may be advantageous to ensure uniformity of the hydrocarbon breakdown pattern and, especially, to encourage degradation of the toxic polycyclic-hydrocarbon fraction.

4.2.2 Hydrocarbon degradation: metabolic specificity

The ability to degrade hydrocarbon substrates is present in a wide variety of bacteria and fungi. The specificity of the process can ultimately be related to the genetic potential of the particular microorganism to introduce molecular oxygen into the hydrocarbon and, with relatively few reactions, to generate the intermediates that subsequently enter the general energy-yielding catabolic pathways of the cell. The specific genetic capacity is expressed in the hydrocarbon substrate specificity of the oxygenase and in the ability of the carbon source to induce the various enzyme activities necessary for its biodegradation.

Alkanes

Oxidation of n-paraffins proceeds, in general, via terminal oxidation to the corresponding alcohol, aldehyde and fatty acid (McKenna & Kallio, 1965).

Hydroperoxides may serve as unstable intermediates in the formation of the alcohol (Watkinson & Morgan, 1990). Fatty acids derived from alkanes are then further oxidized to acetate and propionate (odd-chain alkanes) by inducible β-oxidation systems. The group specificity of the alkane oxygenase system is different in various bacterial species. For example, *Pseudomonas putida* PbG6 (Oct) grows on alkanes of six to ten carbons in chain length (Nieder & Shapiro, 1975), whereas *Acinetobacter* sp. HO1-N is capable of growth on long-chain alkanes (Singer & Finnerty, 1984a). The ability of *P. putida* to grow on C_6–C_{10} alkanes was shown to be plasmid encoded (Chakrabarty, Chou & Gunsalas, 1973). In contrast, all activities necessary for growth of *Acinetobacter* sp. HO1-N and *A. calcoaceticus* BD413 on longer chain hydrocarbons appear to be coded by chromosomal genes (Singer & Finnerty, 1984b).

Subterminal alkane oxidation apparently occurs in some bacterial species (Markovetz, 1971). This type of oxidation is probably responsible for the formation of long-chain secondary alcohols and ketones. Pirnik (1977) and Perry (1984) have reviewed the microbial oxidation of branched and cyclic alkanes, respectively. Interestingly, none of the cyclohexane or cyclopentane compounds seems to be metabolized by pure cultures. Rather, non-specific oxidases present in many bacteria convert the cyclic alkanes into cyclic ketones, which are then oxidized by specific bacteria.

Aromatics

Both procaryotic and eukaryotic microorganisms have the enzymatic potential to oxidize aromatic hydrocarbons that range in size from a single ring (e.g., benzene, toluene and xylene) to polycyclic aromatics (PCAs), such as naphthalane, anthracene, phenanthrene, benzo [*a*] pyrene and benz [*a*] anthracene (Table 4.4). However, the molecular mechanisms by which bacteria and higher microorganisms degrade aromatic compounds are fundamentally different.

Bacteria initially oxidize aromatic compounds by incorporating both atoms of molecular oxygen into the aromatic substrate to form a *cis*-dihydrodiol (Gibson, 1977). The reaction is catalyzed by a multicomponent dioxygenase, consisting of a flavoprotein, an iron–sulfur protein and a ferredoxin (Yeh, Gibson & Liu, 1977). Further oxidation of the *cis*-dihydrodiol leads to the formation of catechols that are substrates for a different dioxygenase, leading to enzymatic fission of the aromatic ring (Dagley, 1971). Fungi and other eukaryotes oxidize aromatic hydrocarbons with a cytochrome P-450 mono-oxygenase system, leading to *trans*-dihydrodiols. The intermediate formed is an arene oxide that is toxic, mutagenic and carcinogenic, because it can bind to nucleophilic sites on DNA. Once the catechol is formed, it can be oxidized via the *ortho* pathway, which involves cleavage of the bond between the two carbon atoms bearing the hydroxyl groups to form *cis, cis*-muconic acid; or via the *meta* pathway, which involves cleavage between one of the carbons with a hydroxyl group and the adjacent carbon.

Substituted aromatics, such as toluene, can either be initially oxidized by bacteria at the methyl group to form benzoic acid, or on the aromatic ring to form a

Table 4.4. *Microorganisms that metabolize aromatic hydrocarbons*

Organisms	Organisms
Bacteria	Fungi
Pseudomonas	*Chytridomycetes*
Aeromonas	*Oomycetes*
Moraxella	*Zygomycota*
Beijerinckia	*Ascomycota*
Flavobacteria	*Basidiomycota*
Achrobacteria	*Deuteromycota*
Nocardia	Microalgae
Corynebacteria	*Porphyridium*
Acinetobacter	*Petalonia*
Alcaligenes	*Diatoms*
Mycobacteria	*Chlorella*
Rhodococci	*Dunaliella*
Streptomyces	*Chlamydomonas*
Bacilli	*Ulva*
Arthrobacter	
Aeromonas	
Cyanobacteria	

From Cerniglia (1992). Reprinted with permission of Kluwer Academic Publishers.

substituted dihydrodiol. Benzoic acid is converted to catechol and the substituted diol is converted to 3-methylcatechol. Ring fission then proceeds as with benzene.

Although the metabolic pathway for the degradation of naphthalene is well understood (e.g., Ensley & Gibson, 1983), little information is available on the molecular mechanisms of microbial degradation of higher PAHs. Naphthalene is oxidized by a multicomponent enzyme system, designated naphthalene dioxygenase, forming *cis*(1R, 2S)-dihydroxy-1, 2,-dihydronaphthalene. The genes for the individual components of the naphthalene dioxygenase in *P. putida* are encoded on a plasmid (Yen & Gunsalus, 1982) that has been cloned in *E. coli* (Ensley *et al.*, 1987) and sequenced.

The recalcitrance of PAHs to biodegradation is directly proportional to molecular weight (Table 4.5). The slow degradation of the high molecular weight PAHs is probably due to low solubility and less availability for biological uptake (Aronstein, Calvillo, & Alexander, 1991). Information on the biochemical pathway for microbial catabolism of higher molecular weight PAHs is scarce. Cerniglia and coworkers (reviewed in Cerniglia, 1992) have demonstrated that a *Mycobacterium* sp. can significantly degrade fluoranthene and pyrene, both in the laboratory and in sediments. In some cases, dead-end intermediates such as aromatic catechols and ketones were formed, raising the question of the toxicity of PAH intermediates in the environment.

Table 4.5. *Half-lives for the microbial degradation of PAHs in soil*[a]

Aromatic compound	Molecular weight	Half-life (weeks)
Naphthalene	128	2.4–4.4
Phenanthrene	178	4–18
2-Methylnaphthalene	142	14–20
Pyrene	202	34–90
3-Methylcholanthrene	226	87–200
Benzo[*a*]pyrene	252	200–300

[a]Data from Cerniglia (1992). Reprinted with permission of Kluwer Academic Publishers.

4.3 Genetics

The genes coding for oxygenases have been studied in a wide variety of hydrocarbon-degrading microorganisms. They are usually organized in inducible operons and can be located either on the chromosomes or on plasmids.

In *Pseudomonas*, the best-studied bacterial group, there are several families of catabolic plasmids involved in hydrocarbon degradation. These include the group of plasmids controlling the degradation of alkanes (CAM-OCT), the NAH plasmids for oxidation of naphthalene and salicylate, and the TOL plasmids for the degradation of toluene and xylene. The plasmids are large, generally conjugative, and present in low copy number. The degradative plasmids of *Pseudomonas* undergo gene rearrangements and shuffling.

The study of catabolic plasmids involves analysis of both structural and regulatory genes and the enzymatic activities of the gene products. Catabolic plasmids that specify similar enzymatic activities often have similar restriction endonuclease fragmentation patterns, and regions of homologous DNA.

The ability of *Pseudomonas putida* to grow on octanol is due to two alkane-inducible enzymes – a hydroxylase (this strain also contains a chromosomal gene coding for a hydroxylase) and a dehydrogenase, located on a CAM plasmid (Chakrabarty *et al.*, 1973).

In other bacteria, such as *Acinetobacter*, alkane oxidation genes are chromosomal (Singer & Finnerty, 1984b). Moreover, some of the genes that participate in alkane oxidation probably comprise indispensable constituents of the bacterial cell. For example, in *P. oleovorans*, one of the gene products that catalyzes the initial oxidation of aliphatic substrates, the alkane hydroxylase, is an integral cytoplasmic membrane protein, and constitutes after induction 1.5–2% of the total cell protein (Nieboer *et al.*; 1993).

The naphthalene catabolic genes are located in most cases on plasmids. In this group the best-studied plasmid is NAH7 of *P. putida* PpG7. It carries two operons, one of which enables the utilization of naphthalene and the other salicylate. Both operons are turned on by the product of another NAH7 gene, *nahR*, in the presence of salicylate (Schell & Poser, 1989; You *et al.*, 1988). The product of the *nahR* gene is a 36-kilodalton polypeptide that is a salicylate-dependent activator of transcription of the *nah* and *sal* hydrocarbon degradation operons (Schell & Wender, 1986).

The coordinated induction of the two degradative operons is not a universal characteristic of the system. In *Rhodococcus* sp. strain B4, isolated from a soil sample contaminated with polycyclic aromatic hydrocarbons, salicylate does not induce the genes of the naphthalene-degradative pathway (Grund *et al.*, 1992).

The plasmid-encoded genes of the naphthalene-degradation pathway of NAH7 have recently been proven to be involved in degradation of PAHs other than naphthalene (Sanseverino *et al.*, 1993). A strain of *P. fluorescens*, 5RL, carries a plasmid with regions highly homologous to upper and lower NAH7 catabolic genes and has been shown capable of mineralization of [9-^{14}C]phenanthrene and [U-^{14}C]anthracene, as well as [1-^{14}C]naphthalene. A mutant, which has the complete lower pathway inactivated by transposon insertion in *nahG*, accumulated a metabolite from phenanthrene and anthracene degradation.

Several strains have been isolated from PAH-contaminated soil that have the ability to metabolize naphthalene as the sole carbon source and have been shown to contain plasmids similar to NAH7, including a group of several plasmids present in strain PpG63 (Ogunseitan *et al.*, 1991). Although the plasmids apparently coded for similar enzymatic activities, they exhibited differences in electrophoretic mobility and endonuclease restriction patterns. NAH2 and NAH3, for example, show different restriction patterns but are highly homologous, as shown by Southern hybridization. It therefore appears that NAH is a family of highly homologous plasmids that have accumulated differences probably due to deletions and insertions (Connors & Barnsley, 1982).

The family of TOL plasmids (such as pWWO of *P. putida*) specifies enzymes for the oxidative catabolism of toluene and xylenes (Frantz & Chakrabarty, 1986). The conversion of the aromatic hydrocarbons to aromatic carboxylic acids via corresponding alcohols and aldehydes is carried out by the upper pathway, and the subsequent conversion to central metabolism intermediates proceeds through the *meta*-cleavage pathway (Franklin *et al.*, 1981). The upper pathway involves three enzymes: xylene oxygenase, benzyl alcohol dehydrogenase and benzaldehyde dehydrogenase. The synthesis of these enzymes is positively regulated by the product of *xylR* (Harayama *et al.*, 1986). The first enzyme in the upper pathway, xylene monooxygenase, catalyzes the oxidation of toluene and xylenes and consists of two different subunits encoded by *xylA* and *xylM*. The *xylM* sequence has a 25% homology with alkane hydroxylase, which catalyzes a similar reaction (the omega-hydroxylation of fatty acids and the terminal hydroxylation of alkanes) (Suzuki *et al.*, 1991).

As previously discussed for the NAH plasmids, the TOL plasmids also constitute a family of related plasmids, with differences that can be attributed to DNA rearrangements. One such rearrangement is a duplication of catabolic genes that occurs in many TOL plasmids (Osborne *et al.*, 1993). In *P. putida* R5-3 it has been shown that plasmid DNA content was altered in specific ways depending on the particular aromatic hydrocarbon utilized as the sole carbon source (Carney & Leary, 1989).

There are several cases in which the genes conferring the ability to grow on toluene are located on the chromosome. In *P. putida* MW1000 this ability is found in a 56-kilobase chromosomal segment that is almost identical to that found in the TOL plasmid pWWO (Sinclair *et al.*, 1986). The chromosomal location of these TOL genes is probably a result of transposition (Tsuda & Iino, 1987). Another example of chromosomal location of an operon coding for aromatic hydrocarbon degradation is in *P. pseudoalcaligenes* KF707. An operon sequence shows 96% homology at the level of nucleotides with the plasmid gene of *P. putida* KF715 (Furukawa *et al.*, 1991). In *Pseudomonas* sp. TA8 (Yano & Nishi, 1980) the TOL phenotype is on an R plasmid.

Comparison of different plasmids with overlapping activities, such as the lower pathways encoded by the TOL and NAH plasmids, indicates a considerable similarity in terms of gene order and DNA sequence (Harayama *et al.*, 1987; Assinder & Williams, 1988; Assinder *et al.*, 1993), suggesting a common origin.

The plasmids and operons described above represent the most studied ones, but probably constitute a small fraction of the catabolic operons in bacteria. In one study, 43 bacterial strains (mostly *Pseudomonas* spp.) from different sources, shown to possess the ability to degrade aromatic and PAHs, were hybridized with probes of NAH and TOL plasmids as well as with genomic DNA of bacteria known to degrade a wide variety of PAHs. Only 14 strains that mineralized naphthalene and phenanthrene showed homology to one of the probes. The remaining isolates mineralized and/or oxidized various PAHs and hybridized with neither pure plasmids nor genomic DNA (Foght & Westlake, 1991).

Although most of the strains capable of degrading hydrocarbons are gram-negative bacteria, this ability has been reported in gram-positive bacteria as well. One such strain, designated F199, utilized toluene, naphthalene, dibenzothiophene, salicylate, benzoate, *p*-cresol, and all isomers of xylene as a sole carbon and energy source. It carries two plasmids larger than 100 kb. No hybridization with the toluene (pWWO) or naphthalene (NAH7) catabolic plasmid DNA probes could be observed with either plasmid or genomic DNA (Fredrickson *et al.*, 1991).

The characterization of bacterial genes involved in degradation pathways has enabled the development of molecular probes for the study of microbial ecology of contaminated environments using DNA hybridization techniques and polymerase chain reaction (PCR). In addition, several of these genes derived from the OCT, NAH, and TOL plasmids were used for detection of pollution. For this purpose they were fused to *lacZ* (lactose utilization) or *lux* (*Vibrio fischeri* luciferase) as reporter genes (King *et al.*, 1990; Greer *et al.*, 1993), enabling specific pollutants to be detected by the appearance of color or light, respectively.

4.3.1 Physical interactions of microorganisms with hydrocarbons

The low solubility of hydrocarbons in water, coupled to the fact that the first step in hydrocarbon degradation involves a membrane-bound oxygenase, makes it essential for bacteria to come into direct contact with the hydrocarbon substrates. Two general biological strategies have been proposed for enhancing contact between bacteria and water-insoluble hydrocarbons: (1) specific adhesion/desorption mechanisms, and (2) emulsification of the hydrocarbon.

In order to understand the special cell-surface properties of bacteria that allow them to grow on hydrocarbons, it is useful to consider the dynamics of petroleum degradation in natural environments. Following an oil spill in the sea, the hydrocarbons rise to the surface and come into contact with air. Some of the low molecular weight hydrocarbons volatilize; the remainder are metabolized relatively rapidly by microorganisms which take up soluble hydrocarbons. These bacteria do not adhere to oil and do not have a high cell-surface hydrophobicity (Rosenberg & Rosenberg, 1985). The next stage of degradation involves microorganisms with high cell-surface hydrophobicity, which can adhere to the residual high molecular weight hydrocarbons.

Adhesion/desorption

Adhesion of microorganisms to the hydrocarbon/water interface is the first step in the growth cycle of microorganisms on water-insoluble hydrocarbons. Adhesion is caused by hydrophobic interactions. The driving force for these interactions, the so-called 'hydrophobic effect', is the transfer of ordered water from the two interacting surfaces to the bulk water phase, leading to an overall increase in entropy. It follows from thermodynamic considerations that hydrophobic interactions are stronger at elevated temperatures, increased ionic strengths of the bulk fluid, and increased apolarity of the interacting surfaces. In considering the adhesion of microorganisms to hydrocarbons via hydrophobic interactions, cell-surface hydrophobicity is the key element. Those components on the cell surface that contribute to cell-surface hydrophobicity have been referred to as *hydrophobins*, while those that reduce cell-surface hydrophobicity have been termed *hydrophilins* (Rosenberg & Kjellerberg, 1986). Bacterial hydrophobins include a variety of fimbriae and fibrils, outer membrane and other surface proteins and lipids, and certain small cell-surface molecules such as gramicidin S of *Bacillus brevis* spores and prodigiosin of pigmented *Serratia* strains. In the case of *A. calcoaceticus* RAG-1, adhesion is due to thin hydrophobic fimbriae (Rosenberg *et al.*, 1982). Mutants lacking these fimbriae failed to adhere to hydrocarbons and were unable to grow on hexadecane. The major hydrophilins of bacteria appear to be anionic exopolysaccharides. Bacterial capsules and other anionic exopolysaccharides appear to inhibit adhesion to hydrocarbons (Rosenberg *et al.*, 1983).

Desorption from the hydrocarbon is a critical part of the growth cycle of petroleum-degrading bacteria. Petroleum is a mixture of thousands of different hydrocarbon molecules. Any particular bacterium is only able to use a part of the petroleum. As the bacteria multiply at the hydrocarbon/water interface of a droplet, the relative amount of nonutilizable hydrocarbon continually increases, until the cells can no longer grow. For bacteria to continue to multiply, they must be able to

Table 4.6. *Microbial surfactants*

Emulsifier	Example	Producing species	References
Low molecular weight			
Glycolipid	Trehalose mycolate	*Rhodococcus*	Kretschmer *et al.* (1982)
	Sophorolipid	*Torulopsis*	Lang *et al.* (1984)
	Rhamnolipid	*Pseudomonas*	Hitsatsuka *et al.* (1971)
Fatty acid	2-Hydroxy-octadecanoic acid	Widespread	
	Corynomycolic acid	*Corynebacterium*	MacDonald *et al.* (1981)
Phospholipid	Phosphatidyl/ethanolamine	Widespread	
Lipopeptide	Surfactin	*B. subtilis*	Arima *et al.* (1968)
	Arthrofactin	*Arthrobacter*	Morikawa *et al.* (1993)
Surfactant	BL-86	*B. lichenformis*	Horowitz & Griffin (1991)
High molecular weight			
Polysaccharide–lipid	Emulsan	*A. calcoaceticus* RAG-1	Rosenberg & Kaplan (1985)
	Lipomannan	*C. tropicalis*	Kappeli *et al.* (1978)
Protein–polysaccharide	Emulsan BD4	*A. calcoaceticus* BD4	Rosenberg & Kaplan (1984)
	Liposan	*C. lipolytica*	Kappeli *et al.* (1978)
	Emulcyan	*Phormidium*	Fattom & Shilo (1984)
Protein	Protein PA	*P. aeroginosa*	Hitsatsuka *et al.* (1971)
Protein–lipid	Emulsifier PG-1	*Pseudomonas* PG-1	Reddy *et al.* (1983)
	AP-6	*P. fluorescens*	Persson *et al.* (1988)

move from the depleted droplet to another one. *A. calcoaceticus* RAG-1 has an interesting mechanism for desorption and for ensuring that it only re-attaches to a droplet of fresh oil (Rosenberg *et al.*, 1989). When cells become starved on the 'used' hydrocarbon droplet or tar ball, they release their capsule. The capsule is composed of an anionic heteropolysaccharide, with fatty acid side-chains, referred to as *emulsan* (Rosenberg, 1986). The extracellular, amphipathic emulsan attaches avidly to the hydrocarbon/water interface, thereby displacing the cells to the aqueous phase. Each 'used' oil droplet or tar ball is then covered with a monomolecular film of emulsan. The hydrophilic outer surface of the emulsan-coated hydrocarbon prevents re-attachment of the RAG-1 cells. The released capsule-deficient bacteria are hydrophobic and readily adhere to fresh hydrocarbon substrate.

An additional microbial surface property that is of critical importance to successful bioremediation of subsurface pollution is transport through soil. Studies on the movement of bacteria through soil with moving water indicated that several properties are involved (Gannon, Manilal & Alexander, 1991), including cell size, hydrophobicity, and net surface electrostatic charge.

Emulsifiers

In any heterogenous system, such as microorganisms growing on water-insoluble hydrocarbons, boundaries are of fundamental importance to the behavior of the system as a whole. Therefore, it is not surprising that hydrocarbon-degrading microorganisms produce a wide variety of surface-active agents. The general topic of microbial surfactants has been reviewed by Zajic & Seffens (1984), Rosenberg (1986), and Desai & Desai (1992). Reviews are also available on the surface-active properties of *Acinetobacter* exopolysaccharides (Rosenberg & Kaplan, 1985), the biosynthesis of lipid surfactants (Zajic & Mahomedy, 1984) and biosurfactants in relation to environmental pollution (Mueller-Hurtig *et al.*, 1993).

Microbiologically derived surfactants can be divided into low and high molecular weight products (Table 4.6). The low molecular weight surfactants are a mixture of glycolipids, fatty acids, phospholipids and lipopeptides. The high molecular weight surfactants are amphipathic polymers and complexes of hydrophobic and hydrophilic polymers. The production of acylated sugars, in which fatty-acid residues are bound directly to a monosaccharide or disaccharide, is often associated with the growth of microorganisms on water-insoluble hydrocarbons.

It has been suggested that the natural role of emulsans is to enhance the growth of bacteria on petroleum by two mechanisms: (1) increasing the hydrocarbon surface area, and (2) desorbing the bacteria from 'used' oil droplets. The growth of microorganisms on water-insoluble hydrocarbons is restricted to the hydrocarbon/water interface because the hydrocarbon oxygenases are always membrane-bound, never extracellular. After the bacteria adhere to the hydrocarbon surface, they begin to multiply on the surface. After a short time, the surface becomes saturated with bacteria, and growth becomes limited by the available surface. If the bacteria can split the oil droplets (emulsification), new surfaces become available for growth.

This is a large effect, as the surface area is inversely proportional to the square of the radius of the drop. For example, consider the minimum time it would take bacteria with a cross-sectional area of $1 \, \mu m^2$ to degrade 1 ml of hydrocarbon, assuming a doubling time of 4 h. If the oil was not emulsified ($r = 1$ mm), then the oil surface would be $3 \times 10^9 \, \mu m^2$ and the degradation time would be 800 h. However, if the oil was emulsified ($r = 2 \, \mu m$) then the oil surface would be $1.5 \times 10^{12} \, \mu m$ and the degradation time would be only 1.6 h. This theoretical argument illustrates both the significance of emulsification for hydrocarbon-degrading microorganisms and the potential of emulsifiers in the bioremediation of petroleum pollution.

An even more critical function of emulsifiers for the producing microorganisms is desorption from 'used' oil droplets. Consider A. calcoaceticus RAG-1 growing on the surface of crude oil in sea water. During the growth phase, the bacteria multiply at the expense of n-alkanes, accumulating emulsan in the form of a minicapsule. After the alkanes in the petroleum droplet are consumed, RAG-1 becomes starved, although it is still attached to the hydrocarbon enriched in aromatic and cyclic paraffins. Starvation of RAG-1 causes release of the minicapsule of emulsan. This released emulsan forms a polymeric film on the n-alkane-depleted oil droplet, thereby desorbing the starved cell. In effect, the 'emulsifier' frees the cell to find a fresh substrate. At the same time, the depleted oil droplet has been 'marked' as being used by having a hydrophilic outer surface to which RAG-1 cannot attach. Presumably, other microorganisms which can attach to hydrophilic surfaces and degrade aromatic and cyclic hydrocarbons continue the process of petroleum biodegradation.

Surfactants are used widely in industry, agriculture and medicine. The materials currently in use are produced primarily by chemical synthesis, or as by-products of industrial processes. For a microbial surfactant to penetrate the market, it must provide a clear advantage over the existing competing materials. The major considerations are: (1) safety, i.e., low toxicity and biodegradability; (2) cost; (3) selectivity; and (4) specific surface modifications. Biosurfactants exhibit low toxicity and good biodegradability, properties that are essential if the surfactant is to be released into the environment.

An emulsifying or dispersing agent not only causes a reduction in the average particle size, but also changes the surface properties of the particle in a fundamental manner. For example, oil droplets that have been emulsified with emulsan contain an outer hydrophilic shell of spatially oriented carboxyl and hydroxyl groups. These emulsan-stabilized oil droplets bind large quantities of uranium (Zosim, Gutnick & Rosenberg, 1983). In an elegant experiment, Pines & Gutnick (1984) demonstrated that emulsan on the surface of hexadecane droplets acts as a receptor for A. calcoaceticus bacteriophage ap3. Clearly, each bioemulsifier is going to alter, in a characteristic manner, the interface to which it adheres. Usually, 0.1–1.0% of a polymeric bioemulsifier or biodispersant is enough to saturate the surface of the dispersed material. Thus, small quantities of a dispersant can alter dramatically the surface properties of a material such as its surface charge, hydrophobicity and, most interestingly, its pattern recognition, based on the three-dimensional structure of the adherent polymer.

Nutrient requirements

As mentioned above, microorganisms that have the genetic potential to bind, emulsify, transport and degrade hydrocarbons are widely distributed in nature. Thus, the rate-limiting step in biodegradation of hydrocarbons is generally not the lack of appropriate microorganisms. Depending upon the particular environmental situation, the extent of degradation depends on the availability of moisture, oxygen, and utilizable sources of nitrogen and phosphorus. Thus, the effectiveness of a bioremediation program depends on defining the limitations and overcoming them – in a practical way.

The requirements for oxygen and moisture are not a problem in oil spills in aquatic environments. However, on land, oxygen and water are often rate-limiting. If oil has not penetrated too deeply into the ground, then watering and tilling are often advantageous. These problems have been discussed theoretically by Atlas (1991), and Bartha (1990) has provided some recent useful data on the subject. In some circumstances, hydrogen peroxide has been used as a source of oxygen (Huling *et al.*, 1991). Nitrate has also been tested as an alternative electron acceptor (Werner, 1985).

The major limitation in the biodegradation of hydrocarbons on land and water is available sources of nitrogen and phosphorus. In theory, approximately 150 mg of nitrogen and 30 mg of phosphorus are consumed in the conversion of 1 g of hydrocarbon to cell material. The nitrogen and phosphorus requirements for maximum growth of hydrocarbon oxidizers can generally be satisfied by ammonium phosphate. Alternatively, these requirements can be met with a mixture of other salts, such as ammonium sulfate, ammonium nitrate, ammonium chloride, potassium phosphate, sodium phosphate and calcium phosphate. When ammonium salts of strong acids are used, the pH of the medium generally decreases with growth. This problem can often be overcome by using urea as the nitrogen source. All of these compounds have a high water solubility, which reduces their effectiveness in open systems because of rapid dilution. In principle, this problem can be solved by using oleophilic nitrogen and phosphorus compounds with low C:N and C:P ratios. For example, it was found that a combination of paraffinized urea and octyl phosphate was able to replace nitrate and inorganic phosphate, respectively (Atlas & Bartha, 1973a).

Recently, commercial nitrogen- and phosphorus-containing fertilizers that have an affinity for hydrocarbons are being developed for treating oil pollution. The most extensively studied example is the oleophilic fertilizer Inipol EAP 22, an oil-in-water microemulsion, containing a N:P ratio of 7.3:2.8 (Glaser, 1991). Inipol EAP 22 is composed of oleic acid, lauryl phosphate, 2-butoxy-1-ethanol, urea, and water. The fertilizer was used in large quantities to treat the oil-contaminated shoreline following the 1989 oil spill in Prince William Sound, Alaska. There are at least three problems with Inipol EAP 22. First, it contains large amounts of oleic acid, which serves as an alternative carbon source, thereby further increasing the C:N ratio in the environment. Second, it contains an emulsifier that could be harmful to the environment. Third, as soon as the fertilizer comes into contact with water, the emulsion breaks, releasing the urea to the water phase where it is not available for the microorganisms. A newly developed controlled release (not slow release) nitrogen fertilizer, which depends on microbial-catalyzed hydrolysis of an oleophilic polymer, will be discussed in the case study of the Haifa Oil Spill.

4.4 Case studies

4.4.1 Exxon Valdez oil spill

On March 24, 1989, the oil tanker *Exxon Valdez* ran aground near Bligh Island, spilling approximately 40 000 tons of crude oil into Prince William Sound, Alaska. During the next few days, the oil spread onto the shoreline of numerous islands in the Sound and into the Gulf of Alaska. The immediate response was to offload the remaining cargo and collect as much oil as possible with skimmers from the sea and washed beaches. Subsequently, about 100 miles of shoreline were treated with fertilizers, in what was probably the largest marine bioremediation project ever undertaken. In spite of the large number of articles published on the project (e.g., Crawford, 1990; Chianelli *et al.*, 1991; Pritchard & Costa, 1991; Lindstrom *et al.*, 1991; Mueller *et al.*, 1992; Pritchard *et al.*, 1992; Prince, 1992; Button *et al.*, 1992), it is still difficult to evaluate how effective the treatment actually was, compared with untreated controls.

The bioremediation treatment consisted only of addition of fertilizers, since preliminary experiments indicated a high level of indigenous hydrocarbon-degrading bacteria and adequate oxygen levels as a result of vigorous tidal flushing. The fertilizer chosen for most of the treatment was Inipol EAP22, a microemulsion of urea, oleic acid, lauryl phosphate, 2-butoxy-1-ethanol and water. Inipol is claimed by its manufacturers to be an oleophilic fertilizer, but we have observed that as soon as it comes into contact with water, the emulsion breaks, releasing urea into the water.

The initial results of treatment were visual changes that showed an apparently remarkable removal of oil from test plots in two weeks. Aerial photographs showed a 'clean rectangle' on the area where Inipol was applied, surrounded by a darker area of oiled cobblestones. These photographs were used extensively in Exxon public relations. The problem with this nonquantitative approach is that it fails to demonstrate biodegradation. For example, physical changes on the exposed surface resulting from interactions with the fertilizer, or the emulsifier it contains, could change the appearance of the surface. Furthermore, considerable amounts of oil remained under the cobblestones and in the underlying sand and gravel. After eight weeks, the contrast between the treated and untreated areas had decreased.

Table 4.7. *Effectiveness of the bioremediation treatment of cobblestone beaches following the Exxon Valdez oil spill*

Parameter measured	Treated[a]	Control
Residual oil (mg/g)		
Day 0[b]	0.6	0.5
Day 18	0.4	0.4
Day 50	0.3	0.3
Days 80–90	0.15	0.25
C18/phytane ratio		
Day 0[b]	1.3	1.1
Day 18	1.4	0.9
Day 50	0.6	0.5
Days 80–90	0.15	0.1

Data from Pritchard *et al.* (1992). Reprinted by permission of Kluwer Academic Publishers.
[a]Inipol fertilizer treated.
[b]Days following fertilizer addition (8 June 1989).

Quantitative measurements demonstrating the effectiveness of the bioremediation treatment of cobblestone samples from Snug Harbor are summarized in Table 4.7. There were large standard deviations in the analyses due largely to the highly heterogeneous distribution of oil on the beach. The averages of several determinations are presented in the table. It should be pointed out that the data are for the cobblestones and not the underlying gravel. Residual oil measurements indicated an approximately 75% removal of the extractable oil in the treated plots, compared with 50% in the control plots, during the three-month summer of 1989. Although the bioremediation effect was not statistically significant ($p = 0.05$), the researchers suggest that bioremediation treatment enhanced the oil removal by 2–3 times, with a half-life of 44 days for the Inipol-treated area, compared with 124 days for the control plots. These calculations are based on the assumption of first-order decay rates, which is certainly an incorrect assumption for a heterogeneous substrate like crude oil.

The ratio of residual straight-chain (C18) to branch-chain (phytane) hydrocarbons clearly demonstrated that biodegradation was largely responsible for the surface beach cleaning. Since the molecular weights of octadecane and phytane are similar, one would expect that removal of these hydrocarbons by physical processes, such as evaporation and solubilization, would occur at similar rates. On the other hand, it is known that branched alkanes biodegrade much more slowly than *n*-alkanes (Singer & Finnerty, 1984b). Thus, the decrease in C18/phytane ratio during the summer of 1989 indicated that biodegradation was occurring. However, the control plots also showed a similar decrease in the ratio. In an independent study supported by the Alaska Sea Grant Program (Button *et al.*, 1992), no differences were observed in visible or quantitatively determined oil or microbial populations on the treated gravel beaches, compared with control plots.

In conclusion, there does not appear to be any compelling evidence that the Inipol fertilizer treatment had any long-term effect on removing bulk oil from the beaches following the *Exxon Valdez* oil spill. However, the treatment did accelerate, during the first few weeks, the biodegradation of oil from the exposed upper surface of the cobblestones. Our interpretation of the published data is that the urea and oleic acid present in the fertilizer had an immediate effect in stimulating the bacterial population, but that most of the urea was removed by tidal flushing within a couple of days. Thereafter, natural biodegradation occurred at similar rates in the treated and control plots. Notwithstanding the inconclusive results of the bioremediation treatment, many techniques were refined, lessons learned and new analytical procedures developed by the researchers involved in the Alaska bioremediation cleanup project.

4.4.2 Bioremediation of an oil-polluted refinery site

In the bioremediation of an oil-polluted refinery site reported by Balba, Ying & McNeice (1991), laboratory experiments performed before the field trial demonstrated:

1. The composition of the hydrocarbon in the contaminated soil was 62.5% total saturates (mainly C_{14}–C_{25}), 25% aromatics, and 12.5% resins and asphaltines.

2. The soil contained hydrocarbon-degrading *Pseudomonas, Rhodococci, Acinetobacter, Mycobacterium*, and *Arthrobacter*. Only the isolated *Mycobacterium* was able to degrade the asphaltene fraction.

3. After screening more than 50 compounds, Cyanamer P70 (Cyanamid) was selected as the surfactant to be used in the trial, because it removed a relatively high proportion of the contaminating oil from the soil (47%) and was not toxic to the isolated bacteria.

4. Over 95% of the hydrocarbon was degraded in 23 weeks in soil microcosm experiments (Table 4.8) that contained added nutrients, Cyanamer P70, and isolated bacteria.

Based on the successful laboratory soil microcosm experiment, the most heavily contaminated soil was excavated from the surface (2 m) layer and treated *ex situ*. The soil was placed in an enclosed rectangular bed and its moisture content maintained at 15% (w/w). The soil was aerated with an agricultural spading machine. Surfactant, organic nutrients, and inorganic nutrients were applied to maintain an optimum C:N:P ratio (data not presented). The isolated microorganisms were inoculated into the soil as dilute suspensions on five separate occasions.

Initially, the excavated soil had an average of 13 g TPH per kg soil. After 34 weeks of treatment, TPH was reduced by about 90% to 1.27 g per kg. Gas chromatography showed a large reduction in the alkane fraction. No data were presented on the aromatic and PAH fractions. The residue remaining in the soil after treatment was immobile, as determined by repeatedly washing the soil with water and testing for hydrocarbons.

Table 4.8. *Degradation of hydrocarbons in oil-polluted soil*

Treatment	Residue TPH (ppm)		
	5 wk	12 wk	23 wk
Control	2000	1700	900 ± 100
Nutrients	1900	1300	400 ± 100
Nutrient + isolate	1200	750	450 ± 100
Nutrient + isolate + surfactant	1300	500	100 ± 50

From Balba, Ying & McNeice (1991).
The initial concentration of total petroleum hydrocarbon (TPH) was 2.4 g per kg soil. All samples were kept moist and mixed regularly.

4.4.3 Bioremediation of Haifa beach (Rosenberg et al., 1992)

In August 1991, approximately 100 tons of heavy crude oil were accidentally spilled about 3 km north of Zvulon Beach, between Haifa and Akko in Israel. The oil-contaminated sand was collected into piles and subsequently spread over the beach.

The technology that was chosen to clean the sand was based on the use of a new controlled-release, polymeric, hydrophobic fertilizer, F-1, and selected bacteria. These bacteria can both utilize hydrocarbons as their carbon and energy source and enzymatically hydrolyze the fertilizer to derive utilizable nitrogen and phosphorus compounds. Since bacteria that can both degrade hydrocarbons and hydrolyze F-1 are rare, it is necessary to inoculate the soil with selected bacteria when using F-1 for bioremediation. The major advantage of this system is that the added nitrogen is used exclusively by the hydrocarbon-degrading bacteria.

In the field trial the experimental plot was inoculated with 20 liters of a mixture of three selected bacteria (5×10^8 cells/ml). Fertilizer F-1 (38 kg, in the form of a fine powder) was then added as the source of nitrogen and phosphorus. The experimental plot was watered from the adjacent sea (water temperature was 27 °C). The plot was then tilled by hand with the help of a simple rake to the depth of about 5 cm. The control plot was left undisturbed. The experimental data are presented in Table 4.9. On day 0 (September 1, 1991), the core samples were taken prior to any treatment. At the beginning of the experiment, the experimental plot contained significantly more hydrocarbon (3.80 mg/g sand) than the control plot (2.30 mg/g sand). There was only a small decrease in the first day. However, by the fourth day, 30% of the hydrocarbon had been degraded. The biodegradation continued, reaching 50% on day 9 and 84.5% on day 25, when the experiment was concluded. The control plot showed an 18% degradation by day 9 which remained relatively constant until day 25.

Gas chromatographic analyses of the alkane fractions demonstrated that the C_{20}–C_{32} fraction, which was the most abundant at day 0, was degraded to the

Table 4.9. *Bioremediation of hydrocarbon-contaminated sand at Haifa beach*

Day	TPH (g/kg)			
	Control	(%)	Experiment	(%)
0	2.30	(100)	3.80	(100)
4	2.53	(110)	2.76	(73)
9	1.88	(82)	1.89	(50)
14	1.70	(74)	0.88	(23)
21	1.94	(84)	1.40	(37)
25	1.95	(85)	0.95	(25)

From Rosenberg *et al.* (1992).

extent of 94%, whereas the C_{14}–C_{18} and C_{36}–C_{40} fractions were degraded to lesser degrees, 80% and 75% respectively. Visual examination of the beach sand following the treatment, in addition to the analytical data, indicated that the F-1/bacteria technology was applicable for bioremediation of the remaining sand.

In the cleanup $30\,000\,m^2$ of polluted sand was treated essentially as in the field trial, except that agricultural equipment was used for applying the bacteria and fertilizer and for tilling. Since the cleanup took place in winter, it was not necessary to water. It should be noted that the winter of 1992 was unusually hard for Israel, and temperatures were around 5–10 °C for a couple of months. The results of the experiment indicated that even under these conditions 88% of the oil was degraded after 4 months, whereas no significant degradation was measured in control plots.

Acknowledgements

This work was supported by Makhteshim Chemical Works Ltd and the Pasha Gol Chair for Applied Microbiology.

References

Arima, K., Kakinuma, A. & Tamura, G. (1968). Surfactin, a crystalline peptidolipid surfactant produced by *Bacillus subtilis*: isolation, characterization and its inhibition of fibrin clot formation. *Biochemical and Biophysical Research Communication*, 31, 488–94.

Aronstein, B. A., Calvillo, Y. M. & Alexander, M. (1991). Effect of surfactants at low concentrations on the desorption and biodegradation of sorbed aromatic compounds in soil. *Environmental Science & Technology*, 25, 1728–31.

Assinder, S., & Williams, X. X. (1988). Comparison of the meta pathway operons on NAH plasmid pWW60-22 and TOL plasmid pWW53-4 and its evolutionary significance. *Journal of General Microbiology*, 134, 2769–78.

Assinder, S., De Marco, P., Osborne, D., Poh, C., Shaw, L. E., Winson, M. K. & Williams,

P. A. (1993). A comparison of the multiple alleles of xylS carried by TOL plasmids pWW53 and pDK1 and its implications for their evolutionary relationship. *Journal of General Microbiology*, 139, 557–68.

Atlas, R. M. (ed.) (1984). *Petroleum Microbiology*. New York: Macmillan Publishing. 692 pp.

Atlas, R. M. (1991). Microbial hydrocarbons degradation-bioremediation of soil spills. *Journal of Chemical Technology and Biotechnology*, 52, 149–56.

Atlas, R. M. & Bartha, R. (1973a). Stimulated biodegradation of oil slicks using oleophilic fertilizers. *Environmental Science & Technology*, 7, 538–41.

Atlas, R. M. & Bartha, R. (1973b). Abundance, distribution and oil biodegradation potential of microorganisms in Raritan Bay. *Environmental Pollution*, 4, 291–300.

Balba, M. T., Ying, A. C. & McNeice, T. G. (1991). Bioremediation of contaminated land: bench scale to field applications. In *Bioreclamation*, ed. R. E. Hinchee & R. F. Olfenbuttel, pp. 464–76. Stoneham, MA: Butterworth–Heinemann Publishers.

Bartha, R. (1990). Bioremediation potential of terrestrial fuel spills. *Applied and Environmental Microbiology*, 56, 652–6.

Bazylinski, D. A., Wirsen, C. O. & Jannasch, H. W. (1989). Microbial utilization of naturally occurring hydrocarbons at the Guaymas Basin hydrothermal vent site. *Applied and Environmental Microbiology*, 55, 2832–6.

Button, D. K., Robertson, B. R., McIntosh, D. & Juttner, F. (1992). Interactions between marine bacteria and dissolved-phase and beached hydrocarbons after the Exxon Valdez oil spill. *Applied and Environmental Microbiology*, 58, 243–51.

Carney, B. F. & Leary, J. V. (1989). Novel alterations in plasmid DNA associated with aromatic hydrocarbon utilization by *Pseudomonas putida* R5-3. *Applied and Environmental Microbiology*, 55, 1523–30.

Cerniglia, C. E. (1992). Biodegradation of polycyclic aromatic hydrocarbons. *Biodegradation*, 3, 351–68.

Chakrabarty, A. M., Chou, G. & Gunsalas. (1973). Genetic regulation of octane dissimilation plasmid in *Pseudomonas*. *Proceedings of National Academy of Science*, 70, 1137–40.

Chianelli, R. R., Aczel, T., Bare, R. E., George, G. N., Genowitz, M. W., Grossman, M. J., Haith, C. E., Kaiser, F. J., Lessard, R. R., Liotta, R., Mastracchio, R. L., Minal-Bernero, V., Prince, R. C., Robbins, W. K., Stiefel, E. I., Wilkinson, J. B., Hinton, S. M., Bragg, J. R., McMillem, S. J. & Atlas, R. M. (1991). Bioremediation technology development and application to the Alaskan spill. In *Proceedings 1991 Oil Spill Conference*, pp. 545–55. Washington, DC: American Petroleum Institute.

Connors, M. A. & Barnsley, E. A. (1982). Naphthalene plasmids in pseudomonads. *Journal of Bacteriology*, 149, 1096–101.

Crawford, M. (1990). Bacteria effective in Alaska cleanup. *Science*, 247, 1537.

Currier, H. B. & Peoples, S. A. (1954). Phytotoxicity of hydrocarbons. *Hilgardia*, 23, 155–74.

Dagley, S. (1971). Catabolism of aromatic hydrocarbons by microorganisms. *Advances in Microbial Physiology*, 6, 1–46.

Davis, J. B. (1967). *Petroleum Microbiology*. New York: Elsevier.

Desai, J. D. & Desai, A. J. (1992). Production of biosurfactants. In *Biosurfactants*, ed. N. Kosaric, pp. 65–97. New York: Marcel Dekker.

Energy Information Administration. (1992). *International Energy Annual*, p. 1990. Washington, DC: Department of Energy.

Ensley, B. D. & Gibson, D. T. (1983). Oxidation of naphthalene by a multicomponent enzyme system from *Pseudomonas* sp. strain NCIB 9816. *Journal of Bacteriology*, 155, 505–11.

Ensley, B. D., Osslund, T. D., Joyce, M. & Simon, M. J. (1987). Expression and complementation of naphthalene dioxygenase activity in *Escherichia coli*. In *Microbial Metabolism and the Carbon Cycle*, ed. S. R. Hagedorn, R. S. Hanson & D. A. Kunz, pp. 437–55. New York: Harwood Academic Publishers.

Fattom, A. & Shilo, M. (1984). Hydrophobicity as an adhesion mechanism of benthic cyanobacteria. *Applied and Environmental Microbiology*, 47, 135.

Foght, J. M. & Westlake, D. W. S. (1991). Cross hybridization of plasmid and genomic DNA from aromatic and polycyclic. *Canadian Journal of Microbiology*, 37, 924–32.

Franklin, F. CH., Bagdasarian, M. M., Bagdasarian, M. & Timmis, K. N. (1981). Molecular and functional analysis of the TOL plasmid pWWO from *Pseudomonas putida* and cloning of genes for the entire regulated aromatic ring meta cleavage pathway. *Proceedings of the National Academy of Science*, 78, 7458–62.

Frantz, B. & Chakrabarty, A. M. (1986). Degradative plasmids in *Pseudomonas*. In *The Biology of* Pseudomonas, ed. J. R. Sokatch, *The Bacteria: A Treatise on Structure and Function*, Vol. 10, pp. 295–323.

Fredrickson, J. K., Brockman, F. J., Workman, D. J., Li, S. W. & Stevens, T. O. (1991). Isolation and characterization of a subsurface bacterium capable of growth on toluene, naphthalene, and other aromatic compounds. *Applied and Environmental Microbiology*, 57, 796–803.

Furukawa, K., Nakagawa, K., Yuyama, N., Sato, A., Kokean, Y., Kato, H., Yamashita, S., Hayase, N., Nishikawa, S. & Taira, K. (1991). Highly conserved coding sequences in polychlorinated biphenyl (PCB)-degraders of *Pseudomonas pseudoalcaligenes* KF707 and *Pseudomonas putida* KF715. *Nucleic-Acids – Symposium Series*, 25, 115–16.

Gannon, J. T., Manilal, V. B. & Alexander, M. (1991). Relationship between cell surface properties and transport of bacteria through soil. *Applied and Environmental Microbiology*, 57, 190–3.

Gibson, D. T. (1977). Biodegradation of aromatic hydrocarbons. In *Fate and Effect of Petroleum in Marine Ecosystems and Organisms*, ed. D. A. Wolfe, pp. 34–46. New York: Pergamon Press.

Glaser, J. A. (1991). Nutrient-enhanced bioremediation of oil-contaminated shoreline: the Valdez experience. In *On Site Bioreclamation*, ed. R. E. Hinchee & R. F. Olfenbuttel, pp. 336–84. Stoneham, MA: Butterworth–Heinemann.

Greer, C., Masson, L., Comeau, Y., Brousseau, R. & Samson, R. (1993). Application of molecular biology techniques for isolating and monitoring. *Water Pollution Research Journal of Canada*, 28, 275–87.

Grund, E., Denecke, B. & Eichenlaub, R. (1992). Naphthalene degradation via salicylate and gentisate by *Rhodococcus* sp. strain B4. *Applied and Environmental Microbiology*, 58, 1874–7.

Harayama, S., Leppik, R. A., Rekik, M., Mermod, N., Lehrbach, P. R., Reineke, W. & Timmis, K. N. (1986). Gene order of the TOL catabolic plasmid upper pathway operon. *Journal of Bacteriology*, 167, 455–61.

Harayama, S., Rekik, M., Wasserfallen, A. & Bairoch, A. (1987). Evolutionary relationship between catabolic pathways for aromatics: conservation of gene order and nucleotide sequences for catechol oxidation genes of pWWO and NAH7 plasmids. *Molecular General Genetics*, 210, 241–7.

Hisatsuka, K., Nakahara, T., Sano, N. & Yamada, K. (1971). Formation of rhamnolipid by *Pseudomonas aeruginosa* and its function in hydrocarbon fermentation. *Agricultural and Biological Chemistry*, 35, 686.

Hood, M. A., Bishop, W. S. Jr., Bishop, F. W., Meyers, S. P. & Whelan T, III (1975). Microbial indicators of oil-rich salt marsh sediments. *Applied and Environmental Microbiology*, 30, 982–7.

Horowitz, A. & Atlas, R. M. (1977a). Continuous open flow-through system as a model for oil degradation in the Arctic Ocean. *Applied and Environmental Microbiology*, 33, 647–53.

Horowitz, A. & Atlas, R. M. (1977b). Response of microorganisms to an accidental gasoline spillage in an Arctic freshwater ecosystem. *Applied and Environmental Microbiology*, 33, 1252–8.

Horowitz, S. & Griffin, M. W. (1991). Structural analysis of *Bacillus licheniformis* 86 surfactant. *Journal of Industrial Microbiology*, 7, 45–52.

Huling, S. G., Bledsoe, B. C. & White, M. G. (1991). The feasibility of utilizing hydrogen peroxide as a source of oxygen in bioremediation. In *Bioreclamation*, ed. R. E. Hinchee & R. F. Olfenbuttel, pp. 83–103. Stoneham, MA: Butterworth–Heinemann.

Kappeli, O., Mueller, M. and Fiechter, A. (1978). Chemical and structural alterations of the cell surface of *Candida tropicalis*, induced by hydrocarbon substrate. *Journal of Bacteriology*, 133, 952.

King, J. M. H., DiGrazia, P. M., Applegate, B., Burlage, R., Sanseverino, J. & Sayler, G. S. (1990). Rapid, sensitive bioluminescent reporter technology for naphthalene exposure. *Science*, 249, 778–81.

Kretschmer, A., Bock, H. & Wagner, F. (1982). Chemical and physical characterization of interfacial-active lipids from *Rhodococcus erythropolis* grown on n-alkanes. *Applied and Environmental Microbiology*, 44, 864.

Lang, S., Gilbon, A., Syldatk, C. & Wagner, F. (1984). Comparison of interfacial active properties of glycolipids from microorganisms. In *Surfactants in Solution*, ed. K. L. Mittal & B. Lindman, p. 1365. New York: Plenum Press.

Leahy, J. G. & Colwell, R. R. (1990). Microbial degradation of hydrocarbons in the environment. *Microbiological Reviews*, 54, 305–15.

Lindstrom, J. E., Prince, R. C., Clark, J. C., Grossman, M. J., Yeager, T. R., Brown, J. F. & Brown, E. J. (1991). Microbial populations and hydrocarbon biodegradation potential in fertilized shoreline sediments affected by the t/v Exxon Valdez oil spill. *Applied and Environmental Microbiology*, 57, 2514–22.

MacDonald, C. R., Cooper, D. G. & Zajic, J. E. (1981). Surface-active lipids from *Nocardia erythropolis* grown on hydrocarbons. *Applied and Environmental Microbiology*, 14, 117.

Markovetz, A. J. (1971). Subterminal oxidation of aliphatic hydrocarbons by microorganisms. *CRC Critical Reviews in Microbiology*, 1, 225–38.

McKenna, E. J. & Kallio, R. E. (1965). The biology of hydrocarbons. *Annual Reviews in Microbiology*, 19, 183–208.

Miller, E. C. & Miller, J. A. (1981). Searches for the ultimate chemical carcinogens and their reactions with cellular macromolecules. *Cancer*, 47, 2327–45.

Morgan, P. & Watkinson, R. J. (1989). Hydrocarbon degradation in soils and methods for biotreatment. *Critical Reviews in Biotechnology*, 8, 305–33.

Morikawa, M., Daido, H., Takao, T., Murata, S., Shimonishi, Y. & Imanaka, T. (1993). A new lipopeptide biosurfactant produced by *Arthrobacter* sp. strain MIS38. *Journal of Bacteriology*, 175, 6459–66.

Mueller, J. G., Resnick, S. M., Shelton, M. E. & Pritchard, P. H. (1992). Effect of inoculation on the biodegradation of weathered Prudhoe Bay crude oil. *Journal of Industrial Microbiology*, 10, 95–102.

Mueller-Hurtig, G., Wagner, F., Blazczyk, R. & Kosaric, N. (1993). Biosurfactants. *Surfactant Science Series*, pp. 447–69.

National Research Council (1985). *Oil in the Sea: Inputs, Fates and Effects*. Washington, DC: National Academy of Sciences.

Nieboer, M., Kingma, J. & Witholt, B. (1993). The alkane oxidation system of *Pseudomonas oleovorans*: induction of the alk genes in *Escherichia coli* W3110 (pGEc47) affects membrane biogenesis and results in overexpression of alkane hydroxylase in a distant cytoplasmic membrane subfraction. *Molecular Microbiology*, 8, 1039–51.

Nieder, M. & Shapiro, J. (1975). Physiological function of *Pseudomonas putida* PpG6 (*Pseudomonas oleovarans*) alkane hydroxylase: monoterminal oxidation of alkanes and fatty acids. *Journal of Bacteriology*, 122, 93–8.

Ogunseitan, O., Delgado, I., Tsai, Y. & Olson, B. (1991). Effect of 2-hydroxybenzoate on the maintenance of naphthalene-degrading pseudomonads in seeded and unseeded soil. *Applied and Environmental Microbiology*, 57, 2873–9.

Osborne, D. J., Pickup, R. W. & Williams, P. A. (1993). The presence of two complete homologous meta pathway operons on TOL plasmid. *Journal of General Microbiology*, 134, 2965–75.

Perry, J. J. (1984). Microbial metabolism of cyclic alkanes. In *Petroleum Microbiology*, ed. R. M. Atlas, pp. 61–98. New York: Macmillan Publishing Company.

Persson, A., Osterberg, E. & Dostalek, M. (1988). Biosurfactant production by *Pseudomonas fluorescens* 378: growth and product characteristics. *Applied Microbiology and Biotechnology*, 29, 1–4.

Pines, O. & Gutnick, D. L. (1984). Specific binding of a bacterio-phage at a hydrocarbon–water interface. *Journal of Bacteriology*, 157, 179–83.

Pirnik, M. P. (1977). Microbial oxidation of methyl branched alkanes. *CRC Critical Reviews in Microbiology*, 5, 413–22.

Prince, R. C. (1992). Bioremediation of oil spills, with particular reference to the spill from the Exxon Valdez. In *Microbial Control of Pollution*, ed. J. C. Fry, G. M. Gadd, R. A. Herbert, C. W. Jones & I. A. Watson-Craik, pp. 5–34. Cambridge: Cambridge University Press.

Pritchard, P. H. & Costa, C. F. (1991). EPA's Alaskan oil spill bioremediation project. *Environmental Science & Technology*, 25, 372–9.

Pritchard, P. H., Mueller, J. G., Rogers, J. C., Kremer, F. V. & Glaser, J. A. (1992). Oil spill bioremediation: experiences, lessons and results from the Exxon Valdez oil spill in Alaska. *Biodegradation*, 3, 315–35.

Raymond, R. L., Hudson, J. O. & Jamison, V. W. (1976). Oil degradation in soil. *Applied and Environmental Microbiology*, 31, 522–35.

Reddy, P. G., Singh, H. D., Roy, P. K. & Baruah, J. N. (1983). Isolation and functional characterization of hydrocarbon emulsifying and solubilizing factors by a *Pseudomonas* species. *Biotechnology and Bioengineering*, 25, 387.

Rosenberg, E. (1986). Microbial surfactants. *CRC Critical Reviews in Biotechnology*, 3, 109–32.

Rosenberg, E. (1991). Hydrocarbon-oxidizing bacteria. In *The Prokaryotes*, ed. A. Ballows, pp. 441–59. Berlin: Springer-Verlag.

Rosenberg, E. (1993). Exploiting microbial growth on hydrocarbons: new markets. *Trends in Biotechnology*, 11, 419–23.

Rosenberg, E. & Kaplan, N. (1985). Surface active properties of *Acinetobacter* exopolysaccharides. In *Bacterial Outer Membranes as Model Systems*, ed. M. Inouye, Chapter 12, pp. 311–42. New York: John Wiley and Sons.

Rosenberg, E., Kaplan, N., Pines, O., Rosenberg, M. & Gutnick, D. (1983). Capsular polysaccharides interfere with adherence of *Acinetobacter*. *FEMS Microbiology Letters*, 17, 157–61.

Rosenberg, E., Rosenberg, M., Shoham, Y., Kaplan, N. & Sar, N. (1989). Adhesion and desorption during growth of *Acinetobacter calcoaceticus* on hydrocarbons. In *Microbial Mats*, ed. Y. Cohen & E. Rosenberg, pp. 218–26. Washington, DC: ASM Publications.

Rosenberg, M. & Kjellerberg, S. (1986). Hydrophobic interactions: role in bacterial adhesion. *Advances in Microbial Ecology*, 9, 353–93.

Rosenberg, M. & Rosenberg, E. (1985). Bacterial adherence at the hydrocarbon–water interface. *Oil and Petrochemical Pollution*, 2, 155–62.

Rosenberg, M., Bayer, E. A., Delaria, J. & Rosenberg, E. (1982). Role of thin fimbriae in adherence and growth of *Acinetobacter calcoaceticus* RAG-1 on hexadecane. *Applied and Environmental Microbiology*, 44, 929–37.

Rosenberg, E., Legmann, R., Kushmaro, A., Taube, R., Adler, E. & Ron, E. (1992). Petroleum bioremediation: a multiphase problem. *Biodegradation* 3, 337–50.

Sanseverino, J., Applegate, B., King, J. & Sayler, G. (1993). Plasmid-mediated mineralization of naphthalene, phenanthrene, and anthracene. *Applied and Environmental Microbiology*, 59, 1931–7.

Schell, M. A. & Poser, E. F. (1989). Demonstration, characterization and mutational analysis of NahR protein in binding to nah and sal promoters. *Journal of Bacteriology*, 171, 837–46.

Schell, M. A. & Wender, P. E. (1986). Identification of the nahR gene product and nucleotide sequence required for its activation of the sal operon. *Journal of Bacteriology*, 166, 9–14.

Sinclair, M. I., Maxwell, P. C., Lyon, B. R. & Halloway, B. W. (1986). Chromosomal location of TOL plasmid DNA in *Pseudomonas putida*. *Journal of Bacteriology*, 168, 1302–8.

Singer, M. E. & Finnerty, W. R. (1984a). Genetics of hydrocarbon-utilizing microorganisms. In *Petroleum Microbiology*, ed. R. M. Atlas, pp. 299–354. New York: Macmillan.

Singer, M. E. & Finnerty, W. R. (1984b). Microbial metabolism of straight-chain and branched alkanes. In *Petroleum Microbiology*, ed. R. M. Atlas, pp. 1–60. New York: Macmillan.

Song, H-G. & Bartha, R. (1990). Effects of jet fuel spills on the microbial community in soil. *Applied and Environmental Microbiology*, 56, 646–51.

Suzuki, M., Hayakawa, T., Shaw, J., Rekik, M. & Harayama, S. (1991). Primary structure of xylene monooxygenase: similarities to and differences from the alkane hydroxylation system. *Journal of Bacteriology*, 173, 1690–5.

Tsuda, M. & Iino, I. (1987). Genetic analysis of a transposon carrying toluene degrading genes on TOL plasmid pWWO. *Molecular and General Genetics*, 210, 270–6.

Walker, J. D. & Colwell, R. R. (1976a). Measuring potential activity of hydrocarbon degrading bacteria. *Applied and Environmental Microbiology*, 31, 189–97.

Walker, J. D. & Colwell, R. R. (1976b). Enumeration of petroleum-degrading microorganisms. *Applied and Environmental Microbiology*, 31, 198–207.

Watkinson, R. J. & Morgan, P. (1990). Physiology of aliphatic hydrocarbon degrading microorganisms. *Biodegradation*, 1, 79–92.

Werner, P. (1985). A new way for the decontamination of polluted aquifers by biodegradation. *Water Supply*, 3, 41–7.

Yano, K. & Nishi, T. (1980). pKJ1, a naturally occurring conjugative plasmid coding for toluene degradation and resistance to streptomycin and sulfonamides. *Journal of Bacteriology*, 143, 552–60.

Yeh, W. K., Gibson, D. T. & Liu, Te-Ning (1977). Toluene dioxygenase: a multicomponent enzyme system. *Biochemical and Biophysical Research Communication*, 78, 401–10.

Yen, K. M. & Gunsalus, I. C. (1982). Plasmid gene organization: naphthalene/salicylate oxidation. *Proceedings of the National Academy of Sciences USA*, 79, 874–8.

You, I. S., Ghosal, D. & Gunsalus, I. C. (1988). Nucleotide sequence of plasmid NAH7 gene nahR and DNA binding of the nahR. *Journal of Bacteriology*, 170, 5409–15.

Zajic, E. C. & Mahomedy, A. Y. (1984). Biosurfactants: intermediates in the biosynthesis of amphipathic molecules in microbes. In *Petroleum Microbiology*, ed. R. M. Atlas, pp. 221–97. New York: Macmillan.

Zajic, E. C. & Seffens, W. (1984). Biosurfactants. *CRC Critical Reviews in Biotechnology*, 1, 87–107.

Zosim, Z., Gutnick, D. L. & Rosenberg, E. (1983). Uranium binding by emulsan and emulsanosols. *Biotechnology and Bioengineering*, 25, 1725–35.

5

Bioremediation of environments contaminated by polycyclic aromatic hydrocarbons

James G. Mueller, Carl E. Cerniglia and P. Hap Pritchard

5.1 Foreword

Bioremediation is described herein as the process whereby organic wastes are biologically degraded under controlled conditions to an innocuous state, or to levels below concentration limits established by regulatory authorities. In past reviews, we have focused on the basic biochemical and microbiological principles of polycyclic aromatic hydrocarbon (PAH) biodegradation (Cerniglia, 1984, 1992, 1993; Cerniglia et al., 1992; Cerniglia & Heitkamp, 1989), and the applicability of these processes toward the bioremediation of PAH-contaminated environments (Mueller et al., 1989a, 1993b; Pritchard, 1992). In this review, we have attempted to combine our perspectives on PAH biodegradation to yield a comprehensive overview of myriad factors impacting the efficacy of bioremediation of environments contaminated by PAHs. As such, this review helps unite technical insight and practical experience from several scientific disciplines including biochemistry, environmental chemistry, environmental microbiology, microbial ecology, and toxicology. Together, these factors interact to influence the scientifically valid, safe and effective use of bioremediation technologies for restoration of PAH-contaminated environments.

5.2 Introduction

Polycyclic aromatic hydrocarbons (PAHs) are chemicals made up of two or more fused benzene rings and/or pentacyclic moieties in linear, angular and cluster arrangements. The structures of several of these hydrocarbons are shown in Figure 5.1. Being derivatives of the benzene ring, they are thermodynamically stable due to their large negative resonance energy. In addition, PAHs often have low aqueous solubilities and tend to be associated with particle surfaces in the environment. For these reasons they are less likely to be affected by different fate processes, such as

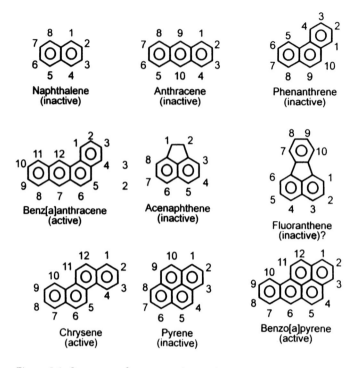

Figure 5.1. Structures of representative carcinogenic (active) and noncarcinogenic (inactive) polycyclic aromatic hydrocarbons.

volatilization, photolysis, and biodegradation. As a result, they are generally persistent under many natural conditions. This persistence, coupled with their potential carcinogenicity, makes PAHs problematic environmental contaminants.

The biodegradability of PAHs, particularly those of low molecular weight (LMW), has been recognized for some time (see Section 5.4). This has been shown through a variety of pure culture studies, as well as from many different observations of PAH disappearance in complex systems such as soils, sediments, and aquifer material when conditions favoring high microbial activity (adequate oxygen, nutrients, biomass) are present. As research has progressed, the considerable metabolic breadth of bacteria and natural microbial communities for degrading high-molecular-weight (HMW) PAHs, those defined herein as containing four or more fused rings, has also been revealed (see Section 5.4).

Axenic bacterial cultures that grow on HMW PAHs of four or more fused rings have been reported. Many studies also suggest that this breadth includes cometabolism or fortuitous metabolism. This is a process in which enzymes responsible for the initial oxidation of a particular PAH fortuitously oxidize other, higher-molecular-weight PAHs, even though these latter substrates may not ultimately be used as a source of carbon and energy for the cells. In addition, methods are continually being developed to improve the bioavailability of PAHs to the degrading microorganisms, thus increasing the rate at which the PAHs are degraded. This will be described in more detail in the following sections.

All this basic information has set the stage for bioremediation as an inexpensive and effective approach for cleaning contaminated soils, waters, and sediments. Considerable efforts are currently underway to develop the necessary field application techniques that make this a wholesale commercial reality. However, the complexities of the biodegradation processes involved make the efforts challenging if not, in some cases, frustrating. This is especially true for larger-scale field studies because increased variability and heterogeneity cause difficulties in data interpretation.

This chapter provides an overview of the current activities associated with PAH bioremediation, with emphasis on the problems, progress, future directions, and regulatory aspects of the technology. We argue that there is considerable potential for the technology but that it must handled with scientific dexterity, considerable care, imagination, and patience. With the rapid move toward bioremediation field applications for PAH-contaminated sites, attention must also be given to integrating basic research thoroughly into the applications and regulatory concerns so that experimental questions in the laboratory are driven by the problems encountered in the field as well as by scientific curiosity. Our chapter will emphasize this integration and provide an overview of how bioremediation technologies for PAHs are evolving and how they are successfully being used in the field.

5.3 Fate and effects of PAHs in the environment

5.3.1 Sources of PAH in the environment

Historically, PAHs have been released into the environment from three sources: biosynthetic (biogenic), geochemical, and anthropogenic (National Research Council, 1983). Anthropogenic sources are of two general types. One is the result of accidental spillage and intentional dumping of such materials as creosote, coal tar, and petroleum products. The other type is from the incomplete combustion of organic material such as from wood burning, municipal incineration, automobile emissions, and industrial discharges. It is these anthropogenic sources of PAHs that are the current focus of many environmental cleanup programs and consequently the basis for the development of effective bioremediation technologies. However, to appreciate further the fate of PAHs in nature, it is important as background information also to understand their origin in the environment from natural biogenic and geochemical sources.

Natural sources of PAHs in the environment

Hydrocarbons have been produced in the environment throughout geological time.

They are of diverse structure and are widely distributed in the biosphere, predominantly as components of surface waxes of leaves, plant oils, cuticles of insects, and the lipids of microorganisms (Millero & Sohn, 1991). In general, many of the alkanes and alkenes are of biogenic origin, being produced from a variety of terrestrial plants and aquatic algae (Millero & Sohn, 1991). In marine systems, good correlations between the presence of certain *n*-alkanes and branched alkanes (e.g., pristine) and blooms of primary productivity have been documented (Gordon *et al.*, 1978). Alkanes appear to be formed in these organisms from fatty acids.

Straight-chain alkanes with carbon number maxima in the C_{17} to C_{21} range are typical of aquatic algae, while terrestrial plant sources typically produce alkanes with C_{25} to C_{33} maxima (Millero & Sohn, 1991; Nishimura & Baker, 1986; Grimalt *et al.*, 1985). Studies in Antarctica, where pollution is minimal, have shown a remarkable consistency in the types of alkanes present and their concentrations, those of C_{15} to C_{30} accounting for most of the alkanes observed (Cripps, 1992). These alkanes are largely from algal biomass. Thus, alkanes would appear to be relatively common natural sources of carbon in the environment and certainly many microorganisms have evolved to use these hydrocarbons as growth substrates.

Petroleum and coal provide the largest source of mononuclear and polynuclear aromatic compounds. In addition, these aromatic chemicals are commonly found in the environment as a result of biosynthesis by plants. Included among these are aromatic carotenoids, lignin, alkaloids, terpenes, and flavinoids (Hopper, 1978). As most of these biosynthetic products are cycled through the biosphere, we can assume that many bacteria and fungi have developed the catabolic capabilities to derive carbon and energy from them. These capabilities will likely be oriented toward alkyl-benzenes, varying from single methyl substitutions to long side chains, many of which are olefinic in nature.

Surveys in terrestrial and marine environments show that PAHs can also occur from geochemical origin. They are formed wherever organic substances are exposed to high temperatures, a process called pyrolysis. In this process, the aromatic compounds formed are more stable than their precursors, usually alkylated benzene rings (Blumer, 1976). The alkyl groups can be of sufficient length to allow cyclization and then, with time, these cyclized moieties become aromatized. The temperature at which this process occurs determines the degree of alkyl substitution on the PAH; the higher the temperature, the more unsubstituted the resulting PAHs become.

The coking of coal (high temperature), for example, produces a material that is composed of a relatively simple mixture of unsubstituted hydrocarbons. With the slow burning of wood (lower temperatures) on the other hand, alkylated hydrocarbons survive, producing a more complex mixture of hydrocarbons. Crude oil is formed over millions of years at low temperatures (100 to 150 °C), and consequently the alkylated PAHs far exceed the unsubstituted PAHs. In this case, the time is adequate to accomplish the energetically favored aromatization but the temperature is not high enough to fragment the carbon–carbon bonds in the alkyl chains (Blumer, 1976). Clearly, microbial communities have been exposed to PAHs as carbon sources for eons, and this is probably reflected in the large number of genera that are capable of aromatic hydrocarbon degradation.

It is interesting to note that the arrangement of the aromatic rings affects their environmental stability and hence their natural distribution (Blumer, 1976). For example, linearly arranged benzene rings, such as in anthracene and tetracene, are the least stable and generally do not survive in nature unless sequestered into certain organic or inorganic matrices. Clusters of benzene are commonly found in pyrolysates, especially where the pyrolysis products are rapidly cooled (as in the generation of wood tar) or distributed; thus, they commonly appear in soils and sediments, as well as in smoke and automobile exhaust gases (Andersson *et al.*, 1983; Partridge *et al.*, 1990). They also persist in petroleum because it has never reached temperatures capable of eliminating or rearranging the less stable ring systems. The most stable arrangement is that of the annular types seen in phenanthrene, chrysene (Figure 5.1), and picene. Such PAHs abound where organics have been exposed to elevated temperatures (Blumer, 1976).

Formation of five-member rings on benzene ring clusters are also common in the pyrolysis of organic materials (e.g., acenaphthene, fluoranthene) (Figure 5.1). These rings create strained carbon–carbon bonds in PAH compounds, making the PAH inherently more unstable. In addition, the five-member rings cannot be converted to aromatic rings, thus preventing the further growth of the aromatic ring cluster. In petroleum, where the high temperatures that can readily break the strained carbon–carbon bond have not been experienced, PAHs with an accumulation of five-member rings around an aromatic nucleus are abundant.

Anthropogenic sources of PAH in the environment

Contemporary anthropogenic inputs of PAHs into the environment originate from two primary sources: (1) point sources such as spills or mismanaged industrial operations (waste treatment lagoons, leaking tanks and holding ponds, and sludge disposal), or (2) low-level inputs such as from atmospheric depositions. Large-scale spillage and disposal of oily wastes and sludges also contribute. Pollution by PAHs is also the result of processing, combustion and disposal of other fossil fuels. For example, PAHs are present in low concentrations in waste sludges originating from the treatment of refinery wastes. In refined petroleum products such as diesel fuel, the LMW PAHs dominate, but HMW PAHs can also be readily detected.

Modern day contamination of soils, sediments, and groundwater by PAHs has originated from four primary waste sources: creosote, coal tar, petroleum, and industrial effluents and gases. Creosote is a product mainly of coal tar, but also of wood tar, and is characterized as a brown–black oily liquid having a significant fraction of mixed phenolics (10%) and heterocyclics (5%), along with the PAHs (85%) (Mueller *et al.*, 1989a). It is a distillate produced by high-temperature carbonization of the tars. Although physical and chemical characteristics vary according to coal source, distillation temperature, and type of equipment used, recent standardization has led to the generation of relatively uniform creosote compositions (Mueller *et al.*, 1989a).

Coal-tar creosote has been widely used throughout the industrialized world for impregnating wood to prevent it from rot and worms. Traditionally, wood-preserving facilities were relatively small operations involving a plant for impregnating the wood under pressure, an area to allow the wood to drip dry, and usually holding lagoons for the creosote wastes. The U.S. EPA (1981) reported that there were 415 creosoting operations in the United States in 1981. Subsequent surveys report an estimated 550 (Micklewright, 1986) and 540 (Burton *et al.*, 1988) wood-preserving sites. Collectively, these plants consumed approximately 4.5×10^7 kg or 454 000 metric tons of creosote annually (Mattraw & Franks, 1986). We would estimate that between 5 and 10% of this volume ended up as residual material that was discharged in a manner that now represents a source of PAH contamination.

Coal gasification has been used for many years, especially in Europe, to generate a fuel source (town gas). The process involves thermal stripping of coal to produce a combustible gas, and yields a coal-tar residue. The residues were typically stored or disposed of at the coal gasification sites themselves. Hence, today many abandoned, defunct sites are responsible for extensive contamination of soils, sediment, and groundwater by PAHs. Concentrations of PAHs in soils at these sites are often very high, ranging into percent concentrations (greater than 10 000 ppm). In addition to HMW PAHs, other chemicals such as BTEXs, cyanide, phenol, and ammonia commonly co-contaminate these sites.

Atmospheric deposits originate primarily from high-temperature, incomplete combustion of fuels (petroleum and coal), typically from industrial activities and automobile exhaust. Some input also comes from the burning of wood and from incineration operations, but depending on the local economic base, it represents a considerably smaller input. For example, Freeman & Cattell (1990) demonstrated that airborne particulate matter resulting from the incomplete combustion of fossil fuels (e.g., coal) and other organic materials (e.g., wood, natural vegetation, paper products) contained significant amounts of PAHs. These data showed that benzo[a]pyrene concentrations ranged from 60 to 3000 μg/m^3 of air from the exhaust stack. Values for 11 other PAHs were also reported, ranging from 0.1 μg of coronene per gram of particulate matter to 240 μg of pyrene per gram of particulate matter.

Chemical analysis of the odorous fraction of diesel engine exhaust showed that these environmental discharges also contained significant amounts of PAHs, with values ranging from 3.5 to 25.7 μg of PAH per liter of exhaust (Partridge *et al.*, 1990). Similarly, Andersson *et al.* (1983) showed that $>424 \mu$g/m^3 particulate PAHs and $>1108 \mu$g/m^3 gaseous PAHs were collected after 2 h exposure to a coke oven atmosphere. Notably, the particulate PAH profile was heavy in HMW PAHs (4 or more fused rings), whereas gaseous emissions were primarily LMW PAHs. Somewhat lower values were reported after 2 hours of operations at an aluminum plant, and 60 minutes' exposure to creosote impregnation operations. Predictably, these extraneous sources of PAHs would significantly impact receptor soils, thus impacting remedial goals.

Atmospheric PAH depositions are usually from very dispersed sources but they cover significant amounts of land surface. PAH concentrations from these sources are typically quite low in soil and they are adsorbed strongly to soil particles. Consequently, there is minimal leaching into the soil below and the adsorbed PAHs tend to resist biodegradation, volatilization, and/or photolysis. If low concentrations of HMW PAHs, such as benzo[*a*]pyrene, need to be reduced below some established risk threshold (often determined in a site-specific manner by the regulatory officials), bioremediation of these low concentrations will likely be a desired alternative because of the scale and magnitude of the problem.

Another type of broad, non-point-source introduction of PAHs into soils is through land treatment procedures. For example, in the U.S. and Europe, sewage sludge is applied to agricultural land as fertilizer, and this has been shown to contain significant concentrations of PAHs (Wild *et al.*, 1990a,b). Presumably, soil particles contaminated from atmospheric depositions are present in surface soil runoff water, and since this water is routinely flushed through sewage treatment facilities, it collects in the sewage sludge. PAH concentrations in the sludge range from 1–10 mg/kg. Studies of the fate of these PAHs after sludge application in the field have shown that the PAHs of three or fewer fused rings were susceptible to biodegradation, especially when sludges were mixed with soils having a previous exposure to PAHs (Wild & Jones, 1993). However, HMW PAHs will apparently remain in the soil for several years (Wild *et al.*, 1991a,b).

5.3.2 Environmental distribution and attenuation of PAH contamination

Point source contamination of soils is exceedingly common in many industrially developed countries and is often the focus of significant environmental concern and regulatory complications. The contaminated areas are generally small (a few hectares) but the PAH concentrations are high and often associated with many other types of hydrocarbons, xenobiotic chemicals, and heavy metals. This produces an oily-phase-contaminated zone that can remain virtually unchanged chemically for years. Because of the high chemical concentrations, fate processes are only significant at the periphery of the contaminated zone and consequently have only a minor effect on the overall disposition of PAHs. However, attempts can be made to redistribute and dilute the contaminated soil (such as tilling, relocation, and land treatment processes) which often allows certain fate processes to have more of an effect.

Such redistribution of contaminated soil has been tried in a variety of contexts reviewed later in this chapter, but it is helpful to illustrate two potential results, as noted in the work of Wang *et al.* (1990) and Park *et al.* (1990). In the former case, extensive removal of PAHs from soil contaminated with a diesel fuel was observed in lysimeter test systems following fertilizer addition, liming and tilling. Even HMW PAHs were substantially removed by this process, making this study noteworthy because of the high extent of degradation. These results represent the type of effect that might come from a dilution/mixing or land treatment approach, but success will likely depend on the history of the contamination, the co-contaminants, and the nature of the soil.

In the latter case, 'interphase transfer' between solid and gas phases in soil appeared to be potentially significant only for two-ring PAH compounds in test systems that represented solid phase treatment of PAH-contaminated soils. Essentially no loss of HMW PAHs occurred. Any mixing or tilling of the soils will therefore likely enhance some volatilization, but by no means affect all of the PAHs.

If point sources of PAH contamination are left undisturbed, which has often been the case in the past, redistribution can occur, albeit slowly. Distribution of PAHs in surface soils is always very heterogenous, with some of the oily phase spread on soil particle surfaces and some existing as free product droplets lodged in the interstices of the soil. Their eventual distribution or movement will depend heavily on the local environmental conditions. But in general, vertical penetration of the oily phase can occur. Because concentrations remain relatively high during any penetration, further effects due to biodegradation, volatilization, and/or photolysis remain minimal. Extensive monitoring of the American Creosote Works in Pensacola, Florida by the U.S. Geological Survey illustrates this penetration potential (Godsy & Gorelitz, 1986).

Because PAHs are found mixed with other aliphatic hydrocarbons, phenolic compounds, and heterocyclic chemicals, the ready biodegradability of these co-contaminants, as well as the LMW PAHs, creates a biological oxygen demand, often severely depleting available oxygen concentrations. In general, PAHs especially the HMW PAHs, are not readily degraded under anaerobic conditions (Mihelcic & Luthy, 1988), thus leading to environmental persistence. In addition, the available nitrogen and phosphorus nutrients are quickly consumed by the indigenous microflora, further limiting biodegradation of the less readily biodegradable HMW PAHs.

A variety of factors affect the horizontal and vertical migration of PAHs, including contaminant volume and viscosity, temperature, land contour, plant cover, and soil composition (Morgan & Watkinson, 1989). Vertical movement occurs as a multiphase flow that will be controlled by soil chemistry and structure, pore size, and water content. For example, non-reactive small molecules (i.e., not PAHs) penetrate very rapidly through dry soils and migration is faster in clays than in loams due to the increased porosity of the clays. Once intercalated, however, sorbed PAHs are essentially immobilized. Mobility of oily hydrophobic substances can potentially be enhanced by the biosurfactant-production capability of bacteria (Zajic *et al.*, 1974) but clear demonstrations of this effect are rare. This is discussed below in more detail (see Section 5.5).

During the passage through dry soil, there is considerable potential for leaching with rain water and it is common to find PAH-contaminated vadose-zone soils depleted of the naphthalenes because of their relatively high water solubility (Bossert & Bartha, 1986). The leached material will ultimately appear in the groundwater, where oxygen and nutrient concentrations may again be plentiful enough to allow for biodegradation of these fairly degradable PAHs. Several studies have demonstrated the biodegradability of PAHs under these conditions (Aelion *et al.*, 1989; Madsen *et al.*, 1992).

Eventually, the PAH-containing material *per se* (e.g., creosote) will reach the water table, where leaching will become more pronounced. It is even possible that small droplets or aggregates resulting from emulsions can be created and transported away from source material. Humic substances in soil may enhance this mobility through emulsification and sorption to particle surfaces. This may account for the observation that significant concentrations of 'dissolved' PAHs often occur in water taken from certain groundwater wells. In addition, the hydrophobic components will tend to spread along the surface via light non-aqueous-phase liquid (LNAPL), or along the bottom of the groundwater table via dense-non-aqueous-phase liquid (DNAPL) to produce a pancake-type flow and lateral distribution of the pollutants. The rise and fall of the water table results in a smearing effect throughout the capillary fringe, with hydrocarbons being distributed vertically through the subsoil. Again, under these conditions PAHs are likely to experience low oxygen and nutrient concentrations, thereby limiting biodegradation.

PAH contamination in aquatic sediments has resulted primarily from atmospheric inputs, surface runoff, spills, and effluents from industrial areas (Bates *et al.*, 1984; LaFlamme & Hites, 1984). Of these, the atmospheric inputs are the most significant on a broad geographical scale (Pahm *et al.*, 1993). These inputs usually involve the PAHs associated with particulate materials of various kinds. As these materials settle through the water column, they are generally stripped of the LMW PAHs through dissolution and volatilization. Concentration of HMW PAHs in sediment ranges from $100 \, \mu g/g$ (such as in abyssal plains) to greater than $100 \, 000 \, \mu g/g$ in urban estuaries (LaFlamme & Hites, 1984; Shiaris & Jambard-Sweet, 1986).

As the PAHs become incorporated into the sediments, biodegradation is greatly reduced because of the general anoxic conditions of all but the upper few millimeters of the sediment environment. Turnover times of PAHs in estuarine and stream sediments (generally based on the mineralization of added radioactive compounds) were approximately 0.3 to 290, 8 to 60, and 50 to $> 20 \, 000$ days for naphthalene, phenanthrene and benzo[*a*]pyrene, respectively (Shiaris, 1989a). These values are based on laboratory assays in which sediment slurries were not extensively mixed and therefore the rates probably represent those that might be expected in actual sediment. However, the large variation in the rates attest to the complex set of environmental conditions (and test methodologies) that can affect PAH biodegradation in sediments.

In general, turnover rates tend to be faster in soils and sediments with long-term exposure to PAHs, but Shiaris (1989a,b) made the interesting observation that high rates of benzo[*a*]pyrene degradation in Boston Harbor sediments – with a significant history of PAH exposure – resulted in the production of considerable partial oxidation products, suggesting the possible involvement of co-oxidation processes. In areas of significant bioturbation, more oxygenated conditions may be present and may possibly yield faster turnover rates for PAHs (Bauer *et al.*, 1988; Bauer & Capone, 1988; Gardner *et al.*, 1979).

It is interesting to note that PAHs at low concentrations in soil may be susceptible to photolysis-enhanced biodegradation if exposure to light is possible (such as through soil reworking or tilling). Work by Miller *et al.* (1988) has shown that exposure of benzo[*a*]pyrene in soil to UV radiation produces partially oxidized products that are much more susceptible to biodegradation. This pretreatment effect has in fact been proposed as a possible bioremediation strategy. Developing bioremediation technologies that in fact will allow for efficient and economical cleanup of low initial concentrations of PAHs is an important research need, requiring creative new approaches.

5.4 PAH biodegradation processes

Given the myriad natural sources of PAHs in the environment, it follows that over eons many microorganisms would adapt toward the use and exploitation of these naturally occurring potential growth substrates. Accordingly, many bacterial, fungal and algal strains have been shown to degrade a wide variety of PAHs containing from two to five aromatic rings (Tables 5.1–5.3). The microbial degradation of PAHs has been studied and the biochemical pathways and mechanisms of enzymatic oxidation have been elucidated (Gibson *et al.*, 1968, 1990; Dagley, 1971, 1975; Gibson & Subramanian, 1984; Williams, 1981; Zylstra & Gibson, 1991). Comprehensive reviews have been published on the microbial metabolism of PAHs (Atlas, 1991; Cerniglia, 1984, 1992, 1993; Pothuluri & Cerniglia, 1994).

In general, HMW PAHs are slowly degraded by indigenous microorganisms and may persist in soils and sediments (Herbes & Schwall, 1978; Bauer & Capone, 1985, 1988; Cerniglia & Heitkamp, 1989; Shiaris, 1989a,b; Heitkamp & Cerniglia, 1987, 1989; Wild *et al.*, 1991a,b). As discussed below, the recalcitrance of these priority pollutants is due in part to the strong adsorption of higher molecular weight PAHs to soil organic matter and in part to low solubility, which decreases their bioavailability for microbial degradation (Manilal & Alexander, 1991; Aronstein *et al.*, 1991; Means *et al.*, 1980; Park *et al.*, 1990) (Figure 5.2).

There are fundamental differences in the mechanisms of PAH metabolism used by microorganisms (Figure 5.3). Bacteria and some green algae initiate the oxidation of PAHs by incorporating both atoms of molecular oxygen, catalyzed by a dioxygenase, into the aromatic ring to produce a *cis*-dihydrodiol, which is then dehydrogenated to give a catechol. Since PAHs such as phenanthrene, pyrene, benzo[*a*]pyrene, and benz[*a*]anthracene are complex fused ring structures, bacteria metabolize PAHs at multiple sites to form isomeric *cis*-dihydrodiols (Table 5.1). The aromatic ring dioxygenases are multi-component enzyme systems consisting of at least three proteins. One example is the naphthalene dioxygenase of *Pseudomonas putida*, which consists of a flavoprotein (ferredoxin reductase), a ferredoxin (ferredoxin$_{NAP}$) and a terminal oxidase (ISP$_{NAP}$) (Ensley *et al.*, 1982; Ensley & Gibson, 1983; Haigler & Gibson, 1990a,b).

Further metabolism of *cis*-dihydrodiols by bacteria involves NAD$^+$-dependent dehydrogenation reactions that produce catechols. The important step in the catabolism of PAHs involves ring-fission by dioxygenase enzymes, which cleave the aromatic ring to give aliphatic intermediates (Figure 5.3). Cleavage of these *ortho*-dihydroxylated aromatic compounds occurs between the two hydroxyl groups (intradiol or *ortho*-fission) or adjacent to one of the hydroxyl groups (extradiol or *meta*-fission) (Figure 5.3). Each of these enzymes is specific for one type of ring fission substrate; different enzymes lead to different aliphatic products (Figure 5.3). These intermediates funnel into central pathways of metabolism, where they are oxidized to provide cellular energy or used in biosynthesis of cell constituents.

In contrast to bacteria, non-ligninolytic fungi and prokaryotic algae (cyanobacteria) metabolize PAHs in pathways that are generally similar to those used by mammalian enzyme systems (Cerniglia *et al.*, 1992; Sutherland, 1992; Holland *et al.*, 1986). Many fungi oxidize PAHs via a cytochrome P-450 monooxygenase by incorporating one atom of the oxygen molecule into the PAH to form an arene oxide and the other atom into water. The monooxygenases, like the bacterial aromatic ring dioxygenases, are complex, multi-component systems, usually membrane-bound with broad substrate specificities. Most arene oxides are unstable and can undergo either enzymatic hydration via epoxide hydrolase to form *trans*-dihydrodiols or non-enzymatic rearrangement to form phenols, which can be conjugated with glucose, sulfate, xylose, or glucuronic acid (Figure 5.3). A diverse group of non-ligninolytic fungi have the ability to oxidize PAHs to *trans*-dihydrodiols, phenols, tetralones, quinones, dihydrodiol epoxides, and various conjugates of the hydroxylated intermediates, but only a few have the ability to degrade PAHs to CO_2 (Table 5.2). The best-studied non-ligninolytic fungus is the zygomycete *Cunninghamella elegans*, which metabolizes PAHs of from two to five aromatic rings. Like most fungi, *C. elegans* does not utilize PAHs as the sole source of carbon and energy but biotransforms or cometabolizes these recalcitrant compounds to detoxified products (Cerniglia *et al.*, 1985a,b). Although *C. elegans* metabolizes PAHs in a manner similar to mammalian systems, there are differences in the regio- and stereospecificities of the fungal and mammalian enzymes (Cerniglia *et al.*, 1983, 1990; Cerniglia & Yang, 1984; Sutherland *et al.*, 1993). For example, the absolute stereochemistries of the fungal dihydrodiols are in most cases *S, S*, which are opposite to the *R, R*-dihydrodiols formed in mammalian systems. The bacterial metabolic pathways have shown a greater tendency toward detoxification than the bioactivation pathways found more commonly in mammals.

The ligninolytic fungi that cause white-rots of wood have been screened for their PAH-degrading ability when grown under ligninolytic and non-ligninolytic culture conditions (Bumpus *et al.*, 1985; Haemmerli *et al.*, 1986; Gold *et al.*, 1989;

Table 5.1. *Representative polycyclic aromatic hydrocarbons metabolized by different species of bacteria*

Compound	Organisms	Metabolites	Reference
Naphthalene	*Acinetobacter calcoaceticus*, *Alcaligenes denitrificans*, *Mycobacterium* sp., *Pseudomonas* sp., *Pseudomonas putida*, *Pseudomonas fluorescens*, *Pseudomonas paucimobilis*, *Pseudomonas vesicularis*, *Pseudomonas cepacia*, *Pseudomonas testosteroni*, *Rhodococcus* sp., *Corynebacterium renale*, *Moraxella* sp., *Streptomyces* sp., *Bacillus cereus*.	Naphthalene *cis*-1,2-dihydrodiol, 1,2-dihydroxynaphthalene, 2-hydroxychromene-2-carboxylic acid, *trans*-o-hydroxybenzylidenepyruvic acid, salicylaldehyde, salicylic acid, catechol, gentisic acid, naphthalene *trans*-1,2-dihydrodiol.	Jerina *et al.* (1971), Catterall *et al.* (1971), Ryu *et al.* (1989), Weissenfels *et al.* (1990, 1991), Kelley *et al.* (1990), Dunn & Gunsalus (1973), Davies & Evans (1964), Foght & Westlake (1988), Jeffrey *et al.* (1975), Mueller *et al.* (1990a,b), Kuhm *et al.* (1991), Walter *et al.* (1991), Dua & Meera (1981), Tagger *et al.* (1990), García-Valdes *et al.* (1988), Trower *et al.* (1988), Grund *et al.* (1992), Cerniglia *et al.* (1984), Barnsley (1976a, 1976b, 1983), Eaton & Chapman (1992), Ensley *et al.* (1987), Ghosal *et al.* (1987).
Acenaphthene	*Beijerinckia* sp., *Pseudomonas putida*, *Pseudomonas fluorescens*, *Pseudomonas cepacia*, *Pseudomonas* sp.	1-Acenaphthenol, 1-acenaphthenone, acenaphthene-*cis*-1,2-dihydrodiol, 1,2-acenaphthenedione, 1,2-dihydroxyacenaphthylene, 7,8-diketonaphthyl-1-acetic acid, 1,8-naphthalenedicarboxylic acid, and 3-hydroxyphthalic acid.	Chapman (1979), Schocken & Gibson (1984), Ellis *et al.* (1991), Komatsu *et al.* (1993), Selifonov *et al.* (1993).
Anthracene	*Beijerinckia* sp., *Mycobacterium* sp., *Pseudomonas putida*, *Pseudomonas paucimobilis*, *Pseudomonas fluorescens*, *Pseudomonas cepacia*, *Rhodococcus* sp., *Flavobacterium* sp., *Arthrobacter* sp.	Anthracene *cis*-1,2-dihydrodiol, 1,2-dihydroxyanthracene, *cis*-4-(2-hydroxynaphth-3-yl)-2-oxobut-3-enoic acid, 2-hydroxy-3-naphthaldehyde, 2-hydroxy-3-naphthoic acid, 2,3-dihydroxynaphthalene, salicylic acid, catechol.	Colla *et al.* (1959), Akhtar *et al.* (1975), Jerina *et al.* (1976), Evans *et al.* (1965), Ellis *et al.* (1991), Weissenfels *et al.* (1990), Foght & Westlake (1988), Walter *et al.* (1991), Mueller *et al.* (1990b), Savino & Lollini (1977), Menn *et al.* (1993).

Phenanthrene	Aeromonas sp., Alcaligenes faecalis, Alcaligenes denitrificans, Arthrobacter polychromogenes, Beijerinckia sp., Micrococcus sp., Mycobacterium sp., Pseudomonas putida, Pseudomonas paucimobilis, Rhodococcus sp., Vibrio sp., Nocardia sp., Flavobacterium sp., Streptomyces sp., Streptomyces griseus, Acinetobacter sp.	Phenanthrene cis-1,2,3,4- and 9,10-dihydrodiols, 3,4-dihydroxyphenanthrene, cis-4-(1-hydroxynaphth-2-yl)-2-oxobut-3-enoic acid, 1-hydroxy-2-naphthaldehyde, 1-hydroxy-2-naphthoic acid, 1,2-dihydroxynaphthalene, 2-carboxybenzaldehyde, o-phthalic acid, protocatechuic acid.	Kiyohara et al. (1976, 1982, 1990), Weissenfels et al. (1990, 1991), Keuth & Rehm (1991), Jerina et al. (1976), Colla et al. (1959), West et al. (1984), Kiyohara & Nagao (1978), Heitkamp & Cerniglia (1988), Guerin & Jones (1988, 1989), Treccani et al. (1954), Evans et al. (1965), Foght & Westlake (1988), Mueller et al. (1990a,b), Sutherland et al. (1990), Ghosh & Mishra (1983), Savino & Lollini (1977), Trower et al. (1988). Barnsley (1983), Boldrin et al. (1993).
Fluoranthene	Alcaligenes denitrificans, Mycobacterium sp., Pseudomonas putida, Pseudomonas paucimobilis, Pseudomonas cepacia, Rhodococcus sp., Pseudomonas sp.	7-Acenaphthenone, 7-hydroxyacenaphthylene, 3-hydroxymethyl-4,5-benzocoumarin, 9-fluorenone-1-carboxylic acid, 8-hydroxy-7-methoxyfluoranthene, 9-hydroxyfluorene, 9-fluorenone, 1-acenaphthenone, 9-hydroxy-1-fluorenecarboxylic acid, phthalic acid, 2-carboxybenzaldehyde, benzoic acid, phenylacetic acid, adipic acid.	Kelley & Cerniglia (1991), Kelley et al. (1991, 1993), Walter et al. (1991), Weissenfels et al. (1990, 1991), Foght & Westlake (1990, 1991), Mueller et al. (1989a,b, 1990b), Barnsley (1975).
Pyrene	Alcaligenes denitrificans, Mycobacterium sp., Rhodococcus sp.	Pyrene cis- and trans-4,5-dihydrodiol, 4-hydroxyperinaphthenone, 4-phenanthroic acid, phthalic acid, 1,2- and 4,5-dihydroxypyrene, cis-2-hydroxy-3-(perinaphthenone-9-yl)propenic acid, 2-hydroxy-2-(phenanthrene-5-one-4-enyl)acetic acid, cinnamic acid.	Heitkamp et al. (1988a, 1988b), Heitkamp & Cerniglia (1988), Walter et al. (1991), Weissenfels et al. (1991), Grosser et al. (1991), Boldrin et al. (1993).
Chrysene	Rhodococcus sp.	None determined.	Walter et al. (1991).
Benz[a]anthracene	Alcaligenes denitrificans, Beijerinckia sp., Pseudomonas putida.	Benz[a]anthracene cis-1,2, cis-8,9-, and cis-10,11-dihydrodiols, 1-hydroxy-2-anthranoic acid, 2-hydroxy-3-phenanthroic acid, 3-hydroxy-2-phenanthroic acid.	Gibson et al. (1975), Mahaffey et al. (1988), Weissenfels et al. (1991), Jerina et al. (1984).
Benzo[a]pyrene	Beijerinckia sp., Mycobacterium sp.	Benzo[a]pyrene cis-7,8- and cis-9,10-dihydrodiols.	Gibson et al. (1975), Heitkamp & Cerniglia (1988), Grosser et al. (1991).

Table 5.2. *Representative polycyclic aromatic hydrocarbons metabolized by different species of fungi*

Compound	Organisms	Metabolites	Reference
Naphthalene	*Absidia glauca, Aspergillus niger, Basidiobolus ranarum, Candida utilis, Choanephora campincta, Circinella* sp.*, Claviceps paspali, Cokeromyces poitrasii, Conidiobolus gonimodes, Cunninghamella bainieri, Cunninghamella elegans, Cunninghamella japonica, Emericellopsis* sp.*, Epicoccum nigrum, Gilbertella persicaria, Gliocladium* sp.*, Helicostylum piriforme, Hypochytrium catenoides, Linderina pennispora, Mucor hiemalis, Neurospora crassa, Panaeolus cambodginensis, Panaeolus subbalteatus, Penicillium chrysogenum, Pestalotia* sp.*, Phlyctochytrium reinboldtae, Phycomyces blakesleanus, Phytophthora cinnamomi, Psilocybe cubensis, Psilocybe strictipes, Psilocybe stuntzii, Psilocybe subaeruginascens, Rhizophlyctis harderi, Rhizophlyctis rosea, Rhizopus oryzae, Rhizopus stolonifer, Saccharomyces cerevisiae, Saprolegnia parasitica, Smittium culicis, Smittium culisetae, Smittium simulii, Sordaria fimicola, Syncephalastrum racemosum, Thamnidium anomalum, Zygorhynchus moelleri.*	1-Naphthol, 2-naphthol, naphthalene *trans*-1,2-dihydrodiol, 4-hydroxy-1-tetralone, 1,2-naphthoquinone, 1,4-naphthoquinone, sulfate and glucuronide conjugates.	Ferris *et al.* (1973), Smith & Rosazza (1974), Cerniglia & Gibson (1977, 1978), Cerniglia *et al.* (1978, 1983, 1985a,b), Cerniglia & Crow (1981), Hofmann (1986).
Acenaphthene	*Cunninghamella elegans.*	1-Acenaphthenone, 1,2-acenaphthenedione, *cis*-1,2-dihydroxyacenaphthene, *trans*-1,2-dihydroxyacenaphthene, 1,5-dihydroxyacenaphthene, 1-acenaphthenol, 6-hydroxyacenaphthenone.	Pothuluri *et al.* (1992a).

Anthracene	*Bjerkandera* sp., *Cunninghamella elegans*, *Phanerochaete chrysosporium*, *Ramaria* sp., *Rhizoctonia solani*, *Trametes versicolor*.	Anthracene *trans*-1,2-dihydrodiol, 1-anthol, anthraquinone, phthalate, glucuronide, sulfate and xyloside conjugates.	Cerniglia (1982), Cerniglia & Yang (1984), Field *et al.* (1992), Hammel *et al.* (1991), Sutherland *et al.* (1992).
Phenanthrene	*Cunninghamella elegans*, *Phanerochaete chrysosporium*, *Trametes versicolor*.	Phenanthrene *trans*-1,2-dihydrodiol, phenanthrene *trans*-3,4-dihydrodiol, phenanthrene *trans*-9,10-dihydrodiol, glucoside conjugate of 1-phenanthrol, 3-,4-, and 9-hydroxyphenanthrene, 2,2′-diphenic acid.	Cerniglia & Yang (1984), Cerniglia *et al.* (1989), Morgan *et al.* (1991), Sutherland *et al.* (1991), Bumpus (1989), Hammel *et al.* (1992), Moen & Hammel (1994), Tatarko & Bumpus (1993), George & Neufeld (1989).
Fluoranthene	*Cunninghamella elegans*.	Fluoranthene *trans*-2,3-dihydrodiol, 8- & 9-hydroxyfluoranthene *trans*-2,3-dihydrodiols, glucoside conjugates	Pothuluri *et al.* (1990, 1992b).
Pyrene	*Cunninghamella elegans*, *Phanerochaete chrysosporium*, *Crinipellis stipitaria*.	1-Hydroxypyrene,1,6- and 1,8-dihydroxypyrene, pyrene *trans*-4,5-dihydrodiol, 1,6-pyrenequinone, 1,8-pyrenequinone, glucoside conjugates.	Cerniglia *et al.* (1986), Hammel *et al.* (1986), Lambert *et al.* (1994), Lange *et al.* (1994).
Benz[*a*]anthracene	*Cunninghamella elegans*.	Benz[*a*]anthracene *trans*-3,4-dihydrodiol, benz[*a*]anthracene *trans*-8,9-dihydrodiol, benz[*a*]anthracene *trans*-10,11-dihydrodiol, phenolic and tetrahydroxy derivatives of benz[*a*]anthracene, glucuronide and sulfate conjugates.	Cerniglia *et al.* (1980d, 1994)
Benzo[*a*]pyrene	*Aspergillus ochraceus*, *Bjerkandera adusta*, *Bjerkandera* sp., *Candida maltosa*, *Candida tropicalis*, *Cbrysosporium pannorum*, *Cunninghamella elegans*, *Mortierella verrucosa*, *Neurospora crassa*, *Penicillium* sp., *Phanerochaete chrysosporium*, *Ramaria* sp., *Saccharomyces cerevisiae*, *Trametes versicolor*, *Trichoderma viride*.	Benzo[*a*]pyrene *trans*-4,5-dihydrodiol, benzo[*a*]pyrene *trans*-7,8-dihydrodiol, benzo[*a*]pyrene *trans*-9,10-dihydrodiol, benzo[*a*]pyrene-1,6-quinone, benzo[*a*]pyrene-3,6-quinone, benzo[*a*]pyrene-6,12-quinone, 3-hydroxybenzo[*a*]pyrene, 9-hydroxybenzo[*a*]pyrene, 7β, 8α, 9α, 10β-tetrahydrobenzo[*a*]pyrene, 7β, 8α, 9β, 10α-tetrahydroxy-7,8,9,10-tetrahydrobenzo[*a*]pyrene, benzo[*a*]pyrene 7,8-dihydrodiol-9,10-epoxide, glucuronide and sulfate conjugates.	Cerniglia *et al.* (1980c), Cerniglia & Crow (1981), Cerniglia & Gibson (1979, 1980a, 1980b), Haemmerli *et al.* (1986), Ghosh *et al.* (1983), Lin & Kapoor (1979), Bumpus *et al.* (1985), Wiseman & Woods (1979), Field *et al.* (1992), Sanglard *et al.* (1986).

Table 5.3. *Representative polycyclic aromatic hydrocarbons metabolized by different species of cyanobacteria and algae*

Compound	Organisms	Metabolites	Reference
Naphthalene	*Oscillatoria* sp. (strain JCM), *Oscillatoria* sp. (strain MEV), *Microcoleus chthonoplastes, Nostoc* sp., *Anabaena* sp. (strain CA), *Anabaena* sp. (strain 1F), *Agmenellum quadruplicatum, Coccochloris elabens, Aphanocapsa* sp., *Chlorella sorokiniana, Chlorella autotrophica, Dunaliella tertiolecta, Chlamydomonas angulosa, Ulva fasciata, Cylindrotheca* sp., *Amphora* sp., *Nitzschia* sp., *Synedra* sp., *Navicula* sp., *Porphyridium cruentum.*	1-Naphthol, 4-hydroxy-1-tetralone, naphthalene *cis*-1,2-dihydrodiol.	Cerniglia *et al.* (1979, 1980a, 1982), Narro *et al.* (1992a).
Phenanthrene	*Oscillatoria* sp. strain JCM, *Agmenellum quadruplicatum.*	Phenanthrene *trans*-9,10-dihydrodiol, 1-methoxyphenanthrene.	Narro *et al.* (1992b).
Benzo[*a*]pyrene	*Selenastrum capricornutum.*	Benzo[*a*]pyrene *cis*-4,5-dihydrodiol, benzo[*a*]pyrene *cis*-7,8-dihydrodiol, benzo[*a*]pyrene *cis*-9,10-dihydrodiol, benzo[*a*]pyrene *cis*-11,12-dihydrodiol.	Cody *et al.* (1984), Lindquist & Warshawsky (1985a,b), Warshawsky *et al.* (1988, 1990).

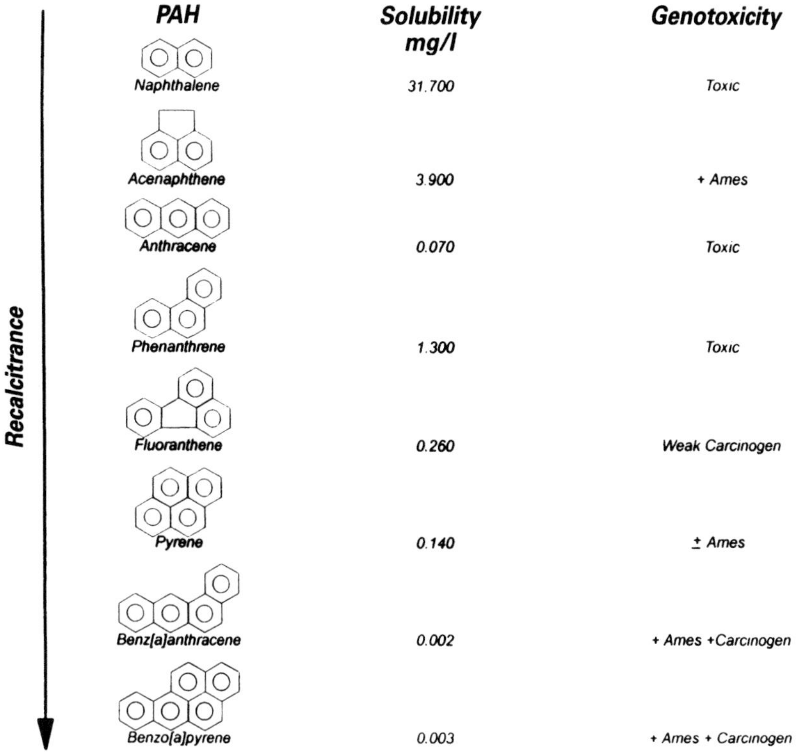

Figure 5.2. Genotoxic activities, physical properties and recalcitrance of polycyclic aromatic hydrocarbons.

Hammel, 1989; Aust, 1990; Morgan *et al.*, 1991, 1993; Dhawale *et al.*, 1992; Sack & Günther, 1993). The best studied strain, *Phanerochaete chrysosporium*, has been used as a model organism to study the potential of white-rot fungi to degrade PAHs (Table 5.2). The initial attack on PAHs by *P. chrysosporium* under culture conditions which favor ligninolysis is catalyzed by extracellular enzymes known as lignin peroxidase and manganese peroxidase, as well as by other H_2O_2-producing peroxidases. Ligninolytic enzymes have been shown *in vitro* to oxidize PAHs with ionization potentials less than 7.35 to 7.55 ev. Lignin peroxidases in the presence of H_2O_2 can remove one electron from a PAH molecule, such as anthracene, pyrene, and benzo[*a*]pyrene, thus creating an aryl cation radical, which undergoes further oxidation to form a quinone (Figure 5.4). Since the earlier studies with *P. chrysosporium*, other white-rot fungi, such as *Trametes versicolor* and *Bjerkandera* sp., have been evaluated and show promise for their ability to degrade PAHs to CO_2 more rapidly than *P. chrysosporium* (Field *et al.*, 1992).

Interestingly, phenanthrene, which is not a substrate for lignin peroxidase, is metabolized by *Phanerochaete chrysosporium* at the C-9 and C-10 positions to give

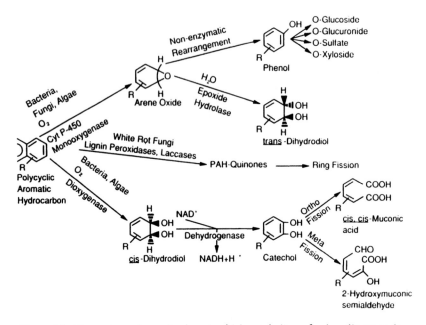

Figure 5.3. The major pathways in the microbial metabolism of polycyclic aromatic hydrocarbons.

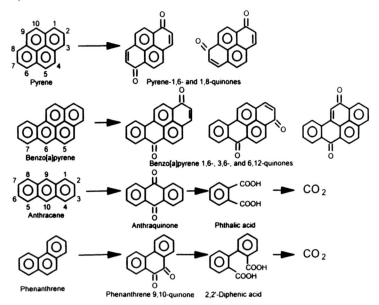

Figure 5.4. Metabolic pathways of polycyclic aromatic hydrocarbon oxidation by white-rot fungi.

2,2'-diphenic acid as a ring cleavage product (Hammel *et al.*, 1992). Further studies have shown that 2,2'-diphenic acid is formed from phenanthrene via an extracellular manganese-peroxidase-mediated lipid peroxidation (Moen & Hammel, 1994). Other enzymes, such as cytochrome P-450 monooxygenase and epoxide hydrolase, appear to be involved in phenanthrene degradation when *P. chrysosporium* is grown under nonligninolytic conditions, since intermediates such as optically active *trans*-dihydrodiols and phenols are isolated from the culture broths (Sutherland *et al.*, 1991).

Cyanobacteria (blue-green algae) and eukaryotic algae oxidize PAHs under photoautotrophic conditions to form hydroxylated intermediates (Table 5.3). Cyanobacteria oxidize naphthalene and phenanthrene to metabolites that are similar to those formed by mammals and fungi (Narro *et al.*, 1992a,b). In contrast, the green alga *Selenastrum capricornutum* oxidizes benzo[*a*]pyrene to isomeric *cis*-dihydrodiols similar to bacterial metabolites (Warshawsky *et al.*, 1988, 1990). Although the level of metabolism is extremely low, both prokaryotic and eukaryotic algae could be important in the overall degradation of PAHs, since they occur in high abundance and are widely distributed in aquatic ecosystems.

All of these studies indicate that the microbial world has developed a wide variety of mechanisms to biodegrade PAHs. However, the specific, fundamental differences in catabolic machinery between microorganisms can be factors to consider in the bioremediation of PAH-contaminated soils and wastewater systems when the objective is detoxification in addition to removal of the parent compound (Middaugh *et al.*, 1991, 1993, 1994). For example, it is clear that further studies are necessary to characterize the enzymes and mechanisms involved in PAH mineralization under ligninolytic and non-ligninolytic conditions when using white-rot fungi.

5.5 Technology-based bioremediation solutions

5.5.1 Factors affecting PAH bioremediation efficacy

The breadth of catabolic activity suggests that its managed and controlled use could affect bioremediation of contaminated environments. But a wide range of physical, chemical, and microbiological factors interact to influence the biodegradation of organic compounds in the environment (Cerniglia, 1992, 1993). Many of these are reviewed extensively in the chapter by Morra (Chapter 2), with special emphasis on the soil environment. These same principles apply to the bioremediation of PAH in water and in the subsurface environment. Below we have identified and described some of the more pertinent factors thought to influence the efficacy of PAH bioremediation efforts.

Bioavailability

Bioavailability may be defined as the complex, interactive effect of myriad physicochemical and microbiological factors on the rate and extent of biodegradation.

As reviewed by Alexander (1985) and Morris & Pritchard (1994), many studies suggest that bioavailability represents one of the most important factors influencing PAH biodegradability in the environment. This is because PAHs, especially the HMW, hydrophobic compounds, possess intrinsically low water solubilities and therefore tend to partition onto soil mineral surfaces and to sorb tightly to available organic materials (Hites *et al.*, 1977; Means *et al.*, 1980; Weber *et al.*, 1993). When PAHs are adsorbed in this way they become physically unavailable to resident microorganisms and are therefore protected from microbial catalysis.

From our perspective, and that of many others, the rate and extent of PAH biodegradation is directly proportional to bioavailability, which will lead to successful bioremediation. Accordingly, much effort has gone towards understanding the relationship between physical adsorption and biodegradability (i.e., bioavailability) (Scow & Alexander, 1992; Scow & Hutson, 1992; Rao *et al.*, 1993; Scow, 1993). Physical sorption of PAHs is one of the primary factors that affect chemical persistence and thus greatly impacts the effectiveness of bioremediation efforts for soil (Jafvert & Weber, 1992), sediments (Coates & Elzerman, 1986; Shimp & Young, 1988), and aquifers (Nau-Ritter *et al.*, 1982; Rostad *et al.*, 1985; Rao *et al.*, 1993). Physical factors that impact bioavailability, such as soil type and texture and organic matter content, are therefore important criteria to assess in the evaluation of any bioremediation strategy.

Sorption/desorption kinetics

Because bioavailability of PAHs is a key factor governing the effectiveness of bioremediation efforts, it is important to understand the forms that nonbiologically available PAHs take in the environment. In experiments with 1,2-dibromomethane, Steinberg *et al.* (1987) showed that entrapment of organic chemicals in the intraparticle micropores of the soil environment was a source of residual materials. Similar studies with PAHs and related chemicals (i.e., PCBs) have suggested that 'irreversible' sorption may occur (Karickhoff & Morris, 1985; Girvin *et al.*, 1990). This is thought to be especially relevant to soils subjected to long-term exposures. Recent studies, however, suggest that PAHs in general are exempt from the phenomenon of 'bulk-sorption' (i.e., sorption by dissolution or partitioning of the sorbed liquids into the macromolecular solid phase organic material naturally present in all soil and aquifers) because size exclusion appears to limit this process to relatively small, polar molecules (Lyon & Rhodes, 1991). Of course, soil type and texture again greatly impact this phenomenon.

While PAHs deeply intercalated into soil matrixes are not thought generally to represent a significant problem for PAH bioremediation applications, residual organic contaminant phases have been described as high-affinity sorptive sinks for these compounds in soil and groundwater (Boyd & Sun, 1990; Sun & Boyd, 1991). Thus, non-aqueous-phase liquids (NAPL) often represent a source of contaminants, with the release of attenuated PAHs being governed by desorption kinetics, which may be very slow. It is therefore important to consider the presence of dense

(DNAPL) and/or light (LNAPL) NAPL during the design of any bioremediation technologies, but especially with *in situ* strategies (MacDonald & Kavanaugh, 1994; U.S. EPA, 1990a, 1992c).

Surface active agents

To increase the usefulness of bioremediation as an effective field remedial tool, significant investments have been made towards the development of means to remove sorbed PAHs, attack sources of NAPL, and subsequently increase the aqueous solubility/bioavailability, and thus the biodegradability, of targeted compounds. To date, one of the most effective ways to accomplish these tasks involves the use of surface active agents (i.e., surfactants). A variety of synthetic surfactants have been shown effective in increasing the bioavailability of PAHs and other hydrophobic contaminants (Kile & Chiou, 1989, 1990; Edwards *et al.*, 1991; Liu *et al.*, 1991). Although the solubilization process is not completely understood, these studies showed that a variety of ionic and nonionic surfactants could significantly increase the water solubility of monitored chemicals.

Synthetic surfactants can be effectively employed to increase bioavailability, but their use in the field must be carefully considered. For example, studies by Laha & Luthy (1991) showed that nonionic synthetic surfactants may actually inhibit the biodegradation process when used at concentrations exceeding the critical micelle concentration. This inhibition was presumably due to extensive micellation, which resulted in a decrease in the equilibrium of free-phase (bioavailable) PAH.

In our studies, the use of microorganisms specially adapted to the presence of high surfactant concentrations showed that PAH biodegradation kinetics can be significantly enhanced in the presence of surfactant concentrations as high as 2% (Pritchard *et al.*, 1994a,b). From the perspective of practical application, the use of low concentrations of surfactants may be more cost effective, depending on the application strategy. But it also appears that the PAH biodegradation process may be enhanced even in the absence of surfactant-enhanced desorption (Aronstein *et al.*, 1991). Here, a low concentration of surfactants enhanced PAH mineralization, although surfactant-induced desorption was not observed. So, surfactants do not need to be added at very high concentrations to enhance biodegradation.

A number of studies have shown that in addition to synthetic surfactants, microbial action itself can lead to enhanced desorption of bound substrate and hence increased rates of biodegradation (Wszolek & Alexander, 1979; Finnerty & Singer, 1984; Ogram *et al.*, 1985). These and other findings led to a large, continuing area of work that describes the biological production of microbial surfactants (i.e., biosurfactants) to enhance the desorption, solubilization, and biodegradation of hydrophobic chemicals (Lang & Wagner, 1987; Rosenberg, 1986; Zhang & Miller, 1992). Biosurfactants have several potential advantages over their synthetic counterparts, including: (1) more rapid PAH biodegradation, (2) a wider range of environmental resilience, and (3) the possibility that they may be naturally produced by resident microflora. It thus appears that stimulation of indigenous PAH-degrading, surfactant-producing microbes represents an attractive bioremediation scenario, especially for *in situ* applications.

It logically followed from these findings that remedial systems designed around the use of surface active agents have emerged. Such systems include *in situ* soil flushing (Vigon & Rubin, 1989; West & Harwell, 1992), direct soil application of surfactants (Graves & Leavitt, 1991), inoculation with biosurfactant-producing microorganisms (Pritchard, 1992), and soil washing with the aid of surfactants (McDermott *et al.*, 1989). However, most of the applications in the remediation industry have met with limited success. For example, Graves & Leavitt (1991) concluded that '. . . application of surfactants directly to soil for purposes of solubilizing hydrophobic compounds was not successful in achieving greater levels of petroleum removal'. Moreover, the results of an extensive field study of the use of surfactants to facilitate *in situ* soil washing concluded that the process 'did not measurably work' (U.S. EPA, 1987) despite favorable results from preliminary laboratory studies (U.S. EPA, 1985).

One of the major focuses of the *in situ* use of surfactants is to accelerate the removal or degradation of free-phase products (DNAPL or oil globules) as variations of technologies used in the oil industry for enhanced tertiary oil recovery as described by Hill *et al.* (1973). Such *in situ* remediation efforts often fail, however, primarily because of differences in the goals and expectations of the applications. For example, the enhanced oil recovery industry is often satisfied with >30% enhanced removal whereas the remediation industry often strives for >99% removal in order to meet remedial guidelines. Unfortunately, removal of about 50% of the free product (i.e., DNAPL) at a PAH-contaminated site appears to represent the full extent of practical field expectations.

The most successful uses of surfactants to accelerate desorption and enhanced biodegradation has clearly been associated with the soil-washing industry (U.S. EPA, 1989) and bioslurry reactors (Oberbremer *et al.*, 1990; Mueller *et al.*, 1991b,c). Here too, however, problems have been encountered at the bench-, pilot-, and full-scale levels with inefficient removal of materials from soils and sediments containing appreciable amounts of clay (*c.* >25%) or organic matter (*c.* >10%).

In general, surfactant-based technologies being developed in the remediation industry to enhance the efficacy of conventional technologies, such as pump-and-treat, have not produced the desired responses, and field data to date do not provide the anticipated successes (Vigon & Rubin, 1989). Of course the technology *per se* continues to be refined with the promise of enhancing overall performance (MacDonald & Kavanaugh, 1994). But it seems that the expectations of the *in situ* applications in particular need to be revised prior to successful full-scale implementation.

146

Table 5.4. *Seven potentially carcinogenic PAHs (pcPAHs) and their benzo{a}pyrene equivalence*

Chemical number	Chemical name	Number of fused rings	Benzo[a]pyrene equivalence[a]
1	Benz[a]anthracene	4	0.1
2	Chrysene	4	0.001
3	Benzo[a]pyrene	5	1.0
4	Benzo[b]fluoranthene	5	0.1
5	Benzo[k]fluoranthene	5	0.01
6	Dibenz[a,b]anthracene	5	1.0
7	Indeno(1,2,3-c,d)pyrene	6	0.1

[a]Benzo[a]pyrene equivalence according to U.S. EPA (1993b).

Regulatory factors

Risk-based remedial goals

Of the many regulatory factors influencing the full-scale implementation of PAH bioremediation technologies, variability in the desired remedial goal (i.e., cleanup criteria) represents one of today's greatest challenges. In many cases, environmental cleanup criteria consider the cumulative concentration of several potentially carcinogenic PAHs (pcPAHs) (Table 5.4). Even with this rather defined unit of measurement, remedial goals are variable in the United States; for example, soils containing pcPAHs originating from wood-preserving operations (K001 or K035 wastes as defined by 40 CFR 261.7) range from 0.19 mg pcPAHs/kg soil for an industrial area in northern California to 700 mg pcPAH/kg soil for a similar site in Texas. In general, other cleanup criteria lie within these ranges (U.S. EPA, 1992d).

This much variability in remedial goals for PAHs makes it difficult to comparatively evaluate bioremediation efforts and to ensure consistency of bioremediation performance in the field. To help reduce this variability it has been recently suggested to regulate the remediation of PAH-contaminated materials according to risk-based cleanup criteria that consider the benzo[a]pyrene equivalence (U.S. EPA, 1993b) of the contaminants (Table 5.4). However, current values are offered only as interim guidance for quantitative risk assessment of PAHs. Obviously, many factors are considered when establishing risk-based cleanup standards. With such complexity, remedial goals are viewed herein as a separate topic and will not be considered further.

European (Dutch) and Canadian standards

As compared with the existing procedure in the United States, both Europe and Canada have a more uniform set of criteria established as the 'Dutch' and MENVIQ standards, respectively. The Dutch A, B, and C standards for select PAHs in soil and groundwater are summarized in Tables 5.5 and 5.6, respectively: the Canadian

Table 5.5. *List of European (Dutch), Canadian, and United States (proposed) cleanup standards for select PAHs in soil*

| Chemical | European (Dutch) standards (mg/kg soil dry weight) | | | United States standards (mg/kg soil dry weight) | |
	Dutch A	Dutch B	Dutch C	LDR	Proposed UTS
Naphthalene	0.1	5	50	3.1 (1.5)	
Anthracene	0.1	10	100	4.0	3.4
Phenanthrene	0.1	10	100	1.5	5.6
Fluoranthene	0.1	10	100	8.2	3.4
Pyrene	ND	ND	ND	8.2 (1.5)	8.2
Chrysene	0.1	5	50	8.2	3.4
Benz[*a*]anthracene	0.1	5	50	8.2	3.4
Benzo[*a*]pyrene	0.05	1	10	8.2	3.4
Benzo[*b*]fluoranthene	ND	ND	ND	3.4	6.8
Benzo[*k*]fluoranthene	0.1	5	50	3.4	6.8
Benzo[*g,h,i*]perylene	0.1	10	100	1.5	1.8
Dibenz[*a,h*]anthracene	ND	ND	ND	8.2	8.2
Indeno[1,2,3-*c,d*]pyrene	0.1	5	50	8.2	3.4
Total PAHs	10	500	1000	—	—
Total pcPAHs	1	20	200	—	—

LDR: Based on constituent concentrations in wastes (40 CFR Section 268.43) for waste codes
F039 and K001 (brackets) where appropriate.
ND – no data.

standards are essentially the same. The Dutch A, B and C values are indicative of natural background levels (Dutch A), threshold levels warranting further investigation (Dutch B) and 'action' levels indicating a need for remedial action (Dutch C). These same categories may be used to establish cleanup criteria which better facilitate the more effective use of bioremediation technologies. For example, bioremediation would be seldom employed to achieve Dutch A standards for PAH-impacted soils (Leidraod bodem sanering, afl. 4, November 1988).

Treatment standards in the United States
Currently, there is much variability in the United States in establishing treatment standards for PAHs in soil and groundwater. For example, Land Disposal Restrictions (LDR) govern the placement of materials destined for any land disposal including landfill, surface impoundments, waste pits, injection wells, land treatment facilities, salt domes, underground mines or caves, and vaults or bunkers. Accordingly, treatment standards for all listed and characteristic hazardous wastes destined for land disposal have been defined (U.S. EPA, 1991). These values thus represent one potential set of treatment standards for PAHs. However, for PAHs,

Table 5.6. *List of European (Dutch), Canadian, and United States (proposed) cleanup standards for select PAHs in groundwater*

Chemical	European (Dutch) standard (mg/l)			United States standards (mg/l)	
	Dutch A	Dutch B	Dutch C	LDR	Proposed UTS
Naphthalene	0.1	5	50	0.059 (0.031)	0.059
Anthracene	0.1	10	100	0.059	0.059
Phenanthrene	0.1	10	100	0.059 (0.031)	0.059
Fluoranthene	0.1	10	100	0.068	0.068
Chrysene	0.1	5	50	0.059	0.059
Pyrene	ND	ND	ND	0.067 (0.028)	0.067
Benz[*a*]anthracene	0.1	5	50	0.059	0.059
Benzo[*a*]pyrene	0.05	1	10	0.061	0.061
Benzo[*b*]fluoranthene	ND	ND	ND	0.055	0.11
Benzo[*k*]fluoranthene	0.1	5	50	0.059	0.11
Benzo[*g,h,i*]perylene	0.1	10	100	0.0055	0.0055
Dibenz[*a,b*]anthracene	ND	ND	ND	0.055	0.055
Indeno[1,2,3-*c,d*]pyrene	0.1	5	50	0.0055	0.0055
Total PAHs	10	500	1000	—	—
Total pcPAHs	1	20	200	—	—

LDR: Based on constituent concentrations in wastes (40 CFR Section 268.43) for waste codes F039 and K001 (brackets) where appropriate.
ND – no data.

they vary depending on the identified source of the material (F039 wastes versus K001 wastes) as shown in Tables 5.5 and 5.6. The use of Alternative or Relevant and Appropriate Requirements (ARARs) and the establishment of Alternate Concentration Limits (ACLs) add even more variability, often leading to confusion in remedial objectives.

Correspondingly, remediation decisions fall by default to the most stringent cleanup criteria in order to meet the mandate for treatment to the maximum extent practical. In practice this means that the more conventional, and usually the more costly, remedial approaches are employed to meet the mandate for permanence of hazardous waste treatment. In turn, more innovative remedial technologies such as bioremediation are not employed despite their potential advantages. Ultimately, this results in increased remediation costs.

Fortunately, efforts are underway in the United States to equilibrate the chaos in the form of establishing Universal Treatment Standards (UTS) for PAHs and other hazardous wastes in non-wastewater (soil, sediments) and wastewater (groundwater) (Federal Register, 1993). The proposed goals for select PAHs are summarized in Tables 5.5 and 5.6. In the proposed UTS scenario, the U.S. EPA offers three options for establishing alternative PAH cleanup criteria for soil or water: (1) a reduction of PAH to $< 10 \times$ the UTS value, (2) a $> 90\%$ reduction in the initial PAH concentration, or (3) a reduction of PAH to $< 10 \times$ the UTS and $> 90\%$ reduction in the initial PAH concentration. The overall impact of such legislation remains to be determined but it would seem to facilitate better the use of bioremediation as a treatment technology.

Background PAH concentrations

Successful bioremediation of PAH-contaminated sites is being more frequently defined as achieving a reduction of PAHs to levels governed by the background concentrations (for example, < 2 standard deviations above background levels). Evidence for this reduction is usually offered by analytical chemistry data generated with replicate soil or water samples extracted and analyzed according to standard procedures such as GC-MS (U.S. EPA Method SW-846 8270) or HPLC (U.S. EPA Method SW-846 8310).

Remedial goals governed by background PAH concentrations offer the advantages of a site-specific cleanup criteria while accounting for natural variability and error associated with the acquisition and analysis of field samples. Such cleanup criteria inherently seem more reasonable because a variety of non-point sources of PAHs in the environment have been scientifically identified (National Research Council, 1983).

For example, Jones *et al.* (1989) conducted long-term studies (> 100 years) to discern the amount of PAH deposition in semi-rural soils. Using an agricultural soil geographically removed from major industries, it was shown that PAH concentrations increased fourfold since the late 1800s (1880s), with some high-molecular-weight PAHs showing substantially greater increases. The authors provided supporting documentation to suggest that regional fallout of anthropogenically generated PAHs derived from the combustion of fossil fuels was the principal source of deposition. Predictably, sources more proximal to these sources are even more extensively impacted. Given the myriad extraneous sources of PAHs, the background concentrations of these compounds is influenced greatly by regional activities. This, in turn, will impact the remedial goal.

5.5.2 Bioremediation technologies: conventional and innovative

Requirement for bioremediation

In 1992 total cleanup costs for all U.S. sites was estimated between $750 billion and $1.5 trillion dollars (U.S. EPA, 1993d). Remediation of 17 000 U.S. Department of Defense sites alone was estimated between $20 and $50 billion dollars in 1991. These estimates were based on the use of conventional technologies to meet the current treatment standards. Obviously, the finances to meet these estimates do not exist.

From a practical perspective, conventional remedial methods (e.g., pump-and-treat) for hydrophobic contaminants in groundwater have generally failed to produce a solution to meet health-based cleanup goals (MacDonald & Kavanaugh, 1994). Moreover, from an end-user's perspective, many conventional remedial technologies are prohibitively expensive to implement and technically inadequate.

From our perspective there are two ways to address these problems: (1) to establish and support less stringent cleanup standards, and/or (2) to develop and implement more cost-efficient, effective remediation technologies. The scope and magnitude of the environmental contamination problems in the U.S. are such that a combination of these two options will be required.

Bioremediation represents one technology that meets these challenges (U.S. EPA, 1992d). Although many site-specific characteristics must be considered, biotreatment was generally recognized in recent U.S. EPA documents (U.S. EPA, 1990b, 1993c) as a 'presumptive remedy' for solids (soil, sediment, and sludges) and water (surface and ground) contaminated by PAHs (e.g., creosote). Incineration (for soils) and activated carbon or oxidation (for water) were also recognized as presumptive remedies. However, the technology performance database used for PAHs in soils, sediments and sludges did not contain sufficient data to list a range of values for predicted performance. Updated versions of these guidelines incorporating other available performance data (U.S. EPA, 1992a, 1993a) will presumably allow such listings for each of the bioremediation strategies discussed below.

PAH bioremediation strategies

Basically, there are three general strategies for implementing bioremediation to treat soil and groundwater contaminated by PAHs: (1) solid-phase, (2) *in situ*, and (3) bioreactor operations (Sims *et al.*, 1989; U.S. EPA, 1990b). Examples of each of these strategies are provided in Table 5.7, along with a summary of the inherent benefits, limitations and challenges associated with each technology.

Each PAH bioremediation strategy has its own unique characteristics. In general, it can be stated that PAH biodegradation will be most rapid and extensive with bioslurry reactors, followed by solid-phase applications and *in situ* strategies, respectively. Of course there is a direct relationship between rate and extent of PAH biodegradation and cost of the remedial technology. As described below, site-specific criteria impact the applicability of each strategy as well as the predicted performance criteria.

The U.S. EPA has compiled what is probably the most comprehensive survey of PAH bioremediation applications available (U.S. EPA, 1994b). This information is available electronically through the Vendor Information System for Innovative Treatment Technologies (VISITT). The 1994 version of this database includes information on 277 treatment technologies offered by over 170 technology vendors. As such, useful information on the performance of myriad bioremediation strategies applied at the bench-, pilot- and full-scale levels is offered. It is important to note, however, that the data provided in the VISITT database was provided by the vendors directly; hence, EPA has neither validated nor approved this technology.

Table 5.7. *Summary of bioremediation strategies potentially applicable to materials contaminated by polycyclic aromatic hydrocarbons*

Technology	Examples	Benefits	Limitations	Factors to consider
Solid phase	• Landfarming • Composting • Engineered soil cell • Soil treatment	• Cost efficient • Modestly effective treatment of persistent HMW PAHs • Low O&M costs • Can be done on-site, in-place	• Space requirements • Extended treatment time • Need to control abiotic loss • Mass transfer problems • Bioavailability limitation	• Catabolic abilities of indigenous microflora • Presence of metals and other inorganics • Physicochemical parameters (pH, temperature, moisture) • Biodegradability
Bioreactors	• Aqueous reactors • Soil slurry reactors	• Most rapid biodegradation kinetics • Optimized physicochemical parameters • Effective use of inoculants and surfactants • Enhanced mass transfer	• Material input requires physical removal • Relatively high capital costs • Relatively high operating costs	• See above • Toxicity of amendments • Toxic concentrations of contaminants possible
In situ	• Biosparging • Bioventing • Groundwater circulation (UVB) • *In situ* bioreactors	• Most cost efficient • Non-invasive • Relatively passive • Complements natural attenuation processes • Treats soil and water simultaneously	• Physicochemical constraints • Extended treatment time • Monitoring difficulties	• See above • Chemical solubility • LNAPL/DNAPL present • Geological factors • Regulatory aspects of groundwater management

Solid-phase bioremediation

Solid-phase bioremediation describes the treatment of contaminated soil or other solid material (e.g., dredged sediment or sludge) in a contained, aboveground system using conventional soil management practices (i.e., tilling, aeration, irrigation, fertilization) to enhance the microbial degradation of organic contaminants. Such treatment systems can be designed to reduce abiotic losses of targeted contaminants through processes such as leaching and volatilization.

Three solid-phase application strategies are considered here: (1) landfarming, (2) composting, and (3) engineered soil cells. The benefits and limitations of these processes are summarized comparative to other implementation strategies in Table 5.7.

Landfarming typically describes the process of treating contaminated surface soil in-place (on-site) via conventional agricultural practices (tilling, irrigation, fertilization). The goal of landfarming is to cultivate catabolically relevant, indigenous microorganisms and facilitate their aerobic catabolism of organic contaminants. In general, the practice is limited to the treatment of the surficial 6–12 inches of soil, with some more recent machinery reportedly capable of tilling soil to a somewhat greater depth. Another method of implementing landfarming involves the placement of excavated soil onto a prepared bed. This is done typically in 6 to 8 inch lifts, with additional (contaminated) lifts being placed on top of a previously placed lift upon its successful biotreatment.

Composting describes a technology very similar in practice to landfarming with the exception that the materials are not necessarily aerated or tilled. As such, compost piles can be constructed of much greater depth (4 to 6 feet) without concern for gas transfer kinetics.

The *engineered soil cell* is a hybrid of the landfarming and the composting processes. Engineered soil cells are constructed, essentially, as aerated compost piles. Such systems offer the best of both processes and are therefore preferred whenever the site-specific application warrants their use. Key factors to consider include: (1) ability to excavate and process material, (2) total volume of material to be treated, (3) expected treatment time, and (4) desired cleanup level.

Probably the greatest advantage of solid-phase bioremediation technology is that the materials can be treated on-site and possibly in-place (conventional landfarming). As such, treated materials are exempt from LDR standards and often qualify for less stringent cleanup levels. Many of the advantages of *in situ* technologies (see below) may also be realized in terms of remediation cost savings. Compared with *in situ* processes, the kinetics of PAH biodegradation in *ex situ* bioremediation should be more favorable and more timely treatment of the HMW PAHs should be realized.

Some technical limitations to solid-phase PAH bioremediation exist. As compared with bioreactor operations, extended periods of treatment time will predictably be required with solid-phase PAH bioremediation operations. In many cases this is quite acceptable, especially when solid-phase systems are designed to provide containment and minimize exposure. When conventional landfarming technologies are employed, there exists a need for sufficient space to accommodate the handling and management of contaminated soils.

Historically, well-documented examples of successful site bioremediation and subsequent closure have been lacking, along with field documentation that the more persistent contaminants (i.e., higher-molecular-weight constituents) were biodegraded during solid-phase bioremediation processes (Mueller *et al.*, 1989a, 1993b). Actually, this is true of all the bioremediation strategies presented herein, but new field data continue to fill these gaps (see Section 5.5.3 below). Furthermore, with all bioremediation approaches, care must be taken to minimize abiotic losses (adsorption, volatilization), and to maximize biodegradation of the more recalcitrant pollutants. Also, the activities of microorganisms can be severely limited by the presence of toxic concentrations of contaminants and/or inhibitory compounds (i.e., heavy metals).

Bioreactor operations

Bioremediation in reactors involves the processing of contaminated solid materials (soil, sediment, sludge) or water through an engineered, contained system. Bioreactors can be specially designed in a variety of configurations to accommodate the physical and chemical characteristics of the contaminated matrix and the targeted pollutant(s) with the objective of maximizing biological degradation while minimizing abiotic losses. Several types of bioreactors are described in Chapter 1 of this volume.

The treatment of PAH-contaminated soil in a reactor environment is basically limited to the use of soil slurry reactors. Conversely, many different bioreactor designs exist for the treatment of water contaminated with PAHs. As reviewed by Grady (1989) and Grady & Lim (1980), these include fixed film reactors, plug flow reactors, and a variety of gas-phase systems, to name a few. Given the depth and magnitude of such a topic, for the purposes of this review discussions will be limited to a generic overview of reactor applications for PAH bioremediation.

In general, the rate and extent of PAH biodegradation is greater in a bioreactor system than in an *in situ* or solid-phase system because the contained environment is made more manageable and hence more controllable and predictable. This is due to a variety of factors including: (1) mixing and intimate contact of microorganisms with targeted pollutants, and (2) maintenance of optimum physicochemical conditions (pH, dissolved oxygen, nutrients, substrate bioavailability) for PAH biodegradation processes. Lastly, because of this contained environment, bioreactors may be inoculated more successfully with specially selected microorganisms often more capable of rapidly and extensively degrading targeted pollutants (Mueller *et al.*, 1993a; Pritchard, 1992).

Despite the obvious potential advantages of reactor systems, significant disadvantages exist which currently limit the implementation of this technology. The greatest of these is the material input requirement which mandates a preliminary 'treatment' process such as groundwater pumping, soil excavation, soil washing, or another form of physical extraction. This requirement adds significantly to the remedial costs. Moreover, the physicochemical properties of PAHs in the environment are often such that the contaminants are not amenable to conventional physical removal processes (e.g., vacuum extraction of semi-volatile PAH is not effective).

Recognizing these limitations, the key challenge is to make bioreactor technology more economical to implement. This is particularly true for soil and sediment applications. As discussed below (see Section 5.5.5), such efforts include the use of *in situ* bioreactors and the conversion of existing lagoons or basins into on-site bioreactors.

In situ bioremediation

In situ bioremediation can be defined generally as the noninvasive stimulation of the native microflora to biodegrade target organics through the passive management of environmental parameters. By and large, the aim of *in situ* bioremediation is to deliver essential co-reagents (principally oxygen) to the indigenous microflora to facilitate the biodegradation of organic compounds. Properly designed *in situ* bioremediation systems will provide effective biodegradation of organic contaminants while simultaneously providing containment of contaminated groundwater.

There are a number of engineering variations of *in situ* bioremediation strategies potentially applicable for soil and/or groundwater contaminated by PAHs. Recent reviews of *in situ* bioremediation technologies by Norris *et al.* (1993) provide excellent sources of references and offer case studies for myriad *in situ* bioremediation applications. These include bioventing vadose zone soil, biosparging saturated zone soil, vacuum-vaporized well (UVB) technology, and *in situ* bioreactors.

Bioventing describes the process of supplying air (oxygen) or other gas to indigenous soil microbes in the unsaturated zone in order to stimulate the aerobic biodegradation of organic contaminants. Early reports of bioventing showed the process to be very effective in stimulating the removal of refined petroleum hydrocarbons (Lee *et al.*, 1988). More recent studies have verified these findings and have applied the technology to treat other contaminants such as chlorinated hydrocarbons, pesticides, and PAHs (see Section 5.5.5).

Biosparging is described by Hinchee (1993) as '. . . the technology of introducing air (or other gases) beneath the water table to promote site remediation'. As with bioventing, biosparging has been most widely used to treat sites contaminated by refined petroleum products and other relatively volatile organic chemicals. As such, removal of organic constituents is due to a combination of physical air stripping as well as oxygen-enhanced *in situ* biodegradation.

Vacuum-vaporized well (German: Unterdruck-Verdampfer-Brunnen; abbreviation UVB) technology for *in situ* treatment of the capillary fringe, phreatic zone, and vadose zone contaminated with volatile organic compounds, including NAPLs, represents to us one of the most promising *in situ* bioremediation technologies (U.S.

Patents 4 892 688; 5 143 606; 5 143 607). The technology consists of a specially adapted groundwater well, a reduced-pressure stripping reactor/oxygen diffuser, an aboveground mounted blower, and a waste air decontamination system (for example, a vapor-phase bioreactor or regenerative activated carbon filters). By adding a support pump to the UVB system, a specific flow direction can be induced which produces a vertical flow either upward (reverse circulation mode) or downward (standard circulation mode). Reverse flow systems would be desired in the presence of LNAPL (prevents smearing of free product through the soil profile) whereas the presence of PAH DNAPL would encourage the use of standard circulation (pulls free product to a DNAPL recovery sump within the UVB well). A generic system diagram of UVB technology operated in standard and reverse-flow circulation is presented in Figure 5.5.

The basic principle of operation is that the water level rises due to a reduced pressure generated by the blower, and fresh air is drawn into the system through a pipe leading to the air diffuser/stripping reactor and pearls up through the raised water column. The rising air bubbles enhance the suction effect at the well bottom thus providing air-lift. The rising air bubbles oxygenate circulating groundwater and supplement the lifting effect of the reduced pressure. The subsequent fall of the groundwater along the walls of the well produces a significant hydraulic pressure. In the surrounding aquifer, a groundwater circulation cell develops with water entering at the base of the well and leaving the upper screened section, or vice versa, depending on the desired flow direction (Mueller & Klingel, 1994). These systems are described in more detail below (see Section 5.5.6).

The potential benefits of *in situ* approaches are that (1) they are usually the most cost efficient technologies to implement; (2) they are noninvasive; and (3) effective bioremediation results in terminal destruction (biodegradation) of the organic contaminant versus 'removal'. For true *in situ* processes, no secondary treatment trains exist for groundwater (since it is never removed from the earth); hence, there are no corresponding discharge problems to confront. However, several *in situ* bioremediation applications do generate off-gas that is potentially contaminated and may require treatment (e.g., passage through activated carbon or a gas-phase bioreactor) prior to terminal discharge.

Two of the recognized limitations of *in situ* technologies are (1) physicochemical restraints (e.g., bioavailability, desorption kinetics), and (2) a need for extended treatment time as compared to *ex situ* biotreatment approaches. Inherent geological parameters such as permeability, vertical and horizontal conductivity, and water depth can also represent constraints that are critically important to recognize and appreciate (Norris *et al.*, 1993; Norris & Falotico, 1994). Another widely recognized limitation inherent to *in situ* processes is that the systems are difficult to monitor and thus effective and complete treatment is difficult to ascertain and validate.

In recognition of the above limitations, a number of factors have been identified as particularly important considerations in the design and implementation of *in situ* technologies (Morgan & Watkinson, 1990). For example, the presence of free

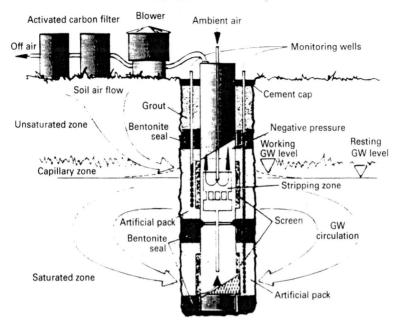

UVB - Standard Circulation

Activated carbon filter Blower Ambient air

Off air

Monitoring wells

Soil air flow

Cement cap

Grout

Unsaturated zone

Bentonite seal

Negative pressure

Working GW level

Resting GW level

Capillary zone

Stripping zone

Artificial pack

Screen

GW circulation

Bentonite seal

Saturated zone

Artificial pack

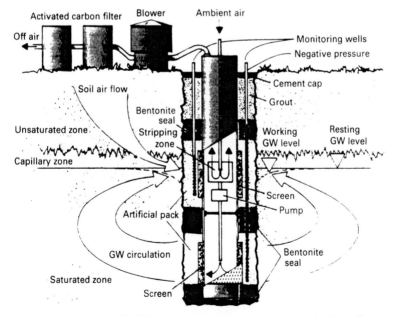

UVB - Reverse Circulation

Activated carbon filter Blower Ambient air

Off air

Monitoring wells

Negative pressure

Soil air flow

Cement cap

Grout

Bentonite seal

Stripping zone

Unsaturated zone

Working GW level

Resting GW level

Capillary zone

Screen

Pump

Artificial pack

GW circulation

Bentonite seal

Saturated zone

Screen

Figure 5.5. Schematic of UVB *in situ* groundwater circulation technology. Reprinted courtesy of IEG Technologies Co., Charlotte, North Carolina.

product as DNAPL or LNAPL represents probable chronic sources of contamination, but they also represent concentrations of contaminants that will be predictably toxic to indigenous cells. Probably the greatest challenge is to deal effectively with natural heterogeneity, anisotrophy and variability intrinsic to geologic formations. This often includes the need to treat simultaneously *in situ* multiple zones of contamination: vadose zone soil, phreatic zone soil, and saturated zone materials. In turn, this generates a requirement for cost-efficient, effective monitoring tools.

A number of co-reagents have been described for *in situ* bioremediation applications. The introduction of oxygen and inorganic nutrients (nitrogen) to facilitate aerobic catalysis of organic contaminants represent the most widely used co-reagents by far (U.S. Patent Office, 1974; Norris *et al.*, 1993; Brown & Crosbie, 1994). Sources of oxygen that have been employed include air, liquid oxygen and solid organic and inorganic peroxides (Britton, 1985; U.S. Patent Office, 1974, 1986; Brown & Norris, 1994). However, multiple economic, practical application, and technical factors appear to be favoring the preferential use of air as the oxygen carrier of choice (Spain *et al.*, 1989; Ward *et al.*, 1992; Hinchee, 1993; Brown & Crosbie, 1994).

Use of alternative electron acceptors such as hydrogen peroxide (Brown & Norris, 1994; Lu, 1994) can be an option, but usually only after careful evaluation of site-specific conditions (Pardieck *et al.*, 1992). Other co-reagents described include: (1) methane to induce methanotrophic cometabolism of chlorinated aliphatics plus oxygen as an electron acceptor (McCarty & Semprini, 1993), (2) phenol or other aromatic hydrocarbon as an inducing substrate and electron donor for chlorinated aromatic catabolism (McCarty & Semprini, 1993; Nelson *et al.*, 1994), (3) tryptophane as an electron donor and an inducing substrate for catalysis of chlorinated aliphatics (U.S. Patent Office, 1990), (4) ammonia as an inducing substrate and nitrogen source (Arciero *et al.*, 1989; Mueller *et al.*, 1994a), and (5) nitrate to induce denitrifying systems and provide a nitrogen source (Reinhard, 1993). As outlined below (see Section 5.5.6), it presently appears that the addition of air as an oxygen source and inorganic nutrients, principally nitrogen, is the desired formula for the most effective *in situ* bioremediation of PAH-contaminated materials.

5.5.3 Use of inoculants

There are several possible situations where inoculation of PAH-contaminated sites with microorganisms possessing unique metabolic capabilities could potentially enhance bioremediation. These include: (1) removal of HMW PAHs, particularly where cometabolism is involved, (2) increasing rates of overall PAH removal, and/or (3) improving bioavailability. Unfortunately, inoculation is a complicated process with relatively little success to encourage its widespread use. Very few scientifically documented examples exist where this process has been successful on a significant scale (Barles *et al.*, 1979; Edgehill & Finn, 1983; Crawford & Mohn, 1985; Briglia *et al.*, 1990; Valo *et al.*, 1990; Pritchard, 1992). Where it has been attempted, performance measures have been mostly inadequate or misleading, with minimal stringency used in determining success. Although potential inoculants often perform well under laboratory conditions where growth and metabolism can be optimized, their performance under conditions that may be more typical of the field and the natural ecosystems is often dismal. However, with more careful attention to selection and application of the inoculants, it is quite reasonable that inoculation could become a major and effective component of biological cleanup methods used for PAHs.

There are recognizably many limitations to the use of inoculation in bioremediation. Only a few have been systematically addressed in an experimental sense (Goldstein *et al.*, 1985; Zaidi *et al.*, 1988, 1989; Ramadan *et al.*, 1990; Greer & Shelton, 1992; Comeau *et al.*, 1993; Pahm & Alexander, 1993). The most common limitation involves the inability to support the growth of the introduced strain, whether it is from inadequate concentrations of growth-supporting substrates and nutrients, competition with indigenous microflora for available substrates and nutrients, or predation by protozoa. Much of the technology considered in inoculation deals with overcoming these growth limitations. Several aspects have been considered.

First, there have been examples in laboratory microcosm studies where inoculation has resulted in significant survival of the introduced microorganism with subsequent improvement in degradation capability. Heitkamp & Cerniglia (1989) demonstrated that the addition of a PAH-degrading *Mycobacterium* species strain Pyr-1 (capable of cometabolizing phenanthrene, pyrene, and fluoranthene) to microcosms containing pristine estuarine sediments produced a PAH-degradation capability that was not present prior to the inoculation. Studies showed that the *Mycobacterium* sp. survived in the microcosms and significantly mineralized multiple doses of pyrene. In addition, although the organisms did not mineralize benzo[*a*]pyrene in pure culture, it did appear to enhance the degradation of this PAH in the microcosms. This unexpected result is not readily explainable but perhaps indicates some of the additional benefits of inoculation. Noteworthy here is the detection of compounds in the microcosms that resemble ring oxidation products of PAHs. Although the chemical nature of these products was not determined, the results are one of the few, if any, verifications that cometabolism of PAHs can possibly occur in natural systems.

In a similar study, Grosser *et al.* (1991) were able to show that the reintroduction of pyrene-degrading bacterial cultures into soil contaminated with coal gasification processing wastes enhanced pyrene mineralization more than 50-fold. This enhancement also occurred in uncontaminated soils, but to a lesser extent. Interestingly, the inoculants used in this study appeared to also be a *Mycobacterium* species, but in contrast to the studies of Heitkamp & Cerniglia (1989), it could grow on pyrene in the absence of supplemental carbon sources.

Møller & Ingvorsen (1993) showed that the degradation of phenanthrene in soil could be enhanced by the addition of an *Alcaligenes* sp. isolated from polluted soil. Inoculation reduced the starting concentration of phenanthrene by 96% in 9 days, whereas non-inoculated soil showed only 12% degradation in 42 days. However, success in this study was probably related to the use of uncontaminated sandy loam soil (no history of significant PAH exposure) that was spiked with phenanthrene. With no previous exposure to PAHs, the indigenous microflora was apparently not acclimated to phenanthrene as a recognizable carbon source, and the authors were unable to detect culturable phenanthrene degraders. In addition, phenanthrene bioavailability was probably much higher than in most polluted soils where oily-phase liquids would likely exist and adsorption of PAHs to soil surfaces would be maximized. Notably, these authors also observed that the *Alcaligenes* sp. did not degrade phenanthrene when sterile soil was used. This was explained by the provision of some growth-stimulating nutrient produced by the indigenous microflora which, if true, is a good example of the precautions that must be taken to achieve successful inoculation. Clearly, using a fastidious inoculant microorganism will reduce the chance of successful inoculation of contaminated soil.

Another possible approach to inoculation is to use an inoculum large enough (either as a single treatment or in multiple treatments) that if the biomass is active immediately after its addition, some degradation enhancement can be realized even though only limited growth occurs (Ramadan *et al.*, 1990; Comeau *et al.*, 1993). In the case of the cometabolism of PAHs, for example, a high but transient biomass might be able to partially oxidize many different PAHs before it decayed into ineffectiveness. If the partially oxidized products are then further degraded by the indigenous microflora, the success of inoculation might be realized. Of course, the necessary application equipment and resources to produce and distribute large bacterial biomasses, perhaps repeatedly, must be available. In addition, the production of biomass should be performed under growth conditions that are as similar to the field as possible. This will help insure that when the inoculum is mixed into the environmental material, the cells will not be shocked (and consequently less active) by an environmental setting that differs considerably from culture conditions.

This approach has been developed in conjunction with the treatment of PAH-contaminated groundwater using a bioreactor system (Mueller *et al.*, 1993a). The design of the bioreactor was based on the premise that a sequential, two-phase process was required (U.S. patent allowed). First, pretreatment of the readily degradable 2- and 3-ring PAHs was performed using a mixture of bacterial cultures that were known to grow on specific LMW PAHs. The activity of this inoculum was probably equivalent to the degradative activity of indigenous soil microflora from PAH-contaminated soil. The pretreatment was then followed by a second phase that involved the addition of *Sphingomonas paucimobilis* strain EPA505 that was known to degrade fluoranthene and cometabolize the more persistent 4- and 5-ring PAHs (Mueller *et al.*, 1989b, 1990b). Since, in this particular study, the source of groundwater contamination was an old creosote-processing site, pentachlorophenol was also present and it too was treated by the addition of a pentachlorophenol-degrading bacterium (*Pseudomonas* sp. strain SR3) (Resnick & Chapman, 1994) in the second treatment phase.

When gas chromatograms of the organic extracts of the original feed into the pilot-scale bioreactors were compared with those for the contents of the two different treatment phases (Figure 5.6), it was apparent that the microorganisms effectively removed most of the PAHs. Without the addition of strains EPA 505 or SR3, the effluents from the first phase of treatment were not acceptable for discharge (Middaugh *et al.*, 1994). Thus, inoculation was effective in improving PAH removal under a controlled bioreactor setting. Since EPA505 did not grow fast enough for the reactor operations, it was apparently having its effects (we speculate largely through cometabolism) before washout. Daily addition of the inocula, of course, helped promote this degradation.

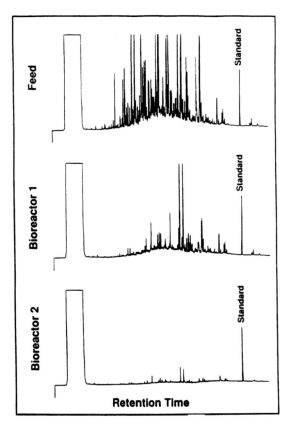

Figure 5.6. Changes in chemical composition of creosote-contaminated groundwater during pilot-scale bioremediation. Gas chromatogram of basic extracts of (1) groundwater bioreactor feed (top panel) and (2) effluent from first-stage bioreactor containing indigenous microbes (middle panel), and the second-stage bioreactor inoculated with HMW PAH degraders. Reprinted with permission from Mueller *et al.*, 1993a. Copyright 1993, American Chemical Society.

For inoculation success outside of a bioreactor, the contaminated environment will almost certainly have to be physically modified in a significant manner, perhaps over an extended period, to optimize the biodegradation process. This generally means establishing conditions in which the availability of oxygen, inorganic nutrients, temperature, degradable substrate, and moisture control are maximized. For *ex situ* or on-site procedures this is frequently accomplished by some mechanical means, including tilling and other homogenization procedures, aeration, and mixing. Probably one of the most severe limitations to *in situ* inoculation involves mass transfer; getting the bacteria to the pollutant and/or getting the pollutant to the bacteria will require significant physical reworking of the contaminated site. Simply sprinkling the inoculum on a contaminated area is not likely to produce optimal results. Thus for *in situ* procedures, we have modified the UVB technology as described below (see Section 5.5.6).

Lajoie *et al.* (1992) have proposed the use of field application vectors in which one creates a temporary niche for the survival and activity of the introduced organism. The niche is created by the addition to the target environment of a selective substrate, in their case several types of detergent, that is readily utilizable by the inoculant but not the indigenous microflora. The potential advantage of using these highly selective substrates is that smaller inoculum sizes can be applied, and, it is argued, that the appropriate degradative enzymes are produced more efficiently because the inoculum is not 'diluted' due to growth on other, less catabolically relevant, carbon sources. The authors were able to demonstrate that the introduction of an organism carrying specific antibiotic resistance markers in soil microcosms supplemented with a growth-supporting detergent, Ipegal CO-720, resulted in considerably improved inoculum survival as compared with unsupplemented systems. It also led to an increase in the number of antibiotic-resistant transformants from gene exchange in the soil from the added microorganisms.

This concept has been applied to the bioremediation of oil-contaminated beach material using a variation on the same theme. Here, Rosenberg *et al.* (1992) selected inoculant microorganisms for their ability to degrade both petroleum hydrocarbons and a unique source of organic nitrogen. The nitrogen source was a hydrophobic fertilizer product made of a modified urea-formaldehyde polymer. Because this source of nitrogen is not readily utilized by indigenous bacteria, it served, in essence, as a 'field application vector', just as Lajoie's detergent did. Rosenberg *et al.* were able to show that the addition of the fertilizer and the hydrocarbon-degrader, which could use the nitrogen from the fertilizer, significantly enhanced oil biodegradation in contaminated beach material. Clearly, this concept could be utilized for the directed degradation of PAHs, especially with the large number of new PAH degraders that are being studied for bioremediation applications.

Another interesting approach to improving the usefulness of inoculation in bioremediation is the development of procedures in which the inoculant can be placed in environmental media in a way that reduces competition from the indigenous microflora and allows expression of the specific introduced metabolic function. The technique of encapsulation is being developed for this purpose. Encapsulation involves the packaging (immobilization) of specific bacterial or fungal cells in polymeric or inert materials in a manner that will allow introduction of viable and active cells into an environment (Lin & Wang, 1991; Trevors *et al.*, 1992). This approach provides a whole series of advantages including longer survival times in soil, better inoculum storage capabilities, and the ability to customize the capsule to enhance growth of the cells after addition to the contaminated soil matrix. A considerable body of information exists on encapsulation/granulation procedures for microorganisms (Paau, 1988; Trevors *et al.*, 1992) but only in a couple of cases has it been applied to bioremediation (O'Reilly & Crawford, 1989; Briglia *et al.*, 1990; Valo *et al.*, 1990; Lin & Wang, 1991; Stormo & Crawford, 1992, 1994; Levinson *et al.*, 1994). In general, there has been relatively little effort to develop procedures and methods for immobilizing or encapsulating inoculants for bioremediation. For *Rhizobium* inoculations, where considerable success has been realized (van Elsas & Heijnen, 1990), extensive research led to the choice of peat moss as the ideal carrier. Similar information is not yet available for many bioremediation inoculants.

Work at the EPA Gulf Breeze Laboratory has demonstrated the potential usefulness of encapsulation in the bioremediation of PAHs. A model system has been developed in which a pure culture capable of degrading fluoranthene (strain EPA505) has been successfully encapsulated in polyurethane foam and polyvinyl alcohol (Baker *et al.*, 1988). The capsules can be stored for several months at 4 °C with only minimal loss of viability. Upon addition of the capsules to moist soil, fluoranthene mineralization commenced in approximately the same way as observed when fresh bacterial cells were added to the soil. These results are shown in Figure 5.7a. Since the same inoculation size was used in all flasks during this experiment, the results suggest that the immobilization process does not significantly affect microbial activity.

Active immobilized cells offer several additional possibilities for further enhancing biodegradation and environmental control. For example, inclusion of adsorbents in the immobilization matrix can result in a more rapid uptake of contaminants from the environment and thereby potentially provide greater accessibility of the adsorbed chemical to the immobilized bacteria (Hu *et al.*, 1994). Two issues need to be addressed, however, when using co-immobilized adsorbents: (1) is microbial activity affected by co-immobilization with adsorbents, and (2) is availability of the adsorbed chemical to the immobilized cells maximal? To study these questions, diatomaceous earth and powdered activated carbon were co-immobilized with strain EPA505 in the polyurethane matrix. In Figure 5.7a, it can be seen that the degrading activity of the cells co-immobilized with the adsorbents was the same as the non-immobilized cells, indicating that the degradation of the adsorbed fluoranthene was complete.

Another possibility involves *in situ* bioremediation situations, where direct addition of nitrogen and phosphorus to soil or water might have a negative

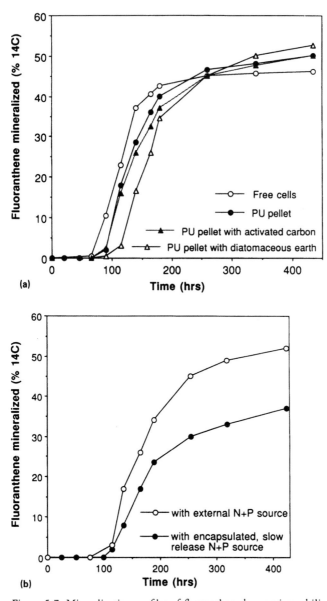

Figure 5.7. Mineralization profiles of fluoranthene by non-immobilized and polyurethane-immobilized cells (PU pellet) of strain EPA505. Results of studies with (a) inorganic nutrients in incubation medium and (b) inorganic co-encapsulated nutrients.

consequence due to either enhancement of the activity of undesired indigenous microflora and/or the leaching of the nutrients to groundwater. By co-immobilizing slow-release formulations of nutrients in the polymer matrix, a major part of the nutrients can be provided to the immobilized cells with considerably less available for leaching into the environment. In our experiments, slow-release formulations of nitrogen and phosphorus were co-immobilized with EPA505 in the polyurethane matrix and then tested in buffer for fluoranthene degradation. As a positive control, immobilized cells with external sources of nitrogen and phosphorus (solution of inorganic salts) were also used. The co-immobilized, slow-release formulations of nutrients supported extensive biodegradation, although the biodegradation rate was slower than that observed with externally supplied nutrients (Figure 5.7b). Further studies on the effect of release rates of the co-immobilized nutrients may provide more in-formation for optimizing this approach to bioaugmentation.

Considerable new research is needed in the future to make inoculation more successful. However, if prudent use of inoculation is encouraged (that is, there is a legitimate need for it relative to the activities of indigenous microorganisms), the potential applications are quite attractive. For example, it can be argued that the wrong microorganisms are being used in the development of inocula because we are too interested in speed rather than remedial 'success'. Laboratory enrichment conditions that are frequently used to obtain microorganisms for an inoculum select for degraders with the fastest growth rate and the highest metabolic rate. When these bacteria are added back to the environment, the enrichment conditions under which they were originally selected are not present and they are easily out-competed by the indigenous microflora. Instead, inoculants should be selected based on their compatibility (environmental competence) first, and their degradative/metabolic capacity second. Selecting in the laboratory for characteristics such as degradation at low concentrations of substrate and or inorganic nutrients, adhesion capability to surfaces where the pollutant chemicals may be concentrated, motility or chemotaxis capability, and high survivability under different environmental conditions, may ultimately produce a better potential inoculant than one capable of very fast degradation. Using inoculants that are very efficient nutrient scavengers, or those that can store extra nutrients into the cell before they are applied in the field, could also greatly improve inoculation procedures.

5.5.4 Solid-phase PAH bioremediation: case studies

Landfarming

McGinnis et al. (1988) conducted some of the earlier work assessing the potential of landfarming technologies to facilitate the biodegradation of PAHs present in soils recovered from wood-treating sites where creosote and/or PCP were used as preservatives. The results of an extensive literature review and detailed laboratory studies indicated that the LMW PAHs such as naphthalene are biodegraded quickly (soil half-life < 10 days), with the HMW PAHs typically exhibiting soil half-lives of < 100 days. The report concluded that all of the PAHs present in soil from creosote-contaminated sites could be biodegraded, with process enhancements capable of reducing the soil half-life for HMW PAHs to less than 10 days in select cases.

A thorough, in-depth assessment and review of solid-phase bioremediation technologies for contaminated surface soils was provided by Sims *et al.* (1989). In this report, a number of factors known to impact on the performance of landfarming and engineered soil cell operations was identified. More recently, Pope & Matthews (1993) essentially updated this data base with particular focus on the application of the landfarming technology and PAH-contaminated soils. To a certain degree, enhanced solid-phase bioremediation has been successful in accelerating the rate, and increasing the extent, of biodegradation of HMW PAHs and other more persistent compounds (Mueller *et al.*, 1991a). As discussed above, these enhancements include the use of inoculants, the addition of inorganic nutrients, and/or physical modification (i.e., tilling, composting, use of plants) to increase bioavailability, biological activity and contaminant biodegradation. Some of these examples are described below.

Relying on the activity of the indigenous microflora, Tremaine *et al.* (1994) comparatively evaluated three solid-phase treatment strategies according to their ability to remove PAHs from creosote-contaminated soil. These efforts culminated in an initial biodegradation rate (k) of 122 mg total PAH per kg of soil per day with a starting soil total PAH concentration of 500–14 000 mg/kg. However, when biodegradation data are considered over the entire 12-week incubation period, PAH removal appears to be less extensive; removal of the HMW PAHs was not described.

Using tilling and inorganic nutrient fertilization, Ellis (1994) conducted full-scale (4600 yd^3) land biotreatment of creosote-contaminated soil. This application followed a series of laboratory treatability studies which suggested that a 59% reduction in total PAHs could be achieved after 28 days incubation at 25 °C (starting soil PAH concentration of about 4300 mg/kg). As would be expected, HMW PAHs were less readily biodegraded, but significant removal of these chemicals was also observed. Subsequent full-scale application resulted in the reduction of total creosote PAHs from 1024 to 324 mg/kg soil dry weight after 88 days (cleanup goal was < 200 mg/kg total PAH). Although the HMW PAHs again persisted, significant reduction (e.g., > 50%) in the soil concentration of these chemicals was again observed.

Field-scale demonstration of solid-phase PAH bioremediation is currently being supported by the U.S. EPA's Bioremediation Field Initiative Program at the Libby Superfund Site, Libby, Montana (U.S. EPA, 1993f). Here, biotreatment of about 45 000 yd^3 of soil contaminated by PCP and PAH is being performed by *ex situ* treatment of soil removed to a treatment facility and placed in 9-inch-thick lifts and treating the material using conventional solid-phase approaches (tilling, fertilization) until the soil meets the targeted cleanup criteria (dry weight basis) of: 88 mg/kg cumulative (10) carcinogenic PAHs, 8 mg/kg naphthalene, 8 mg/kg phenanthrene, 7.3 mg/kg pyrene, 37 mg/kg PCP, and <0.001 mg/kg 2,3,7,8-tetrachlorodioxin equivalent. This reportedly takes between 32 and 163 days to achieve (Sims *et al.*, 1994; U.S. EPA, 1993f), with detoxification to nontoxic levels being documented by MicroTox™ test and Ames Assay (*Salmonella typhimurium* mutagenicity test). At this same site, treatment of groundwater with fixed-film bioreactors or injection of hydrogen peroxide and inorganic nutrients is also being evaluated.

Engineered soil cell

Field demonstration of an aerated soil cell was conducted with soil contaminated by PAH originating from a coal gasification plant (Taddeo *et al.*, 1989). After 78 days of compost treatment, total PAH concentrations were reduced from 6330 to 370 mg/kg compost. However, 5-ring PAHs including benzo[*a*]pyrene were not degraded within this time period.

Cold-temperature composting has been used successfully to treat PAHs in soil from creosote waste sites in Norway (Berg & Eggen, 1991; Eikum *et al.*, 1992). Following a number of process optimization studies performed at the bench-, pilot- and full-scale levels, full-scale remediation of 20 000 tons of soil (containing an average of 500 mg PAH/kg soil) was undertaken. Field data showed that the concentration of total PAHs was successfully reduced from *c*. 500 mg/kg soil to <20 mg/kg soil within 8–10 weeks of compost bioremediation. The Dutch 'B' remedial goal of reducing the concentration of total PAHs in soil to below the level (500 mg/kg soil) was met routinely within 6–8 weeks.

Inoculation with fungi

The use of ligninolytic fungi inoculants in the biodegradation of PAHs has been an active topic for several years (Cerniglia, 1982; Bumpus *et al.*, 1985; Cerniglia & Heitkamp, 1989; Aust, 1990; Aust *et al.*, 1994). These systems offer the potential advantages associated with extracellular enzymatic activity that exerts non-substrate-specific catalysis. Although the very nature of this attractive catalysis is such that there is recognized potential for generating toxic, reactive intermediates (see Section 5.4 above) (Cerniglia, 1992), the controlled and managed use of lignin-degrading fungi may yield effective treatment of soil contaminated by PAHs and related compounds.

For example, Lamar & Glaser (1994) reported that when soil containing 672 mg pentachlorophenol (PCP) plus 4017 mg total PAH/kg was inoculated with *Phanerochaete sordida*, an average 89% PCP removal and 75% PAH removal was observed within 8 weeks of pilot-scale field testing. However, biodegradation of HMW PAHs was generally low (i.e., 6% for chrysene), with the most extensive removal (i.e., 33% for chrysene) being observed in the control plot despite a detailed, statistically valid field sampling and analysis plan supported by the U.S. EPA's Bioremediation in the Field Initiative (U.S. EPA, 1993g).

Additional efforts continue to improve on this process. For example, laboratory studies of solid-phase bioremediation conducted by Baud-Gasset *et al.* (1994) reported somewhat better PAH biodegradation data with *Phanerochaete chrysosporium* and *P. sordida* inoculants. Here, the total PAH concentrations were reduced by 76 to 86% after an 8 week incubation period (starting soil total PAH concentration ranged from 5139 to 6582 mg/kg soil) when soils were inoculated with ligninolytic fungi. This was significantly greater than the removals observed in the corresponding controls, but the use of sterile wood chips alone also yielded significant PAH removal, suggesting that abiotic factors contributed to these figures. Differences between the control and active treatments were more apparent in terms of reduction in 4-ringed PAHs, with 22% removal in the control and up to 79% removal in the fungal treatments. In addition to demonstrating the removal of the parent compounds (i.e., PAHs), a series of biological toxicity tests (plants and Microtox[TM] assays) showed significant reduction in toxicity that correlated with PAH removal.

Bioaugmentation with plants

Another upcoming area of solid-phase bioremediation enhancement entails the managed use of the plant–microbe relationship (Bell, 1992; Kingsley *et al.*, 1994). The use of plants in solid-phase bioremediation offers a number of potential advantages associated with the support of rhizosphere microbial communities that may be active towards soil contaminants (Bolton *et al.*, 1992).

Hsu & Bartha (1979) published one of the earliest reports on the use of plants to stimulate the removal of organic contaminants (organophosphate insecticides) from soils. A number of related studies subsequently showed that an active rhizosphere community could more readily biodegrade other organic chemicals such as parathion (Reddy & Sethunathan, 1983), 2,4-D (Short *et al.*, 1992), chlorinated solvents (Walton & Anderson, 1990) and PAHs (Aprill & Sims, 1990). While all the intricacies of the plant–microbe interactions are not fully understood for any system, efforts are being undertaken to apply these technologies to enhance the bioremediation of PAH-contaminated sites. Such applications can prove especially useful in removing inorganic metal constituents (i.e., copper, chrome, arsenic) at co-contaminated sites. Such co-contamination is common to many wood-preserving operations.

The plant–microbe symbiosis may help facilitate the effective use of inoculants. For example, developed (brady)rhizobial strains or root-colonizing pseudomonads may be more effectively introduced into a contaminated soil environment when they are applied in conjunction with their plant host. Kingsley *et al.* (1994) showed that inoculation of soil with a 2,4-D-degrader protected germinating seeds from the herbicidal effects of residual pesticide. Thus, plants may be used to help restore treated soils that contain residual but biologically active compounds.

5.5.5 Bioreactor PAH bioremediation: case studies

The nature of reactor operations requires that the technology be coupled with another physical extraction operation, such as groundwater removal, soil excavation, and soil washing/classification. In 1993, soil washing was recommended at 20 U.S. Superfund sites, with 17 systems being in the pre-design/design phase and three in the installation/operational phase (U.S. EPA, 1993d, 1994a). At the time of these publications, however, none of the systems was completed.

Since soil washing only serves to reduce the volume of material requiring subsequent treatment, seven of these applications had some form of bioremediation (i.e., slurry reactors) coupled to them as a secondary treatment step. This represents a common combination of technologies: soil washing/separation followed by bioslurry reactor, then drying/polishing in a land treatment unit (Mueller *et al.*, 1990a, 1991b; U.S. EPA, 1990b, 1992b). Based on this concept, Simpkin & Giesbrecht (1994) conducted a series of laboratory-scale (shake-flask) and pilot-scale (60 liter reactor) tests to assess the potential of bioslurry reactors combined with land treatment to treat a silty-clay pond sediment contaminated by PAHs. Although very few data are presented for review, a sediment solids loading rate of 25% and a 3 to 6 day hydraulic retention time yielded PAH biodegradation of 88% (as measured by pyrene removal). No insight into the land treatment unit was provided.

Along these same lines, Berg *et al.* (1994) conducted a series of bench-scale and pilot-scale studies to discern the effectiveness of soil washing by flotation followed by slurry-phase biotreatment. At the pilot-scale, soil washing was performed at a rate of 500–800 kg/h with a 454 liter bioslurry reactor being operated in the batch mode at 14% solids. Laboratory data showed that an average 97% reduction in total PAHs could be realized within a 21-day study with the HMW PAHs being more refractory, as expected. When the pilot-scale reactor was loaded with washed soil material containing 5700 mg total PAH/kg soil, similar results were obtained. Here, a 70% reduction in total PAHs was observed within 6 days of operation, and 95% reduction within 8 days. Although mass balance calculations were attempted, the data were not reported. Presumably, volatilization was insignificant and the HMW PAHs composed a majority of the residual contaminants.

On-site, slurry-phase bioremediation of a lagoon sludge containing a cocktail of aromatic hydrocarbons, including PAHs, was tested at the laboratory- and pilot-scale to assess the ability of bioremediation as a treatment technology (Lee *et al.*, 1994). As with many such applications, biotreatment offered the potential advantages of on-site, *in situ* treatment, thus exempting the sludge from land disposal restrictions and qualifying it for alternative cleanup criteria. These studies showed that a combination of biotic and abiotic processes were responsible for the PAH losses observed. In general, almost all of the volatile constituents were removed via air stripping; PAH removal was variable, ranging from 50 to 100%. However, no attempt was made to mass balance the reactors and physical sorption has been shown to be a potential source of removal (Park *et al.*, 1990; Mueller *et al.*, 1991b, 1993a). Full-scale implementation of the technology was estimated to require 1.5 to 2 years at a cost of $66/m^3 of sludge processes.

Full-scale implementation of a similar system was initiated at the French Limited Superfund Site to treat sludge contaminated with a variety of chemicals, including PAHs (Woodward & Ramsden, 1989). Following four laboratory treatability studies, two on-site, 20 000-gallon bioreactor studies and a 6-month, 0.5 acre field demonstration, full-scale operations commenced by converting the lagoon into an on-site bioreactor. These efforts have been determined to be successful with site closure possibly complete by this time.

Tremaine *et al.* (1994) conducted pilot-scale studies to compare the effect of a suspended-growth reactor and a fixed-film bioreactor with a constructed wetland environment in removing creosote-PAHs from contaminated water recovered from a wood-preserving facility. Mass balanced chemical analysis of 5 PAHs used as model constituents of creosote showed that the wetland yielded between 20 and 84% removal, whereas the fixed-film reactor yielded 90 to 99% PAH removal. Biodegradation accounted for >99% of the losses observed in the fixed-film reactor, but only 1–55% of the compounds removed in the artificial wetland was attributable to biodegradation. Again, physical sorption of PAHs, especially HMW PAHs, was found to be significant.

Full-scale separation/washing and bioslurry reactor operations have been used to treat creosote-contaminated soil at the former Southeastern Wood Preserving Site at Canton, Mississippi (Jerger *et al.*, 1994; Woodhull & Jerger, 1994). Here, an estimated 10 500 yd^3 of soil and sludge were excavated from various process areas, stabilized with kiln dust and stockpiled for subsequent treatment. Based on the results of preliminary bench studies, four 680 000 liter reactors were eventually established to handle 7050 yd^3 of the screened (200-mesh) soil fraction at a solids content of 20–25%. Other oil fractions and waters were handled separately (data and costs not reported).

Operating slurry reactors in the batch mode for 8 to 12 days reportedly yielded a removal efficiency of 95%, with a majority of the degradation being realized within the first few days of operation. Treatment of the pcPAHs ranged between 55 and 85%, leaving an estimated total (data not reported) of 400 mg pcPAHs/kg soil (dry weight). Ultimately, a soil cleanup goal of 950 mg/kg total PAHs and 180 mg/kg pcPAHs was achieved. But biotreatment performance of this magnitude could presumably have been realized using a more cost-efficient technology (i.e., solid-phase treatment).

The available data from this full-scale operation demonstrate some of the difficulties associated with sampling and analysis of such systems. In general, reported PAH removal rates would seem to challenge the expected results of microbiological processes. But optimized engineering studies of similar bioslurry reactors reported by Glaser *et al.* (1994) showed similar results, reducing the slurry concentration of LMW PAHs from 798 to 45.1 mg/kg and the concentration of HMW PAHs from 1249 to 374 mg/kg in a 30% slurry within 7 days of treatment. But no attempt was made to mass balance these removal values (Lauch *et al.*, 1992). While laboratory models of the system by Woodhull & Jerger (1994) suggested that <0.002% of PAHs were sorbed to reactor walls, this is significantly less than results of other bioslurry reactor studies where mass balance chemistry showed that sorption and biomass accumulation accounted for up to 37% of the removal of HMW PAHs (Park *et al.*, 1990; Mueller *et al.*, 1991b, 1993a). Thus, despite the cost and difficulty (impossibility?) of sample acquisition, mass balance (or pseudo-mass balance) PAH biodegradation data under full-scale bioslurry reactor operations would have been very informative.

5.5.6 In situ *PAH bioremediation: case studies*

In 1992, *in situ* bioremediation was recommended for 4% of the U.S. EPA's Superfund sites (20/498 sites) (U.S. EPA, 1992b). As of June 1993, this number increased to 26 applications, with 16 systems in the design phase, nine systems in the installation/operational phases, and one application completed (U.S. EPA, 1994a). *In situ* installation at non-Superfund sites has been even more popular. This growing trend of implementing *in situ* bioremediation technologies is expected to continue, especially with recent field documentation of natural attenuation and intrinsic biodegradation occurring *in situ* with little or no augmentation (Wilson *et al.*, 1994).

Brown & Norris (1994) state that the first reported case of field-scale *in situ* bioremediation was conducted by Raymond in 1972 to treat a gasoline pipeline spill (U.S. Patent Office, 1974). Since this time, most *in situ* bioremediation applications have continued to focus on the removal of refined petroleum products and other relatively volatile organic contaminants, such as TCE (Norris *et al.*, 1993). Historically, few applications have been attempted with PAHs. This is because *in situ* bioremediation technologies effectively address certain factors such as the addition of electron acceptors (oxygen, nitrate) and inorganic nutrients (e.g., nitrogen), but they are typically limited in their ability to overcome other physicochemical factors known to inhibit the biodegradation of PAHs (i.e., bioavailability).

While technology limitations currently inhibit broader implementation, recent studies have demonstrated the potential effectiveness of *in situ* PAH bioremediation. For example, Flyvberg *et al.* (1991) described the use of nitrate to enhance the *in situ* biodegradation of PAHs from creosote and oil, but the extensive removal was generally limited to the lower-molecular-weight PAHs. These findings support many studies showing that unsubstituted aromatic hydrocarbons (such as benzene) tend to persist under denitrifying conditions (Reinhard, 1993). In processes where unsubstituted aromatic hydrocarbons were degraded upon the addition of nitrate, the presence of oxygen as the effective electron acceptor is often suspected (Major *et al.*, 1988; Werner, 1991).

Piontek & Simpkin (1994) recently reported on a long-term (3 year), detailed assessment of multiple *in situ* bioremediation strategies that were tested under field conditions at an abandoned creosote facility in Laramie, Wyoming. Here, as at many wood-preserving sites, creosote DNAPL represented the greatest problem, with an estimated 19 to 30 million liters of creosote representing a continuous source of PAHs, and hence a long-term, high requirement for an electron acceptor to maintain effective *in situ* bioremediation. Bench- and pilot-scale testing of water-flood product (creosote DNAPL) recovery and chemically (surfactant) enhanced soil flushing to reduce the oxygen requirement was combined with three strategies of supplying an electron acceptor to the indigenous microflora: (1) nitrate addition, (2) bioventing (air), and (3) injection of pure oxygen. Hydrogen peroxide was not considered for this application due to chemical reactivity and cost. Each technology was evaluated in terms of full-scale implementability, cost, and ability to meet desired cleanup standards. These studies concluded that water flooding could potentially remove 50% of the DNAPL and that subsequent *in situ* bioventing could reduce the total PAH concentration in soil from > 500 mg/kg to as low as 5 mg/kg soil. Nitrate was generally not effective.

Field testing of injecting hydrogen peroxide and inorganic nutrients to stimulate the biodegradation of PAHs is currently being supported by the U.S. EPA through the Bioremediation in the Field Initiative at the Libby Superfund Site (U.S. EPA, 1993f). Limited data are available to date. This same program is sponsoring a field demonstration of air bioventing to treat PAH in a sandy vadose zone ranging from 2 to 10 feet below ground surface. Over the next three years, an expected 27% reduction in PAHs will calculate to a 10–15 year *in situ* remediation time, assuming first-order kinetics.

Also sponsored by the U.S. EPA's Bioremediation Field Initiative, air bioventing of unsaturated soils contaminated by PAHs from creosote and coal tar is being field tested at the Reilly Tar and Chemical Corporation, St. Louis Park, Minnesota (McCauley *et al.*, 1994; U.S. EPA, 1993h). A 3-year program was initiated in November of 1992, with quarterly soil gas analyses for oxygen and carbon dioxide being conducted. Initially, microbial respiration rates (i.e., oxygen uptake and carbon dioxide production) will be used to calculate PAH biodegradation as described by Hinchee & Ong (1992). Data to date are not conclusive. However, in our use of similar tools during bioremediation studies, stable isotope analysis of the respired carbon provides more accurate identification of the mineralized substrates (Coffin *et al.*, unpublished data).

In situ bioremediation of PAH-contaminated vadose zones via bioventing, and saturated zones via groundwater circulation and/or air sparging, represent areas of great future advancement. As documented by multiple case studies, the UVB technology provides for simultaneous *in situ* remediation of contaminated soil and groundwater, especially those contaminated with volatile organic compounds (VOCs) such as BTEX constituents and chlorinated solvents (Borchert & Sick, 1992; Herrling *et al.*, 1994). Properly designed systems effectively treat the capillary fringe zones where LNAPLs typically accumulate, and technology developments promise to enhance the performance of these systems with PAH-contaminated materials. Total remedial efficacy is accomplished through the combined use of physical and biological processes including: (1) *in situ* circulation of air and water, (2) stimulated *in situ* bioremediation by indigenous microflora via the introduction and distribution of electron acceptor (oxygen) and nutrients (nitrogen), and (3) *in situ* air stripping of volatile organics.

Our recent modification of UVB technology with *in situ* bioreactors (Mueller & Klingel, 1994) housing immobilized cells of known HMW PAH degraders such as *Sphingomonas paucimobilis* strain EPA505 (Mueller *et al.*, 1989b, 1990b) and slow-release inorganic nutrients (Mueller *et al.*, 1994b) is currently being field tested at two PAH-contaminated sites in the United States. Preliminary data suggest that these systems will be effective in reducing PAH concentrations in soil and groundwater in a cost-efficient, true *in situ* manner.

5.6 Conclusions

With varying degrees of success, bioremediation has been used to treat PAHs in creosote wastes, oil field and refinery sludges, and petroleum products (U.S. EPA, 1993d). A common feature to all these efforts is that differential rates of PAH biodegradation typically occur. In general, the lower-molecular-weight, more water-soluble organics such as naphthalene are more readily biodegraded than the higher-molecular-weight PAHs such as benzo[*a*]pyrene. Thus, for all of the bioremediation technologies described above, their demonstrated applicability toward more challenging organic contaminants, such as HMW PAHs, represents an area in which continued scientific advancement is needed. Actually, recognizing these trends in biogeochemical persistence has helped in the conduct of chemical mass balance assessments of bioremediation efficacy (Elmendorf *et al.*, 1994; Butler *et al.*, 1991). Use of such monitoring and assessment tools during field bioremediation studies is becoming more popular, and will aid in the evaluation of bioremediation technologies for a variety of applications.

In general, data on bioremediation performance under actual field conditions represents the most valuable information since the effect of myriad interactive, unknown parameters can only be realized under these test conditions. Efforts such as the U.S. EPA's Bioremediation in the Field Program (U.S. EPA, 1993e), and others, therefore perform a critical role in the continuing development and implementation of bioremediation technologies. Adequately addressing in the field problems of measuring process effectiveness, evaluating site characterizations, establishing long-term process control, and attaining appropriate cleanup levels will help resolve complicated scientific, public, and regulatory issues. With a careful balance between basic and applied research, we believe that environmental biotechnology will rapidly become a routine remedial tool for restoration of PAH-contaminated environments.

References

Aelion, C. M., Dobbins, C. D. & Pfaender, F. K. (1989). Adaptation of aquifer microbial communities to the biodegradation of xenobiotic compounds: Influence of substrate concentration and preexposure. *Environmental Toxicology and Chemistry*, 8, 75–86.

Akhtar, M. N., Boyd, D. R., Thomas, N. J., Koreeda, M., Gibson, D. T., Mahadevan, V. &

Jerina, D. M. (1975). Absolute stereochemistry of the dihydroanthracene-*cis*- and *trans*-1,2-diols produced from anthracene by mammals and bacteria. *Journal of the Chemical Society Perkin Transactions*, 1975, 2506–11.

Alexander, M. (1985). Biodegradation of organic chemicals. *Environmental Science & Technology*, 18, 106–11.

Andersson, K., Levin, J-O. & Nilsson, C-A. (1983). Sampling and analysis of particulate and gaseous polycyclic aromatic hydrocarbons from coal tar sources in the working environment. *Chemosphere*, 12, 197–207.

Aprill, W. & Sims, R. C. (1990). Evaluation of the use of prairie grass for stimulating polycyclic aromatic hydrocarbon treatment in soil. *Chemosphere*, 20, 253–65.

Arciero, D., Vannelli, T., Logan, M. & Hooper, A. B. (1989). Degradation of trichloroethylene by the ammonia-oxidizing bacterium *Nitrosomonas europaea*. *Biochemistry and Biophysics Research Communications*, 159, 640–3.

Aronstein, B. N., Cavillo, Y. M. & Alexander, M. (1991). Effect of surfactants at low concentrations on the desorption and biodegradation of sorbed aromatic compounds in soil. *Environmental Science & Technology*, 25, 1728–31.

Atlas, R. M. (1991). Microbial hydrocarbon degradation: bioremediation of oil spills. *Journal of Chemical Technology and Biotechnology*, 52, 149–56.

Aust, S. D. (1990). Degradation of environmental pollutants by *Phanerochaete chrysosporium*. *Microbial Ecology*, 20, 197–209.

Aust, S. D., Shah, M. M., Barr, D. P. & Chung, N. (1994). Biodegradation of environmental pollutants by white-rot fungi. In *Bioremediation of Chlorinated and Polycyclic Aromatic Hydrocarbon Compounds*, ed. R. E. Hinchee *et al.*, pp. 441–5. Boca Raton, FL: CRC Press.

Baker, C. A., Brooks, A. A., Greenly, R. Z. & Henis, J. M. (1988). *Encapsulation Method*. Monsanto Corp. U.S. Patent #131965.

Barles, R. W., Daughton, G. C. & Hsieh, D. P. (1979). Accelerated parathion degradation in soil inoculated with acclimated bacteria under field conditions. *Archives of Environmental Contamination and Toxicology*, 8, 647–60.

Barnsley, E. A. (1975). The bacterial degradation of fluoranthene and benzo[a]pyrene. *Canadian Journal of Microbiology*, 21, 1004–8.

Barnsley, E. A. (1976a). Naphthalene metabolism by pseudomonads: The oxidation of 1,2-dihydroxynaphthalene to 2-hydroxychromene-2-carboxylic acid and the formation of 2'-hydroxybenzalpyruvate. *Biochemistry and Biophysics Research Communications*, 72, 1116–21.

Barnsley, E. A. (1976b). Role and regulation of the *ortho* and *meta* pathways of catechol metabolism in pseudomonads metabolizing naphthalene and salicylate. *Journal of Bacteriology*, 125, 404–8.

Barnsley, E. A. (1983). Bacterial oxidation of naphthalene and phenanthrene. *Journal of Bacteriology*, 153, 1069–71.

Bates, T. S., Hamilton, S. E. & Cline, J. D. (1984). Vertical transport and sedimentation of hydrocarbons in the central main basin of Puget Sound, Washington. *Environmental Science & Technology*, 18, 299–305.

Baud-Gasset, F., Safferman, S. I., Baud-Gasset, S. & Lamar, R. T. (1994). Demonstration of soil bioremediation and toxicity reduction by fungal treatment. In *Bioremediation of Chlorinated and Polycyclic Aromatic Hydrocarbon Compounds*, ed. R. E. Hinchee *et al.*, pp. 496–500. Boca Raton, FL: CRC Press.

Bauer, J. E. & Capone, D. G. (1985). Degradation and mineralization of the polycyclic aromatic hydrocarbons anthracene and naphthalene in intertidal sediments. *Applied and Environmental Microbiology*, 50, 81–90.

Bauer, J. E. & Capone, D. G. (1988). Effects of co-occurring aromatic hydrocarbons on the degradation of individual polycyclic aromatic hydrocarbons in marine sediment slurries. *Applied and Environmental Microbiology*, 54, 1649–55.

Bauer, J. E., Kerr, R. P., Bautista, M. F., Decker, C. J. & Capone, D. G. (1988). Stimulation of microbial activities and polycyclic aromatic hydrocarbon degradation in marine sediments inhabited by *Capitella capitata*. *Marine Environmental Research*, 25, 63–84.

Bell, R. M. (1992). *Higher Plant Accumulation of Organic Pollutants from Soils*. EPA/600/SR-92/138.

Berg, J. D. & Eggen, T. (1991). Enhanced composting for cold-climate biodegradation of organic contamination in soil. In *Proceedings, Third EPA Forum on Innovative Hazardous Waste Technologies, Dallas, Texas, June 11–13*, pp. 17–36.

Berg, J. D., Nesgard, B., Gunderson, R., Lorentsen, A. & Bennett, T. E. (1994). Washing and slurry-phase biotreatment of creosote-contaminated soil. In *Bioremediation of Chlorinated and Polycyclic Aromatic Hydrocarbon Compounds*, ed. R. E. Hinchee *et. al.*, pp. 489–95. Boca Raton, FL: CRC Press.

Blumer, M. (1976). Polycyclic aromatic compounds in nature. *Scientific American*, 234, 35–45.

Boldrin, B., Tiehm, A. & Fritzsche, C. (1993). Degradation of phenanthrene, fluorene, and fluoranthene, and pyrene by a *Mycobacterium* sp. *Applied and Environmental Microbiology*, 59, 1927–30.

Bolton, H., Fredrickson, J. K. & Elliot, L. F. (1992). Microbial ecology of the rhizosphere. In *Soil Microbial Ecology, Applications in Agriculture, Forestry and Environmental Management*, ed. F. B. Metting, pp. 27–63. New York: Marcel-Dekker.

Borchert, S. & Sick, M. (1992). The vacuum-vaporizer-well (UVB) technology for *in situ* groundwater remediation: Brief description of technology and three case studies. In *Proceedings, International Symposium on Environmental Contamination in Central and Eastern Europe, Budapest, Hungary, October 12–16, 1992*.

Bossert, I. & Bartha, R. (1986). Structure-biodegradability relationships of polycyclic aromatic hydrocarbons in soil. *Bulletin of Environmental Contamination and Toxicology*, 37, 490–5.

Boyd, S. A. & Sun, S. (1990). Residual petroleum and polychlorinated oils as sorptive phases for organic contaminants in soils. *Environmental Science & Technology*, 24, 142–4.

Briglia, M., Nurmiaho-Lassila, E. L., Vallini, G. & Salkinoja-Salonen, M. (1990). The survival of pentachlorophenol-degrading *Rhodococcus chlorophenolicus* and *Flavobacterium* sp. in natural soil. *Biodegradation*, 1, 273–81.

Britton, L. N. (1985). *Feasibility Studies on the Use of Hydrogen Peroxide to Enhance Microbial Degradation of Gasoline*. American Petroleum Institute Publication 4389.

Brown, R. A. & Crosbie, J. R. (1994). Oxygen sources for *in situ* bioremediation. In *Bioremediation: Field Experience*, ed. P. Flathman *et al.*, pp. 311–31. Boca Raton, FL: CRC Press.

Brown, R. A. & Norris, R. D. (1994). The evaluation of a technology: Hydrogen peroxide in *in situ* bioremediation. In *Hydrocarbon Bioremediation*, ed. R. E. Hinchee *et al.*, pp. 148–62. Boca Raton, FL: CRC Press.

Bumpus, J. A. (1989). Biodegradation of polycyclic aromatic hydrocarbons by *Phanerochaete chrysosporium. Applied and Environmental Microbiology*, 55, 54–158.

Bumpus, J. A., Tien, M., Wright, D. & Aust, S. D. (1985). Oxidation of persistent environmental pollutants by a white rot fungus. *Science*, 228, 1434–6.

Burton, M. B., Martinson, M. M. & Barr, K. D. (1988). Microbiology treatment of contaminated water and soils. In *Biotechnology USA, 5th Annual Industrial Conference, Nov 14–16, San Francisco, CA.*

Butler, E. L., Prince, R. C., Douglas, G. S., Aczel, T. Hsu, C. S., Bronson, M. T., Clark, J. R., Lindstrom, J. E. & Steinhauer, W. G. (1991). Hopane, a new chemical tool for the measurement of oil biodegradation. In *On-Site Bioreclamation: Processes for Xenobiotic and Hydrocarbon Treatment*, ed. R. E. Hinchee & R. F. Olfenbuttel, pp. 515–21. Boston, MA: Butterworth–Heinemann.

Catterall, F. A., Murray, K. & Williams, P. A. (1971). The configuration of the 1,2-dihydroxy-1,2-dihydronaphthalene formed by the bacterial metabolism of naphthalene. *Biochimica et Biophysica Acta*, 237, 361–4.

Cerniglia, C. E. (1982). Initial reactions in the oxidation of anthracene by *Cunninghamella elegans. Journal of General Microbiology*, 128, 2055–61.

Cerniglia, C. E. (1984). Microbial metabolism of polycyclic aromatic hydrocarbons. In *Advances in Applied Microbiology*, Vol. 30, ed. A. Laskin, pp. 31–71. New York: Academic Press.

Cerniglia, C. E. (1992). Biodegradation of polycyclic aromatic hydrocarbons. *Biodegradation*, 3, 351–68.

Cerniglia, C. E. (1993). Biodegradation of polycyclic aromatic hydrocarbons. *Current Opinion in Biotechnology*, 4, 331–8.

Cerniglia, C. E. & Crow, S. A. (1981). Metabolism of aromatic hydrocarbons by yeasts. *Archives of Microbiology*, 129, 9–13.

Cerniglia, C. E. & Gibson, D. T. (1977). Metabolism of naphthalene by *Cunninghamella elegans. Applied and Environmental Microbiology*, 34, 363–70.

Cerniglia, C. E. & Gibson, D. T. (1978). Metabolism of naphthalene by cell extracts of *Cunninghamella elegans. Archives of Biochemistry and Biophysics*, 186, 121–7.

Cerniglia, C. E. & Gibson, D. T. (1979). Oxidation of benzo(a)pyrene by the filamentous fungus *Cunninghamella elegans. Journal of Biological Chemistry*, 254, 12 174–80.

Cerniglia, C. E. & Gibson, D. T. (1980a). Fungal oxidation of benzo(a)pyrene and (±)-trans-7,8-dihydrodroxy-7,8-dihydrobenzo(a)pyrene: evidence for the formation of a benzo(a)pyrene 7,8-diol-9,10-epoxide. *Journal of Biological Chemistry*, 255, 5159–63.

Cerniglia, C. E. & Gibson, D. T. (1980b). Fungal oxidation of (±)-9,10-dihydroxy-9,10-dihydrobenzo(a)pyrene: formation of diastereomeric benzo(a)pyrene 9,10-diol 7,8-epoxides. *Proceedings of the National Academy of Science USA*, 77, 4554–8.

Cerniglia, C. E. & Heitkamp, M. A. (1989). Microbial degradation of polycyclic aromatic hydrocarbons in the aquatic environment. In *Metabolism of Polycyclic Aromatic Hydrocarbons in the Aquatic Environment*, ed. U. Varanasi, pp. 41–68. Boca Raton, FL: CRC Press.

Cerniglia, C. E. & Yang, S. K. (1984). Stereoselective metabolism of anthracene and phenanthrene by the fungus *Cunninghamella elegans. Applied and Environmental Microbiology*, 47, 119–24.

Cerniglia, C. E., Hebert, R. L., Dodge, R. H., Szaniszlo, P. J. & Gibson, D. T. (1978). Fungal transformation of naphthalene. *Archives of Microbiology*, 117, 135–43.

Cerniglia, C. E., Gibson, D. T. & Van Baalen, C. (1979). Algal oxidation of aromatic hydrocarbons: formation of 1-naphthol from naphthalene by *Agmenellum quadruplicatum* strain PR-6. *Biochemistry and Biophysics Research Communications*, 88, 50–8.

Cerniglia, C. E., Van Baalen, C. & Gibson, D. T. (1980a). Metabolism of naphthalene by the cyanobacterium *Oscillatoria* sp., strain JCM. *Journal of General Microbiology*, 116, 485–95.

Cerniglia, C. E., Gibson, D. T. & Van Baalen, C. (1980b). Algal oxidation of naphthalene. *Journal of General Microbiology*, 116, 495–500.

Cerniglia, C. E., Mahaffey, W. & Gibson, D. T. (1980c). Fungal oxidation of benzo(a)pyrene: formation of (-)-*trans*-7,8-dihydroxy-7,8-dihydrobenzo(a)pyrene by *Cunninghamella elegans*. *Biochemistry and Biophysics Research Communications*, 94, 226–32.

Cerniglia, C. E., Dodge, R. H. & Gibson, D. T. (1980d). Studies on the fungal oxidation of polycyclic aromatic hydrocarbons. *Botanica Marina*, 23, 121–4.

Cerniglia, C. E., Gibson, D. T. & Van Baalen, C. (1982). Aromatic hydrocarbon oxidation by diatoms isolated from the Kachemak Bay region of Alaska. *Journal of General Microbiology*, 128, 987–90.

Cerniglia, C. E., Althaus, J. R., Evans, F. E., Freeman, J. P., Mitchum, R. K. & Yang, S. K. (1983). Stereochemistry and evidence for an arene-oxide-NIH shift pathway in the fungal metabolism of naphthalene. *Chemico-Biological Interactions*, 44, 119–32.

Cerniglia, C. E., Freeman, J. P. & Evans, F. E. (1984). Evidence for an arene oxide-NIH shift pathway in the transformation of naphthalene to 1-naphthol in *Bacillus cereus*. *Archives of Microbiology*, 138, 283–6.

Cerniglia, C. E., White, G. L. & Heflich, R. H. (1985a). Fungal metabolism and detoxification of polycyclic aromatic hydrocarbons. *Archives of Microbiology*, 143, 105–10.

Cerniglia, C. E., Freeman, J. P., White, G. L., Heflich, R. H. & Miller, D. W. (1985b). Fungal metabolism and detoxification of the nitropolycyclic aromatic hydrocarbon, 1-nitropyrene. *Applied and Environmental Microbiology*, 50, 649.

Cerniglia, C. E., Kelly, D. W., Freeman, J. P. & Miller, D. W. (1986). Microbial metabolism of pyrene. *Chemical Biological Interactions*, 57, 203–16.

Cerniglia, C. E., Campbell, W. L., Freeman, J. P. & Evans, F. E. (1989). Identification of a novel metabolite in phenanthrene metabolism by the fungus *Cunninghamella elegans*. *Applied and Environmental Microbiology*, 55, 2275–9.

Cerniglia, C. E., Campbell, W. L., Fu, P. P., Freeman, J. P. & Evans, F. E. (1990). Stereoselective fungal metabolism of methylated anthracenes. *Applied and Environmental Microbiology*, 56, 661–8.

Cerniglia, C. E., Sutherland, J. B. & Crow, S. A. (1992). Fungal metabolism of aromatic hydrocarbons. In *Microbial Degradation of Natural Products*, ed. G. Winkelmann, pp. 193–217. Weinheim: VCH Press.

Cerniglia, C. E., Gibson, D. T. & Dodge, R. H. (1994). Metabolism of benz[a]anthracene by the filamentous fungus *Cunninghamella elegans*. *Applied and Environmental Microbiology*, 60, 3931–8.

Chapman, P. J. (1979). Degradation mechanisms. In *Proceedings of the Workshop: Microbial Degradation of Pollutants in Marine Environments*, ed. A. W. Bourquin & P. H. Pritchard, pp. 28–66. Gulf Breeze, FL: U.S. Environmental Protection Agency.

Coates, J. T. & Elzerman, A. W. (1986). Desorption kinetics for selected PCB congeners from river sediments. *Journal of Contaminant Hydrology*, 1, 191–210.

Cody, T. E., Radike, M. J. & Warshawsky, D. (1984). The phytotoxicity of benzo[a]pyrene in the green alga, *Selenastrum capricornutum*. *Environmental Research*, 35, 122–31.

Colla, C., Fiecchi, A. & Treccani, V. (1959). Ricerche sul metabolismo ossidativo microbico dell'antracene e del fenantrene. Nota II. Isolamento e caratterizzazione del 3,4-diidro-3,4-diossifenantrene. *Annali di Microbiologia ed Enzimologia*, 9, 87–91.

Comeau, Y., Greer, C. W. & Samson, R. (1993). Role of inoculum preparation and density on the bioremediation of 2,4-D contaminated soil by bioaugmentation. *Applied and Environmental Microbiology*, 38, 681–7.

Crawford, R. L. & Mohn, W. W. (1985). Microbiological removal of pentachlorophenol from soil using a *Flavobacterium*. *Enzyme and Microbial Technology*, 7, 617–20.

Cripps, G. C. (1992). Baseline levels of hydrocarbons in seawater of the southern ocean: natural variability and regional patterns. *Marine Pollution Bulletin*, 24, 109–14.

Dagley, S. (1971). Catabolism of aromatic compounds by microorganisms. *Advances in Microbial Physiology*, 6, 1–46.

Dagley, S. (1975). A biochemical approach to some problems of environmental pollution. *Essays in Biochemistry*, 11, 81–138.

Davies, J. I. & Evans, W. C. (1964). Oxidative metabolism of naphthalene by soil pseudomonads: The ring-fission mechanism. *Biochemical Journal*, 91, 251–61.

Dhawale, S. W., Dhawale, S. S. & Dean-Ross, D. (1992). Degradation of phenanthrene by *Phanerochaete chrysosporium* occurs under ligninolytic as well as nonligninolytic conditions. *Applied and Environmental Microbiology*, 58, 3000–6.

Dua, R. D. & Meera, S. (1981). Purification and characterization of naphthalene oxygenase from *Corynebacterium renale*. *European Journal of Biochemistry*, 120, 461–5.

Dunn, N. W. & Gunsalus, I. C. (1973). Transmissible plasmid coding early enzymes of naphthalene oxidation in *Pseudomonas putida*. *Journal of Bacteriology*, 114, 974–9.

Eaton, R. W. & Chapman, P. J. (1992). Bacterial metabolism of naphthalene: Construction and use of recombinant bacteria to study ring cleavage of 1,2-dihydroxynaphthalene and subsequent reactions. *Journal of Bacteriology*, 174, 7542–54.

Edgehill, R. U. & Finn, R. K. (1983). Microbial treatment of soil to remove pentachlorophenol. *Applied and Environmental Microbiology*, 45, 1122–7.

Edwards, D. A., Luthy, R. G. & Liu, Z. (1991). Solubilization of polycyclic aromatic hydrocarbons in micellar nonionic surfactant solutions. *Environmental Science & Technology*, 25, 127–33.

Eikum, A. S., Berg, J. D., Selfors, H. & Eggen, T. (1992). Simple composting system for biotreatment of PAH-contaminated soil. In *Proceedings, International Symposium on Environmental Contamination in Central and Eastern Europe, Budapest, Hungary*. October 12–16, 1992. 5 pp.

Ellis, B. (1994). Reclaiming contaminated land: *in situ/ex situ* remediation of creosote- and petroleum hydrocarbons-contaminated sites. In *Bioremediation: Field Experience*, ed. P. Flathman *et al.*, pp. 107–43. Boca Raton, FL: CRC Press.

Ellis, B., Harold, P. & Kronberg, H. (1991). Bioremediation of a creosote contaminated site. *Environmental Technology*, 12, 447–59.

Elmendorf, D. L., Haith, C. E., Douglas, G. S. & Prince, R. C. (1994). Relative rates of biodegradation of substituted polycyclic aromatic hydrocarbons. In *Bioremediation of Chlorinated and Polycyclic Aromatic Hydrocarbon Compounds*, ed. R. E. Hinchee *et al.*, pp. 188–208. Boca Raton, FL: CRC Press.

Ensley, B. D. & Gibson, D. T. (1983). Naphthalene dioxygenase: purification and properties of a terminal oxygenase component. *Journal of Bacteriology*, 155, 505–11.

Ensley, B. D., Gibson, D. T. & Laborde, A. L. (1982). Oxidation of naphthalene by a multicomponent enzyme system from *Pseudomonas* sp. strain NCIB 9816. *Journal of Bacteriology*, 149, 948–54.

Ensley, B. D., Osslund, T. D., Joyce, M. & Simon, M. J. (1987). Expression and complementation of naphthalene dioxygenase activity in *Escherichia coli*. In *Microbial Metabolism and the Carbon Cycle*, ed. S. R. Hagedorn, R. S. Hanson & D. A. Kunz, pp. 437–55. New York: Harwood Academic Publishers.

Evans, W. C., Fernley, H. N. & Griffiths, E. (1965). Oxidative metabolism of phenanthrene and anthracene by soil pseudomonads: the ring-fission mechanism. *Biochemical Journal*, 95, 819–31.

Federal Register (1993). *Land Disposal Restrictions for Newly Identified and Listed Hazardous Wastes and Hazardous Soil: Proposed Rule*. Federal Register, Tuesday, September 14, 1993.

Ferris, J. P., Fasco, M. J., Stylianopoulou, F. L., Jerina, D. M., Daly, J. W. & Jeffrey, A. M. (1973). Mono-oxygenase activity in *Cunninghamella bainieri*: Evidence for a fungal system similar to liver microsomes. *Archives of Biochemistry and Biophysics*, 156, 97–103.

Field, J. A., de Jong, E., Feijoo Costa, G. & de Bont, J. A. M. (1992). Biodegradation of polycyclic aromatic hydrocarbons by new isolates of white rot fungi. *Applied and Environmental Microbiology*, 58, 2219–26.

Finnerty, W. R. & Singer, M. E. (1984). A microbial biosurfactant: physiology, biochemistry and applications. *Developments in Industrial Microbiology*, 25, 311–40.

Flyvberg, J., Arvin, E., Jensen, B. K. & Olson, S. K. (1991). Bioremediation of oil- and creosote-related aromatic compounds under nitrate-reducing conditions. In *In Site Bioreclamation; Applications and Investigations for Hydrocarbon and Contaminated Site Remediation*, ed. R. E. Hinchee & R. F. Olfenbuttel, pp. 471–9. Boston, MA: Butterworth-Heinemann.

Foght, J. M. & Westlake, D. W. S. (1988). Degradation of polycyclic aromatic hydrocarbons and aromatic heterocycles by a *Pseudomonas* species. *Canadian Journal of Microbiology*, 34, 1135–41.

Freeman, D. J. & Cattell, F. C. R. (1990). Woodburning as a source of atmospheric polycyclic aromatic hydrocarbons. *Environmental Science & Technology*, 24, 1581–5.

García-Valdés, E., Cozar, E., Rotger, R., Lalucat, J. & Ursing, J. (1988). New naphthalene-degrading marine *Pseudomonas* strains. *Applied and Environmental Microbiology*, 54, 2478–85.

Gardner, W. S., Lee, R. F., Tenore, K. R. & Smith, L. W. (1979). Degradation of selected polycyclic aromatic hydrocarbons in coastal sediments: importance of microbes and polychaete worms. *Water, Air, and Soil Pollution*, 11, 339–47.

George, E. J. & Neufeld, R. D. (1989). Degradation of fluorene in soil by fungus *Phanerochaete chrysosporium*. *Biotechnology and Bioengineering*, 33, 1306–10.

Ghosal, D., You, I. S. & Gunsalus, I. C. (1987). Nucleotide sequence and expression of gene *nahH* of plasmid NAH7 and homology with gene *xylE* of TOL pWWO. *Gene*, 55, 19–28.

Ghosh, D. K. & Mishra, A. K. (1983). Oxidation of phenanthrene by a strain of *Micrococcus*: Evidence of protocatechuate pathway. *Current Microbiology*, 9, 219–24.

Ghosh, D. K., Dutta, D., Samanta, T. B. & Mishra, A. K. (1983). Microsomal benzo[a]pyrene hydroxylase in *Aspergillus ochraceus* TS: Assay and characterization of the enzyme system. *Biochemistry and Biophysics Research Communications*, 113, 497–505.

Gibson, D. T. & Subramanian, V. (1984). Microbial degradation of aromatic hydrocarbons.

In *Microbial Degradation of Organic Compounds*, ed. D. T. Gibson, pp. 181–252. New York: Marcel Dekker.

Gibson, D. T., Koch, J. R. & Kallio, R. E. (1968). Oxidative degradation of aromatic hydrocarbons by microorganisms. I. Enzymatic formation of catechol from benzene. *Biochemistry*, 7, 2653–61.

Gibson, D. T., Mahadevan, V., Jerina, D. M., Yagi, H. & Yeh, H. J. C. (1975). Oxidation of the carcinogens benzo[a]pyrene and benz[a]anthracene to dihydrodiols by a bacterium. *Science*, 189, 295–7.

Gibson, D. T., Zylstra, G. J. & Chauhan, S. (1990). Biotransformations catalyzed by toluene dioxygenase from *Pseudomonas putida* F1. In *Pseudomonas: Biotransformations, Pathogenesis, and Evolving Biotechnology*, ed. S. Silver, A. M. Chakrabarty, B. Iglewski & S. Kaplan, pp. 121–32. Washington, DC: American Society for Microbiology.

Girvin, D. C., Skalrew, D. S., Scott, A. J. & Zipperer, J. P. (1990). *Release and Attenuation of PCB Congeners: Measurement of Desorption Kinetics and Equilibrium Sorption Partition Coefficients*. Palo Alto, CA. Electric Power Research Institute, EPRI GS-6875.

Glaser, J. A., McCauley, P. T., Dosani, M. A., Platt, J. S. & Krishan, E. R. (1994). Engineering optimization of slurry bioreactors for treating hazardous wastes. In *Proceedings, Symposium on Bioremediation of Hazardous Wastes: Research, Development and Field Evaluations*, pp. 109–15. EPA/600/R-94/075.

Godsy, E. M. & Gorelitz, D. F. (1986). *USGS Water Supply Paper no. 2285*, ed. A. C. Mattraw & B. J. Franks. Denver, CO: US Geological Survey. 36 p.

Gold, M. H., Wariishi, H. & Valli, K. (1989). Extracellular peroxidases involved in lignin degradation by the white rot basidiomycete *Phanerochaete chrysosporium*. In *Biocatalysis in Agricultural Biotechnology*, ed. J. R. Whitaker & P. E. Sonnet, pp. 127–40. Washington, DC: American Chemical Society.

Goldstein, R. M., Mallory, L. M. & Alexander, M. (1985). Reasons for possible failure of inoculation to enhance biodegradation. *Applied and Environmental Microbiology*, 50, 977–83.

Gordon, D. C., Keizer, P. D. & Dale, J. (1978). Temporal variations and probable origins of hydrocarbons in the water column of Bedford Basin, Nova Scotia. *Estuarine and Coastal Marine Science*, 7, 243–56.

Grady, C. P. L., Jr (1989). The enhancement of microbial activity through bioreactor design. In *Biotechnology Applications in Hazardous Waste Treatment*, ed. G. Lewandowski *et al.*, pp. 81–93. New York: Engineering Foundation.

Grady, C. P. L., Jr & Lim, H. C. (1980). *Biological Wastewater Treatment: Theory and Applications*. New York: Marcel-Dekker.

Graves, D. & Leavitt, M. (1991). Petroleum biodegradation in soil: the effect of direct application of surfactants. *Remediation*, 2, 147–66.

Greer, L. E. & Shelton, D. R. (1992). Effect of inoculant strain and organic matter content on kinetics of 2,4-dichlorophenoxyacetic acid degradation in soil. *Applied and Environmental Microbiology*, 58, 1459–65.

Grimalt, J., Albaiges, J., Al Saad, H. T. & Douabul, A. A. (1985). N-alkane distributions in surface sediments from the Arabian Gulf. *Naturwissenschaften*, 72, 35–7.

Grosser, R. J., Warshawsky, D. & Vestal, J. R. (1991). Indigenous and enhanced mineralization of pyrene, benzo[a]pyrene and carbazole in soils. *Applied and Environmental Microbiology*, 57, 3462–9.

Grund, E., Denecke, B. & Eichenlaub, R. (1992). Naphthalene degradation via salicylate and gentisate by *Rhodococcus* sp. strain B4. *Applied and Environmental Microbiology*, 58, 1874–7.

Guerin, W. F. & Jones, G. E. (1988). Mineralization of phenanthrene by a *Mycobacterium* sp. *Applied and Environmental Microbiology*, 54, 937–44.

Guerin, W. F. & Jones, G. E. (1989). Estuarine ecology of phenanthrene-degrading bacteria. *Estuarine Coastal and Shelf Science*, 29, 115–30.

Haemmerli, S. D., Leisola, M. S. A., Sanglard, D. & Fiechter, A. (1986). Oxidation of benzo[a]pyrene by extracellular ligninases of *Phanerochaete chrysosporium*: veratryl alcohol and stability of ligninase. *Journal of Biological Chemistry*, 261, 6900–3.

Haigler, B. E. & Gibson, D. T. (1990a). Purification and properties of NADH-ferredoxin$_{NAP}$ reductase, a component of naphthalene dioxygenase from *Pseudomonas* sp. strain NCIB 9816. *Journal of Bacteriology*, 172, 457–64.

Haigler, B. E. & Gibson, D. T. (1990b). Purification and properties of ferredoxin$_{NAP}$, a component of naphthalene dioxygenase from *Pseudomonas* sp. strain NCIB 9816. *Journal of Bacteriology*, 172, 465–8.

Hammel, K. E. (1989). Organopollutant degradation by ligninolytic fungi. *Enzyme and Microbial Technology*, 11, 776–7.

Hammel, K. E., Kalyanaraman, B. & Kirk, T. K. (1986). Oxidation of polycyclic aromatic hydrocarbons and dibenzo[*p*]dioxins by *Phanerochaete chrysosporium* ligninase. *Journal of Biological Chemistry*, 261, 16 948–52.

Hammel, K. E., Green, B. & Gai, W. Z. (1991). Ring fission of anthracene by a eukaryote. *Proceedings of the National Academy of Science USA*, 88, 10 605–8.

Hammel, K. E., Gai, W. Z., Green, B. & Moen, M. A. (1992). Oxidative degradation of phenanthrene by the ligninolytic fungus *Phanerochaete chrysosporium*. *Applied and Environmental Microbiology*, 58, 1832–8.

Heitkamp, M. A. & Cerniglia, C. E. (1987). The effects of chemical structure and exposure on the microbial degradation of polycyclic aromatic hydrocarbons in freshwater and estuarine ecosystems. *Environmental Toxicology and Chemistry*, 6, 535–46.

Heitkamp, M. A. & Cerniglia, C. E. (1988). Mineralization of polycyclic aromatic hydrocarbons by a bacterium isolated from sediment below an oil field. *Applied and Environmental Microbiology*, 54, 1612–14.

Heitkamp, M. A. & Cerniglia, C. E. (1989). Polycyclic aromatic hydrocarbon degradation by a *Mycobacterium* sp. in microcosms containing sediment and water from a pristine ecosystem. *Applied and Environmental Microbiology*, 55, 1968–73.

Heitkamp, M. A., Franklin, W. & Cerniglia, C. E. (1988a). Microbial metabolism of polycyclic aromatic hydrocarbons: isolation and characterization of a pyrene-degrading bacterium. *Applied and Environmental Microbiology*, 54, 2549–55.

Heitkamp, M. A., Freeman, J. P., Miller, D. W. & Cerniglia, C. E. (1988b). Pyrene degradation by a *Mycobacterium* sp.: identification of ring oxidation and ring fission products. *Applied and Environmental Microbiology*, 54, 2556–65.

Herbes, S. E. & Schwall, L. R. (1978). Microbial transformation of polycyclic aromatic hydrocarbons in pristine and petroleum contaminated sediments. *Applied and Environmental Microbiology*, 35, 306–16.

Herrling, B., Stamm, J., Alesi, E. J., Bott-Breuning, G. & Diekmann, S. (1994). *In situ* bioremediation of groundwater containing hydrocarbons, pesticides, or nitrate using vertical circulation flows (UVB/GZB technique). In *Air Sparging for Site Remediation*, ed. R. Hinchee, pp. 56–80. Boca Raton, FL: CRC Press.

Hill, H. J., Reisberg, J. & Stegemeier, G. L. (1973). Aqueous surfactant systems for oil recovery. *Journal of Petroleum Technology*, 25, 186–94.

Hinchee, R. (1993). Bioventing of Petroleum Hydrocarbons. In In Situ *Bioremediation of Ground Water and Geological Material: A Review of Technologies*, ed. R. D. Norris *et al.*, pp. 3.1–3.21. EPA/600/R-93/124. Section 3. NTIS Document No. PB93-215564. Washington, DC.

Hinchee, R. E. & Ong, S. K. (1992). A rapid *in situ* respiration test for measuring aerobic biodegradation rates of hydrocarbons in soils. *Journal of the Air and Waste Management Association*, 42, 1305–12.

Hites, R. A., LaFlamme, R. E. & Farrington, J. W. (1977). Sedimentary polycyclic aromatic hydrocarbons: the historical record. *Science*, 198, 829–31.

Hofmann, K. H. (1986). Oxidation of naphthalene by *Saccharomyces cerevisiae* and *Candida utilis*. *Journal of Basic Microbiology*, 26, 109–11.

Holland, H. L., Khan, S. H., Richards, D. & Riemland, E. (1986). Biotransformation of polycyclic aromatic compounds by fungi. *Xenobiotica*, 16, 733–41.

Hopper, D. J. (1978). Microbial degradation of aromatic hydrocarbons. In *Developments in Biodegradation of Hydrocarbons - 1*, ed. R. J. Watkinson, pp. 85–112. London: Applied Science Publishers.

Hsu, T-S. & Bartha, R. (1979). Accelerated mineralization of two organophosphate insecticides in the rhizosphere. *Applied and Environmental Microbiology*, 37, 36–41.

Hu, Z-C., Korus, R. A., Levinson, W. E. & Crawford, R. L. (1994). Adsorption and biodegradation of pentachlorophenol by polyurethane-immobilized *Flavobacterium*. *Environmental Science & Technology*, 28, 491–6.

Jafvert, C. T. & Weber, E. J. (1992). *Sorption of Ionizable Organic Compounds to Sediments and Soils*. EPA/600/S3-91/017.

Jeffrey, A. M., Yeh, H. J. C., Jerina, D. M., Patel, T. R., Davey, J. F. & Gibson, D. T. (1975). Initial reactions in the oxidation of naphthalene by *Pseudomonas putida*. *Biochemistry*, 14, 575–84.

Jerger, D. E., Cady, D. J. & Exner, J. H. (1994). Full-scale slurry-phase biological treatment of wood-preserving wastes. In *Bioremediation of Chlorinated and Polycyclic Aromatic Hydrocarbon Compounds*, ed. R. E. Hinchee *et al.*, pp. 490–3. Boca Raton, FL: CRC Press.

Jerina, D. M., Daly, J. W., Jeffrey, A. M. & Gibson, D. T. (1971). *cis*-1,2-Dihydroxy-1,2-dihydronaphthalene: A bacterial metabolite from naphthalene. *Archives of Biochemistry and Biophysics*, 142, 394–6.

Jerina, D. M., Selander, H., Yagi, H., Wells, M. C., Davey, J. F., Mahadevan, V. & Gibson, D. T. (1976). Dihydrodiols from anthracene and phenanthrene. *Journal of the American Chemical Society*, 98, 5988–96.

Jerina, D. M., van Bladeren, P. J., Yagi, H., Gibson, D. T., Mahadevan, V., Neese, A. S., Koreeda, M., Sharma, N. D. & Boyd, D. R. (1984). Synthesis and absolute configuration of the bacterial *cis*-1,2-, *cis*-8,9-, and *cis*-10,11-dihydrodiol metabolites of benz[a]anthracene formed by a strain of *Beijerinckia*. *Journal of Organic Chemistry*, 49, 3621–8.

Jones, K. C., Stratford, J. A., Waterhouse, K. S., Furlong, E. T., Giger, W., Hites, R. A., Schaffner, C. & Johnston, A. E. (1989). Increases in the polynuclear aromatic hydrocarbon content of an agricultural soil over the last century. *Environmental Science & Technology*, 23, 95–101.

Karickhoff, S. W. & Morris, K. R. (1985). Sorption dynamics of hydrophobic pollutants in sediment suspensions. *Environmental Toxicology and Chemistry*, 4, 462–79.

Kelley, I. & Cerniglia, C. E. (1991). The metabolism of fluoranthene by a species of *Mycobacterium. Journal of Industrial Microbiology*, 7, 19–26.

Kelley, I., Freeman, J. P., Cerniglia, C. E. (1990). Identification of metabolites from degradation of naphthalene by a *Mycobacterium* sp. *Biodegradation*, 1, 283–90.

Kelley, I., Freeman, J. P., Evans, F. E. & Cerniglia, C. E. (1991). Identification of a carboxylic acid metabolite from the catabolism of fluoranthene by a *Mycobacterium* sp. *Applied and Environmental Microbiology*, 57, 636–41.

Kelley, I., Freeman, J. P., Evans, F. E. & Cerniglia, C. E. (1993). Identification of metabolites from the degradation of fluoranthene by *Mycobacterium* sp. strain PYR-1. *Applied and Environmental Microbiology*, 59, 800–6.

Keuth, S. & Rehm, H. J. (1991). Biodegradation of phenanthrene by *Arthrobacter polychromogenes* isolated from a contaminated soil. *Applied Microbiology and Biotechnology*, 34, 804–8.

Kile, D. E. & Chiou, C. T. (1989). Water solubility enhancement of DDT and trichlorobenzene by some surfactants below and above the critical micelle concentration. *Environmental Science & Technology*, 23, 832–8.

Kile, D. E. & Chiou, C. T. (1990). Effect of some petroleum sulfonate surfactants on the apparent water solubility of organic compounds. *Environmental Science & Technology*, 24, 205–8.

Kingsley, M. T., Fredrickson, J. K., Metting, F. B. & Seidler, R. J. (1994). Environmental restoration using plant-microbe bioaugmentation. In *Bioremediation of Chlorinated and Polycyclic Aromatic Hydrocarbon Compounds*, ed. R. E. Hinchee *et al.*, pp. 287–92. Boca Raton, FL: CRC Press.

Kiyohara, H. & Nagao, K. (1978). The catabolism of phenanthrene and naphthalene by bacteria. *Journal of General Microbiology*, 105, 69–75.

Kiyohara, H., Nagao, K. & Nomi, R. (1976). Degradation of phenanthrene through o-phthalate by an *Aeromonas* sp. *Agricultural and Biological Chemistry*, 40, 1075–82.

Kiyohara, H., Nagao, K., Kuono, K. & Yano, K. (1982). Phenanthrene degrading phenotype of *Alcaligenes faecalis* AFK2. *Applied and Environmental Microbiology*, 43, 458–61.

Kiyohara, H., Takizawa, N., Date, H., Torigoe, S. & Yano, K. (1990). Characterization of a phenanthrene degradation plasmid from *Alcaligenes faecalis* AFK2. *Journal of Fermentation and Bioengineering*, 69, 54–6.

Komatsu, T., Omori, T. & Kodama, T. (1993). Microbial degradation of the polycyclic aromatic hydrocarbons acenaphthene and acenaphthylene by a pure bacterial culture. *Bioscience Biotechnology and Biochemistry*, 57, 864–5.

Kuhm, A. E., Stoiz, A. & Knackmuss, H. J. (1991). Metabolism of naphthalene by the biphenyl-degrading bacterium *Pseudomonas paucimobilis* Q1. *Biodegradation*, 2, 115–20.

LaFlamme, R. E. & Hites, R. A. (1984). The global distribution of polycyclic aromatic hydrocarbons. *Geochimica et Cosmochimica Acta*, 43, 289–303.

Laha, S. & Luthy, R. G. (1991). Inhibition of phenanthrene mineralization by non-ionic surfactants in soil-water systems. *Environmental Science & Technology*, 25, 1920–30.

Lajoie, C. A., Chen, S. Y., Oh, K. C., & Strom, P. F. (1992). Development and use of field application vectors to express non-adaptive foreign genes in competitive environments. *Applied and Environmental Microbiology*, 58, 655–63.

Lamar, R. T. & Glaser, J. A. (1994). Field evaluations of the remediation of soil contaminated with wood-preserving chemicals using lignin-degrading fungi. In *Bioremediation of Chlorinated and Polycyclic Aromatic Hydrocarbon Compounds*, ed. R. E. Hinchee *et. al.*, pp. 239–47. Boca Raton, FL: CRC Press.

Lambert, M., Kremer, S., Sterner, O. & Anke, H. (1994). Metabolism of pyrene by the basidiomycete *Crinipellis stipitaria* and identification of pyrenequinones and their hydroxylated precursors in strain JK375. *Applied and Environmental Microbiology*, **60**, 3597–601.

Lang, S. & Wagner, F. (1987). Structure and properties of biosurfactants. *Surfactant Science Series*, **25**, 21–45.

Lange, B., Kremer, S., Sterner, O. & Anke, H. (1994). Pyrene metabolism in *Crinipellis stipitaria*: Identification of *trans*-4,5-dihydro-4,5-dihydroxypyrene and 1-pyrenylsulfate in strain JK364. *Applied and Environmental Microbiology*, **60**, 3602–7.

Lauch, R. P., Herrmann, J. G., Mahaffey, W. R., Jones, A. B., Dosani, M. & Hessling, J. (1992). Removal of creosote from soil by bioslurry reactors. *Environmental Progress*, **11**, 265–71.

Lee, M. D., Ward, J. M., Borden, R. C., Bedient, P. B., Ward, C. H. & Wilson, J. T. (1988). Biorestoration of aquifers contaminated with organic compounds. *CRC Critical Reviews in Environmental Control*, **18**, 29–89.

Lee, M. D., Butler, W. A., Mistretta, T. F., Zanikos, I. J. & Perkins, R. E. (1994). Biological treatability studies on a surface impoundment sludge from a chemical manufacturing facility. In *Bioremediation of Chlorinated and Polycyclic Aromatic Hydrocarbon Compounds*, ed. R. E. Hinchee *et al.*, pp. 129–43. Boca Raton, FL: CRC Press.

Levinson, W. E., Stormo, K. E., Tao, H-I. & Crawford, R. L. (1994). Hazardous waste treatment using encapsulated or entrapped microorganisms. In *Biological Degradation and Bioremediation of Toxic Chemicals*, ed. G. R. Chaudhry, pp. 455–69. Portland, OR: Timber Press.

Lin, J. & Wang, H. Y. (1991). Degradation of pentachlorophenol by non-immobilized, immobilized and co-immobilized *Arthrobacter* cells. *Journal of Fermentation and Bioengineering*, **72**, 311–14.

Lin, W. S. & Kapoor, M. (1979). Induction of aryl hydrocarbon hydroxylase in *Neurospora crassa* by benzo[a]pyrene. *Current Microbiology*, **3**, 177–81.

Lindquist, B. & Warshawsky, D. (1985a). Identification of the 11,12-dihydroxybenzo[a]pyrene as a major metabolite produced by the green alga, *Selenastrum capricornutum*. *Biochemistry and Biophysics Research Communications*, **130**, 71–5.

Lindquist, B. & Warshawsky, D. (1985b). Stereospecificity in algal oxidation of the carcinogen benzo[a]pyrene. *Experientia*, **41**, 767–9.

Liu, Z., Laha, S. & Luthy, R. G. (1991). Surfactant solubilization of polycyclic aromatic hydrocarbon compounds in soil-water suspensions. *Water Science and Technology*, **23**, 475–85.

Lu, C. J. (1994). Effects of hydrogen peroxide on the *in situ* biodegradation of organic chemicals in a simulated groundwater system. In *Hydrocarbon Bioremediation*, ed. R. E. Hinchee *et al.*, pp. 140–7. Boca Raton, FL: CRC Press.

Lyon, W. G. & Rhodes, D. E. (1991). *The Swelling Properties of Soil Organic Matter and their Relation to Sorption of Non-ionic Compounds*. EPA/600/S2-91/033. Cincinnati, OH: US EPA.

MacDonald, J. A. & Kavanaugh, M. C. (1994). Restoring contaminated groundwater: an achievable goal? *Environmental Science & Technology*, **28**, 362A–8A.

Madsen, E. L., Winding, A., Malachowsky, K., Thomas, C. T. & Ghiorse, W. C. (1992).

Contrasts between subsurface microbial communities and their metabolic adaptation to polycyclic aromatic hydrocarbons at a forested and an urban coal-tar disposal site. *Microbial Ecology*, 24, 199–213.

Mahaffey, W. R., Gibson, D. T. & Cerniglia, C. E. (1988). Bacterial oxidation of chemical carcinogens: Formation of polycyclic aromatic acids from benz[a]anthracene. *Applied and Environmental Microbiology*, 54, 2415–23.

Major, D. W., Barker, C. I. & Barker, J. F. (1988). Biotransformation of benzene by denitrification on aquifer sand. *Ground Water*, 26, 8–14.

Manilal, V. B. & Alexander, M. (1991). Factors affecting the microbial degradation of phenanthrene in soil. *Applied Microbiology and Biotechnology*, 35, 401–5.

Mattraw, H. C. & Franks, B. J. (eds.) (1986). *USGS Water Supply Paper No. 2285.* Washington, DC: US Geological Survey. pp. 1–8.

McCarty, P. L. & Semprini, L. (1993). Ground water treatment for chlorinated solvents. In In Situ *Bioremediation of Ground Water and Geological Material: A Review of Technologies,* ed. R. D. Norris *et al.* EPA/600/R-93/124. Section 5. NTIS Document No. PB93-215564, Washington DC.

McCauley, P. T., Brenner, R. C., Kremer, F. V., Alleman, B. C. & Beckwith, D. C. (1994). Bioventing soils contaminated with wood preservatives. In *Proceedings, Symposium on Bioremediation of Hazardous Wastes: Research, Development and Field Evaluations,* pp. 40–5. EPA/600/R-94/075. Cincinnati, OH: US EPA.

McDermott, J. B., Unterman, R., Brennan, M. J., Brooks, R. E., Mobley, D. P., Schwartz, C. C. & Dietrich, D. K. (1989). Two strategies for PCB soil remediation: Biodegradation and surfactant extraction. *Experimental Progress*, 8, 46–51.

McGinnis, G. D., Borazjani, H., McFarland, L. K., Pope, D. F., Strobel, D. A. & Matthews, J. E. (1988). *Characterization and Laboratory Soil Treatability Studies for Creosote and Pentachlorophenol Sludges and Contaminated Soil.* EPA/600/2-88/055. Cincinnati, OH: US EPA.

Means, J. C., Ward, S. G., Hassett, J. J. & Banwart, W. L. (1980). Sorption of polycyclic aromatic hydrocarbons by sediments and soils. *Environmental Science & Technology*, 14, 1524–8.

Menn, F. M., Applegate, B. M. & Sayler, G. S. (1993). NAH plasmid-mediated catabolism of anthracene and phenanthrene to naphthoic acids. *Applied and Environmental Microbiology*, 59, 1938–42.

Micklewright, J. T. (1986). *Contract Report to the American Wood-Preserver's Institute.* Washington, DC: International Statistics Council.

Middaugh, D. P., Mueller, J. G., Thomas, R. L., Lantz, S. E., Hemmer, M. J., Brooks, G. T. & Chapman, P. J. (1991). Detoxification of creosote- and PCP-contaminated groundwater by physical extraction: chemical and biological assessment. *Archives of Environmental Contamination Toxicology*, 21, 233–44.

Middaugh, D. P., Resnick, S. M., Lantz, S. E., Heard, C. S. & Mueller, J. G. (1993). Toxicological assessment of biodegraded pentachlorophenol: Microtox™ and fish embryos. *Archives of Environmental Contamination Toxicology*, 24, 165–72.

Middaugh, D. P., Lantz, S. E., Heard, C. S. & Mueller, J. G. (1994). Field-scale testing of a two-stage bioreactor for removal of creosote and pentachlorophenol from ground water: chemical and biological assessment. *Archives of Environmental Contamination Toxicology*, 26, 320–8.

Mihelcic, J. R. & Luthy, R. G. (1988). Degradation of polycyclic aromatic hydrocarbon compounds under various redox conditions in soil-water systems. *Applied and Environmental Microbiology*, 54, 1182–7.

Miller, R. M., Singer, G. M., Rosen, J. D. & Bartha, R. (1988). Photolysis primes biodegradation of benzo(a)pyrene. *Applied and Environmental Microbiology*, 54, 1724–30.

Millero, F. J. & Sohn, M. L. (1991). *Chemical Oceanography*, Boca Raton, FL: CRC Press. 531 pp.

Moen, M. A. & Hammel, K. E. (1994). Lipid peroxidation by the manganese peroxidase of *Phanerochaete chrysosporium* is the basis for phenanthrene oxidation by the intact fungus. *Applied and Environmental Microbiology*, 60, 1956–61.

Morgan, P. & Watkinson, P. J. (1989). Hydrocarbon degradation in soils and methods for soil biotreatment. *CRC Critical Reviews in Biotechnology*, 8, 305–53.

Morgan, P. & Watkinson, R. J. (1990). Assessment of the potential for *in situ* biotreatment of hydrocarbon-contaminated soils. *Water Science and Technology*, 22, 63–8.

Morgan, P., Lewis, S. T. & Watkinson, R. J. (1991). Comparison of abilities of white-rot fungi to mineralize selected xenobiotic compounds. *Applied Microbiology and Biotechnology*, 34, 693–6.

Morgan, P., Lee, S. A., Lewis, S. T., Sheppard, A. N. & Watkinson, R. J. (1993). Growth and biodegradation by white-rot fungi inoculated into soil. *Soil Biology and Biochemistry*, 25, 279–87.

Morris, P. J. & Pritchard, P. H. (1994). Concepts in improving polychlorinated biphenyl bioavailability to bioremediation strategies. In *Bioremediation of Chlorinated and Polycyclic Aromatic Hydrocarbon Compounds*, ed. R. E. Hinchee *et al.*, pp. 359–67. Boca Raton, FL: Lewis Publishers.

Mueller, J. G. & Klingel, E. J. (1994). Combined use of ground water circulation (uvb) technology and *in situ* bioreactors for enhanced bioremediation of subsurface contamination. In *Joint Army, Navy, NASA, Air Force (JANNAF) Workshop on Environmentally Benign Cleaning and Degreasing Technology, June 14–15, Indian Head, MD*. Washington, DC: US Department of Defense.

Mueller, J. G., Chapman, P. J. & Pritchard, P. H. (1989a). Creosote-contaminated sites: their potential for bioremediation. *Environmental Science & Technology*, 23, 1197–201.

Mueller, J. G., Chapman, P. J. & Pritchard, P. H. (1989b). Action of a fluoranthene-utilizing bacterial community on polycyclic aromatic hydrocarbon components of creosote. *Applied and Environmental Microbiology*, 55, 3085–90.

Mueller, J. G., Chapman, P. J. & Pritchard, P. H. (1990a). *Development of a Sequential Treatment System for Creosote-contaminated Soil and Water: Bench Studies*, pp. 42–5. EPA/600/9-90/041. Cincinnati, OH: US EPA.

Mueller, J. G., Chapman, P. J., Blattmann, B. O. & Pritchard, P. H. (1990b). Isolation and characterization of a fluoranthene-utilizing strain of *Pseudomonas paucimobilis*. *Applied and Environmental Microbiology*, 56, 1079–86.

Mueller, J. G., Lantz, S. E., Blattmann, B. O. & Chapman, P. J. (1991a). Bench-scale evaluation of alternative biological treatment processes for the remediation of pentachlorophenol- and creosote-contaminated materials: Solid-phase bioremediation. *Environmental Science & Technology*, 25, 1045–55.

Mueller, J. G., Lantz, S. E., Blattmann, B. O. & Chapman, P. J. (1991b). Bench-scale evaluation of alternative biological treatment processes for the remediation of pentachlorophenol- and creosote-contaminated materials: slurry-phase bioremediation. *Environmental Science & Technology*, 25, 1055–61.

Mueller, J. G., Middaugh, D. P., Lantz, S. E. & Chapman, P. J. (1991c). Biodegradation of creosote and pentachlorophenol in groundwater: chemical and biological assessment. *Applied and Environmental Microbiology*, 57, 1277–85.

Mueller, J. G., Resnick, S. M., Shelton, M. E. & Pritchard, P. H. (1992). Effect of inoculation on the biodegradation of weathered Prudhoe Bay crude oil. *Journal of Industrial Microbiology*, 10, 95–105.

Mueller, J. G., Lantz, S. E., Colvin, R. J., Ross, D., Middaugh, D. P. & Pritchard, P. H. (1993a). Strategy using bioreactors and specially selected microorganisms for bioremediation of ground water contaminated with creosote and pentachlorophenol. *Environmental Science & Technology*, 27, 691–8.

Mueller, J. G., Lin, J-E., Lantz, S. E. & Pritchard, P. H. (1993b). Recent developments in cleanup technologies: implementing innovative bioremediation technologies. *Remediation*, 3, 369–91.

Mueller, J. G., Brissette, K. M., Becker, A. P. & Van De Steeg, G. E. (1994a). Evaluation of *in situ* bioventing and solid-phase bioremediation strategies for creosote-contaminated soils: laboratory test results. In *Emerging Technologies in Hazardous Waste Management VI*, ed. D. W. Tedder, p. 1017. Atlanta, GA: American Chemical Society.

Mueller, J. G., Pritchard, P. H., Montgomery, M. T. & Schultz, W. W. (1994b). *In situ* treatment technologies for PAH-contaminated sediments. *National Research Council's Committee on Contaminated Marine Sediments, Technology Review Workshop, April 21–22, 1994, Chicago IL.* Cincinnati, OH: US EPA.

Narro, M. L., Cerniglia, C. E., Van Baalen, C. & Gibson, D. T. (1992a). Evidence of NIH shift in naphthalene oxidation by the marine cyanobacterium, *Oscillatoria* species strain JCM. *Applied and Environmental Microbiology*, 58, 1360–3.

Narro, M. L., Cerniglia, C. E., Van Baalen, C. & Gibson, D. T. (1992b). Metabolism of phenanthrene by the marine cyanobacterium *Agmenellum quadruplicatum*, strain PR-6. *Applied and Environmental Microbiology*, 58, 1351–9.

National Research Council (1983). *Polycyclic Aromatic Hydrocarbons: Evaluation of Sources and Effects*. Washington, DC: National Academy Press.

Nau-Ritter, G. M., Wurster, C. F. & Rowland, R. G. (1982). Partitioning of ^{14}C-PCB between water and particulate with various organic contents. *Water Research*, 16, 1615–18.

Nelson, M., Mills, D. & Downs, L. (1994). Application of cometabolism for remediation of chloroethenes at industrial sites. In *Emerging Technologies in Hazardous Waste Management VI*, ed. D. W. Tedder, pp. 849–51. Atlanta, GA: American Chemical Society.

Nishimura, M. & Baker, E. W. (1986). Possible origin of n-alkanes with a remarkable even to odd predominance in recent marine sediments. *Geochimica et Cosmochimica Acta*, 50, 299–305.

Norris, R. D. & Falotico, R. J. (1994). Modeling of hydrogeological field data for design and optimization of *in situ* bioremediation of contaminated aquifers. In *Bioremediation: Field Experience*, ed. P. Flathman *et al.*, pp. 287–307. Boca Raton, FL: CRC Press.

Norris, R. D., Hinchee, R. E., Brown, R., McCarty, P. L., Semprini, L., Wilson, J. T., Kampbell, D. H., Reinhard, M., Bouwer, E. J., Borden, R. C., Vogel, T. M., Thomas, J. M. & Ward, C. H. (1993). In situ *Bioremediation of Ground water and Geological Material: A Review of Technologies*. EPA/600/R-93/124. Cincinnati, OH: US EPA.

Oberbremer, A., Muller-Hurtig, R. & Wagner, F. (1990). Effect of the addition of microbial surfactants on hydrocarbon degradation in a soil population in a stirred reactor. *Applied Microbiology and Biotechnology*, 32, 485–9.

Ogram, A. V., Jessup, R. E., Ou, L. T. & Rao, P. S. C. (1985). Effects of sorption on biological degradation rates of 2,4-dichlorophenoxyacetic acid in soils. *Applied and Environmental Microbiology*, 49, 582–7.

O'Reilly, K. T. & Crawford, R. L. (1989). Degradation of pentachlorophenol by polyurethane-immobilized *Flavobacterium* cells. *Applied and Environmental Microbiology*, 9, 2113–18.

Paau, A. S. (1988). Formulations useful in applying beneficial microorganisms to seeds. *Trends in Biotechnology*, 6, 276–9.

Pahm, A. M. & Alexander, M. (1993). Selecting inocula for the bioremediation of organic compounds at low concentrations. *Microbial Ecology*, 25, 275–86.

Pahm, T., Lum, K. & Lemieux, C. (1993). Sources of PAHs in the St. Lawrence River (Canada) and their relative importance. *Chemosphere*, 27, 1137–49.

Pardieck, D. L., Bouwer, E. J. & Stone, A. T. (1992). Hydrogen peroxide use to increase oxidant capacity for *in situ* bioremediation of contaminated soils and aquifers: a review. *Journal of Contaminant Hydrology*, 9, 221–45.

Park, K. S., Sims, R. C., DuPont, R. R., Doucette, W. J. & Matthews, J. E. (1990). Fate of PAH compounds in two soil types: influence of volatilization, abiotic loss and biological activity. *Environmental Toxicology and Chemistry*, 9, 187–95.

Partridge, P. A., Shala, F. J., Cernasky, N. P. & Suffet, I. H. (1990). Comparison of diesel engine exhaust using chromatographic profiling techniques. *Environmental Science & Technology*, 24, 189–94.

Piontek, K. R. & Simpkin, T. J. (1994). Practicability of *in situ* bioremediation at a wood-preserving site. In *Bioremediation of Chlorinated and Polycyclic Aromatic Hydrocarbon Compounds*, ed. R. E. Hinchee *et al.*, pp. 117–28. Boca Raton, FL: CRC Press.

Pope, D. F. & Matthews, J. E. (1993). *Environmental Regulations and Technology: Bioremediation Using the Land Treatment Concept.* EPA/600/R-93/163. Cincinnati, OH: US EPA.

Pothuluri, J. V. & Cerniglia, C. E. (1994). Microbial metabolism of polycyclic aromatic hydrocarbons. In *Biological Degradation and Bioremediation of Toxic Chemicals*, ed. G. R. Chaudhry, Portland, OR: Dioscorides Press.

Pothuluri, J. V., Freeman, J. P., Evans, F. E. & Cerniglia, C. E. (1990). Fungal transformation of fluoranthene. *Applied and Environmental Microbiology*, 56, 2974–83.

Pothuluri, J. V., Freeman, J. P., Evans, F. E. & Cerniglia, C. E. (1992a). Fungal metabolism of acenaphthene by *Cunninghamella elegans*. *Applied and Environmental Microbiology*, 58, 3654–9.

Pothuluri, J. V., Heflich, R. H., Fu, P. P. & Cerniglia, C. E. (1992b). Fungal metabolism and detoxification of fluoranthene. *Applied and Environmental Microbiology*, 58, 937–41.

Pritchard, P. (1992). Use of inoculation in bioremediation. *Current Opinion in Biotechnology*, 3, 232–43.

Pritchard, P. H., Lantz, S. E., Lin, J-E. & Mueller, J. G. (1994). Metabolic and ecological factors affecting the bioremediation of PAH- and creosote-contaminated soil and water. In *U.S. EPA Annual Symposium on Bioremediation of Hazardous Wastes: Research, Development and Field Evaluations.* San Francisco, California, June 28–30, 1994, pp. 129–38. EPA/600/R-94/075.

Pritchard, P. H., Lantz, S. E., Lin, J-E., & Mueller, J. G. (1996). Influencing mechanisms of operational factors on the degradation of fluoranthene by *Sphingomonas paucimobilis* strain EPA505. *Environmental Science & Technology* (submitted).

Ramadan, M. A., El-Tayeb, O. M. & Alexander, M. (1990). Inoculum size as a factor limiting the success of inoculation in bioremediation. *Applied and Environmental Microbiology*, 56, 1392–6.

Rao, P. S. C., Bellin, C. A. & Brusseau, M. L. (1993). Coupling biodegradation of organic chemicals to sorption and transport in soils and aquifers: paradigms and paradoxes. In *Sorption and Degradation of Pesticides and Organic Chemicals in Soil*, pp. 1–26. Madison, WI: Soil Science Society of America.

Reddy, B. R. & Sethunathan, N. (1983). Mineralization of parathion in the rice rhizosphere. *Applied and Environmental Microbiology*, 45, 826–9.

Reinhard, M. (1993). *In situ* bioremediation technologies for petroleum-derived hydrocarbons based on alternate electron acceptors (other than molecular oxygen). In *In Situ Bioremediation of Ground Water and Geological Material: A Review of Technologies*, ed. R. D. Norris *et al.*, section 7, pp. 7.1–7.7. EPA/600/R-93/124. NTIS Document No. PB93-215564, Washington, DC.

Resnick, S. M. & Chapman, P. J. (1994). Physiological properties of substrate specificity of a pentachlorophenol-degrading *Pseudomonas* sp. *Biodegradation*, 5, 47–54.

Rosenberg, E. (1986). Microbial surfactants. *CRC Critical Reviews in Biotechnology*, 3, 109–32.

Rosenberg, E., Legmann, R., Kushmaro, A., Taube, R., Adler, E. & Ron, E. (1992). Petroleum biodegradation: a multiphase problem. *Biodegradation*, 3, 337–50.

Rostad, C. E., Pereira, W. E. & Hult, M. F. (1985). Partitioning studies of coal-tar constituents in a two-phase contaminated ground water system. *Chemosphere*, 14, 1023–36.

Ryu, B. H., Oh, Y. K. & Bin, J. H. (1989). Biodegradation of naphthalene by *Acinetobacter calcoaceticus* R-88. *Journal of the Korean Agricultural Chemical Society*, 32, 315–20.

Sack, U. & Günther, T. (1993). Metabolism of PAH by fungi and correlation with extracellular enzymatic activities. *Journal of Basic Microbiology*, 33, 269–77.

Sanglard, D., Leisola, M. S. A. & Fiechter, A. (1986). Role of extracellular ligninases in biodegradation of benzo[α]pyrene by *Phanerochaete chrysosporium*. *Enzyme Microbial Technology*, 8, 209–12.

Savino, A. & Lollini, M. N. (1977). Identification of some fermentation products of phenanthrene in microorganisms of the genus *Arthrobacter*. *Bolletino Societa Italiana Biologa Sperimentale*, 53, 916–21.

Schocken, M. J. & Gibson, D. T. (1984). Bacterial oxidation of the polycyclic aromatic hydrocarbons acenaphthene and acenaphthylene. *Applied and Environmental Microbiology*, 48, 10–16.

Scow, K. M. (1993). Effects of sorption-desorption and diffusion processes on the kinetics of biodegradation of organic chemicals in soil. In *Sorption and Degradation of Pesticides and Organic Chemicals in Soil*, ed. D. M. Linn *et al.*, pp. 73–114. Madison, WI: Soil Science Society of America.

Scow, K. M. & Alexander, M. (1992). Effect of diffusion on the kinetics of biodegradation: experimental results with synthetic aggregates. *Soil Science Society of America Journal*, 56, 128–34.

Scow, K. M. & Hutson, J. (1992). Effect of diffusion on the kinetics of biodegradation: theoretical considerations. *Soil Science Society of America Journal*, 56, 119–27.

Selifonov, S. A., Slepenkin, A. V., Adanin, V. M., Grechkina, G. M. & Starovoitov, I. I.

(1993). Acenaphthene catabolism by strains of *Alcaligenes eutrophus* and *Alcaligenes paradoxus*. *Microbiology* (Engl. Transl.), 62, 85–92.

Shiaris, M. P. (1989a). Seasonal biotransformation of naphthalene, phenanthrene and benzo[a]pyrene in surficial estuarine sediments. *Applied and Environmental Microbiology*, 55, 1391–9.

Shiaris, M. P. (1989b). Phenanthrene mineralization along a natural salinity gradient in an urban estuary, Boston Harbor, Massachusetts. *Microbial Ecology*, 18, 135–46.

Shiaris, M. P. & Jambard-Sweet, D. (1986). Polycyclic aromatic hydrocarbons in surficial sediments of Boston Harbour, Massachusetts, USA. *Marine Pollution Bulletin*, 17, 469–72.

Shimp, R. J. & Young, R. L. (1988). Availability of organic chemicals for biodegradation in settled bottom sediments. *Ecotoxicology and Environmental Safety*, 15, 31–45.

Short, K. A., King, R. J., Seidler, R. J. & Olsen, R. H. (1992). Biodegradation of phenoxyacetic acid in soil by *Pseudomonas putida* PP0301 (pRO103), a constitutive degrader of 2,4-dichlorophenoxyacetate. *Molecular Ecology*, 1, 89–94.

Simpkin, T. J. & Giesbrecht, G. (1994). Slurry bioremediation of polycyclic aromatic hydrocarbons in sediments from an industrial complex. In *Bioremediation of Chlorinated and Polycyclic Aromatic Hydrocarbon Compounds*, ed. R. E. Hinchee *et al.*, pp. 484–8. Boca Raton, FL: CRC Press.

Sims, J. L., Sims, R. C. & Matthews, J. E. (1989). *Bioremediation of Contaminated Surface Soils*. EPA/600/9-89/073. Cincinnati, OH: US EPA.

Sims, R. C., Sims, J. L., Sorensen, D. L., Stevens, D. K., Huling, S. G., Bledsoe, B. E., Matthews, J. E. & Pope, D. (1994). Performance evaluation of full-scale *in situ* and *ex situ* bioremediation of creosote wastes in ground water and soils. In *Proceedings, Symposium on Bioremediation of Hazardous Wastes: Research, Development and Field Evaluations*, pp. 35–9. EPA/600/R-94/075. Cincinnati, OH: US EPA.

Smith, R. V. & Rosazza, J. P. (1974). Microbial methods of mammalian metabolism. Aromatic hydroxylation. *Archives of Biochemistry and Biophysics*, 161, 551–8.

Spain, J. C., Milligan, J. D., Downey, D. C. & Slaughter, J. K. (1989). Excessive bacterial decomposition of H_2O_2 during enhanced biodegradation. *Ground Water*, 27, 163–7.

Steinberg, S. M., Pignatello, J. J. & Sawhney, B. L. (1987). Persistence of 1,2-dibromoethane in soils: entrapment in intraparticle micropores. *Environmental Science & Technology*, 21, 1201–8.

Stormo, K. E. & Crawford, R. L. (1992). Preparation of encapsulated microbial cells for environmental applications. *Applied and Environmental Microbiology*, 58, 727–30.

Stormo, K. E. & Crawford, R. L. (1994). Pentachlorophenol degradation by microencapsulated *Flavobacterium* and their enhanced survival for in situ aquifer bioremediation. In *Applied Biotechnology for Site Remediation*, ed. R. E. Hinchee *et al.*, pp. 422–8. Boca Raton, FL: Lewis Publishers.

Sun, S. & Boyd, S. A. (1991). Sorption of polychlorobiphenyl (PCB) congeners by residual PCB-oil phases in soils. *Journal of Environmental Quality*, 20, 557–61.

Sutherland, J. B. (1992). Detoxification of polycyclic aromatic hydrocarbons by fungi. *Journal of Industrial Microbiology*, 9, 53–62.

Sutherland, J. B., Freeman, J. P., Selby, A. L., Fu, P. P., Miller, D. W. & Cerniglia, C. E. (1990). Stereoselective formation of a K-region dihydrodiol from phenanthrene by *Streptomyces flavovirens*. *Archives of Microbiology*, 154, 260–6.

Sutherland, J. B., Selby, A. L., Freeman, J. P., Evans, F. E. & Cerniglia, C. E. (1991).

Metabolism of phenanthrene by *Phanerochaete chrysosporium*. *Applied and Environmental Microbiology*, 57, 3310–16.

Sutherland, J. B., Selby, A. L., Freeman, J. P., Fu, P. P., Miller, D. W. & Cerniglia, C. E. (1992). Identification of xyloside conjugates formed from anthracene by *Rhizoctonia solani. Mycology Research*, 96, 509–17.

Sutherland, J. B., Fu, P. P., Yang, S. K., Von Tungein, L. S., Casillas, R. P., Crow, S. A. & Cerniglia, C. E. (1993). Enantiomeric composition of the *trans*-dihydrodiols produced from phenanthrene by fungi. *Applied and Environmental Microbiology*, 59, 2145–9.

Taddeo, A., Findlay, M., Dooley-Dana, M. & Fogel, A. (1989). Field demonstration of a forced aeration composting treatment for coal tar. In *Proceedings of the 2nd National Conference Superfund '89, November 27–29, Washington, DC*, pp. 57–62. Hazardous Materials Control Research Institute.

Tagger, S., Truffaut, N. & Le Petit, J. (1990). Preliminary study on relationships among strains forming a bacterial community selected on naphthalene from marine sediment. *Canadian Journal of Microbiology*, 36, 676–81.

Tatarko, M. & Bumpus, J. A. (1993). Biodegradation of phenanthrene by *Phanerochaete chrysosporium*: on the role of lignin peroxidase. *Letters in Applied Microbiology*, 17, 20–4.

Treccani, V., Walker, N. & Wiltshire, G. H. (1954). The metabolism of naphthalene by soil bacteria. *Journal of General Microbiology*, 11, 341–8.

Tremaine, S. C., McIntire, P. E., Bell, P. E., Siler, A. K., Matolak, N. B., Payne, T. W. & Nimo, N. A. (1994). Bioremediation of water and soils contaminated with creosote: suspension and fixed-film bioreactors vs. constructed wetlands and plowing vs. solid peroxygen treatment. In *Bioremediation of Chlorinated and Polycyclic Aromatic Hydrocarbon Compounds*, ed. R. E. Hinchee *et al.*, pp. 172–87. Boca Raton, FL: CRC Press.

Trevors, J. T., van Elsas, J. D., Lee, H. & Overbeck, L. S. (1992). Use of alginate and other carriers for encapsulation of microbial cells for use in soil. *Microbial Releases*, 1, 61–9.

Trower, M. K., Sariaslani, F. S. & Kitson, F. G. (1988). Xenobiotic oxidation by cytochrome P-450-enriched extracts of *Streptomyces griseus. Biochemistry and Biophysics Research Communications*, 157, 1417–22.

U.S. EPA (1981). *Development Document for Effluent Limitations Guidelines and Standards for Timber Products*. EPA/440/1-81/023. January 1981. Cincinnati, OH: US EPA.

U.S. EPA (1985). *Treatment of Contaminated Soils with Aqueous Surfactants*. EPA/600/2-85/129.

U.S. EPA (1987). *Field Studies of in situ Soil Washing*. EPA/600/2-87/110.

U.S. EPA (1989). *Cleaning Excavated Soil Using Extraction Agents: A State-of-the-art Review*. EPA/600/2-89/034.

U.S. EPA (1990a). *Basics of Pump-and-treat Ground-water Remediation Technology*. EPA/600/8-90/003.

U.S. EPA (1990b). *Approaches for Remediation of Uncontrolled Wood Preserving Sites*. EPA/625/7-90/011.

U.S. EPA (1991). *Land Disposal Restrictions: Summary of Requirements*. OSWER 9934.0-1A. February 1991.

U.S. EPA (1992a). *Accessing Federal Data Bases for Contaminated Site Clean-up Technologies*, 2nd edn. EPA/542/B-92/002.

U.S. EPA (1992b). *Innovative Treatment Technologies: Semi-annual Status Report*. EPA/542/R-92/011.

U.S. EPA (1992c). *Dense Non-Aqueous Phase Liquids: A Workshop Summary*. EPA/600/R-92/030.

U.S. EPA (1992d). *Contaminants and Remedial Options at Wood Preserving Sites.* EPA/600/R-92/182.

U.S. EPA (1993a). *Demonstration of Remedial Action Technologies for Contaminated Land and Groundwater.* EPA/600/R-93/012a.

U.S. EPA (1993b). *Provisional Guidance for Quantitative Risk Assessment of Polycyclic Aromatic Hydrocarbons.* EPA/600/R-93/089.

U.S. EPA (1993c). *Technology Selection Guide for Wood Treater Sites.* EPA/540/F-93/020.

U.S. EPA (1993d). *Cleaning up the Nation's Waste Sites: Markets and Technology Trends.* EPA/542/R-92/012.

U.S. EPA (1993e). *Bioremediation Field Initiative.* EPA/540/F-93/510.

U.S. EPA (1993f). *Bioremediation Field Initiative Site Profile: Ground Water Superfund Site.* EPA/540/F-93/510A.

U.S. EPA (1993g). *Bioremediation Field Initiative Site Profile: Escambia Wood Preserving Site – Brookhaven.* EPA/540/F-93/510G.

U.S. EPA (1993h). *Bioremediation Field Initiative Site Profile: Reilly Tar and Chemical Corporation Superfund Site.* EPA/540/F-93/510H.

U.S. EPA (1994a). *Tech Trends.* February Issues. EPA/542/N-94/001.

U.S. EPA (1994b). *VISITT Vendor Information System for Innovative Treatment Technologies: User Manual (Version 3.0).* EPA/542/R-94/003.

U.S. Patent Office (1974). *Reclamation of Hydrocarbon Contaminated Ground Waters.* Raymond, R. L. US Patent No. 3 846 290.

U.S. Patent Office (1986). *Stimulation of Biooxidation Processes in Subterranean Formations.* Raymond, R. L., Brown, R. A., Norris, R. D. & O'Neill, E. T. US Patent No. 4 588 506.

U.S. Patent Office (1990). *Method For Stimulating Biodegradation of Halogenated Aliphatic Hydrocarbons.* Nelson, M. J. K. & Bourguin, A. US Patent No. 4 925 802.

Valo, R., Haggblom, M. M. & Salkinoja-Salonen, M. (1990). Bioremediation of chlorophenol-containing simulated ground water by immobilized cells. *Water Research*, 24, 253–8.

van Elsas, J. D. & Heijnen, C. E. (1990). Methods for the introduction of bacteria into soil: a review. *Biology and Fertility of Soils*, 10, 127–33.

Vigon, B. W. & Rubin, A. J. (1989). Practical considerations in the surfactant-aided mobilization of contaminants in aquifers. *Journal of the Water Pollution Control Federation*, 61, 1233–40.

Walter, U., Beyer, M., Klein, J., Rehm, H. J. (1991). Degradation of pyrene by *Rhodococcus* sp. UW1. *Applied Microbiology and Biotechnology*, 34, 671–6.

Walton, B. T. & Anderson, T. A. (1990). Microbial degradation of trichloroethylene in the rhizosphere: potential application to biological remediation of waste sites. *Applied and Environmental Microbiology*, 56, 1012–16.

Wang, X., Yu, X. & Bartha, R. (1990). Effect of bioremediation on polycyclic aromatic hydrocarbon residues in soil. *Environmental Science & Technology*, 24, 1086–9.

Ward, C. H., Wilson, J. T., Kampbell, D. H. & Hutchins, S. (1992). Performance and cost evaluation of bioremediation techniques for fuel spills. In *Proceedings, Bioremediation Symposium, 1992.* Niagara-On-The Lake, Ontario, Canada, September 20–24, 1992, pp. 15–21.

Warshawsky, D., Radike, M., Jayasimhulu, K. & Cody, T. (1988). Metabolism of benzo[a]pyrene by a dioxygenase system of the freshwater green alga *Selenastrum capricornutum. Biochemistry and Biophysics Research Communications*, 152, 540–4.

Warshawsky, D., Keenan, T. H., Reilman, R., Cody, T. E. & Radike, M. J. (1990). Conjugation of benzo[a]pyrene metabolites by freshwater green alga *Selenastrum capricornutum*. *Chemico-Biological Interactions*, 74, 93–105.

Weber, J. B., Best, J. A. & Gonese, J. U. (1993). Bioavailability and bioactivity of sorbed organic chemicals. In *Sorption and Degradation of Pesticides and Organic Chemicals in Soil*, pp. 153–96. Madison, WI: Soil Science Society of America.

Weissenfels, W. D., Beyer, M. & Klein, J. (1990). Degradation of phenanthrene, fluorene and fluoranthene by pure bacterial cultures. *Applied Microbiology and Biotechnology*, 32, 479–84.

Weissenfels, W. D., Beyer, M., Klein, J. & Rehm, H. J. (1991). Microbial metabolism of fluoranthene: isolation and identification of ring fission products. *Applied Microbiology and Biotechnology*, 34, 528–35.

Werner, P. (1991). German experiences in the biodegradation of creosote and gaswork-specific substances. In *In Situ Bioreclamation: Applications and Investigations for Hydrocarbon and Contaminated Site Remediation*, ed. R. E. Hinchee & R. F. Olfenbuttel, pp. 496–517. Boston, MA: Butterworth–Heinemann.

West, C. C. & Harwell, J. H. (1992). Surfactants and subsurface remediation. *Environmental Science & Technology*, 26, 2324–30.

West, P. A., Okpokwasili, G. C., Brayton, P. R., Grimes, D. J. & Colwell, R. R. (1984). Numerical taxonomy of phenanthrene-degrading bacteria isolated from the Chesapeake Bay. *Applied and Environmental Microbiology*, 48, 988–93.

Wild, S. R. & Jones, K. C. (1993). Biological and abiotic losses of polynuclear aromatic hydrocarbons (PAHs) from soils freshly amended with sewage sludge. *Environmental Toxicology and Chemistry*, 12, 5–12.

Wild, S. R., Jones, K. C., Waterhouse, K. S. & McGrath, S. P. (1990a). Organic contaminants in an agricultural soil with a history of sewage sludge amendments: Polynuclear aromatic hydrocarbons. *Environmental Science & Technology*, 24, 1706–11.

Wild, S. R., McGrath, S. P. & Jones, K. C. (1990b). The polynuclear aromatic hydrocarbon (PAH) content of archives sewage sludge. *Chemosphere*, 20, 703–16.

Wild, S. R., Berrow, M. L. & Jones, K. C. (1991a). The persistence of polynuclear aromatic hydrocarbons (PAHs) in sewage sludge amended agricultural soils. *Environmental Pollution*, 72, 141–57.

Wild, S. R., Obbard, J. P., Munn, C. I., Berrow, M. L. & Jones, K. C. (1991b). The long-term persistence of polynuclear aromatic hydrocarbons (PAHs) in an agricultural soil amended with metal-contaminated sewage sludges. *Science of the Total Environment*, 101, 235–53.

Williams, P. A. (1981). Genetics of biodegradation in microbial degradation of xenobiotics and recalcitrant compounds. In *Microbial Degradation of Xenobiotics and Recalcitrant Compounds*, ed. T. Leisinger, R. Hutter, A. M. Cook & J. Nuesch, pp. 97–130. New York: Academic Press.

Wilson, J. T., Weaver, J. W. & Kampbell, D. H. (1994). Intrinsic bioremediation to TCE in ground water at an NPL Site in St. Joseph, Michigan. In *Symposium on Bioremediation of Hazardous Wastes: Research, Development and Field Evaluations*, pp. 3–10. EPA/600/R-94/075.

Wiseman, A. & Woods, L. F. J. (1979). Benzo[a]pyrene metabolites formed by the action of yeast cytochrome P-450/P-448. *Journal of Chemical Technology and Biotechnology*, 29, 320–4.

Woodhull, P. M. & Jerger, D. E. (1994). Bioremediation using a commercial scale slurry-phase biological treatment system. *Remediation*, 4, 353–62.

Woodward, R. E. & Ramsden, D. K. (1989). Update on *in situ* bioremediation of the French Limited Superfund site. *Proceedings of the 2nd National Conference Superfund '89*, Nov. 27–29, Washington, DC, pp. 29–32. Hazardous Material Control Research Institute.

Wszolek, P. C. & Alexander, M. (1979). Effect of desorption rate on the biodegradation of n-alkylamines bound to clay. *Journal of Agricultural and Food Chemistry*, 27, 410–14.

Zaidi, B. R., Stucki, G. & Alexander, M. (1988). Low chemical concentration and pH as factors limiting the success of inoculation in bioremediation. *Environmental Toxicology and Chemistry*, 7, 143–51.

Zaidi, B. R., Murakami, Y. & Alexander, M. (1989). Predation and inhibitors in lake water affect the success of inoculation to enhance biodegradation of organic materials. *Environmental Science & Technology*, 23, 859–63.

Zajic, J. E., Supplisson, B. & Volesky, B. (1974). Bacterial degradation and emulsification of no. 6 fuel oil. *Environmental Science & Technology*, 8, 664–7.

Zhang, Y. & Miller, R. M. (1992). Enhanced octadecane dispersion and biodegradation by a *Pseudomonas* rhamnolipid surfactant (biosurfactant). *Applied and Environmental Microbiology*, 58, 3276–82.

Zylstra, G. J. & Gibson, D. T. (1991). Aromatic hydrocarbon degradation: a molecular approach. In *Genetic Engineering: Principles and Methods*, Vol. 13, ed. J. K. Setlow, pp. 183–203. New York: Plenum Press.

6

Bioremediation of nitroaromatic compounds

Stephen B. Funk, Don L. Crawford and Ronald L. Crawford

6.1 Introduction

Explosives and related compounds have become widely recognized as serious environmental contaminants. Among the nitrosubstituted aromatic compounds causing particular concern are 2,4,6-trinitrotoluene (TNT), 2,4,6-trinitrophenol (picric acid), and many nitro- and/or amino-substituted aromatics that result from the manufacture and transformation of explosives. The threat posed by the presence of these compounds in soil and water is the result of their toxicity and is compounded by their recalcitrance to biodegradation.

Contamination by nitroaromatic compounds, especially TNT, stems primarily from military activities (Boopathy et al., 1994). During the manufacture of explosives and the disposal of old munitions, large quantities of water became contaminated. This wash water was typically disposed of in unlined lagoons that facilitated the slow release of the explosives from the soil in the lagoons into groundwater, lakes, and rivers.

The mutagenicity of TNT (Kaplan & Kaplan, 1982a; Won et al., 1976), as well as its toxic effects on algae and fish, humans, and other vertebrates, make it an environmental hazard (Hudock & Gring, 1970; Smock et al., 1976; Won et al., 1976). TNT has been listed as a priority pollutant by the U.S. Environmental Protection Agency (Keither & Telliard, 1979). Tan et al. (1992) compared the mutagenicity of the biologically reduced forms of TNT (2-amino-4,6-dinitrotoluene and 4-amino-2,6-dinitrotoluene) to TNT, hexahydro-1,3,5-trinitro-1,3,5-triazine (RDX), octahydro-1,3,5,7-tetranitro-1,3,5,7-tetrazocine (HMX), and N-methyl-N,2,4,6-tetranitroaniline (Tetryl) using the *Salmonella*/mammalian microsome plate incorporation method. They found that the reduction of the nitro groups of TNT to the amino products caused a decrease in the mutagenic activity of the compounds proportional to the number of amino groups formed. They also noted that a mononitrotoluene was mutagenic only if the nitro group was in the *para* position. Tetryl was found to be 30 times more mutagenic than TNT, while RDX and HMX were not mutagenic up to 1 mg/plate. La & Froines (1993) found that the nitroaromatic compound 2,4-dinitrotoluene (24DNT) was toxic and carcinogenic to rats. This compound formed adducts with the DNA in liver cells and resulted in 50% lethality at doses of 0.3 mmol/kg. In contrast, 2,4-diaminotoluene (24DAT) did not form DNA adducts and was not lethal at doses up to 1.2 mmol/kg. These data suggest that either the nitro, nitroso, or hydroxylamino substituent of nitroaryl compounds disrupts DNA, while the fully reduced amino substituent does not. This suggestion is supported by the work of Heflich *et al.* (1985), who found that the enzymatic two-electron reduction of 1-nitropyrene to 1-nitrosopyrene was the rate-limiting step in the formation of specific DNA adducts in *S. typhimurium* exposed to this compound.

Until very recently, explosives-contaminated soils have been remediated by incineration, a process whose high cost has stimulated the search for a more economical cleanup method (Roberts *et al.*, 1993). Microbially mediated degradation of explosives is a promising technology. Many researchers have studied microbial consortia and various pure cultures for their ability to degrade TNT and other nitroaromatic compounds (for a review see Crawford, 1995), bringing about the development of bioremediation processes that can remove TNT and other explosives from contaminated soil and water (Funk *et al.*, 1995; Williams *et al.*, 1992).

6.2 Nitroaromatic degradation by aerobic and microaerophilic microorganisms

Nitroaromatic transformations have been observed in aerobic environments. Aerobic composting of explosives has been investigated to address the feasibility of bioremediating soils contaminated with TNT, RDX, and/or HMX (Kaplan & Kaplan, 1982b; Isbister *et al.*, 1984; Williams & Marks, 1991; Williams *et al.*, 1992). Through this process, TNT and other explosives have been significantly depleted from soil, but the fate of the biotransformed TNT molecule is not completely known. In a bench-scale compost experiment with *Bacillus* spp. and *Actinomyces* spp., Kaplan & Kaplan (1982b) showed neither significant mineralization of ^{14}C-TNT nor evidence for cleavage of the ring. The recovered ^{14}C-label was shown to partially consist of azoxy dimers, TNT, and mono- and diaminonitrotoluenes. Furthermore, the humic fraction was thought to contain larger insoluble TNT conjugates linked into the humus.

In a treatability study, Craig & Sisk (1994) composted 30 cubic yards of highly contaminated soil by forced aeration in a two-window configuration. After 40 days of treatment, the concentration of explosives decreased >99% for TNT and RDX and 96.9% for HMX. The toxicity of the compost leachate was reduced 87% to 92%, as shown by a *Ceriodaphnia dubia* assay, and a 99.3% to 99.6% reduction in mutagenicity was indicated by the Ames assays. The composting process used native thermophilic microbes, eliminating the need for inoculation. However, the degradation pathways and ultimate fate of the explosives were not determined.

Duque *et al.* (1993) described a *Pseudomonas* sp. capable of cometabolizing TNT, 2,4- and 2,6-dinitrotoluene, and 2-nitrotoluene. Nitrite was formed from the nitroaromatic compounds when the organism was grown on fructose as the cosubstrate. The pseudomonad was able to use these compounds as a nitrogen source, forming toluene as an end-product through successive removal of the nitro groups (nitro groups were also reduced on the ring to their amino analogs via a hydroxylamine intermediate). The exact mechanism for the elimination of the nitro substituent from the ring remains unknown. It was also confirmed that nitrite reductase activity was induced by the presence of TNT. Upon transfer into this strain of the TOL plasmid pWWo-Km, which codes for enzymes necessary for toluene degradation, namely dioxygenases, the transconjugate was able to grow slowly on TNT as a sole source for carbon and nitrogen. Interestingly, this *Pseudomonas* sp. was also able to tolerate high concentrations of TNT (in excess of 1 g/l).

In another study using *Pseudomonas* spp., Boopathy *et al.* (1994) described a consortium of bacteria isolated from a soil that was chronically contaminated with TNT. Four pseudomonad isolates were purified from the soil and were examined for their ability to degrade TNT. None of these cultures could degrade TNT without utilizing a cosubstrate such as succinate, either independently or as a consortium. When the consortium was incubated with ^{14}C-TNT, 3.1% of the label was recovered as $^{14}CO_2$, 8% of the label was associated with biomass, and the remainder was recovered as either TNT or the mono amino metabolites. In this study, as in many other studies of aerobic TNT biotransformation, the formation of aminodinitrotoluenes appears to be the rate-limiting step. This may be due to the toxic or mutagenic effect of the nitroso, hydroxylamino, or amino adducts formed on the aromatic ring.

Phanerochaete chrysosporium is a basidiomycete white-rot fungus capable of transforming and mineralizing TNT. Stahl & Aust (1993b) observed a correlation between TNT mineralization and the presence of a manganese-dependent peroxidase. Furthermore, they attributed the initial reduction of the TNT molecule to a plasma-membrane-dependent redox system (Stahl & Aust, 1993a). This reduction required live mycelia and was inhibited by a number of compounds. However, Rieble *et al.* (1994) observed the reduction of TNT in cell-free extracts of *P. chrysosporium* by an NAD(P)H-dependent nitroreductase which proceeded in a stepwise manner through nitroso and hydroxylamino intermediates. Two hydroxyl-aminodinitrotoluene isomers formed from the reduction of TNT by *P. chrysosporium* were found to directly inhibit lignin peroxidases, which may inhibit further TNT metabolism (Michels & Gottschalk, 1994). This work confirmed a previous report on TNT toxicity to *P. chrysosporium* (Spiker *et al.*, 1992), but also shed light on the mechanism of toxicity.

All of these results show that aerobic biotransformations of TNT are initially similar regardless of the microorganism. The initial transformation generally involves the reduction or removal of one of the nitro substituents, giving way to an amino derivative or free nitrite.

The formation of an unstable hydride–Meisenheimer complex has been observed in a culture containing a strain of *Mycobacterium* growing in the presence of TNT (Vorbeck *et al.*, 1994). Here, a hydride ion attaches to C-3 of TNT and provides a way of releasing nitrite without the need for an oxygenase. Lenke & Knackmuss (1992) originally observed this complex in cultures of *Rhodococcus erythropolis* growing in the presence of picric acid. Again, the nucleophilic attack of the hydride ion occurred on the C-3.

Many aerobic microorganisms have been examined for their ability to degrade nitroaromatic compounds other than TNT. Several aerobes can utilize mononitrophenols as carbon and nitrogen sources. Typically, the nitro group is eliminated as nitrite, followed by ring cleavage through the action of monooxygenases or dioxygenases (Gorontzy *et al.*, 1994). An example is *Arthrobacter aurescens* TW17, which was able to convert *p*-nitrophenol to nitrite and hydroquinone (Hanne *et al.*, 1993). Apparently, the elimination of the nitro substituent as nitrite may be a mechanism used by some cells to avoid the ill effects of the reduced amino derivative. While it is commonly seen during the aerobic degradation of mono- and dinitrophenols, its relative importance in explosives degradation remains unclear.

Rhys-Williams *et al.* (1993) reported on the catabolism of 4-nitrotoluene (4NT) by several *Pseudomonas* spp. that used 4NT as a carbon and nitrogen source. A pathway for the destruction of 4NT was proposed which proceeds through 4-nitrobenzyl alcohol, 4-nitrobenzaldehyde, 4-nitrobenzoate, and finally to protocatechuate, with concomitant release of ammonia. The protocatechuate ring is then oxidized and broken open by the action of a dioxygenase. The authors hypothesized that the nitro substituent was initially reduced through nitroso and hydroxylamino intermediates, with the elimination of the hydroxylamino group as ammonia. A similar pathway was observed by Grownewegen *et al.* (1992), in which *Comomonas acidovorans* catabolized 4-nitrobenzoate to protocatechuate with elimination of a hydroxylamine group as ammonium.

6.3 Nitroaromatic degradation by anaerobic microorganisms

Reduction of aromatic nitro groups is the primary initial activity observed when anaerobes metabolize nitroaromatics. For example, Preuss *et al.* (1993) examined the sequential reduction of the nitro substituents on TNT by a *Desulfovibrio*. This sulfate-reducer was able to use TNT as the sole nitrogen source, while anaerobically respiring pyruvate; sulfate served as an energy source. TNT was completely reduced to triaminotoluene (TAT) and then further degraded to unknown intermediates. 2,4-diamino-6-nitrotoluene (2,4-DANT) was found to be the limiting step in the formation of TAT, a reduction proceeding through 2,4-diamino-6-hydroxylaminotoluene (DAHAT) as an intermediate. The conversion of DAHAT to TAT was inhibited by CO and NH_2OH. DAHAT was also an intermediate in the conversion of 2,4-DANT to TAT by two strains of *Clostridium*, *C. pasteurianum* and *C. thermoaceticum*.

Gorontzy *et al.* (1993) observed the reduction of several mono- and dinitroaromatic compounds by methanogenic bacteria, sulfate-reducing bacteria, and clostridia. Here, the anaerobic cultures reduced aromatic nitro groups to their corresponding amino groups without further conversion of the molecule. The methanogenic bacteria were unable to grow in the presence of these highly oxidized compounds; however, whole cell suspensions as well as crude cell extracts were able to carry out the reductions.

The two nonaromatic but related nitramine explosives, RDX and HMX, are sometimes found with TNT and other nitroaromatics in explosives-contaminated soils (Funk *et al.*, 1993b). Kitts *et al.* (1994) degraded RDX and HMX by means of three aerotolerant bacteria isolated from nitramine-contaminated soil. These were *Providencia rettgeri* B1, *Morganella morganii*, and *Citrobacter freundii* NS2. Initially, the cultures were grown aerobically in the presence of RDX and HMX without notable degradation of the nitramines. However, upon O_2 starvation, *P. rettgeri* and *M. morganii* were able to completely transform RDX, while *C. freundii* was able to partially transform it. Mono- and dinitroso derivatives of RDX transiently accumulated before being further transformed. *C. freundii* accumulated the highest concentration of intermediates, while *P. rettgeri* and *M. morganii* removed the intermediates to less than 5% of the original RDX concentration after 45 days. The authors suggested that the nitroso groups were further reduced to unstable hydroxylamino groups, implying ring cleavage. The three bacteria also reduced HMX to mono- and dinitroso intermediates, but to a lesser extent than they reduced RDX. When both HMX and RDX were present in the medium, RDX inhibited HMX degradation in *P. rettgeri* and *M. morganii*, but not in *C. freundii*.

Clostridium bifermentans is an anaerobe isolated from a four-liter bioreactor enriched from municipal waste and munitions-contaminated soil (Funk *et al.*, 1993a; Regan & Crawford, 1994; Funk *et al.*, 1993b; Shin & Crawford, 1995). This organism was capable of transforming 50 mg/liter TNT to the transient intermediates 4A26DNT and 24DA6NT in the presence of yeast extract or trypticase soy and a fermentable cosubstrate. *C. bifermentans* was not able to grow on TNT as a sole source of carbon or nitrogen, but degraded it cometabolically. No other aromatic intermediates were detectable after the disappearance of 24DA6NT. The benzene ring may have been cleaved, leading to the formation of volatile organic acids, a conclusion that is not yet confirmed but is supported by indirect evidence (Shin & Crawford, 1995).

6.4 Consortia versus pure cultures

In determining pathways, pure cultures are easier to work with than are consortia, since consortia make it difficult to determine which organism is responsible for individual steps in the transformation process. Furthermore, consortia are usually found in environmental samples and may require unknown factors and minerals present in sediment in order for individual strains to coexist. If such factors are absent, a single strain that may not perform the complete transformation could dominate a culture. However, we have found that an anaerobic consortium kept under a selective pressure of TNT in a minimal medium was able to thrive on TNT in a bench-top bioreactor without the addition of a carbon cosubstrate (Funk *et al.*, 1993b). When pure strains from this consortium were cultured, none of the isolates could grow in the same minimal medium with TNT as the only carbon source. Furthermore, the consortium was able to tolerate > 100 mg/liter TNT while the isolates alone became somewhat inhibited at concentrations > 50 mg/liter TNT.

We have followed the pathway of TNT degradation in the consortium using reverse phase high performance liquid chromatography (HPLC) and mass spectroscopy (MS). Analysis of the supernatant in the consortium showed sequential reduction of TNT to TAT; however, TAT only occasionally accumulated, and even then only in very low concentrations. In addition, methylphloroglucinol (MPG) and *p*-cresol transiently appeared in the supernatant. All compounds were identified using HPLC by UV spectra and retention times as compared to authentic standards chromatographed under the same conditions. The mono- and diamino intermediates were also confirmed by the MS results.

Regan & Crawford (1994) examined a strain of *C. bifermentans* (KMR-1) isolated from our anaerobic bioreactor, which was capable of transforming RDX and TNT in a rich medium. TNT was selectively removed by KMR-1 prior to RDX degradation.

C. bifermentans strain CYS-1 was also isolated from our anaerobic bioreactor. Shin & Crawford (1995) examined the ability of CYS-1 to degrade TNT cometabolically in various defined media. This strain could overcome the toxicity of and degrade > 150 ppm TNT in liquid media supplemented with a rich cosubstrate such as yeast extract or trypticase soy, given an appropriate inoculum ($\approx 10^7$ CFU/ml). Furthermore, it was found that CYS-1 could degrade TNT which contaminated a sandy loam soil. The degradation of TNT proceeded through the transient intermediates 4-amino-2,6-dinitrotoluene and 2,4-diamino-6-nitrotoluene.

Overall, the results obtained from research with anaerobic consortia, with pure cultures derived from the consortia, and with other pure bacterial cultures are indicative of a complex TNT biodegradation process in soil that involves multiple organisms acting synergistically and probably sequentially.

6.5 Enzymology of nitroaromatic degradation

Knowledge of the enzymes used by microorganisms in the transformation of nitroaromatic compounds is limited. Blasco & Castillo (1993) characterized an inducible nitrophenol reductase from *Rhodobacter capsulatus* that catalyzed the reduction of 2,4-dinitrophenol (DNP) to 2-amino-4-nitrophenol. This enzyme was a dimer that contained flavin mononucleotide and possibly nonheme iron as prosthetic groups. NAD(P)H was an electron donor and DNP acted as an electron acceptor in the presence of light under anaerobic conditions.

Rafii & Cerniglia (1993) purified an azoreductase and a nitroreductase from *Clostridium perfringens* which had similar physical properties and which were thought to be the same protein. Nitroreductase activity was measured colorimetrically by the reduction of *p*-nitrobenzoic acid; *p*-aminobenzoic acid was converted to a purple dye by N-(1-naphthyl)ethylenediamine dihydrochloride. The physiological role of nitroreductase in clostridial metabolism is not known. However, other data from the same authors suggest that this enzyme may be a dehydrogenase used by this anaerobe for electron transport (Rafii & Cerniglia, 1990). This would imply that the nitroaromatic compounds may function as nonspecific alternate electron acceptors and that clostridia may be capable of anaerobic respiration under certain conditions. Angermaier & Simon (1983) also observed this phenomenon in clostridia where *p*-nitrobenzoate accepted electrons from NAD(P)H via ferrodoxin-NAD(P)$^+$ reductase.

The enteric bacterium *Enterobacter cloacae* produces a nitroreductase that reduces nitrofurans, nitroimidazoles, nitrobenzene derivatives, and quinones (Bryant & DeLuca, 1991). This oxygen-insensitive enzyme has been purified and is known to require FMN to transfer reducing equivalents from NAD(P)H to the nitroaromatic compounds, TNT being the preferred substrate. Aerobically, this enzyme reduces nitrofurazone through the hydroxylamine intermediate, which then tautomerizes to yield an oxime end-product. Anaerobically, however, the reduction proceeds to the fully reduced amine adduct. When *E. cloacae* was grown in the presence of TNT, the nitroreductase activity increased five- to tenfold.

Generally speaking, the reduction of the nitrosubstituted compounds mentioned here appears to be the result of broad-substrate-specificity enzymes using these compounds as electron-acceptors. This mechanism may account for the observation by Preuss *et al.* (1993) that the reduction of 2,4-diamino-6-hydroxylaminotoluene to TAT was inhibited by CO, a known inhibitor of hydrogenase enzymes in clostridial bacteria. The electron-accepting function may also explain why some nitro compounds, in the presence of others, are preferentially reduced first. Kitts *et al.* (1994) showed that RDX reduction was preferred over HMX reduction. It would appear that if these nitramines are being reduced from electron sinks made by these organisms (such as reduced ferrodoxins or other iron/sulfur-containing proteins) and not substrate-specific enzymes, then HMX requires a lower redox value for reduction than RDX. This same principle may hold true for TNT as well. Funk *et al.* (1993b) showed the reduction of TNT to 4-amino-2,6-dinitrotoluene (4A26DNT) and then to 2,4-diamino-6-nitrotoluene by a mixed anaerobic consortium in a medium that contained TNT and RDX. It was shown that the RDX concentration remained constant until all of the TNT was converted to the diamino compound. Hence, RDX may require a lower redox potential for reduction than TNT and 4A26DNT. Similar results were observed by Hoffsommer *et al.* (1978) and Regan & Crawford (1994).

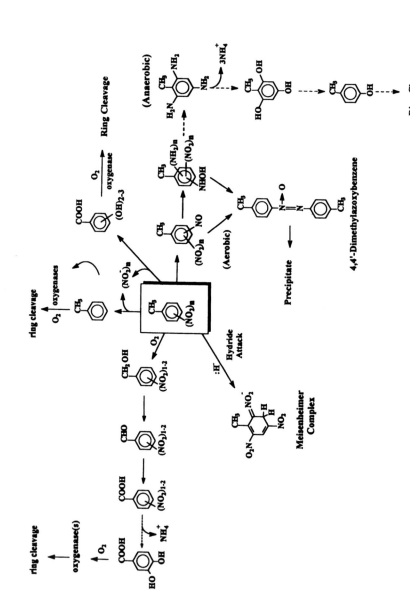

Figure 6.1. Representative biotransformations of nitrotoluenes.

The hypothesis stating that these various nitro-substituted compounds are reduced using reducing power stored in proteins such as ferrodoxins needs to be confirmed. Santangelo *et al.* (1991) developed a genetic system that tentatively identified several genes in *Clostridium acetobutylicum* which transfer electrons to nitro substituents of 5-nitroimidazole (metronidazole). This compound is typically reduced to 5-aminoimidazole, which has broad-spectrum antimicrobial activity. A nitrate-reductase-deficient mutant of *E. coli* was used as a host for inserting plasmids containing DNA fragments from *C. acetobutylicum*. Recombinants that became sensitive to metronidazole were further investigated. One particular insert was characterized and was found to encode a flavodoxin protein capable of reducing the nitro group, thus making the recombinant sensitive to metronidazole. Additional tests such as this still need to be done to confirm which proteins are involved in the reduction of nitroaromatic compounds. Possibilities are oxidoreductase, ferrodoxin, flavodoxin, rubiboxin, and hydrogenase, to name just a few. These proteins may also require cofactors such as iron, FMN, FADH, and NAD(P)H that allow them to function.

Figure 6.1 depicts a general summary of the various known nitroaromatic biodegradation and biotransformation pathways. This pathway may involve mono, di, or trinitrated toluene as the parent compound. Formation of the Meisenheimer complex and the anaerobic portion of the pathway were observed only when TNT was the starting substrate. Included here are examples of both oxidative and reductive mechanisms for biological metabolism of generic nitroaromatic compounds. Cleavage of the aromatic ring may be implied.

6.7 Current technologies for bioremediation of nitroaromatic-contaminated soils and waters

An anaerobic environment alone (e.g., sediments, soils, and aquifers) may not be enough to facilitate biotransformation of nitroaromatic compounds. Limiting factors such as pH, temperature, and nutrient availability can prevent the natural destruction of these contaminants in the environment. Therefore, *ex situ* treatment strategies are used to bioremediate contaminated soil. Research performed at the University of Idaho has enabled the J. R. Simplot Company of Boise, Idaho, to develop and implement such a process, termed the 'Simplot Anaerobic Bioremediation Ex Situ' or SABRE^{TM} process (U.S. patent 5 387 271, February 1995; Biological system for degrading nitroaromatics in water and soils; EPA Technology Profiles, 1993). Here, the nitroaromatic-contaminated soil is excavated, augmented with buffer and nutrients, and mixed into a slurry within a lined pit. Temperature and pH are maintained while the slurry batch is periodically mixed. The soil slurry is allowed to incubate under strictly anaerobic conditions (redox potential of $< 250\,mV$). The nitroaromatics in the soil are removed and degraded by an enriched anaerobic bacterial consortium dominated by clostridia and perhaps sulfate-reducing bacteria. Once the contaminant is completely eliminated, the soil is replaced in the original site. This treatment procedure typically costs only a fraction of the cost for incineration.

Slurry reactors have been used to treat municipal sewage for years. Innovative research is now showing that slurry reactors can be modified to treat many kinds of contaminated water and soil. Kleijntjens *et al.* (1987) have, for example, modified a bioreactor to decontaminate soil. Slurry reactors enhance bioremediation by allowing close contact between the microorganisms and the contaminant, improving desorption of the contaminant from soil, and increasing degradation rates to shorten incubation times. However, a batch slurry reactor is limited in the amount of soil it can treat. Typically, soil is mixed with $> 30\%$ aqueous medium in a slurry reactor equipped with controls that maintain pH, temperature, moisture content, and proper mixing. These reactors may be used for aerobic or anaerobic treatments, depending on the target compound and the microorganisms that degrade it. Recently, Stormo & Deobald (1995) have developed a novel hydrolytically propelled slurry bioreactor that can mix soil–water slurries of $> 50\%$ solids while consuming little energy. Such a reactor may prove useful in the bioremediation of explosives-contaminated soils.

Composting of explosives-contaminated soils has also been demonstrated. Here, contaminated soil is mixed with a bulking agent such as straw or wood chips to increase the porosity of the material and facilitate air exchange. Typically, windrow configurations are constructed and are mechanically turned over periodically. However, some static compost heaps are aerated through an internal network of piping. Alternatively, mechanical in-vessel composting offers the highest level of control (aeration, pH, temperature, moisture) but considerably raises the costs of remediation. The metabolism of carbon by microorganisms creates heat in the compost heaps. The elevated temperatures favor thermophilic microorganisms and increase the solubility of some explosive contaminants. For example, the solubility of HMX in an aqueous medium is only 6.6 mg/liter at 20 °C. Williams *et al.* (1992) conducted field demonstrations in which soil contaminated with TNT, RDX, HMX, Tetryl, and nitrocellulose was composted. By volume, 80–90% of the compost consisted of bulking agents; fertilizer was added to achieve a C : N ratio of 30 : 1. Total explosives in mesophilic piles were reduced from 17 870 to 74 mg/kg, and in thermophilic piles, from 16 460 to 326 mg/kg over 153 days' incubation. Transient accumulation of initial TNT degradation products (mono- and diaminonitrotoluenes) were observed during the incubation, but the exact fate of the explosives was not determined. The explosives and initial intermediates may have been bound up in the compost matrix and humic acids. Although the explosives were not extractable from the compost, it is unclear whether these explosives can 'bleed' back into the environment. Consequently, depositing compost back into the environment, at 5 to 30 times the original volume of the contaminated soil, raises some concerns.

The treatment of wastewater contaminated with explosive is rather limited. Brooks & Livingston (1994) developed a novel membrane bioreactor which was capable of extracting nitrobenzene (NB) and 3-chloronitrobenzene (3-CNB) from industrial wastewater with pH of < 1.0 and an inorganic salt concentration of $> 5\%$ w/w. A selective membrane was used in conjunction with specialized microorganisms enriched to degrade these compounds. The novel reactor was uniquely designed to prevent direct contact of the wastewater with the microorganisms, thus avoiding the need for pretreatment of the wastewater. NB and 3-CNB were over 98% removed with wastewater residence times of 30 min. Approximately 75% of the total organic carbon entering this system was transformed to CO_2.

The purification of TNT during its manufacture produces a wastewater known as 'red water'. Red water contains a variety of nitro-substituted toluene isomers as well as their sulfonated derivatives. Hao *et al.* (1993) used wet air oxidation (WAO) at temperatures up to 340 °C to remove much of the total organic carbon. WAO-treated red water was tested against nontreated red water for the toxicity to *Nitrosomonas* and activated sludge cultures. Despite some lingering effects, a significant decrease in toxicity was observed, raising the possibility that the abiotic WAO pretreatment may be a viable process for further remediation of red water.

Currently, no technology exists for an *in situ* treatment process that remediates nitroaromatic compounds in soil, but one could potentially be developed, depending on many factors. *In situ* treatment, or bioreclamation, does not require the excavation of contaminated soil. Instead, nutrients are added directly into the contaminated zone to enhance the indigenous population of microorganisms. Aqueous medium is recovered by use of extraction wells and can be reapplied to the contaminated site through injection wells. In situations where a viable population of microorganisms capable of carrying out the degradation of the target compound is not present, microorganisms may need to be added. The addition of encapsulated bacteria or spores via injection wells is currently being investigated in our laboratory. Some potential problems for *in situ* treatment include the recalcitrance of the contaminant, the effect of soil type (e.g., clay soil may be nonporous), location of the contaminant, and the presence of a water table or aquifer that could become contaminated. Therefore, a complete understanding of the hydrogeology of the contaminated site is necessary before *in situ* approaches are even considered. These problems, of course, are not unique to nitroaromatic contaminants.

6.8 Conclusion

Extensive research has contributed to the recent development of treatment processes for the bioremediation of soils and waters contaminated with nitro-substituted explosives. By elucidating the degradative pathways in both aerobic and anaerobic systems, we can determine the fate of the parent molecule and assess its effects on the environment. Further research into treating soil contaminated with various nitroaromatics is essential, since their incineration is not always a viable option, due to high cost and risk of pollution.

References

Angermaier, L. & Simon, H. (1983). On nitroaryl reductase activities in several clostridia. *Hoppe–Seyler's Zeitschrift für Physiologische Chemie*, 364, 1653–63.

Blasco, R. & Castillo, F. (1993). Characterization of a nitrophenol reductase from the phototrophic bacterium *Rhodobacter capsulatus* E1F1. *Applied and Environmental Microbiology*, 59, 1774–8.

Boopathy, R., Manning, J., Montemagno, C. & Kulpa, C. (1994). Metabolism of 2,4,6-trinitrotoluene by a *Pseudomonas* consortium under aerobic conditions. *Current Microbiology*, 28, 131–7.

Brooks, P. R. & Livingston, A. G. (1994). Biological detoxification of a 3-chloronitrobenzene manufacture wastewater in an extractive membrane bioreactor. *Water Research*, 28, 1347–54.

Bryant, C. & DeLuca, M. (1991). Purification and characterization of an oxygen-insensitive NAD(P)H nitroreductase from *Enterobacter cloacae*. *The Journal of Biological Chemistry*, 266, 4119–25.

Craig, H. & Sisk, W. (1994). The composting alternative to incineration of explosives contaminated soils. *Tech Trends*. EPA Publication 542-N-94-008. November 1994.

Crawford, R. L. (1995). The microbiology and treatment of nitroaromatic compounds. *Current Opinion in Biotechnology*, 6, 329–36.

Duque, E., Haidour, A., Godoy, F. & Ramos, J. (1993). Construction of a *Pseudomonas* hybrid strain that mineralizes 2,4,6-trinitrotoluene. *Journal of Bacteriology*, 175, 2278–83.

EPA Superfund Innovative Technology Evaluation Program (1993). *Technology Profiles Sixth Edition*. EPA publication EPA/540/R-93/526. November 1993.

Funk, S. B., Roberts, D. J., Crawford, D. L. & Crawford, R. L. (1993a). Degradation of trinitrotoluene (TNT) and sequential accumulation of metabolic intermediates by an anaerobic bioreactor during its adaptation to a TNT feed. In *Abstracts, 93rd General Meeting of the American Society of Microbiology*, abstr. Q410, p. 421. Washington, D.C.: American Society for Microbiology.

Funk, S. B., Roberts, D. J., Crawford, D. L. & Crawford, R. L. (1993b). Initial-phase optimization for bioremediation of munition compound-contaminated soils. *Applied and Environmental Microbiology*, 59, 2171–7.

Funk, S. B., Crawford, D. L., Crawford, R. L, Mead, G. & Davis-Hoover, W. (1995). Full-scale anaerobic bioremediation of trinitrotoluene (TNT) contaminated soil. *Applied Biochemistry and Biotechnology*, 51/52, 625–33.

Gorontzy, T., Kuver, J. & Blotevogel, K-H. (1993). Microbial transformation of nitroaromatic compounds under anaerobic conditions. *Journal of General Microbiology*, 139, 1331–6.

Gorontzy, T., Drzyzga, O., Kahl, M. W., Bruns-Nagel, D., Breitung, J., von Loew, E. & Blotevogel, K. H. (1994). Microbial degradation of explosives and related compounds. *Critical Reviews in Microbiology*, 20(4), 265–84.

Grownewegen, P. E. J., Breeuwer, P., van Helvoort, J. M. L. M., Langenhoff, A. A. M., De Vries, F. P. & De Bont, J. A. M. (1992). Novel degradative pathway of 4-nitrobenzoate in *Comomonas acidovorans* NBA-10. *Journal of General Microbiology*, 138, 1599–605.

Hanne, L. L., Kirk, L. L., Appel, S. M., Narayan, A. D. & Bains, K. K. (1993). Degradation and induction specificity in actinomycetes that degrade *p*-nitrophenol. *Applied and Environmental Microbiology*, 59, 2505–8.

Hao, O. J., Phull, K. K., Davis, A. P., Chen, J. M. & Maloney, S. W. (1993). Wet air oxidation of trinitrotoluene manufacturing red water. *Water and Environmental Research*, 65, 213–20.

Heflich, R. H., Howard, P. C. & Beland, F. A. (1985). 1-Nitrosopyrene: An intermediate in the metabolic activation of 1-nitropyrene to a mutagen in *Salmonella typhimurium* TA1538. *Mutation Research*, 149, 25–32.

Hoffsommer, J. C., Kaplan, L. A., Glover, D. J., Kubrose, D. A., Dickinson, C., Goya, H., Kayser, E. G., Groves, C. L. & Sitzmann, M. E. (1978). *Biodegradability of TNT: A Three Year Pilot Plant Study*. NSWC/WOL TR 77-136. White Oak, Silver Spring, MD: Naval Surface Weapons Center.

Hudock, G. A. & Gring, D. M. (1970). *Biological Effects of Trinitrotoluene*. Contract NO0164-69-CO822, Naval Environmental Health Center.

Isbister, J. D., Anspach, G. L., Kitchens, J. F. & Doyle, R. C. (1984). Composting for decontamination of soils containing explosives. *Microbiologica*, 7, 47–73.

Kaplan, D. L. & Kaplan, A. M. (1982a). Mutagenicity of 2,4,6-trinitrotoluene surfactant complexes. *Bulletin of Environmental Contamination and Toxicology*, 28, 33–8.

Kaplan, D. L. & Kaplan, A. M. (1982b). Thermophilic biotransformations of 2,4,6-trinitrotoluene under simulated composting conditions. *Applied and Environmental Microbiology*, 44(3), 757–60.

Keither, L. H. & Telliard, W. A. (1979). Priority pollutants. I. A perspective view. *Environmental Science and Technology*, 13, 416–23.

Kitts, C. L., Cunningham, D. P. & Unkefer, P. J. (1994). Isolation of three hexahydro-1,3,5-trinitro-1,3,5-triazine degrading species of the family Enterobacteriaceae from nitramine explosive-contaminated soil. *Applied and Environmental Microbiology*, 60, 4608–11.

Kleijntjens, R. H., Luyben, K. C. A. M., Bosse, M. A. & Velthuisen, L. P. (1987). Process development for biological soil decontamination in a slurry reactor. In *Proceedings of the European Congress on Biotechnology, 1987*, Vol. 1, ed. O. M. Neijssel, R. R. van der Meer & K. C. A. M. Luyben, pp. 252–5. Amsterdam: Elsevier Science.

La, D. K. & Froines, J. R. (1993). Comparison of DNA binding between the carcinogen 2,6-dinitrotoluene and its noncarcinogenic analog 2,4-diaminotoluene. *Mutation Research*, 301, 79–85.

Leuke, H. & Knackmuss, H-J. (1992). Initial hydrogenase during catabolism of picric acid by *Rhodococcus erythopolis*. *Applied and Environmental Microbiology*, 59, 2933–7.

Michels, J. & Gottschalk, G. (1994). Inhibition of the lignin peroxidase of *Phanerochaete chrysosporium* by the hydroxylaminodinitrotoluene intermediate in the degradation of 2,4,6-trinitrotoluene. *Applied and Environmental Microbiology*, 60, 186–94.

Preuss, A., Fimpel, J. & Diekert, G. (1993). Anaerobic transformation of 2,4,6-trinitrotoluene (TNT). *Archives of Microbiology*, 159, 345–53.

Rafii, F. & Cerniglia, C. E. (1990). An anaerobic nondenaturing gel assay for the detection of azoreductase from anaerobic bacteria. *Journal of Microbiological Methods*, 12, 139–48.

Rafii, F. & Cerniglia, C. E. (1993). Comparison of the azoreductase and nitroreductase from *Clostridium perfringens*. *Applied and Environmental Microbiology*, 59, 1731–4.

Regan, K. M. & Crawford, R. L. (1994). Characterization of *Clostridium bifermentans* and its biotransformation of 2,4,6-trinitrotoluene (TNT) and 1,3,5-triaza-1,3,5-trinitrocyclohexane (RDX). *Biotechnology Letters*, 16(10), 1081–6.

Rhys-Williams, W., Taylor, S. C. & Williams, P. A. (1993). A novel pathway for the catabolism of 4-nitrotoluene by *Pseudomonas*. *Journal of General Microbiology*, 139, 1967–72.

Rieble, A., Joshi, D. K. & Gold, M. H. (1994). Aromatic nitroreductase from the basidiomycete *Phanerochaete chrysosporium*. *Biochemical and Biophysical Research Communications*, 205(1), 298–304.

Roberts, D. J., Kaake, R. H., Funk, S. B., Crawford, D. L. & Crawford, R. L. (1993). Field-scale anaerobic bioremediation of dinoseb-contaminated soils. In *Biotreatment of Industrial and Hazardous Waste*, ed. M. A. Levin & M. A. Gealt, pp. 219–44. New York: McGraw-Hill.

Santangelo, J. D., Jones, D. T. & Woods, D. R. (1991). Metronidazole activation and isolation of *Clostridium acetobutylicum* electron transport genes. *Journal of Bacteriology*, 173, 1088–95.

Shin, C. Y. & Crawford, D. L. (1995). Biodegradation of trinitrotoluene (TNT) by a strain of *Clostridium bifermentans*. In *Bioaugmentation for Site Remediation*, ed. R. E. Hinchee, G. D. Sayles & B. C. Alleman, pp. 57–69. Columbus, OH: Battelle Press.

Smock, L. A., Stoneburner, D. L. & Clark, J. R. (1976). The toxic effects of trinitrotoluene (TNT) and its primary degradation products on two species of algae and the fathead minnow. *Water Research*, 10, 537–43.

Spiker, J., Crawford, D. & Crawford, R. (1992). Influence of 2,4,6-trinitrotoluene (TNT) concentration of the degradation of TNT in explosives-contaminated soils by the white-rot fungus *Phanerochaete chrysosporium*. *Applied and Environmental Microbiology*, 58, 3199–202.

Stahl, J. D. & Aust, S. D. (1993a). Plasma membrane dependent reduction of 2,4,6-trinitrotoluene by *Phanerochaete chrysosporium*. *Biochemical and Biophysical Research Communications*, 192(2), 471–6.

Stahl, J. D. & Aust, S. D. (1993b). Metabolism and detoxification of TNT by *Phanerochaete chrysosporium*. *Biochemical and Biophysical Research Communications*, 192(2), 477–82.

Stormo, K. E. & Deobald, L. A. (1995). Novel slurry bioreactor with efficient operation and intermittent mixing capabilities. In *Bioaugmentation for Site Remediation*, ed. R. E. Hinchee, G. D. Sayles & B. C. Alleman, pp. 129–35. Columbus, OH: Battelle Press.

Tan, E. L., Ho, C. H., Griest, W. H. & Tyndall, R. L. (1992). Mutagenicity of trinitrotoluene and its metabolites formed during composting. *Journal of Toxicology and Environmental Health*, 36, 165–75.

Vorbeck, C., Lenke, H., Fischer, P. & Knackmuss, H.-J. (1994). Identification of a hydride-meisenheimer complex as a metabolite of 2,4,6-trinitrotoluene by a *Mycobacterium* strain. *Journal of Bacteriology*, 176(3), 932–4.

Williams, R. T. & Marks, P. J. (1991). *Optimization of Composting for Explosives Contaminated Soil*. Final Report prepared for U.S. Army Toxic Hazardous Materials Agency. Report no. CETHA-TS-CR-91053. November 1991.

Williams, R. T., Ziegenfuss, P. S. & Sisk, W. E. (1992). Composting of explosives and propellant contaminated soils under thermophilic and mesophilic conditions. *Journal of Industrial Microbiology*, 9, 137–44.

Won, W. D., DiSalvo, L. H. & Ng, J. (1976). Toxicity and mutagenicity of 2,4,6-trinitrotoluene and its microbial metabolites. *Applied and Environmental Microbiology*, 31, 576–80.

7

A history of PCB biodegradation

Ronald Unterman

7.1 Introduction

Polychlorinated biphenyls (PCBs) are a group of related chemicals (congeners) which were produced during the middle of this century as chlorinated derivatives of biphenyl. There are 209 congeners distinguished by the number and position of chlorine atoms on a biphenyl backbone, although many of these congeners are not found in the commercial PCB mixtures (marketed under the tradename 'Aroclor' in the U.S., 'Clophen' in Europe, and 'Kaneclor' in Japan) (Ballschmiter & Zell, 1980; Hutzinger *et al.*, 1983; Erickson, 1992). The chemical and thermal stability of these compounds formed the basis for the widespread use of PCBs as the fire-retardant fluid in electrical equipment, heat exchangers, hydraulic fluids, and compressor fluids. However, this chemical stability also carried with it a degree of biological stability that has resulted in the accumulation of PCBs at sites of its manufacture, use, storage, and disposal.

Although the term PCBs has evoked the specter of the grim reaper, at the outset of this review it should be stated that this visceral response to PCBs has at times bordered on the irrational. As our society moves into the next century and we increasingly address our environmental problems, we must take a realistic view of the true risks posed by each of the chemical targets that we address. It should also be remembered that in their time PCBs were an important life-saving invention. Prior to 1929 electrical equipment contained flammable fluids which often ignited, creating conflagrations that truly posed serious risks to human health. The use of chlorinated organics, including PCBs, as fire-retardant liquids helped create a safer environment for our growing industrial society. It was not until many years later that possible health effects were discovered and in the case of PCBs these have at times been overstated, as documented in more recent health studies (Abelson, 1991). This is not to say that environmental release of halogenated chemicals is innocuous; however, our society must realistically assess, using sound risk analysis, which environmental problems we should tackle using our limited resources. The ultimate goal must be protection of human health and environmental well being.

The environmental release and fate of PCBs has occupied the front row of our environmental consciousness for the better part of a quarter century. This began in the late 1960s as a result of concern over the possible toxic effects of PCBs as first seen in Japan and Taiwan, as well as a perception that many industrial chemicals such as PCBs were immutable and would persist forever. Other early studies also indicated possible health effects, and numerous continuing studies have extended these initial findings. It is also clear from the literature that the publicly perceived toxicity of PCBs almost surely overstates their actual toxicity and health risk, and this controversy continues today. However, it is not the purpose of this review to discuss PCB toxicity and risk and the social issues underlying this topic, but to present an overview and historical perspective of the extensive body of work generated over the past 22 years that has demonstrated the biodegradability of PCBs, and what implications this has for their fate in the environment. Our ultimate goal is to understand better the microbial biodegradation of this family of compounds and the significance this has for natural attenuation, and where necessary to evaluate possible approaches for accelerating these natural processes through bioremediation techniques.

7.2 The early years – demonstrating the biodegradability of PCBs

In 1973, Ahmed & Focht reported the first evidence for PCB biodegradation. They isolated two *Achromobacter* species from sewage, the first by biphenyl enrichment and the second by 4-chlorobiphenyl (4-CB) enrichment, and showed that these strains could degrade several PCB congeners. The products produced by these two species were different, thus suggesting different metabolic pathways for degradation. In addition, they showed that degradation of the unsubstituted aromatic ring was preferred, that no dechlorination occurred, and that this led to a build-up in chlorobenzoic acids.

Baxter *et al.* (1975) measured PCB degradation by a species of *Nocardia* and a species of *Pseudomonas*. With the *Nocardia* species they reported 88% degradation and 95% degradation for 52 and 100 days, respectively. With the *Pseudomonas* species they reported 76 and 85% degradation for the same periods. Both of these studies were done with the PCB mixture Aroclor 1242. In their research with single PCB congeners they found that PCBs with up to six chlorines could be degraded to some degree and that the pattern of degradation for the two species was different; for example, the \Nocardia* could not degrade 4,4'-dichlorobiphenyl (4,4'-CB) in 121 days, whereas the *Pseudomonas* species degraded about 50% of this PCB in 15 days. They also described the activity as cometabolic, meaning in their case that some congeners that degraded slowly when present as the only carbon source were degraded more rapidly when in a mixture of congeners or when biphenyl was present.

Furukawa & Matsumara (1976) isolated an *Alcaligenes* species (designated Y42) from a lake sediment by enrichment with biphenyl. This culture was also able to degrade PCBs. The degradation followed a stepwise pathway ending in chlorinated benzoic acids. PCB congeners with chlorines on only one of the rings were degraded more easily than those with chlorines on both rings. Lower chlorinated congeners were more easily degraded and congeners with up to five chlorines could be degraded.

Subsequent studies by Furukawa and co-workers (Furukawa *et al.*, 1978a, 1979, 1982) with *Alcaligenes* Y42 and an *Acinetobacter* species (designated P6) using 31 different PCB congeners showed similar results. They found that congeners with chlorine at positions 2,6- or 2,2'- (i.e., highly ortho-substituted) were poorly degraded, and that less degradation occurred as the number of chlorine atoms per molecule increased. The two species showed similar degradative competence and several trichlorobiphenyl congeners could be degraded almost completely in less than three hours. *Acinetobacter* sp. P6 (Furukawa *et al.*, 1978b) has been extensively studied by this group and many others during the intervening years, including its taxonomic reassignment as the gram positive *Corynebacterium* sp. strain MB1 (Bedard *et al.*, 1984) and later as a *Rhodococcus globerulus* (Asturias & Timmis, 1993; Envirogen, Inc., unpublished results). P6 (MB1) has been demonstrated to be one of a select group of PCB-degrading bacterial strains that can degrade highly chlorinated PCBs (Bedard *et al.*, 1984, 1986; Kohler *et al.*, 1988a).

Sayler and co-workers isolated a species of *Pseudomonas* from sea water that degraded PCBs. They obtained between 9 and 39% degradation of Aroclor 1254 in 22 days, the percentage depending on the starting concentration. In 60 days they found between 63 and 84% degradation, again the percentage depending on the starting concentration (Sayler *et al.*, 1977, 1978).

Tucker *et al.* (1975), working with a continuously fed activated sludge unit at a feed level of 1 mg/48 hours, reported 100% degradation of biphenyl, 81% degradation of Aroclor 1221, 33% degradation of Aroclor 1016, 26% degradation of Aroclor 1242, and 15% degradation of Aroclor 1254. With Aroclor 1221, only biphenyl and the mono-chlorobiphenyls were biodegraded; the di-, tri-, tetra-, and pentachlorobiphenyl congeners were not degraded.

During these early years a number of laboratories postulated similar pathways for PCB (Ahmed & Focht, 1973; Furukawa *et al.*, 1978a; Yagi & Sudo, 1980; Furukawa *et al.*, 1983) and biphenyl (Lunt & Evans, 1970; Gibson *et al.*, 1973; Catelani *et al.*, 1973) degradation. The initial enzyme was described as a 2,3-dioxygenase with subsequent attack by a dihydrodiol dehydrogenase to produce (chlorinated) 2,3-dihydroxybiphenyl. The dihydroxybiphenyl then underwent ring cleavage via another dioxygenase reaction between carbon atoms 1 and 2 (*meta*-cleavage) to produce (chlorinated) 2-hydroxyl-6-oxo-6-phenylhexa-2,4-dienoic acid (the so-called '*meta*-cleavage product') which itself was oxidized in a fourth reaction to produce (chlorinated) benzoic acid and a smaller aliphatic fragment. Although additional PCB metabolites and pathway branches have since been described, this original 'upper' PCB/biphenyl pathway is the accepted main route for aerobic PCB biodegradation.

It was also during the late 1970s that the first work into the microbial genetics of PCB biodegradation took place. These studies were focused on describing the plasmids which encoded PCB-degradative enzymes (Kamp & Chakrabarty, 1979; Farrell & Chakrabarty, 1979; Christopher et al., 1981; Chatterjee et al., 1981; Chakrabarty, 1982). Furukawa & Chakrabarty (1982) described a 53.7-megadalton plasmid from *Acinetobacter* sp. P6 and *Arthrobacter* sp. strain M5 (M5 was later shown to be presumably identical to P6 and both were reclassified as *Rhodococcus globerulus*, as discussed above) which appeared to harbor the biphenyl degradative pathway in that biphenyl-minus segregants lost a small piece of the plasmid and this loss correlated with the loss of biphenyl and 4-CB utilization.

Thus, it was well documented during this early period that many different genera of bacteria were capable of degrading mono-, di-, and trichlorobiphenyls when presented as single compounds (Ahmed & Focht, 1973; Baxter et al., 1975; Tulp et al., 1978; Reichardt et al., 1981; Furukawa, 1982) or in a commercial mixture such as Aroclor 1242 (Kaiser & Wong, 1974; Tucker et al., 1975; Clark et al.,1979; Yagi & Sudo, 1980; Liu, 1981). However, there were few reports of degradation of the more highly chlorinated PCBs, as in Aroclor 1254 (Sayler et al., 1977; Liu, 1980). Throughout all of these studies it was generally observed that PCB congeners containing more than three chlorine atoms were markedly resistant to biodegradation, and that with the exception of monochlorobiphenyls, degradation was cometabolic, requiring biphenyl or 4-CB as carbon and energy source; higher chlorinated PCBs could not support bacterial growth. Most importantly, there was now a clear body of evidence that bacteria did exist that were capable of degrading these 'difficult' chlorinated aromatics. The groundwork was laid, and names such as Focht, Furukawa, Sayler, Gibson, and Chakrabarty are to this day still contributing to their pioneering findings from the 1970s.

7.3 The microbial expansion

7.3.1 New bacterial strains and activities – aerobic metabolism

The level of PCB biodegradation research into aerobic processes expanded during the 1980s as a growing number of bacterial systems were identified that were capable of oxidatively attacking PCBs. During this time several new biological pathways were identified which demonstrated the breadth of catalytic capability that existed naturally. This work included research by a group at the General Electric Research Laboratories which identified more than 30 new environmental isolates from PCB-contaminated soil and river silt, and which characterized several exceptional strains of aerobic bacteria that could biodegrade a broad suite of PCB congeners as well as attack some highly chlorinated PCBs including hexa- and heptachlorobiphenyls (Bedard et al., 1984; Unterman et al., 1985; Bedard et al., 1986; Bopp, 1986; Unterman et al., 1988). A congener-specific analysis of these transformations demonstrated that some of these organisms preferentially attacked different groups of PCB congeners (Bedard et al., 1984; Unterman et al., 1985;

Bedard *et al.*, 1987a). Further studies using two of these complementary strains (originally designated Types I and II for so-called 2,3- and 3,4-dioxygenase activity) demonstrated the utility of using a mixed culture approach to effect more extensive PCB degradation. (Unterman *et al.*, 1987a, 1988).

The early work of Bedard and co-workers resulted in the isolation of a novel PCB-degrading bacterial strain. This conclusion was based initially on its specificity for degrading characteristic types of PCB congeners; in particular, those PCBs with a 2,5-substituted phenyl ring (Bedard *et al.*, 1984). Extensive characterization of this new strain, designated *Alcaligenes eutrophus* H850, showed that H850 could attack PCBs that had no unchlorinated 2,3 sites and could hydroxylate at positions 3,4 on the phenyl ring (Bedard *et al.*, 1987a; Nadim *et al.*, 1987). The strain also showed a preference to degrade congeners by attack on 2-; 2,4-; 2,5-; or 2,4,5-chlorophenyl rings, and was also shown to produce, in addition to chlorobenzoic acids, a novel metabolite, 2′,3′-dichloroacetophenone (Bedard *et al.*, 1987a). Adding to the uniqueness of this strain, H850 had exceptional PCB degradative competence in terms of its breadth of congener specificity, including many tetra- and pentachlorobiphenyls and several hexachlorobiphenyls, and showed an extensive ability to degrade Aroclor mixtures (Bedard *et al.*, 1987b). In the intervening years, it has turned out that very few bacterial strains have ever been isolated which have the biochemical specificity exhibited by H850, these being LB400, ENV307 and ENV391 as described below.

The production of chloroacetophenones as metabolites of PCB biodegradation was eventually shown to be more common than just from H850 metabolism of PCBs (Barton & Crawford, 1988; Bedard & Haberl, 1990). The study by Bedard & Haberl (1990) on the influence of chlorine substitution pattern on the degradation of PCBs by eight bacterial strains demonstrated several patterns of reactivity and specificity of ring attack. From an analysis of these strains it was concluded that the collection of strains characterized to date could be divided into four classes based on their mode of attack and preference for chlorine substitution pattern. From the continuing work of these researchers and others, it now appears that the variety of PCB degradative competence even exceeds the initial four classes as described by Bedard and Haberl and in some ways may represent a continuum of genetic and biochemical variability (Unterman *et al.*, 1991).

Following the isolation of H850, a second exceptional PCB-degrading strain designated *Pseudomonas* sp. LB400 (Bopp, 1986) was isolated at the General Electric laboratories. This strain was similar to H850 in biochemical and PCB congener specificity, but was a distinct strain as later detected by flanking DNA sequences and microbial characterization. Like H850, LB400 could also degrade higher chlorinated congeners and showed a breadth of PCB-degradative capability beyond those that had been previously demonstrated. Only two other strains have as yet been characterized that have the unique congener specificity and degradative capabilities of H850 and LB400. One of these cultures, *Pseudomonas* sp. strain ENV307 (Sharma *et al.*, 1991; Shannon *et al.*, 1994), was isolated from the upstate

New York region as were LB400 and H850, and the fourth culture, *Pseudomonas* sp. strain ENV391, was isolated from a PCB-contaminated site in Pennsylvania (Rothmel *et al.*, 1993). It is interesting to note the scarcity of this degradative competence, considering the extensive efforts that have occurred over the past 22 years to isolate novel and superior PCB-degrading strains.

Because of its exceptional PCB-degradative competence, LB400 was chosen by several laboratories for further biochemical and genetic characterization. The multi-component enzyme system from LB400 was purified and characterized by Gibson and co-workers (Haddock *et al.*, 1993, 1995) and shown to be very similar to other multi-component aromatic hydrocarbon dioxygenases. This work showed that a single dioxygenase was responsible for the previously demonstrated (Bedard *et al.*, 1984, 1987a; Nadim *et al.*, 1987) attack at both 2,3 and 3,4 ring positions, and that following dihydroxylation of a 2-chlorophenyl ring spontaneous dechlorination occurred. In addition, the third enzyme in the biphenyl pathway of LB400 (2,3-dihydroxybiphenyl-1,2-dioxygenase) has recently been purified and crystallized, allowing for further characterization of the enzymes of this PCB pathway (Eltis *et al.*, 1993). Finally, extensive genetic studies of LB400 have been conducted, as described below.

Following on the initial work from 1977, Sayler and colleagues expanded their PCB biodegradation studies initially to characterize mixed bacterial cultures for their ability to biodegrade monochlorobiphenyls (Shiaris & Sayler, 1982; Kong & Sayler, 1983). In subsequent work, Shields *et al.* (1985) isolated and characterized a bacterial strain capable of mineralizing monohalogenated biphenyls and showed that it harbored a 35-megadalton plasmid which encoded the complete pathway for 4-chlorobiphenyl degradation. This plasmid, pSS50, was the first plasmid ever reported which encoded the complete mineralization of a model PCB (Hooper *et al.*, 1989; Layton *et al.*, 1992). Other work by this group has focused on the use of DNA probes for isolating PCB-degradative strains (Pettigrew & Sayler, 1986), the population dynamics of PCB degradation (Pettigrew *et al.*, 1990), as well as novel application vectors (as described below) and a planned field demonstration of PCB biodegradation using both naturally occurring and genetically engineered strains.

In the study of Pettigrew *et al.* (1990) a bacterial consortium was shown to mineralize 4-CB and dehalogenate 4,4'-CB. It included three isolates: a *Pseudomonas testosteroni* which catalyzed the breakdown of the chlorinated biphenyls to 4-chlorobenzoic acid (the so-called 'upper pathway'); an *Arthrobacter* species that mediated 4-chlorobenzoic acid mineralization (the so-called 'lower pathway'); and a third strain from the consortium with a role that has not been determined. This pattern of co-culture degradation for upper and lower pathway degradation has been observed generally in the field of PCB biodegradation. Few strains have been shown with the capability to catalyze both upper and lower pathway degradation.

In the early 1980s Sylvestre and co-workers began their studies first by isolating a facultative anaerobe, strain B-206, which was able to grow on 4-chlorobiphenyl, but could not grow on or degrade 4-chlorobenzoic acid. Strain B-206 could also grow on biphenyl, 3-CB, 2-CB, and benzoic acid, but it was unable to grow on any of the monochlorinated benzoic acids. Furthermore, although B-206 could not grow on Aroclor 1254 it was reported to partially transform this Aroclor (Sylvestre & Fauteux, 1982). Subsequent work with strain B-206 demonstrated nitration of 4-chlorobiphenyl dependent on the nitrogen source in the medium. Strain B-206 generated the novel metabolites of 4-CB (hydroxy nitrochlorobiphenyls), probably as products of side reactions involving intermediates in the 4-CB biodegradation pathway. It was presumed that these compounds were the result of a reaction with an arene oxide intermediate and therefore indicated the presence of a monooxygenase in strain B-206 (Sylvestre et al., 1982).

Continued work in this laboratory focused on two newly isolated strains, an *Achromobacter* sp. strain B-218 and a *Bacillus brevis* strain B-257. This work involved the characterization of 4-chlorobiphenyl degradation and the identification of the intermediates through 4-chlorobenzoic acid, and confirmed the work of others which had previously postulated the dioxygenase pathway for biphenyl and PCBs. In the case of both strains, 4-chlorobenzoic acid was not further metabolized and was found to accumulate in the growth medium. In addition, this study identified several novel PCB metabolites and postulated that 3,4-dioxygenase attack was a possible source for some of these (Masse et al., 1984). In continuing work, Sylvestre et al., (1985) demonstrated that a two-member bacterial culture could overcome the accumulation of 4-chlorobenzoic acid as seen in previous studies. Under these mixed culture conditions, the degradation of 4-chlorobiphenyl was more rapid and complete. Finally, Sondossi et al., (1991) showed that the metabolism of the hydroxylated biphenyls and chlorohydroxybiphenyl was catalyzed by the same set of enzymes used for biphenyl and PCB degradation. Thus, these enzymes were shown to have a very broad substrate specificity allowing them to degrade many biphenyl analogs carrying various substituents.

It was during the mid-1980s that the first demonstration of PCB biodegradation by white-rot fungi was demonstrated (Bumpus et al., 1985; Eaton, 1985). This work sparked a tremendous interest in the broader substrate specificity of the white-rot fungi for many environmental pollutants. However, the limitations of this enzyme system were evident in terms of both the ability to grow these organisms for bioremediation applications as well as the breadth and extent of PCB congener degradation. The early results demonstrated only low levels of PCB degradation and a narrower breadth of specificity as compared to bacterial strains. However, progress on both of these fronts has been made during the intervening years; therefore, further research and commercialization of fungal bioremediation are continuing today and hold promise as a bioremediation approach (Zeddel et al., 1994).

Following on his pioneering work from 1973, Focht and colleagues expanded their research to new bacterial strains and applications of these cultures. This work initially focused on the *Acinetobacter* P6 strain of Furukawa and utilized ^{14}C-labeled PCBs to demonstrate the extent of mineralization that could be observed with Aroclors under conditions of biphenyl metabolism and bacterial inoculation of soils

(Brunner *et al.*, 1985; Focht & Brunner, 1985). The metabolism of PCBs (Aroclor 1242) was greatly enhanced by the addition of biphenyl to soils. After 49 days only 25–35% of the original PCBs remained in the soil and 48–49% was converted to $^{14}CO_2$ in treatments enriched with P6. In contrast, 92% of the PCBs remained and less than 2% was converted to $^{14}CO_2$ in unenriched controls. Inoculation with strain P6 did not increase mineralization of PCBs, but did show a higher rate and extent of degradation of the more highly chlorinated PCBs. These results are consistent with our understanding of PCB cometabolism and the requirement for biphenyl as both a growth substrate as well as a genetic inducer for the oxidation of PCBs. Further work with strain P6 and an *Arthrobacter* strain B1B demonstrated the excellent capabilities of P6 for degrading even the higher chlorinated congeners contained in Aroclor 1254. Both strains showed significantly higher levels of degradation when the cells were actively growing as compared with being held in a resting state (Kohler *et al.*, 1988a).

A novel PCB-degradative bacterial strain was isolated by Barton & Crawford (1988) and was identified as a *Pseudomonas* sp. (strain MB86). This strain was isolated from a 4-chlorobenzoate enrichment and was able to grow on 4-chlorobenzoic acid and 4-chlorobiphenyl as sole carbon and energy sources, thus demonstrating the rare case of a single organism harboring both the upper and lower PCB-degradative pathways (this had only been seen once before by Shields *et al.*, 1985). This strain also produced 4'-chloroacetophenone, which as described above had previously been identified from the General Electric (GE) cultures. However, in contrast to the GE cultures, large amounts of the chloroacetophenone accumulated in this culture despite the fact that it could grow on 4-chlorobenzoic acid. It was presumed that by enriching on 4-chlorobenzoic acid, in contrast to the traditional route of biphenyl enrichment, Barton & Crawford were able to isolate this novel strain. Interestingly, although the strain could grow on 4-chlorobenzoate it could not grow on benzoate or other chlorobenzoates. In addition, growth on 4-chlorobiphenyl was superior to growth on biphenyl, which was further evidence for the uniqueness of this strain.

Continued work by Focht and co-workers expanded the capabilities of bacterial PCB degradation by the development of co-culture techniques for combining upper- and lower-pathway degradation to facilitate the complete destruction of PCBs (Adriaens *et al.*, 1989; Adriaens & Focht, 1990). This work demonstrated the advantages of co-culture degradation of 4,4'-CB and also described a continuous fixed-film bioreactor for developing stable co-cultures for degrading di- and tetrachlorobiphenyls. In this system the PCB-degradative capability of strain P6 was combined with the chlorobenzoate-degradative ability of *Acinetobacter* strain 4-CBA, which utilized 4-chlorobenzoate as its sole source of carbon and energy. This research approach also led to the construction of a 3-chlorobiphenyl-utilizing strain that was produced using a multiple chemostat system (Krockel & Focht, 1987) which resulted in an intergeneric mating between a 3-CB-utilizing strain and a 3-chlorobenzoate strain. The recombinant isolate was able to grow on both

3-CB and 3-chlorobenzoate and therefore could mineralize 3-CB (Adams *et al.*, 1992). Ultimately, this approach led to the construction of another novel PCB bacterial isolate using the continuous chemostat approach of Krockel & Focht (1987). This recombinant bacterium has the unique capacity to utilize the PCB congener 3,4'-CB, which is chlorinated on both rings, as a sole carbon and energy source (McCullar *et al.*, 1994). This work of Focht and co-workers has been an important contribution to our understanding and development of hybrid pathways for PCB and chlorobenzoate biodegradation (Brenner *et al.*, 1994). Finally, the studies of Focht and co-workers led to the analysis of bacterial metabolism of hydroxylated biphenyls as potentially important pathways for the degradation of eukaryotic PCB by-products, and further demonstrated the breadth of substituted biphenyl degradability (Kohler *et al.*, 1988b; Higson & Focht, 1989).

The work of Reineke and colleagues in the early 1990s also made important contributions to demonstrating the capability of microorganisms to mineralize PCBs. Building upon the observation that very few strains had been isolated which harbor both the upper and lower PCB-degradative pathways (Shields *et al.*, 1985; Barton & Crawford, 1988), this group constructed hybrid *Pseudomonas* strains by combining the upper PCB-degradative pathways of a PCB-degrading *Pseudomonas putida* strain (BN10) as isolated from a biphenyl enrichment, with the lower pathway strain, *Pseudomonas* species B13. The *in vivo*-constructed hybrid organisms were able to mineralize 3-CB and to convert some of the congeners of Aroclor 1221 (mono- and dichlorobiphenyls) (Mokross *et al.*, 1990). In subsequent work, Havel & Reineke (1991) expanded on their initial research, first by assessing the stability of co-cultures to mineralize a broader group of PCB congeners, and then, through conjugative transfer, they generated a hybrid strain that could mineralize 2-CB, 3-CB, 4-CB, 2,4-CB, and 3,5-CB. Other work by this group (Havel & Reineke, 1992) further assessed the limitation of partial PCB degradation in soil microcosms and the limitations that one observes due to production of chlorobenzoate intermediates. Thus, these studies by Reineke and colleagues clearly demonstrate the potential for expanding the degradative capability of PCB strains for their use in bioremediation, as well as the potential for natural genetic exchange to occur, thereby facilitating complete destruction of PCBs in soil environments. The research of Fulthorpe & Wyndham (1992) using lake water and sediment flow-through microcosms further demonstrated that microbial communities can selectively adopt chlorobenzoate catabolic genes when dosed with 3-chlorobiphenyl, thus indicating that natural adaptation to complete PCB mineralization is observed in natural systems.

Other work through the 1980s focused on the identification and characterization of various PCB degraders and included work on the biodegradation of mono-chlorobiphenyls in river water (Bailey *et al.*, 1983), and the biodegradation of PCBs by mixed bacterial cultures grown on naphthalene and 2-methyl, 4-chloro-phenoxyacetic acid (Kilpi *et al.*, 1988). The work of Parsons & Sijm (1988) demonstrated the degradation of mono-, di- and tetrachlorobiphenyls in chemostat cultures using a *Pseudomonas* strain (JB1) that was isolated from soil. Finally, the work of Minoda and co-workers has further characterized the enzymes and biochemistry of biphenyl biodegradation (Ishigooka *et al.*, 1986; Omori *et al.*, 1986a, b) and growth of a *Pseudomonas cruciviae* strain S93B1 on 10 biphenyl-related compounds, including 4-CB (Takase *et al.*, 1986).

Recently, the PCB-biodegradative capabilities of methanotrophs have been demonstrated (Adriaens, 1994). In this study, 2-CB was oxidized by a methanotrophic culture (CSC1) to a hydroxylated chlorobiphenyl intermediate identified as 2-hydroxy-3-chlorobiphenyl. This intermediate indicated that the metabolite was formed via a concerted oxidation involving an arene oxide which rearranges spontaneously via an NIH shift. No studies have shown, however, that methanotrophs can degrade more highly chlorinated PCBs, and their utility for bioremediation processes does not seem promising.

The ultimate outcome of this microbial expansion (which perhaps should be called an explosion) was a greater understanding of the breadth of PCB-degradative activity that exists naturally. Prior to the work of Ahmed & Focht (1973) PCBs were perceived as immutable, and this perception lingered even into the 1970s and early 1980s despite the early research. It was because of the extensive body of work as described above that the aerobic biodegradability of PCBs was finally accepted generally, and indeed the dogma of PCB recalcitrance was shattered. It is now clear that a broad spectrum of naturally occurring aerobic bacteria exist which can degrade at least the lower chlorinated PCBs, and this fact has become broadly accepted, as evidenced by a 1987 article in *Science* entitled 'Discovering microbes with a taste for PCBs' (Roberts, 1987).

7.3.2 Discovery of anaerobic dechlorination and natural attenuation

Concurrent with the research being conducted on the isolation and characterization of aerobic PCB-degrading strains, an exciting and extremely important discovery was made by two groups regarding gas chromatographic (GC) analysis of Aroclors from Hudson River sediment (Bopp *et al.*, 1984; Brown *et al.*, 1984, 1986, 1987a, b). Both groups observed that the GC profile of sediment PCBs indicated an altered pattern from the Aroclor 1242 that had been released from the General Electric capacitor plants at Fort Edward, New York. The composition of the Aroclor in these samples was totally unlike any commercial mixture in that the PCB distribution was substantially enriched in the lower-chlorinated congeners, in particular 2-CB, 2,2'-CB, and 2,6-CB, and this altered pattern was observed only in the deeper sediments. Although others had observed 'weathered' PCB GC profiles in Hudson River and many other environmental samples, it required the insight of Brown and Bopp and their colleagues to see beyond the dogma that all weathering was the result of selective physical and chemical redistribution (e.g., sorption, volatilization, dissolution). They postulated that the weathering was a result of the anaerobic conditions of the lower sediments and further proposed that this altered GC peak distribution was the result of reductive dechlorination, probably biologically mediated. This thinking was a paradigm shift of far-reaching importance, as evidenced by the research and discoveries that followed.

Different GC patterns were recognized in the Hudson River sediments and each was presumably the result of a different population of anaerobic bacteria each with its own distinctive pattern of PCB congener specificity (Brown *et al.*, 1987a, b). It was proposed that these unusual congeners arose from anaerobic microbial dehalogenation of more highly chlorinated PCBs through a series of stepwise dehalogenations. The observed reductive dechlorination patterns in the Hudson River demonstrated that this process favored the reductive dehalogenation of *meta*- and *para*-chlorines and resulted in the accumulation of *ortho*-substituted PCBs. A similar observation of an altered PCB pattern was observed for a site in Pittsfield, Massachusetts, which was contaminated with Aroclor 1260 (Brown *et al.*, 1987a,b). The data from this site also suggested sequential reductive dechlorination with *meta*-.and *para*-dechlorination specificity.

The first reports and conclusions from the General Electric research group (J.F. Brown, Jr. and colleagues) and the Bopp group were not immediately accepted as a biologically mediated process, especially by the regulatory community, which attempted to explain the altered PCB congener distribution in terms of other physical and chemical processes (M.P. Brown *et al.*, 1988). However, the response of J.F. Brown, Jr. *et al.* (1988) refuted these claims. Ultimately, the large body of information that developed both in the laboratory and from the identification of other sites of natural dechlorination convinced even the initial skeptics of the widespread existence of microbial reductive dechlorination as well as its importance as a process for PCB attenuation in the environment.

The dramatic inferences made from PCB chromatograms were subsequently verified through the work of many researchers. The first report confirming the postulated anaerobic microbial activity in the Hudson River was from a group at Michigan State University (Quensen *et al.*, 1988). This work compared the activity of autoclaved versus live Hudson River sediments over a 16-week time course and demonstrated a characteristic shift in the congener distribution of added PCBs; a large increase in lower-chlorinated congeners and concomitant decrease of higher congeners. As seen in the original Hudson River samples, this activity was specific for *meta*- and *para*-dechlorination. This work was followed by a broader examination of four different Aroclor mixtures using anaerobic sediments from the Hudson River and Silver Lake, Massachusetts. The differences in dechlorination patterns suggested that different organisms were responsible for the dechlorination at the different sites and that each had its characteristic PCB dechlorination specificity (Quensen *et al.*, 1990).

Following the initial observations and confirmation of microbial reductive dechlorination, a large number of laboratories began to conduct research into this area. Some of this work focused on confirmation and further elucidation of the Hudson River dechlorination (Chen *et al.*, 1988; Rhee *et al.*, 1989, 1993a, b; Abramowicz, 1990; Abramowicz *et al.*, 1993). Other work sought evidence for reductive dechlorination at other environmental sites and found this to be broadly based in many anaerobic environments (Alder *et al.*, 1993; Sokol *et al.*, 1994; Ofjord *et al.*, 1994; Bedard & Quensen, 1995 (this reference is an excellent and comprehensive review of microbial dechlorination of PCBs); Bedard & May, 1996).

Further laboratory work also focused on elucidating the biochemical mechanism of PCB dechlorination. This work generally utilized pure PCB congeners and helped develop an understanding of the mechanism of PCB dechlorination and induction, as well as the alternative pathways for chlorine removal (Abramowicz, 1990; Nies & Vogel, 1990, 1991; Assaf-Anid *et al.*, 1992; Williams, 1994; Bedard & Quensen, 1995). In addition, the first experimental demonstration of biologically mediated *ortho*-dechlorination was published (Van Dort & Bedard, 1991). Furthermore, Bedard and Van Dort (1992) made the important discovery that in methanogenic sediment slurries brominated biphenyls, as well as other brominated aromatics, could be used to stimulate reductive dechlorination of Aroclor 1260 (Bedard *et al.*, 1993; Stokes *et al.*, 1994). Following the laboratory demonstration of this capability, a field demonstration was initiated in Woods Pond, Massachusetts, and has been underway for over a year with very successful dechlorination results (discussed below).

Continuing research has been focusing on attempts to enrich and isolate pure strains which can dechlorinate PCBs. To date these attempts have met with only limited success, although some work has been able to develop subpopulations of either anaerobic spore formers (Dingyi *et al.*, 1992), or cultures that have a selective *meta*-dechlorination activity (Morris *et al.*, 1992). A team from Celgene Corporation was able to subculture an anaerobic PCB dechlorination system on solid media while maintaining dechlorination of tri- and tetra-chlorobiphenyls (May *et al.*, 1992), as well as employ an enrichment method that allowed for a significant increase in the rate of reductive dechlorination of PCBs (Boyle *et al.*, 1993).

It has become clear over the past ten years that the 1980s demonstration of anaerobic microbial dechlorination of PCBs is probably the most important discovery in the field of PCB biotransformations since Ahmed and Focht first demonstrated in 1973 that PCBs were biodegradable. Many new anaerobic microbial activities have been enriched and characterized from anaerobic fresh water and marine environments and heavily polluted industrial sediments. These anaerobic cultures are capable of dechlorinating PCBs, thereby transforming highly chlorinated Aroclors to lower-chlorinated mixtures. This natural attenuation process is an important contributor to PCB degradation and detoxification in the environment and can form the basis for intrinsic remediation of many PCB-contaminated sites.

At least three environmental benefits are derived from PCB dechlorination. First the *meta*- and *para*-dechlorination observed in most anaerobic systems removes those chlorines that confer upon PCBs the 'dioxin-like' structure and potential toxicity. Second, the products of anaerobic dechlorination are more soluble and have a decreased tendency to bioaccumulate. Third, the lower-chlorinated products from PCB dechlorination are more readily biodegradable aerobically as documented above, irrespective of whether this process occurs through natural attenuation or active bioremediation. Furthermore, anaerobic cultures also have the potential to be used in anaerobic microbial bioremediation systems that could address even Aroclor mixtures as highly chlorinated as Aroclor 1260, which is essentially nonbiodegradable by aerobic cultures. In addition, these anaerobic cultures also exhibit substantial activity on the lower-chlorinated Aroclors (1242, 1248) and facilitate the degradation of these PCBs as well. Thus, the anaerobic transformation of PCBs plays a critical role for both natural attenuation as well as future development of PCB bioremediation systems.

7.4 Recent history

7.4.1 Genes and genetic recombination

The tools of genetic engineering have now been applied to PCB-degrading organisms and many research laboratories have isolated and cloned the genes for PCB biodegradation from various bacterial strains. It is ultimately hoped that this line of research will result in the construction of microbes with exceptional PCB-degradative competence and simpler process control (Unterman *et al.*, 1987b; Timmis *et al.*, 1994). In any event, this research is helping to elucidate the mechanisms, regulation, and diversity of PCB biodegradation.

As a continuation of their comprehensive and pioneering work of the 1970s and early 1980s into the microbial characterization of PCB-degradative strains, Furukawa and his colleagues began a focused effort in the mid-1980s to isolate, clone, and characterize the genes from many of their PCB-degradative strains. This work included the cloning of a gene cluster from *Pseudomonas pseudoalcaligenes* KF707 which encoded the first three enzymes in the biphenyl pathway: *bphA* (encoding biphenyl dioxygenase), *bphB* (encoding dihydrodiol dehydrogenase), and *bphC* (encoding 2,3-dihydroxybiphenyl dioxygenase). The fourth gene, *bphD*, which encodes the last enzyme in the upper pathway, the hydrolase, was missing from this fragment (Furukawa & Miyazaki, 1986). This was followed in the next year by the first sequence of a PCB-degradative gene, that being the *bphC* gene of KF707 encoding the 2,3-dihydroxybiphenyl dioxygenase (Furukawa *et al*, 1987), and subsequently the purification and characterization of the protein as isolated from the recombinant KF707 clone (Furukawa & Arimura, 1987). The cloning of the *bphC* gene from *Pseudomonas paucimobilis* Q1 permitted the first comparison of the two genes from different PCB-degradative strains. These showed only moderate homology to each other and were the first indication of the genetic variability of PCB degradative strains that exist in nature. Indeed, polyclonal antibodies raised against the protein from Q1 failed to cross react with the previously isolated 2,3-dihydroxybiphenyl dioxygenase from KF707 (Taira *et al.*, 1988).

One of the most extensive genetic comparisons of a broad group of PCB-degrading strains was performed by Furukawa *et al.* (1989). Fifteen strains were compared with gene probes from KF707 and Q1. The KF707 probe demonstrated essentially identical homology between seven strains despite a dissimilarity in the flanking chromosomal regions. Three other strains were homologous but not identical to KF707, and five other strains showed no significant genetic homology with KF707. The Q1 probe (*bphC* gene only) lacked genetic as well as immunological homology with any of the other 15 biphenyl/PCB degraders tested. The unique strain, *Arthrobacter* sp. M5 (identical to P6 as discussed above) also showed no homology to any of the other strains. This study showed at least three different PCB type strains based on genetic and immunological homology; however, it also demonstrated nearly identical chromosomal genes among various strains, thus suggesting a mechanism for gene transfer of the *bphABC* locus (Furukawa, 1994). This research team further expanded their genetic library by cloning and characterizing the *bphABCD* genes from yet another PCB degrading strain, *Pseudomonas putida* KF715 (Hayase *et al.*, 1990).

Furukawa *et al.* (1993) extended the homology analysis to the toluene–benzene operon of *P. putida* F1 ('*tod*' genes). Despite their discrete substrate ranges for aromatic metabolism, the two genes showed very similar gene organization as well as size and homology of the corresponding enzymes. Through the construction of a *bph* and *tod* chimera it was demonstrated that these two gene sets could complement each other for significant enzyme components and confer cross reactivity between these two metabolic pathways. This cross reactivity between the toluene and PCB-catabolic pathways was also shown by Zylstra *et al.* (1990). This and other evidence indicates that the various aromatic ring dioxygenases probably evolved from a common ancestor and that the biphenyl/PCB degradative pathways are one branch in this aromatic degradative gene family (Neidle *et al.*, 1991; Haddock *et al.*, 1993; Furukawa, 1994).

During this same period, Mondello cloned and expressed the upper pathway genes from strain LB400 in *E. coli* and achieved the first isolation of the whole gene set for upper pathway metabolism of PCBs (to the chlorobenzoic acids). The recombinant strain had the same exceptional and characteristic congener specificity of LB400, yet did not require growth on biphenyl for full induction of the PCB pathway (Mondello, 1989). A probe from the LB400 genes was used to compare homology of this strain with seven other PCB-degrading strains and only demonstrated significant homology to the *Alcaligenes eutrophus* H850, which is a strain that has the same unique PCB-degrading competence. No homology was found with the other strains, nor with the flanking region of strain H850 (Yates & Mondello, 1989). Thus, at least two different classes of PCB genes were demonstrated in this study, consistent with the biochemical and congener specificity evidence as previously described (Bedard *et al.*, 1984, 1986; Nadim *et al.*, 1987). Further characterization of the LB400 genes demonstrated the structure of this operon and its homology to components of the toluene dioxygenase genes from *P. putida* F1 (Erickson & Mondello, 1992).

In a fascinating comparison of the gene sequences from LB400 and KF707, two groups independently drew some important conclusions and demonstrated findings as to the DNA sequence variability that determines the significant differences in PCB congener substrate specificity between these two strains. In a comparison of the published DNA sequences and a direct comparison of the bacterial congener specificity of these two strains, Gibson *et al*. (1993) showed that despite the very significant differences in congener specificity, there were only very slight differences in nucleotide sequences. Indeed, it was postulated that single amino acid substitutions at two positions in the KF707 large subunit could play the determining role in substrate specificity for each enzyme.

Concurrently, Erickson & Mondello (1993) deduced from a comparison of the nucleotide and amino acid sequences of the biphenyl dioxygenases of LB400 and KF707 that despite the dramatic differences in substrate specificity the sequences were nearly identical; over a 4750-bp region an overall similarity in excess of 97% was demonstrated. Through resting cell assay analysis of LB400 and KF707, Erickson and Mondello also demonstrated the significant differences in PCB congener specificity between these two strains. By targeting four amino acids for site directed mutagenesis, Erickson and Mondello were able to show that the modified LB400 *bphA* gene product exhibited a broader congener specificity with increased activity against several congeners characteristic of KF707. This work elegantly demonstrated the extremely subtle differences within the *bphA* gene that confer such wide differences in PCB congener specificity. These results also demonstrate the possibility of expanding the range of congeners degradable by a single bacterial culture and the potential use of developing even more effective biocatalysts for bioremediation for PCB-contaminated sites.

During this same time period, Walia and colleagues cloned and expressed in *E. coli* the gene for 2,3-dihydroxybiphenyl dioxygenase from a 4-chlorobiphenyl-degrading *Pseudomonas putida* strain, OU83, and characterized this dioxygenase from the *E. coli* recombinant (Khan *et al*., 1988). In subsequent work the full upper pathway for PCB degradation in OU83 was cloned and the operon further characterized in terms of both mapping and expression (Khan & Walia, 1989, 1990, 1991), and production of metabolites from the cloned genes (Khan & Walia, 1992).

Throughout the early 1990s other research laboratories also isolated and characterized the PCB-degradative genes from various bacterial strains. Kimbara *et al*. (1989) cloned and sequenced the last two genes in the biphenyl pathway from *Pseudomonas* sp. strain KKS102. Kikuchi *et al*. (1994) later cloned and characterized the complete upper pathway from KKS102 and demonstrated the presence of 12 genes which encode this pathway, including four open reading frames.

Sylvestre and co-workers cloned the genes from their *Pseudomonas testosteroni* strain B-356, which had been demonstrated to biodegrade 4-CB into 4-chlorobenzoic acid. In these studies, they demonstrated the homology between many different PCB-degrading bacteria isolated from different geographic locations (Ahmad *et al*.,

1990). Furthermore, they described the limitations of expression of *bphA* and *bphB* genes in *E. coli* as well as the dehalogenation of chlorobiphenyls during initial oxygenase attack (Ahmad *et al.*, 1991). Work by other colleagues demonstrated the first successful cloning of the genes from Furukawa's P6/M5 culture and compared these with analogous genes from gram-negative strains (Peloquin & Greer, 1993).

The work of Timmis and co-workers resulted in the cloning of the seven structural genes of LB400 and the complete DNA sequences for *bphB*, *bphC*, and *bphD*. This information, when combined with the DNA sequence of *bphA* from Erickson & Mondello (1992), produced the first DNA sequence for all of the genes encoding the metabolism from biphenyl to benzoate (Hofer *et al.*, 1993). In addition, Asturias & Timmis (1993; Asturias *et al.*, 1994) cloned and characterized the *bphB*, *bphC*, and *bphD* genes from *Rhodococcus globerulus* P6 (the original Furukawa strain) as part of their goal to isolate genes from complementary PCB-degradative strains as a first step to the possible genetic combination of complementary degradative activities (Timmis *et al.*, 1994). The genetic uniqueness of this strain was demonstrated first by the presence of three *bphC* genes with narrow substrate specificity, as well as the fact that none of the P6 genes showed hybridization homology to other *bphC* genes from previously characterized strains. This finding was consistent with the uniqueness of the PCB-degradative activity of P6. A further understanding of the genetic differences of P6 may eventually help to elucidate the genetic basis for the capabilities of this extremely competent and unique PCB-degradative strain. Subsequent work by Maeda *et al.* (1995) characterized the genes from a second gram-positive strain, *Rhodococcus erythropolis* TA421, and demonstrated the presence of four *bphC* genes, suggesting that the *Rhodococcus* genus may contain a diverse family of *bphC* genes as also seen by Asturias *et al.* (1994).

In a collaboration between groups at Rutgers University and Envirogen, Inc., the PCB-degradative genes from *Pseudomonas* sp. strain ENV307 (previously isolated at Envirogen, Inc., unpublished data) were transferred into *Pseudomonas* strains which utilize a nontoxic, water-soluble surfactant as a selective growth substrate. The goal of this work was to demonstrate the utility of field application vectors. These vectors are a combination of a selective substrate host and a cloning vector developed for the purpose of expressing foreign genes in nonsterile competitive environments in which the genes confer no advantage to the host, as for example with PCB cometabolism. This technique involves the addition to the targeted environment of a selective substrate readily utilizable by the host microorganism, but not utilizable by most of the indigenous species, thereby conferring a selective advantage for the added bacterial strain (Lajoie *et al.*, 1992). The recombinant constructs were demonstrated to be able to degrade many PCB congeners in soil environments in the absence of the normally utilized biphenyl substrate (Lajoie *et al.*, 1993). This field application vector research was then expanded to include the full spectrum of PCB-degradative competence as encoded by the ENV307 donor strain, as well as demonstrate the stability and selective growth advantages in soil systems (Lajoie *et al.*, 1994).

Finally, the utility of PCB-degradative genes as probes for detection and monitoring of bacterial populations has also been demonstrated. These research programs have shown that DNA probes can be effective monitoring tools: (1) for both soil and water samples (Packard *et al.*, 1989); (2) to assess the abundance and diversity of PCB-degrading bacteria in polluted soil environments (Walia *et al.*, 1990); (3) for the detection of DNA sequences encoding PCB-degradative catabolic pathways in aquatic sediments without prior cultivation of the degradative microorganisms by using direct extraction of total DNA and PCR amplification, thereby detecting a chromosomally encoded single-copy gene from a highly specialized subpopulation within a total microbial community in a natural sediment (Erb & Wagner-Dobler, 1993); and (4) for a comparison of two soils for the presence of specific PCB-degradative strains (Layton *et al.*, 1994).

7.4.2 The lower pathway

Although the initial work on PCB biodegradation focused on the upper pathway from chlorinated biphenyls through the chlorinated benzoic acid by-products, more recent studies by some PCB research groups have focused on the lower pathway as well as the aliphatic fragment generated from the last upper pathway hydrolytic step to chlorinated benzoic acid. In model studies, Kohler-Staub & Kohler (1989) demonstrated the biodegradation of chlorobutyric and chlorocrotonic acids as model substrates for the chlorinated aliphatic five-carbon acid produced during the last upper pathway step of PCB biodegradation (hydrolysis of the aliphatic side chain following *meta*-cleavage of the first PCB ring). The *Alcaligenes* sp. strain CC1 was able to grow on several α-chlorinated aliphatic acids as well as the β-chlorinated 4-carbon aliphatic acids as sole carbon and energy sources and dehalogenation was observed. This work further confirmed the generally accepted view that the chlorinated aliphatic PCB-degradation products are not refractile to microbial attack.

Because of the potential inhibition of PCB biodegradation through accumulation of chlorinated benzoic acids, Focht and co-workers conducted extensive studies on the accumulation and biodegradation of many of the various chlorinated benzoic acids that are formed from PCB biodegradation (Hickey & Focht, 1990; Hernendez *et al.*, 1991; Adriaens & Focht, 1991; Arensdorf & Focht, 1994). They showed that many of these chlorinated benzoic acids are biodegraded by enrichment strains that can utilize them as sole carbon and energy source and in some cases cometabolically. This work has recently demonstrated the complete biodegradation of 4-chlorobiphenyl through a *meta*-cleavage pathway of the 4-chlorobenzoate intermediate. This mineralization of 4-chlorobiphenyl by *Pseudomonas cepacia* P166 is another of the rare examples of a PCB bacterial strain harboring both the upper and lower degradative pathways (Arensdorf & Focht, 1995). Higson & Focht (1990) also demonstrated that one of the alternative metabolites of PCB biodegradation, the ring-chlorinated acetophenones (Bedard *et al.*, 1987a; Barton & Crawford, 1988; Bedard & Haberl, 1990) can be cometabolized by two characterized bacterial strains.

As a logical extension of their earlier studies, the Focht group demonstrated in soil microcosms that the addition of chlorobenzoate-degrading bacteria enhanced the mineralization of PCBs in soil which contained no chlorobenzoate utilizer prior to inoculation. From this soil they also isolated pure cultures able to utilize biphenyl and 2,5-dichlorobenzoate, thus suggesting the possibility that genetic transfer may have occurred from the added chlorobenzoate degrader into the indigenous population of biphenyl utilizers (Hickey *et al.*, 1993).

Just as the Focht group has expanded their PCB biodegradation studies to assess the lower pathway degradation of PCB metabolites, so too have Rieneke and co-workers focused on these PCB intermediates. Although their studies on the degradation of chlorinated aromatics by *Pseudomonas* sp. strain B13 are obviously generally applicable to the biodegradation of various chlorinated aromatics, one must consider the importance of these studies to PCB metabolism as well (Kaschabek & Reineke, 1992, 1993). Likewise, the work of Havel & Reineke (1993) on the microbial degradation of chlorinated acetophenones by a defined mixed culture of *Arthrobacter* and *Micrococcus* species is important in light of the observed production of mono-, di-, and trichlorinated acetophenones from PCBs.

Thus, although most PCB-degrading bacteria have been shown to harbor only the upper pathway genes for PCB degradation and to produce single-ring metabolites, there is extensive evidence of lower pathway degradation of these metabolites. Therefore, the upper pathway biodegradation of PCBs to these and other intermediates is the critical first step in aerobic PCB destruction, and it is evident that complete mineralization is possible depending on the indigenous organisms that are present at contaminated sites, as well as the potential to add co-cultures of these strains. Finally, although chlorobenzoate metabolites have recently been demonstrated in aerobic Hudson River sediments they are generally only observed at very low concentrations and are believed to be transient intermediates (Flanagan & May, 1993).

7.5 Bioremediation alternatives

7.5.1 General applicability of bioremediation

In many cases biological approaches for remediating polluted environments provide a significant advantage over alternative technologies in both cost savings and reduction of environmental and human exposure. Biological treatment technologies can facilitate the complete destruction of hazardous chemicals without the generation of toxic emissions or by-products. Cost is reduced by lower capital expenditure and operating expenses. *In situ* treatment systems provide an added advantage of destroying the contaminant in place, thereby further reducing human exposure. Cost is also reduced because contaminated media do not have to be excavated, relocated, or otherwise disposed of. Of course, if applicable, natural attenuation is the most cost-effective 'bioremediation'.

Many laboratories and biotreatment companies have demonstrated that bioremediation can be an effective treatment technology for the destruction of a wide range of environmental contaminants including hydrocarbons (BTEX, jet fuels, alkanes, alkenes, etc.), substituted aromatics (2,4-D, chlorobenzenes, styrene, aniline, etc.), polyaromatic hydrocarbons (PAHs), chlorinated solvents (TCE, DCE, chloroform, methylene chloride, TCA, etc.), nitroaromatics (TNT, DNT, nitrobenzenes, etc.), carbon disulfide, and a wide variety of other compounds. The target compounds can serve as a sole source of carbon and energy for the degradative organisms, or, as for PCBs, they may be cometabolic substrates for which supplementary carbon sources must be present to stimulate degradative activity. Although cometabolic bioremediation systems are more challenging, they are feasible, as demonstrated for chlorinated solvents and high molecular weight PAHs. The extent to which compounds are destroyed is controlled by a number of factors, including chemical structure and composition, concentration, solubility, hydrophobicity and environmental conditions (pH, temperature, etc.). The versatility of naturally occurring degradative microorganisms, coupled with proper system design and treatment strategies, provides the possibility to expand greatly the current scope of bioremediation as a cost-effective and efficient treatment alternative for remediating many of the nation's environmental pollution problems. As one of the more difficult chemical pollutants, PCBs are just now being considered as a candidate for commercial bioremediation.

Because of the mixtures of contaminants present in the environment, the matrices containing the contamination, and the unique physical/chemical parameters of contaminated sites, a variety of biological treatment alternatives is needed to address individual problem definitions, and these require a critical integration of biological catalysts and engineered treatment systems. The various soil bioremediation technologies can be roughly categorized into aboveground systems, including a variety of slurry bioreactors, 'landfarming' techniques, and engineered soil piles, and subsurface *in situ* technologies focused on remediation of groundwater and the unsaturated zone of soils, including biosparging/bioventing technologies. Because PCBs are so insoluble they are generally not found as a significant groundwater problem.

7.5.2 A hierarchy of options

When one considers the approaches that can be utilized to biologically treat PCB-contaminated sites, a hierarchy is apparent based upon ease of cleanup, time frame, and cost, and this hierarchy can be summarized as follows.

(a) The first choice for 'biotreatment' is intrinsic bioremediation (natural attenuation) whereby the indigenous microflora and microbial conditions exist which demonstrate that natural attenuation is already occurring and is continuing. This is generally the first choice of remediation because quite obviously it requires no intervention, just monitoring of the natural progress of degradation. This important concept has evolved over the past ten years in terms of hydrocarbon biodegradation in general, and PCB biodegradation in particular. Starting with the observations of Brown *et al.* (1984) and Bopp *et al.* (1984), the intrinsic anaerobic biodegradation of PCBs has become an important remediation strategy to consider. More recently, the less common case of intrinsic aerobic PCB biodegradation has also been documented and this logically completes the strategy for total intrinsic PCB mineralization (Unterman *et al.*, 1991; Flanagan & May, 1993).

(b) The second choice in the hierarchy of biological treatment is the biostimulation of indigenous microbial populations to remediate the target chemical. This alternative is chosen when a natural degradative population exists within the contaminated zone, but available oxygen, nitrogen, or other nutrients are insufficient to support microbial activity. For example, through the introduction of an oxygen source such as hydrogen peroxide or air, or a nitrogen source such as fertilizer, the natural population can be induced to degrade the target chemicals. This is generally the most cost-effective approach for active bioremediation (Rothmel *et al.*, 1994; Smith *et al.*, 1995).

(c) The third choice in the hierarchy, bioaugmentation, is for cases where intrinsic bioremediation or biostimulation does not work because of insufficient or unacclimated bacterial cultures. In this case selected strains of bacteria with the desired catalytic capabilities and other characteristics can be injected directly into the contaminated zone, along with nutrients if necessary, to effect the biodegradation of the target chemicals. There are also cases where although the natural population may be sufficient to achieve a biostimulation remediation, it may be desirable to choose a bioaugmentation approach to increase the rate of degradation and therefore shorten the time frame for full-scale remediation (Shannon *et al.*, 1994). Ultimately, the use of superior genetically engineered strains harboring characteristics such as high rates of activity, broader substrate specificity, uncoupled cometabolism, optimized biocatalyst and process control, or various resistances, will become candidates for bioremediation systems (Erickson & Mondello, 1993; Timmis *et al.*, 1994).

7.5.3 Development of alternative PCB biotreatment systems

Although the extensive research presented above clearly demonstrates that PCB biodegradation occurs under both aerobic and anaerobic conditions, and that this degradation activity is widespread, the challenge that technologists face is transitioning this microbiology, biochemistry, and genetics into an efficient and cost-effective bioremediation process. There have been many claims of commercial PCB-bioremediation technologies; however, none of these has yet survived rigorous technical and analytical scrutiny. Hydrophobic substrates such as PCBs tend to elude analysts, and many claims by commercial PCB-bioremediation vendors have been based on PCB disappearance, which can most often be attributed to repartitioning and redistribution within the biotreatment system, or to other abiotic losses and artifacts. The characteristic mark of these systems is no congener specificity (Unterman, 1991; Shannon & Unterman, 1993).

There are several defining boundaries that one should consider as a framework in order to assess realistically what performance can be expected from a PCB bioremediation system. First, under strictly aerobic conditions one can anticipate at best between 50 and 85% PCB biodegradation of Aroclors 1242 and 1248, up to concentrations of approximately 1000 ppm (in one rare exception we have seen 75% degradation of 4000 ppm Aroclor 1242 in sediment). Therefore, one must view the potential for utilizing an aerobic PCB-bioremediation system in terms of what the final target concentration is and then calculate back to the upper limit for a starting PCB concentration. In addition, the strictly aerobic biodegradation of PCBs is limited to the lower-chlorinated congeners, although as discussed above, some exceptional strains can degrade penta- and hexachlorobiphenyls to varying extents.

Second, under strictly anaerobic biotreatment conditions, one sees little or no mass reduction in PCB (just the mass of chloride removal), but as discussed above, one can effect a dramatic reduction of higher chlorinated congeners, in particular some of the 'dioxin-like' congeners. Therefore, an anaerobic treatment by itself may effect a sufficient detoxification and decrease in bioaccumulation potential to suffice in terms of a risk-based cleanup target. Third, one can integrate both anaerobic and aerobic biotreatment into a two-stage system whereby under initial anaerobic dechlorination conditions one produces lower molecular weight congeners which in a second-stage aerobic treatment are readily biodegraded by either indigenous or inoculated organisms. This approach can therefore effect a greater extent of degradation, especially of the higher congeners. Finally, one can develop both *ex situ* reactor-based technologies as well as *in situ* PCB biotreatment. These various options are discussed in more detail below.

An aerobic microbial biotreatment system using two complementary PCB-degrading bacterial strains, ENV 307 and ENV 360, has demonstrated up to 80% destruction of PCB in site soils containing 300 ppm Aroclor 1248, using bench-scale bioslurry reactors. These strains demonstrate exceptional ability to degrade not only the lower-, but also higher-chlorinated PCB congeners (tetra-, penta-, hexa-) by using two distinct and complementary dioxygenase enzyme systems that differ in terms of their substrate specificities (Bedard et al., 1984; Unterman et al., 1988; Haddock et al., 1995). There is an additive effect of using these two PCB-degrading cultures, in contrast to using one culture exclusively (Shannon et al., 1994). Additional research for enhancing aerobic biodegradation has been directed at developing techniques to increase the survivability and activity of PCB-degrading bacteria after they have been applied to soils, and a significant increase in the viable population of PCB-degrading microbes has been achieved in treated soils with the addition of carbon (e.g., biphenyl) and co-cultures (Kohler et al., 1988a; Barriault & Sylvestre, 1993; Hickey et al., 1993; Envirogen, Inc., unpublished results).

In situ bench-scale microcosms as well as field pilot studies have been conducted to optimize *in situ* PCB soil biotreatment processes. This work has focused on dilute PCB contamination (less that 100 mg/kg) because of the current rate limitations of *in situ* treatments. Parameters being addressed include bacterial dosing regime, moisture content, nutrient addition, and additives. Results to date have shown 20% to 65% PCB biodegradation in soils contaminated with 10–100 ppm in the laboratory.

Laboratory comparisons of aerobic reactor-based versus *in situ* PCB processes demonstrated significantly higher rates of PCB destruction in soil slurry reactors; however, for many sites the advantages of not excavating continue to favor the *in situ* process configuration as a very viable, albeit slower, alternative. Reactor development is currently assessing many system parameters, including mixing mechanism and rate, percentage of solids, bacterial dosing regime, the use of growing versus stationary growth cells, and the integration of co- and pretreatment steps. The first field pilots have now demonstrated 40–60% PCB biodegradation in systems of up to 1000 gallons, as discussed below (Harkness *et al.*, 1993; *Biotreatment News*, May 1995, pp. 3–14; Envirogen, Inc., study in progress).

Anaerobic PCB dechlorination has been initiated and enhanced by providing co-substrates such as bromobiphenyls and natural polymers (Bedard & Van Dort, 1992; Stokes *et al.*, 1994; Bedard & Quensen, 1995; Envirogen, Inc., unpublished results); however, the major focus for anaerobic PCB treatment is intrinsic bioremediation. Soils that show no stimulation with carbon source addition may undergo dechlorination by adding PCB-dechlorinating bacteria from actively dechlorinating bacterial populations to soils that are depleted or lacking such strains. However, to date no mass-culturing of PCB-dechlorination strains has been demonstrated, so this approach is not yet viable on a large scale.

Two-stage anaerobic/aerobic microbial systems are also a viable biotreatment alternative. There are many cases where halogenated compounds, such as PCBs, PCE or chloroform, that are resistant to aerobic degradation have been partially dehalogenated by anaerobic bacterial enrichments or cultures. After chloride removal aerobic bacteria are then capable of degrading the partially dehalogenated compounds (Bedard *et al.*, 1987b; Abramowicz, 1990; Anid *et al.*, 1991; Fathepure & Vogel, 1991). Using this approach a sequential anaerobic/aerobic process has been demonstrated whereby more than 80% of the PCB from a 1240-ppm Aroclor 1248-contaminated sediment was biodegraded (Shannon *et al.*, 1994).

The results from these and other laboratory studies have now led to the testing of various PCB bioremediation approaches in pilot demonstrations. The field trials are designed to demonstrate the applicability of field-scale treatment of soils, with the ultimate goal of commercial-scale bioremediation of PCB-contaminated sites, or demonstrated efficacy of intrinsic bioremediation.

7.5.4 Field demonstrations

To date there have been only a limited number of field demonstrations of PCB bioremediation technologies and no documented full-scale bioremediation cleanups. Three of the field demonstrations have been conducted by the General Electric

Research Team. The first was in New York in 1987 at the site of a former race track that had been contaminated by the application of spent PCB oils which were applied to the dirt roadways for dust control. The test was performed as a solid-phase land treatment (similar to 'landfarming') whereby the contaminated surface soils were tilled with PCB-degrading bacteria and nutrients. During this study, the application of a highly competent aerobic PCB-degrading strain (*Pseudomonas* sp. LB400, see above; Bopp, 1986) was monitored over a four-month period. The test demonstrated up to 20% PCB biodegradation from a starting concentration that ranged between 50 and 525 ppm. There was no evidence of any PCB biodegradation in the soil samples from an adjacent control plot. This rate and extent of biodegradation was about one-half that seen in prior laboratory experiments and was attributed to problems with control of soil temperature and moisture content. The limited biodegradation in this soil was also shown to be due to poor bioavailability. This demonstration was the first *in situ* PCB biodegradation field test ever conducted and although the results were modest by comparison to laboratory tests, they did demonstrate the possibility of soil inoculation as a PCB bioremediation approach (McDermott *et al.*, 1989; Harkness & Bergeron, 1990).

The GE team subsequently conducted a major bioslurry field study in the Hudson River to demonstrate that indigenous aerobic microorganisms could degrade PCBs present in anaerobically dechlorinated sediments. Over the 73-day 1991 field study, aerobic PCB degradation was stimulated by inorganic nutrients, biphenyl, and oxygen, and even under mixed, but unamended, conditions. The test demonstrated 37–55% loss of PCBs from a starting concentration averaging 39 ppm, with concomitant production of chlorobenzoates. The use of a PCB-degrading bacterial inoculum, *Alcaligenes eutrophus* strain H850 (Bedard *et al.*, 1984; also see discussion above), did not improve the biodegradative activity as compared with that observed with the indigenous strains, probably due to poor survivability in the river sediment. This work demonstrated that natural indigenous populations exist which when given proper conditions can biodegrade at least the lower-chlorinated congeners, as seen in this highly dechlorinated environment. The fact that only partial biodegradation was observed even though the majority of PCBs were mono- and dichlorobiphenyls was attributed to the fact that this sediment contains a PCB fraction that is not bioavailable to microorganisms (over the short 73-day time frame) due to sorption of PCBs to natural organic matter (Harkness *et al.*, 1993, 1994).

Complementary work by the GE team on Hudson River sediments has utilized a microcosm system to understand better both the aerobic and anaerobic transformations that occur in stationary sediments. These studies provided further evidence that the naturally occurring microorganisms shown in the Hudson River field trial have the potential to attack Aroclor 1242 at rapid rates both aerobically and anaerobically. These results are helping to further our understanding of the fate of released PCBs in river sediments (Fish & Principe, 1994).

Following the important results obtained in laboratory studies using bromobiphenyls to stimulate anaerobic dechlorination (Bedard & Van Dort, 1992; Bedard *et al.*, 1993), a field demonstration of stimulated anaerobic PCB dechlorination was conducted in Woods Pond within the Housatonic River (Massachusetts). The PCB concentrations in the experimental and control caissons were 26 and 31 ppm, respectively, and over a 373-day time course no change in PCBs was observed in the control caisson. In contrast, the PCBs in the experimental caisson, which received one dose of 2,6-dibromobiphenyl, were extensively dechlorinated and showed decreases of 70, 57 and 27% in the hexa-, hepta-, and octachlorobiphenyls, respectively, after only 93 days. At the same time the tri- and tetrachlorobiphenyls increased from 18 to 63 mol per cent. By 373 days the hexa- through nonachlorobiphenyls had decreased by approximately 70%. Most of the dechlorination was *meta*-specific with little *para*-dechlorination and no *ortho*-dechlorination. As discussed above, the dechlorination resulted in congeners which show less potential to bioaccumulate in humans and have an expected decrease in 'dioxin-like' toxicity associated with PCBs. The lower-chlorinated congeners which were produced are also more readily biodegradable by aerobic bacteria. Thus, this field trial clearly demonstrated that it is possible to stimulate substantial microbial dechlorination of highly chlorinated PCBs *in situ* with the single application of an inducing chemical even at ambient temperatures (Bedard *et al.*, 1995).

An in-lagoon slurry bioremediation at the French Limited Lagoon (Texas) was conducted at both pilot and eventually full scale and successfully bioremediated many of the aromatic chemicals found at that site, including BTEX and PAHs. This site also contained approximately 100 ppm PCBs; however, it is not clear from the results released to date whether the decrease observed in PCB concentration during the field work can be attributed to biodegradation or some abiotic redistribution or loss (Unterman, 1991; *Biotreatment News*, October, November, December, 1992; Shannon & Unterman, 1993).

Beginning in 1985 a five-year field-scale PCB research study was conducted by the Madison Metropolitan Sewerage District (Wisconsin) in order to assess whether municipal sludge containing approximately 50 ppm PCBs could be biologically treated using land application biotreatment. A series of test plots was established and PCB-contaminated sludge applied by subsurface injection followed by tilling and addition of commercial fertilizer. Two different PCB concentrations of approximately 25 and 75 ppm were tested. The results showed 85% degradation of biodegradable lower-chlorinated congeners and only 6% loss of a marker 'biorecalcitrant' congener (2,5,2',5'-CB). This specificity indicates that a significant pathway for PCB loss was biodegradation and in total 43 PCB congeners had rate constants that were significantly greater than zero. The soil half-life for the dichlorinated PCB congeners was 7–11 months, 5–17 months for the trichlorinated congeners, 11–58 months for the tetrachlorinated PCB congeners and 19 months for total PCBs. The results of this field study support land treatment as an effective management approach for municipal sludge containing low levels of PCB (Gan & Berthouex, 1994).

PCB contamination in the Sheboygan River (Wisconsin) has also been a target of PCB cleanup and demonstration of bioremediation technology through a collaboration between Blasland, Bouck and Lee, Inc. and the U.S. Environmental Protection Agency. The overall objective of the field study was to assess the feasibility of utilizing biodegradation as a remedial treatment of PCBs contained in aquatic sediments. The test evaluated both in-vessel and *in situ* biodegradation. The in-vessel study utilized a confined treatment facility (CTF) that was constructed adjacent to the river. Different cells were established to maintain aerobic and anaerobic conditions. The report for the in-vessel study is currently in preparation and results are as yet unavailable. However, preliminary results characterizing the extent and progression of PCB dechlorination in armored sediments within the river (the '*in situ* study') indicate up to 59% PCB dechlorination relative to Aroclor 1248 and the sum of mono-, di-, and trichlorobiphenyls becoming 87% (Dawn Foster, pers. commun.; 'Sheboygan River and Harbor Biodegradation Pilot Study Workplan', prepared by Blasland, Bouck and Lee, Inc. and U.S. Environmental Protection Agency, September 1992).

For the past three years the Aluminum Company of America (ALCOA) has been involved in an extensive bioremediation evaluation program. In the initial laboratory studies comparing biostimulation versus bioaugmentation of PCB-contaminated lagoon sediments, it was demonstrated that the major site lagoon contained a sufficient indigenous population which could degrade a substantial proportion of the PCBs (30–60%) (and PAHs; greater than 90% reduction) within the sediments (Rothmel *et al.*, 1994). On-site field demonstrations of two bioremediation processes, biological slurry reactor treatment (BSR) and engineered land treatment (LTU), were conducted from August through November, 1994, as a collaboration between ALCOA, Carnegie Mellon University, Camp, Dresser & McKee, and Envirogen, Inc.

With total PCB sludge/sediment concentrations ranging from 100 to 1000 ppm, the BSR treatment achieved between 20 and 50% PCB reduction. The LTUs, which contained contaminated sediment mixed with uncontaminated sand to provide a workable land treatment base, achieved similar contaminant reductions as seen in the BSRs. No degradation was observed in killed control reactors or untreated control samples. Essentially all of the biodegradation was directed at the lower-chlorinated PCBs. Considering the fact that these are the PCB congeners which are more water soluble, more desorbable, and therefore pose the greatest potential for migration through leaching, this aerobic bioremediation demonstrates the possibility of achieving a partially destroyed PCB mixture which can be considered 'biostabilized' in that the biotreatment results in less leachable organics which pose a lower risk for exposure. This bioremediation treatment strategy could prove to be very cost effective when followed by proper containment management of the treated residuals to prevent other routes of exposure (e.g., volatilization, dermal contact, ingestion) which can be more easily limited through capping of sites (*Biotreatment News*, May 1995, pp. 3–14).

Three field demonstrations of PCB biodegradation approaches are currently being tested at a natural gas pipeline site in Pennsylvania (Envirogen, Inc.): *in situ* aerobic biotreatment (solid phase land treatment; 'landfarming'); aerobic soil slurry biotreatment; and *in situ* anaerobic biotreatment. These studies are being conducted over a two-year time frame and will be completed by early 1996. Preliminary results are presented below.

The demonstration of *in situ* aerobic land treatment (tilling of surface soils) is being conducted over a 24-month period using soils having PCB concentrations between 30 and 100 ppm. The biodegradation was initiated by adding two complementary PCB-degrading strains, ENV360 and ENV307, which together degrade a broader range of PCB congeners (Unterman *et al.*, 1988; Shannon *et al.*, 1994). To ensure optimal degradation rates, the soil has been supplemented with additional growth/inducing substrates. A second nondosed control plot is being run in parallel and has shown no PCB loss. After 24 months of operation PCB levels in the experimental test plot have dropped approximately 44%, primarily the lower-chlorinated congeners (Envirogen, Inc., study in progress).

Aerobic PCB biodegradation in a bioslurry reactor is currently being demonstrated (1995) with soils obtained from site areas which have PCB concentrations between 200 and 500 ppm. Two complementary PCB-degrading strains (as discussed above) were added to the soil/slurry mixture with the necessary nutrients to accomplish biodegradation. A series of tests were performed to compare the efficacy of various unit operations and the use of inducing cosubstrates. Preliminary results indicate up to 55% PCB biodegradation (Envirogen, Inc., study in progress).

Demonstration of *in situ* anaerobic biotreatment is being conducted in a heavily forested area with low levels of PCB (<50 ppm). The test is being run over a 14-month period in six test plot areas. Biodegradation is being stimulated by the addition of natural polymer growth substrates for the development of indigenous PCB-dechlorinating soil microorganisms. The test plots are flooded with water to ensure the maintenance of anaerobic conditions throughout the study. This study is of particular interest because the site soils show no evidence of prior dechlorination; however, laboratory experiments had demonstrated that substrate additives can induce indigenous dechlorination. Results are not yet available (Envirogen, Inc., study in progress).

7.5.5 The issues of bioavailability

For biological degradation to be efficient, competent organisms and substrate(s) must come in contact with each other; therefore, the transport of highly insoluble substrates, such as PCBs, can be a rate-limiting factor in a biological process. The hydrophobic property of PCBs results in tight binding (sorption) to soil and/or organic matrices, which greatly reduces the contact between potential PCB-degrading microorganisms and their target. This issue of bioavailability, which is directly related to the solubility, diffusion, and desorption of soil-bound PCBs, has been an important research area, and is critical for understanding both the rates of natural

PCB degradation and for developing a PCB biotreatment process (Morris & Pritchard, 1994).

There are three major physical/chemical characteristics that govern the rate and extent of biodegradation of a compound. These include: (1) affinity of the compound to a sorbent matrix; (2) the solubility of the chemical; and (3) the distribution of the compound between air and water. The most critical factor in developing a biodegradation process for hydrophobic organic compounds (HOCs), such as PCBs is the rate and extent of sorption/desorption of the HOC to an organic matrix such as soil (Scow et al., 1986; Thomas et al., 1986; Stucki & Alexander, 1987). It is this sorption of hydrophobic organic compounds to soil organic matter which limits the diffusion of HOCs. In fact, the concentration of organic matter, which includes both natural organic matter (e.g. humic acids) and residual synthetic co-contaminants, greatly influences the degree of sorption and thus desorption from the soil (Sayler & Colwell, 1976; Carter & Suffet, 1982; Lee et al., 1989; Boyd & Sun, 1990; Murphy et al., 1990; Harkness & Bergeron, 1990; Sun & Boyd, 1991). The higher the concentration of nonaqueous organic matter the more extensive is the sorption of HOCs and the less bioavailable are the HOCs to the microbial population (Wszolek & Alexander, 1979; Subba-Rao & Alexander, 1982; Umbreit et al., 1986; Bedard & Bergeron, 1988; Bedard, 1990). Binding of HOCs on soil can be correlated well with the octanol–water partition coefficient, K_{ow}, of the HOC (Murphy et al., 1990). For PCBs, this coefficient is dependent on the molecular weight of the PCB congener; the more highly chlorinated congeners have a significantly higher K_{ow} (Hawker & Connell, 1988) and thus have a greater affinity to sorb onto a solid phase organic material, such as soil.

The diffusion or desorption of bound PCBs, and HOCs in general, can be modeled as a two-component process. There is a readily available component ('component 1') that is capable of immediately diffusing/desorbing from the soil, which can be rapidly degraded, assuming there is a competent microbial consortia in the system. The second component ('component 2') is not immediately desorbable into the bulk aqueous phase (Di Toro & Horzempa, 1982; Karickhoff & Morris, 1985; Electric Power Research Institute, 1990; Carroll et al., 1994). This component is considered to include substrate that is either sequestered in soil micropores or so strongly adsorbed to the organic matter that it is not readily available to the microbial population (Scow et al., 1986; Chiou et al., 1990). Furthermore, this resistant fraction can result from diffusional limitation from condensed phases of sediment organic matter (Carroll et al., 1994) or PCB binding in the intracrystalline water layers of clay (Uzgiris et al., 1995). This two-component process has been described in a nonequilibrium sorption model in which the diffusion of the hydrophobic organic chemical(s) within the intra-organic matter ('component 2') is responsible for the rate-limited or nonequilibrium sorption/desorption (Brusseau et al., 1991; Angley et al., 1992). Solute (HOC) located within the internal structure of soil is likely to be protected from biodegradation due to both the exclusion of bacteria from the internal soil matrix

(Angley *et al.*, 1992) and the retarded rate of desorption and diffusion from the soil micropore surfaces to the macropore water. In the case of PCBs, the retarded rate of desorption can affect both lower- and higher-chlorinated PCB congeners; however, the effect is likely to be more severe with higher chlorinated congeners.

In addition to soil binding affinity, the concentration of HOC desorbed into the aqueous phase at any one time depends on the solubility of the HOC. This is the second physical/chemical characteristic (as mentioned above) that governs the rate and extent of biodegradation (Alexander *et al.*, 1986). For PCBs, the solubility of individual congeners is generally very low, and decreases with an increase in the molecular weight of the congener (Opperhulzen *et al.*, 1988; Dunnivant *et al.*, 1992). Thus, this characteristic also contributes significantly to poor PCB biodegradation in soils. The third characteristic governing the extent of biodegradation, the air/water distribution, is determined by Henry's Law constants. For the case of PCBs, however, volatilization is quite low and therefore does not significantly compete with bioremediation.

A general model of microbial degradation proposed by Mihelcic & Luthy (1991) can be used to describe the microbial degradation of PCBs. The model can be described as follows. The rate of PCB degradation is initially proportional to the concentration of PCB in the bulk aqueous phase and the active cell population. The maximum aqueous PCB concentration is obviously limited by the solubility of the individual PCB congener. As the soluble PCBs are degraded, a concentration gradient develops between the soil and water, and PCBs that are loosely associated at the surface of the soil readily desorb to reestablish the soil/water equilibrium. With time, the readily desorbable fraction will move into the bulk aqueous phase, where it will be transported onto a second organic phase, namely microorganisms, where biodegradation occurs. Once the more labile PCB fraction is transported from the soil into the bacterium, a nonequilibrium state develops where desorption of PCB from the soil micropores is dependent on the very slow rate of intra-aggregate or intra-organic matter diffusion. As a result of this nonequilibrium desorption, there is an apparent threshold that is reached in terms of PCB degradation, which is partially related to the immediately bioavailable PCBs or readily desorbable congeners. This threshold effect has been observed in numerous PCB treatability studies, where 50% to 65% of total soil PCBs can be readily degraded within 24 to 72 hours. However, after this time only a very slow rate of degradation (if any) is detectable, even though a significant concentration of degradable PCB congeners may remain in the soil. It is believed that this slow rate of degradation is controlled by diffusion of the 'component 2' through the intra-organic matter (Harkness *et al.*, 1993).

PCB biodegradation can be enhanced by increasing the extent and/or rate of the PCB intra-organic matter diffusion, and can therefore lead to an apparent increase in the aqueous PCB concentration available for biodegradation. Treatment methods that have been used to disrupt the sorption of PCBs from the soil matrix include the addition of surfactants, solvents, or caustic. Surfactants have been used to increase the apparent solubility of sorbed organic compounds from soil matrices by entrapping organic molecules into micelles (Electric Power Research Institute, 1985; Liu *et al.*, 1990; Abdul & Gibson, 1991; Bury & Miller, 1993). However, the effect of using surfactants to enhance biodegradation is very variable. In some cases biodegradation can be increased (Green, 1990; Aronstein *et al.*, 1991), but in other cases biodegradation is inhibited (Laha & Luthy, 1991; Envirogen, Inc., unpublished results). Solvents can be used to extract PCBs from soil with an 80 to 99% efficiency; however, they are generally incompatible with microorganisms at the concentrations needed to achieve this extraction efficiency. Co-solvents have also been shown to increase PCB mobility in soils, and presumably bioavailability (Griffin & Chou, 1981). Disruption of the humic material in soil can also increase the partitioning of PCBs from the soil to the aqueous phase. This can be achieved by caustic treatment (Harkness & Bergeron, 1990; Envirogen, Inc., unpublished results) or by microbial degradation of humic material (Dec & Bollag, 1988).

Some of the most problematic PCB-contaminated matrices include lagoon and soil materials that are high in non-PCB oils (e.g., cutting oils and hydraulic fluids). The concentration of these co-contaminating oils can vary significantly, and consequently vary the bioavailability of the PCB. A matrix that contains nonbiodegradable oils at high concentrations is not a good candidate for bioremediation of PCB (Harkness & Bergeron, 1990; Shannon *et al.*, 1992). This was clearly demonstrated in two GE studies where various concentrations of di(2-ethylhexyl)phthalate and mineral oil were shown to dramatically inhibit the biodegradation of 2,4'-CB and Aroclor 1242, respectively (Bedard, 1990; Bedard & Bergeron, 1988).

Microbially produced biosurfactants can also have important effects on increasing bioavailability and biodegradation (Zhang & Miller, 1992). A new PCB-degrading strain was recently isolated and found to produce a bio-emulsifier (Rothmel *et al.*, 1993). Experiments show that the bio-emulsifier stimulates the extent of PCB degradation well beyond the levels seen before with other Type strains of its class, and that the bio-emulsifier can stimulate the extent of PCB biodegradation of added co-cultures as well.

From another viewpoint one can consider the limitations of PCB bioavailability as a potential positive factor in determining PCB-remediation strategies. For example, one can develop a conceptual approach for beneficially evaluating the limited range and extent of PCB bioavailability and degradation in terms of a risk-based cleanup standard. Experiments have shown that at 50% total PCB biodegradation, extensive degradation of the lower-chlorinated PCB homologs results, and this in turn results in a substantial decrease in the leachable fraction of PCBs and PAHs (Smith *et al.*, 1995). Additionally, studies have been conducted to assess the biodegradation of water-soluble PCBs in the low parts per billion range ($\mu g/kg$), and these studies have demonstrated that virtually all of the water-soluble PCBs can be biodegraded (Envirogen, Inc., study in progress). When these observations are combined, one can propose an approach for biodegrading the lower-chlorinated PCB homologs to decrease subsequent migration of the residual PCB fraction and therefore exposure and potential toxicity. An example of this effect with PAHs has been described by Weissenfels *et al.* (1992). In addition, a model can be proposed from biodegradation studies with PAHs (Smith *et al.*, 1995) for a significant rate of degradation for any subsequently solubilized PCBs. This cleanup approach is one embodiment of 'biostabilization' and is being considered as a possible remediation alternative. This mechanism of biostabilization should be distinguished from previously described biostabilization processes (Stott *et al.*, 1983; Shannon & Bartha, 1988), which describe immobilization of the oxidized by-products of biodegradation.

7.5.6 Summary

Microorganisms are responsible for the breakdown of complex organic materials into simpler forms in the environment. However, synthetic compounds, particularly those that are chlorinated, frequently resist microbial degradation (Alexander, 1969). This has been true of PCBs; congeners containing more than three chlorine atoms have been markedly resistant to biodegradation. On the other hand, various microbial cultures capable of degrading many PCB congeners have been characterized over the past 22 years. From laboratory and field studies 50% to 85% biodegradation of Aroclors 1242 and 1248 has been demonstrated. Depending on the target cleanup level, this translates into the ability to bioremediate PCB-contaminated soils aerobically at starting concentrations of up to 500 to 1000 ppm. In terms of the commercial Aroclor mixtures, it is clear that only Aroclors 1221, 1242, and 1248 are currently amenable to direct aerobic biodegradation. Aroclors 1254 and 1260 are too highly chlorinated to be reasonably degraded by existing aerobic bacterial strains. Because the products of anaerobic dechlorination are readily biodegradable by aerobic strains, a two-stage anaerobic/aerobic treatment can destroy a substantial level of PCBs. Thus, by using both anaerobic and aerobic systems in conjunction, one can now anticipate the ability to degrade even higher concentrations of the lower-chlorinated Aroclors (1242, 1248) as well as the higher Aroclors (1254, 1260).

Current PCB bioremediation development programs continue to focus on: (1) optimization of conditions for growth and activity of PCB-degrading bacteria, both aerobic and anaerobic; (2) isolation and characterization of superior and novel PCB degraders; (3) genetic engineering of PCB pathways; (4) development of various physical and chemical pretreatment and cotreatments for improving the effectiveness of biotreatment steps; and (5) design and testing of field pilot systems.

Finally, another major area of focus continues to be the characterization of intrinsic bioremediation as the most cost-effective cleanup alternative. As part of this strategy further work will need to be conducted on developing risk-based cleanup standards using 'environmentally acceptable endpoints', including the potential applicability of biostabilization. These approaches are being developed to help our society implement more reasonable and cost-effective remediation goals and priorities for remediation of PCB-contaminated sites. Ultimately, we must hope that the technical base of information developed over the past 22 years will drive our decision making.

References

Abdul, A. S. & Gibson, T. L. (1991). Laboratory studies of surfactant-enhanced washing of polychlorinated biphenyl from sandy material. *Environmental Science & Technology*, 25(4), 665–71.

Abelson, P. H. (1991). Excessive fear of PCBs. *Science*, 253, 361.

Abramowicz, D. A. (1990). Aerobic and anaerobic biodegradation of PCBs: a review. In *CRC Critical Reviews in Biotechnology*, ed. G. G. Stewart & I. Russel, vol. 10, pp. 241–51. Boca Raton, FL: CRC Press.

Abramowicz, D. A., Brennan, M. J., Van Dort, H. M. & Gallagher, E. L. (1993). Factors influencing the rate of polychlorinated biphenyl dechlorination in Hudson River sediments. *Environmental Science & Technology*, 27, 1125–31.

Adams, R. H., Huang, C-M., Higson, F. K., Brenner, V. & Focht, D. D. (1992). Construction of a 3-chlorobiphenyl-utilizing recombinant from an intergeneric mating. *Applied and Environmental Microbiology*, 58, 647–54.

Adriaens, P. (1994). Evidence for chlorine migration during oxidation of 2-chlorobiphenyl by a type II methanotroph. *Applied and Environmental Microbiology*, 60, 1658–62.

Adriaens, P. & Focht, D. D. (1990). Continuous coculture degradation of selected polychlorinated biphenyl congeners by *Acinetobacter* spp. in an aerobic reactor system. *Environmental Science & Technology*, 24, 1042–9.

Adriaens, P. & Focht, D. D. (1991). Cometabolism of 3,4-dichlorobenzoate by *Acinetobacter* sp. strain 4-CB1. *Applied and Environmental Microbiology*, 57, 173–9.

Adriaens, P., Kohler, H-P. E., Kohler-Staub, D. & Focht, D. D. (1989). Bacterial dehalogenation of chlorobenzoates and coculture biodegradation of 4,4'-dichlorobiphenyl. *Applied and Environmental Microbiology*, 55, 887–92.

Ahmad, D., Masse, R. & Sylvestre, M. (1990). Cloning and expression of genes involved in 4-chlorobiphenyl transformation by *Pseudomonas testosteroni*: homology to polychlorobiphenyl-degrading genes in other bacteria. *Gene*, 86, 53–61.

Ahmad, D., Sylvestre, M. & Sondossi, M. (1991). Subcloning of *bph* genes from *Pseudomonas testosteroni* B-356 in *Pseudomonas putida* and *Escherichia coli*: evidence for dehalogenation during initial attack on chlorobiphenyls. *Applied and Environmental Microbiology*, 57, 2880–7.

Ahmed, M. & Focht, D. D. (1973). Degradation of polychlorinated biphenyls by two species of *Achromobacter*. *Canadian Journal of Microbiology*, 19, 47–52.

Alder, A. C., Haggblom, M. M., Oppenheimer, S. R. & Young, L. Y. (1993). Reductive dechlorination of polychlorinated biphenyls in anaerobic sediments. *Environmental Science & Technology*, 27, 530–8.

Alexander, M. (1969). Soil biology. *Reviews of Research, National Resources Research, UNESCO*, 9, 209–40.

Alexander, M., Michele Thomas, J., Yordy, J. R. & Amador, J. A. (1986). Rates of dissolution and biodegradation of water-insoluble organic compounds. *Applied and Environmental Microbiology*, 52, 290–6.

Angley, J. T., Brusseau, M. L., Miller, W. L. & Delfino, J. J. (1992). Nonequilibrium sorption and aerobic biodegradation of dissolved alkylbenzenes during transport in aquifer material: column experiments and evaluation of a coupled-process model. *Environmental Science & Technology*, 26(7), 1404–10.

Anid, P. J., Nies, L. & Vogel, T. M. (1991). Sequential anaerobic-aerobic biodegradation of PCBs in the river model. In *On-site Bioreclamation*, ed. R. E. Hinchee & R. F. Olfenbuttel, pp. 428–36. Stoneham, MA: Butterworth-Heinemann.

Arensdorf, J. J. & Focht, D. D. (1994). Formation of chlorocatechol *meta* cleavage products by a Pseudomonad during metabolism of monochlorobiphenyls. *Applied and Environmental Microbiology*, 60, 2884–9.

Arensdorf, J. J. & Focht, D. D. (1995). A *meta* cleavage pathway for 4-chlorobenzoate, an intermediate in the metabolism of 4-chlorobiphenyl by *Pseudomonas cepacia* P166. *Applied and Environmental Microbiology*, 61, 443–7.

Aronstein, B. N., Calvillo, Y. M. & Alexander, M. (1991). Effect of surfactants at low concentrations on the desorption and biodegradation of sorbed aromatic compounds in soil. *Environmental Science & Technology*, 25(10), 1728–31.

Assaf-Anid, N., Nies, L. & Vogel, T. M. (1992). Reductive dechlorination of a polychlorinated biphenyl congener and hexachlorobenzene by vitamin B_{12}. *Applied and Environmental Microbiology*, 58, 1057–60.

Asturias, J. A. & Timmis, K. N. (1993). Three different 2,3-dihydroxybiphenyl-1,2-dioxygenase genes in the gram-positive polychlorobiphenyl-degrading bacterium *Rhodococcus globerulus* P6. *Journal of Bacteriology*, 175, 4631–40.

Asturias, J. A., Eltis, J. D., Prucha, M. & Timmis, K. N. (1994). Analysis of three 2,3-dihydroxybiphenyl 1,2-dioxygenase genes found in *Rhodococcus globerulus* P6. *Journal of Biological Chemistry*, 269, 7807–15.

Bailey, R. E., Gonsior, S. & Rhinehart, W. L. (1983). Biodegradation of monochlorobiphenyls and biphenyl in river water. *Environmental Technology*, 17, 617–20.

Ballschmiter, K. & Zell, M. (1980). Analysis of polychlorinated biphenyls by glass capillary gas chromatography; composition of technical Aroclor and Clophen–PCB mixtures. *Fresenius Zeitschrift für Analytische Chemie*, 302, 20–31.

Barriault, D. & Sylvestre, M. (1993). Factors affecting PCB degradation by an implanted bacterial strain in soil microcosms. *Canadian Journal of Microbiology*, 39, 594–602.

Barton, M. R. & Crawford, R. L. (1988). Novel biotransformations of 4-chlorobiphenyl by a *Pseudomonas* sp. *Applied and Environmental Microbiology*, 54, 594–5.

Baxter, R. A., Gilbert, R. E., Lidgett, R. A., Mainprize, J. H. & Vodden, H. A. (1975). The degradation of polychlorinated biphenyls by microorganisms. *Science of the Total Environment*, 4, 53–61.

Bedard, D. L. (1990). Biochemical transformation of polychlorinated biphenyls. In *Biotechnology and Biodegradation*, ed. D. Kamely, A. Chakrabarty & G. S. Omenn, pp. 369–88. Woodlands, TX: Portfolio Publishing Co.

Bedard, D. L. & Bergeron, J. A. (1988). Studies of a PCB-contaminated industrial sludge. In *Seventh Progress Report for the Research and Development Program for the Destruction of PCBs*, pp. 17–21. Schenectady, NY: General Electric Co. Corporate Research and Development.

Bedard, D. L. & Haberl, M. L. (1990). Influence of chlorine substitution pattern on the degradation of polychlorinated biphenyls by eight bacterial strains. *Microbial Ecology*, 20, 87–102.

Bedard, D. L. & May, R. J. (1996). Characterization of the polychlorinated biphenyls (PCBs) in the sediments of Woods Pond: evidence for microbial dechlorination of Aroclor 1260 *in situ*. *Environmental Science & Technology*, 30, 237–45.

Bedard, D. L. & Quensen, J. F. (1995). Microbial reductive declorination of polychlorinated biphenyls. In *Microbial Transformation and Degradation of Toxic Organic Chemicals*, ed. L. Y. Young & C. Cerniglia, pp. 127–216. New York: Wiley-Liss Division, John Wiley & Sons.

Bedard, D. L. & Van Dort, H. M. (1992). Brominated biphenyls can stimulate reductive dechlorination of endogenous Aroclor 1260 in methanogenic sediment slurries. In *Abstracts of the General Meeting of the American Society for Microbiology*. New Orleans, Louisiana #Q-26, p. 339.

Bedard, D. L., Brennan, M. J. & Unterman, R. (1984). Bacterial degradation of PCBs: Evidence of distinct pathways in *Corynebacterium* sp. MB1 and *Alcaligenes eutrophus* H850. In *Proceedings of the 1983 Polychlorinated Biphenyl Seminar*, ed. G. Addis & R. Komai, pp. 4-101–8. Palo Alto, CA: Electric Power Research Institute.

Bedard, D. L., Unterman, R. D., Bopp, L. H., Brennan, M. J., Haberl, M. L. & Johnson, C. (1986). Rapid assay for screening and characterizing microorganisms for the ability to degrade polychlorinated biphenyls. *Applied and Environmental Microbiology*, 51, 761–8.

Bedard, D. L., Haberl, M. L., May, R. J. & Brennan, M. J. (1987a). Evidence for novel mechanisms of PCB metabolism in *Alcaligenes eutrophus* H850. *Applied and Environmental Microbiology*, 53, 1103–12.

Bedard, D. L., Wagner, R. E., Brennan, M. J., Haberl, M. L. & Brown, J. F., Jr (1987b). Extensive degradation of Aroclors and environmentally transformed polychlorinated biphenyls by *Alcaligenes eutrophus* H850. *Applied and Environmental Microbiology*, 53, 1094–102.

Bedard, D. L., Van Dort, H. M., Bunnell, S. C., Principe, J. M., DeWeerd, K. A., May, R. J. & Smullen, L. A. (1993). Stimulation of reductive dechlorination of Aroclor 1260 contaminant in anaerobic slurries of Woods Pond sediment. In *Anaerobic Dehalogenation and its Environmental Implications*, pp. 19–21. Abstract of 1992 American Society of Microbiology Conference, Athens, GA: Office of Research and Development, U.S. EPA.

Bedard, D. L., Smullen, L-A., DeWeerd, K. A., Dietrich, D. K., Frame, G. M., May, R. J., Principe, J. M., Rouse, T. O., Fessler, W. A. & Nicholson, J. S. (1995). Chemical activation of microbially-mediated PCB dechlorination: a field study. *Organohalogen Compounds*, 24, 23–8.

Bopp, L. H. (1986). Degradation of highly chlorinated PCBs by *Pseudomonas* strain LB400. *Journal of Industrial Microbiology*, 1, 23–9.

Bopp, R. F., Simpson, H. J., Deck, B. L. & Kostyk, N. (1984). The persistance of PCB components in sediments of the Lower Hudson. *Northeastern Environmental Science*, 3, 180–4.

Boyd, S. A. & Sun, S. (1990). Residual petroleum and polychlorobiphenyl oils as sorptive phases for organic contaminants in soils. *Environmental Science & Technology*, 24, 142–4.

Boyle, A. W., Blake, C. K., Price, W. A. & May, H. D. (1993). Effects of polychlorinated biphenyl congener concentration and sediment supplementation on rates of methanogenesis and 2,3,6-trichlorobiphenyl dechlorination in an anaerobic enrichment. *Applied and Environmental Microbiology*, 59, 3027–31.

Brenner, V., Arensdorf, J. J. & Focht, D. D. (1994). Genetic construction of PCB degraders. *Biodegradation*, 5, 359–77.

Brown, J. F., Jr, Wagner, R. E., Bedard, D. L., Brennan, M. J., Carnahan, J. C., May, R. J. & Tofflemire, T. J. (1984). PCB transformations in upper Hudson sediments. *Northeastern Environmental Science*, 3, 169–79.

Brown, J. F., Jr, Bedard, D. L., Bopp, L. H., Carnahan, J. C., Lawton, R. W., Unterman, R. D. & Wagner, R. E. (1986). Human and environmental degradation of PCBs. In *Proc. 1985 PCB Seminar, Electric Power Research Institute, Palo Alto, CA*, ed. R. Y. Komai, V. Niemeyer & G. Addis, pp. 10-17 to 10-10. Palo Alto, CA: Electric Power Research Institute.

Brown, J. F., Jr, Bedard, D. L., Brennan, M. J., Carnahan, J. C., Feng, H. & Wagner, R. E. (1987a). Polychlorinated biphenyl dechlorination in aquatic sedments. *Science*, 236, 709–12.

Brown, J. F., Jr, Wagner, R. E., Feng, H., Bedard, D. L., Brennan, M. J., Carnahan, J. C. & May, R. J. (1987b). Environmental dechlorination of PCBs. *Environmental Toxicology and Chemistry*, 6, 579–93.

Brown, J. F., Jr, Wagner, R. E. & Bedard, D. L. (1988). PCB dechlorination in Hudson River sediment: response. *Science*, 240, 1675–6.

Brown, M. P., Bush, B., Rhee, G. Y. & Shane, L. (1988). PCB dechlorination in Hudson River sediment. *Science*, 240, 1674–5.

Brunner, W., Sutherland, F. H. & Focht, D. D. (1985). Enhanced biodegradation of polychlorinated biphenyls in soils by analog enrichment and bacterial inoculation. *Journal of Environmental Quality*, 14, 324–8.

Brusseau, M. L., Jessup, R. E. & Rao, P. S. C. (1991). Nonequilibrium sorption of organic chemicals: elucidation of rate-limiting processes. *Environmental Science & Technology*, 25, 134–42.

Bumpus, J. A., Tien, M., Wright, D. & Aust, S. D. (1985). Oxidation of persistent environmental pollutants by white rot fungus. *Science*, 228, 1434–6.

Bury, S. J. & Miller, C. A. (1993). Effect of micellar solubilization on biodegradation rates of hydrocarbons. *Environmental Science & Technology*, 27(1), 104–10.

Carroll, K. M., Harkness, M. R., Bracco, A. A. & Balcarcel, R. R. (1994). Application of a permeant/polymer diffusional model to the desorption of polychlorinated biphenyls from Hudson River sediments. *Environmental Science & Technology*, 28, 253–8.

Carter, C. W. & Suffet, I. H. (1982). Binding of DDT to dissolved humic materials. *Environmental Science & Technology*, 16(11), 735–40.

Catelani, D., Colombi, A., Sorlini, C. & Treccani, V. (1973). Metabolism of biphenyl. 2-Hydroxy-6-oxo-6-phenylhexa-2,4-dienoate: the meta-cleavage product from 2,3-dihydroxybiphenyl by *Pseudomonas putida*. *Biochemical Journal*, 134, 1063–6.

Chakrabarty, A. M. (1982). Genetic mechanisms in the dissimulation of chlorinated compounds. In *Biodegradation and Detoxification of Environmental Pollutants*, ed. A. M. Chakrabarty, pp. 127–39. Boca Raton, FL: CRC Press.

Chatterjee, D. K., Kellogg, S. T., Watkins, D. R. & Chakrabarty, A. M. (1981). Plasmids in the biodegradation of chlorinated aromatic compounds. In *Molecular Biology, Pathogenicity, and Ecology of Bacterial Plasmids*, ed. S. B. Levy, R. C. Clowes & E. L. Koenig, pp. 519–28. New York: Plenum Press.

Chen, M., Hong, C. S., Bush, B. & Rhee, G-Y. (1988). Anaerobic biodegradation of polychlorinated biphenyls by bacteria from Hudson River sediments. *Ecotoxicology and Environmental Safety*, 16, 95–105.

Chiou, C. T., Lee, J-F. & Boyd, S. A. (1990). The surface area of soil organic matter. *Environmental Science & Technology*, 24(8), 1164–6.

Christopher, F., Franklin, H., Bagdasarian, M. & Timmis, K. N. (1981). Manipulation of degradative genes of soil bacteria. In *Plasmids of Medical, Environmental, and Commercial Importance, Microbial Degradation of Xenobiotics and Recalcitrant Compounds*, ed. T. Leisinger, R. Eutter, A. M. Cook & J. Nuesch, pp. 109–30. New York: Academic Press.

Clark, R. R., Chian, E. S. K. & Griffin, R. A. (1979). Degradation of polychlorinated biphenyls by mixed microbial cultures. *Applied and Environmental Microbiology*, 37, 680–5.

Dec, J. & Bollag, J. M. (1988). Microbial release and degradation of catechol and chlorophenols bound to synthetic humic acid. *Soil Science Society of American Journal*, 52(5), 1366–71.

Di Toro, D. M. & Horzempa, L. M. (1982). Reversible and resistant components of PCB adsorption–desorption: Isotherms. *Environmental Science & Technology*, 16, 594–602.

Dingyi, Y., Quensen, J. F., III, Tiedje, J. S. & Boyd, S. A. (1992). Anaerobic dechlorination of polychlorobiphenyls (Aroclor 1242) by pasteurized and ethanol-treated microorganisms from sediments. *Applied and Environmental Microbiology*, 58, 1110–14.

Dunnivant, F. M., Elzerman, A. W., Jure, P. C. & Hasan, M. N. (1992). Quantitative structure–property relationships for aqueous solubilities and Henry's Law constants of polychlorinated biphenyls. *Environmental Science & Technology*, 26(8), 1567–73.

Eaton, D. C. (1985). Mineralization of polychlorinated biphenyls by *Phanerochaete chrysosporium*: a ligninolytic fungus. *Enzyme and Microbial Technology*, 7, 194–6.

Electric Power Research Institute (1985). *Application of Biotechnology to PCB Disposal Problems*. Report No. EPRI CS-3807, Project No. 1263-16, Palo Alto, CA.

Electric Power Research Institute (1990). *Release and Attenuation of PCB Congeners: Measurement of Desorption Kinetics and Equilibrium Sorption Partitioning Coefficients*. Report No. EPRI GS-67875, Project No. 1263-22, Palo Alto, CA.

Eltis, L. D., Hofmann, B., Hecht, H-J., Lunsdorf, H. & Timmis, K. N. (1993). Purification and crystallization of 2,3-dihydroxybiphenyl 1,2-dioxygenase. *Journal of Biological Chemistry*, 268, 2727–32.

Erb, R. W. & Wagner-Dobler, I. (1993). Detection of polychlorinated biphenyl degradation genes in polluted sediments by direct DNA extraction and polymerase chain reaction. *Applied and Environmental Microbiology*, 59, 4065–73.

Erickson, B. D. & Mondello, F. J. (1992). Nucleotide sequencing and transcriptional mapping of the genes encoding biphenyl dioxygenase, a multicomponent polychlorinated-biphenyl-degrading enzyme in *Pseudomonas* strain LB400. *Journal of Bacteriology*, 174, 2903–12.

Erickson, B. D. & Mondello, F. J. (1993). Enhanced biodegradation of polychlorinated biphenyls after site-directed mutagenesis of a biphenyl dioxygenase gene. *Applied and Environmental Microbiology*, 59, 3858–62.

Erickson, M. D. (1992). *Analytical Chemistry of PCBs*. Chelsea, MI: Lewis Publishers.

Farrell, R. & Chakrabarty, A. M. (1979). Degradative plasmids: molecular nature and mode of evolution. In *Plasmids of Medical, Environmental, and Commercial Importance*, ed. K. N. Timmis & A. Puhler, pp. 97–109. New York: Elsevier/North-Holland Biomedical Press.

Fathepure, B. Z. & Vogel, T. M. (1991). Complete degradation of polychlorinated hydrocarbons by a two-stage biofilm reactor. *Applied and Environmental Microbiology*, 57, 3418–22.

Fish, K. M. & Principe, J. M. (1994). Biotransformations of Aroclor 1242 in Hudson River test tube microcosms. *Applied and Environmental Microbiology*, 60, 4289–96.

Flanagan, W. P. & May, R. J. (1993). Metabolite detection as evidence for naturally occurring aerobic PCB biodegradation in Hudson River sediments. *Environmental Science & Technology*, 27, 2207–12.

Focht, D. D. & Brunner, W. (1985). Kinetics of biphenyl and polychlorinated biphenyl metabolism in soil. *Applied and Environmental Microbiology*, 50, 1058–63.

Fulthorpe, R. R. & Wyndham, R. C. (1992). Involvement of a chlorobenzoate-catabolic transposon, Tn5271, in community adaptation to chlorobiphenyl, chloroaniline, and 2,4-dichlorophenoxyacetic acid in a freshwater ecosystem. *Applied and Environmental Microbiology*, 58, 314–25.

Furukawa, K. (1982). Microbial degradation of polychlorinated biphenyls. In *Biodegradation and Detoxification of Environmental Pollutants*, ed. A. M. Chakrabarty, pp. 33–57. Boca Raton, FL: CRC Press.

Furukawa, K. (1994). Molecular genetics and evolutionary relationship of PCB-degrading bacteria. *Biodegradation*, 5, 289–300.

Furukawa, K. & Arimura, N. (1987). Purification and properties of 2,3-dihydroxybiphenyl dioxygenase from polychlorinated biphenyl-degrading *Pseudomonas pseudoalcaligenes* and *Pseudomonas aeruginosa* carrying the cloned bphC gene. *Journal of Bacteriology*, 169, 924–7.

Furukawa, K. & Chakrabarty, A. M. (1982). Involvement of plasmids in total degradation of chlorinated biphenyls. *Applied and Environmental Microbiology*, 44, 619–26.

Furukawa, K. & Matsumura, F. (1976). Microbial metabolism of polychlorinated biphenyls: Studies on the relative degradability of polychlorinated biphenyl components by *Alkaligenes* sp. *Journal of Agricultural and Food Chemistry*, 24, 251–6.

Furukawa, K. & Miyazaki, T. (1986). Cloning of a gene cluster encoding biphenyl and chlorobiphenyl degradation in *Pseudomonas pseudoalcaligenes*. *Journal of Bacteriology*, 166, 392–8.

Furukawa, K., Tonomura, K. & Kamibayashi, A. (1978a). Effect of chlorine substitution on the biodegradability of polychlorinated biphenyls. *Applied and Environmental Microbiology*, 35, 223–7.

Furukawa, K., Matsumura, F. & Tonomura, K. (1978b). *Alcaligenes* and *Acinetobacter* strains capable of degrading polychlorinated biphenyls. *Agricultural and Biological Chemistry*, 42(3), 543–8.

Furukawa, K., Tomizuka, N. & Kamibayashi, A. (1979). Effect of chlorine substitution on the bacterial metabolism of various polychlorinated biphenyls. *Applied and Environmental Microbiology*, 38, 301–10.

Furukawa, K., Tomizuka, N., Deck, B. L. & Koystyk, N. (1982). Bacterial degradation of polychlorinated biphenyls (PCB) and their metabolites. *Advances in Experimental Medicine and Biology*, 136A, 407–18.

Furukawa, K., Tomizuka, N. & Kamibayashi, A. (1983). Metabolic breakdown of kaneclors (polychlorobiphenyls) and their products by *Acinetobacter* sp. *Applied and Environmental Microbiology*, 46, 140–5.

Furukawa, K., Arimura, N. & Miyazaki, T. (1987). Nucleotide sequence of the 2,3-dihydroxybiphenyl dioxygenase gene of *Pseudomonas pseudoalcaligenes*. *Journal of Bacteriology*, 169, 427–9.

Furukawa, K., Hayase, N., Taira, K. & Tomizuka, N. (1989). Molecular relationship of chromosomal genes encoding biphenyl/polychlorinated biphenyl catabolism: some soil bacteria possess a highly conserved *bph* operon. *Journal of Bacteriology*, 171, 5467–72.

Furukawa, K., Hirose, J., Suyama, A., Zaiki, T. & Hayashida, S. (1993). Gene components responsible for discrete substrate specificity in the metabolism of biphenyl (*bph* operon) and toluene (*tod* operon). *Journal of Bacteriology*, 175, 5224–32.

Gan, D. R. & Berthouex, P. M. (1994). Disappearance and crop uptake of PCBs from sludge-amended farmland. *Water Environment Research*, 66, 54–69.

Gibson, D. T., Roberts, R. L., Wells, M. C. & Kobal, V. M. (1973). Oxidation of biphenyl by a Beijerinckia species. *Biochemistry and Biophysics Research Communications*, 50, 211–19.

Gibson, D. T., Cruden, D. L., Haddock, J. D., Zylstra, G. J. & Brand, J. M. (1993). Oxidation of polychlorinated biphenyls by *Pseudomonas* sp. LB400 and *Pseudomonas pseudoalcaligenes* KF707. *Journal of Bacteriology*, 175, 4561–4.

Green, G. (1990). *The Use of Surfactants in the Bioremediation of Petroleum Contaminated Soils*. USA Report No. EPA/101/F-90/013, NTIS Report No. PB90-256546.

Griffin, R. A. & Chou, S. F. J. (1981). Movement of PCBs and other persistent compounds through soil. *Water Science and Technology*, 13, 1153–63.

Haddock, J. D., Nadim, L. M. & Gibson, D. T. (1993). Oxidation of biphenyl by a multicomponent enzyme system from *Pseudomonas* sp. Strain LB400. *Journal of Bacteriology*, 175, 395–400.

Haddock, J. D., Horton, J. R. & Gibson, D. T. (1995). Dihydroxylation and dechlorination of chlorinated biphyenls by purified biphenyl 2,3-dioxygenase from *Pseudomonas* sp. Strain LB400. *Journal of Bacteriology*, 177, 20–6.

Harkness, M. R. & Bergeron, J. A. (1990). Availability of PCBs in soils and sediments to surfactant extraction and aerobic biodegradation. In *Ninth Progress Report for the Research and Development Program for the Destruction of PCBs*, pp. 109–20. Schenectady, NY: General Electric Co. Corporate Research and Development.

Harkness, M. R., McDermott, J. B., Abramowicz, D. A., Salvo, J. J., Flanagan, W. P., Stephens, M. L., Mondello, F. J., May, R. J., Lobos, J. H., Carroll, K. M., Brennan, M. J., Bracco, A. A., Fish, K. M., Warner, G. L., Wilson, P. R., Dietrich, D. K., Lin, D. T., Morgan, C. B. & Gately, W. L. (1993). *In situ* stimulation of aerobic PCB biodegradation in Hudson River sediments. *Science*, 259, 503–7.

Harkness, M. R., McDermott, J. B., Abramowicz, D. A., Salvo, J. J., Flanagan, W. P., Stephens, M. L., Mondello, F. J., May, R. J., Lobos, J. H., Carroll, K. M., Brennan, M. J., Bracco, A. A., Fish, K. M., Warner, G. L., Wilson, P. R., Dietrich, D. K., Lin, D. T., Morgan, C. B. & Gately, W. L. (1994). Field study of aerobic polychlorinated biphenyl biodegradation in Hudson River sediments. In *Bioremediation of Chlorinated and Polycyclic Aromatic Hydrocarbon Compounds*, ed. R. E. Hinchee, A. Leeson, L. Semprini & S. K. Ong, pp. 368–75. New York: Lewis Publishers.

Havel, J. & Reineke, W. (1991). Total degradation of various chlorobiphenyls by cocultures and in vivo constructed hybrid pseudomonads. *FEMS Microbiology Letters*, 78, 163–70.

Havel, J. & Reineke, W. (1992). Degradation of Aroclor 1221 and survival of strains in soil microcosms. *Applied and Environmental Microbiology*, 38, 129–34.

Havel, J. & Reineke, W. (1993). Microbial degradation of chlorinated acetophenones. *Applied and Environmental Microbiology*, 59, 2706–12.

Hawker, D. W. & Connell, D. W. (1988). Octanol-water partition coefficients of polychlorinated biphenyl congeners. *Environmental Science & Technology*, 22(4), 382–7.

Hayase, N., Taira, K. & Furukawa, K. (1990). *Pseudomonas putida* KF715 bphABCD operon encoding biphenyl and polychlorinated biphenyl degradation: Cloning, analysis, and expression in soil bacteria. *Journal of Bacteriology*, 172, 1160–4.

Hernandez, B. S., Higson, F. K., Kondrat, R. & Focht, D. D. (1991). Metabolism of and inhibition by chlorobenzoates in *Pseudomonas putida* P111. *Applied and Environmental Microbiology*, 57, 3361–6.

Hickey, W. J. & Focht, D. D. (1990). Degradation of mono-, di-, and trihalogenated benzoic acids by *Pseudomonas aeruginosa* JB2. *Applied and Environmental Microbiology*, 56, 3842–50.

Hickey, W. J., Searles, D. B. & Focht, D. D. (1993). Enhanced mineralization of polychlorinated biphenyls in soil inoculated with chlorobenzoate-degrading bacteria. *Applied and Environmental Microbiology*, 59, 1194–200.

Higson, F. K. & Focht, D. D. (1989). Bacterial metabolism of hydroxylated biphenyls. *Applied and Environmental Microbiology*, 55, 946–52.

Higson, F. K. & Focht, D. D. (1990). Bacterial degradation of ring-chlorinated acetophenones. *Applied and Environmental Microbiology*, 56, 3678–85.

Hofer, B., Eltis, L. D., Dowling, D. N. & Timmis, K. N. (1993). Genetic analysis of a *Pseudomonas* locus encoding a pathway for biphenyl/polychlorinated biphenyl degradation. *Gene*, 130, 47–55.

Hooper, S. W., Dockendorff, T. C. & Sayler, G. S. (1989). Characteristics and restriction analysis of the 4-chlorobiphenyl catabolic plasmid, pSS50. *Applied and Environmental Microbiology*, 55, 1286–8.

Hutzinger, O., Safe, S. & Zitko, V. (1983). *The Chemistry of PCBs*. Malabar, FL: Robert E. Krieger Publishing Company.

Ishigooka, H., Yoshida, Y., Omori, T. & Minoda, Y. (1986). Enzymatic dioxygenation of biphenyl-2,3-diol and 3-isopropylcatechol. *Agricultural and Biological Chemistry*, 50, 1045–6.

Kaiser, K. L. E. & Wong, P. T. S. (1974). Bacterial degradation of polychlorinated biphenyls. I. Identification of some metabolic products from Aroclor 1242. *Bulletin of Environmental Contamination and Toxicology*, 11, 291–6.

Kamp, P. F. & Chakrabarty, A. M. (1979). Plasmids specifying p-chlorobiphenyl degradation in enteric bacteria. In *Plasmids of Medical, Environmental, and Commercial Importance*, ed. E. N. Timmis & A. Puhler, pp. 275–85. New York: Elsevier/North Holland Biomedical Press.

Karickhoff, S. W. & Morris, K. R. (1985). Sorption dynamics of hydrophobic pollutants in sediment suspensions. *Environmental Toxicology and Chemistry*, 4, 462–79.

Kaschabek, S. R. & Reineke, W. (1992). Maleylacetate reductase of *Pseudomonas* sp. strain B13: dechlorination of chloromaleylacetates, metabolites in the degradation of chloroaromatic compounds. *Archives of Microbiology*, 158, 412–17.

Kaschabek, S. R. & Reineke, W. (1993). Degradation of chloroaromatics: purification and characterization of meleylacetate reductase from *Pseudomonas* sp. strain B13. *Journal of Bacteriology*, 175, 6075–81.

Khan, A. A. & Walia, S. K. (1989). Cloning of bacterial genes specifying degradation of 4-chlorobiphenyl from *Pseudomonas putida* OU83. *Applied and Environmental Microbiology*, 55, 798–805.

Khan, A. A. & Walia, S. K. (1990). Identification and localization of 3-phenylcatechol dioxygenase and 2-hydroxy-6-oxo-6-phenylhexa-2,4-dienoate hydrolase genes of *Pseudomonas putida* and expression in *Escherichia coli*. *Applied and Environmental Microbiology*, 56, 956–62.

Khan, A. A. & Walia, S. K. (1991). Expression, localization, and functional analysis of polychlorinated biphenyl degradation genes *cbp*ABCD of *Pseudomonas putida*. *Applied and Environmental Microbiology*, 57, 1325–32.

Khan, A. A. & Walia, S. K. (1992). Use of genetically engineered *Escherichia coli* strain to produce 1,2-dihydroxy-4'-chlorobiphenyl. *Applied and Environmental Microbiology*, 58, 1388–91.

Khan, A. A., Tewari, R. & Walia, S. K. (1988). Molecular cloning of 3-phenylcatechol dioxygenase involved in the catabolic pathway of chlorinated biphenyl from *Pseudomonas putida* and its expression in *Escherichia coli*. *Applied and Environmental Microbiology*, 54, 2664–71.

Kikuchi, Y., Yasukochi, Y., Nagata, Y., Fukuda, M. & Takagi, M. (1994). Nucleotide sequence and functional analysis of the *meta*-cleavage pathway involved in biphenyl and polychlorinated biphenyl degradation in *Pseudomonas* sp. strain KKS102. *Journal of Bacteriology*, 176, 4269–76.

Kilpi, S., Himberg, K., Yrjala, K. & Backstrom, V. (1988). The degradation of biphenyl and chlorobiphenyls by mixed bacterial cultures. *FEMS Microbiology Ecology*, 53, 19–26.

Kimbara, K., Hashimoto, T., Fukuda, M., Koana, T., Takagi, M., Oishi, M. & Yano, K. (1989). Cloning and sequencing of two tandem genes involved in degradation of 2,3-dihydroxybiphenyl to benzoic acid in the polychlorinated biphenyl-degrading soil bacterium *Pseudomonas* sp. strain KKS102. *Journal of Bacteriology*, 171, 2740–7.

Kohler, H-P. E., Kohler-Staub, D. & Focht, D. D. (1988a). Cometabolism of polychlorinated biphenyls: enhanced transformation of Aroclor 1254 by growing bacterial cells. *Applied and Environmental Microbiology*, 54, 1940–5.

Kohler, H-P. E., Kohler-Staub, D. & Focht, D. D. (1988b). Degradation of 2-hydroxybiphenyl and 2,2'-dihydroxybiphenyl by *Pseudomonas* sp. strain HBP1. *Applied and Environmental Microbiology*, 54, 2683–8.

Kohler-Staub, D. & Kohler, H-P. E. (1989). Microbial degradation of β-chlorinated four-carbon aliphatic acids. *Journal of Bacteriology*, 171, 1428–34.

Kong, H-L. & Sayler, G. S. (1983). Degradation and total mineralization of monohalogenated biphenyls in natural sediment and mixed bacterial culture. *Applied and Environmental Microbiology*, 46, 666–72.

Krockel, L. & Focht, D. D. (1987). Construction of chlorobenzene-utilizing recombinants by progenative manifestation of a rare event. *Applied and Environmental Microbiology*, 53, 2470–5.

Laha, S. & Luthy, R. G. (1991). Inhibition of phenanthrene mineralization by nonionic surfactants in soil–water systems. *Envionmental Science & Technology*, 25, 1920–30.

Lajoie, C. A., Chen, S-Y., Oh, K-C. & Strom, P. F. (1992). Development and use of field application vectors to express nonadaptive foreign genes in competitive environments. *Applied and Environmental Microbiology*, 58, 655–63.

Lajoie, C. A., Zylstra, G. J., DeFlaun, M. F. & Strom, P. F. (1993). Development of field application vectors for bioremediation of soils contaminated with polychlorinated biphenyls. *Applied and Environmental Microbiology*, 59, 1735–41.

Lajoie, C. A., Layton, A. C. & Sayler, G. S. (1994). Cometabolic oxidation of polychorinated biphenyls in soil with a surfactant-based field application vector. *Applied and Environmental Microbiology*, **60**, 2826–33.

Layton, A. C., Sanseverion, J., Wallace, W., Corcoran, C. & Sayler, G. S. (1992). Evidence for 4-chlorobenzoic acid dehalogenation mediated by plasmids related to pSS50. *Applied and Environmental Microbiology*, **58**, 399–402.

Layton, A. C., Lajoie, C. A., Easter, J. P., Jernigan, R., Sanseverino, J. & Sayler, G. S. (1994). Molecular diagnostics and chemical analysis for assessing biodegradation of polychlorinated biphenyls in contaminated soils. *Journal of Industrial Microbiology*, **13**, 392–401.

Lee, J-F., Crum, J. R. & Boyd, S. (1989). Enhanced retention of organic contaminants by soils exchanged with organic cations. *Environmental Science & Technology*, 23(11), 1365–72.

Liu, D. (1980). Enhancement of PCBs biodegradation by sodium lignin sulfonate. *Water Research*, **14**, 1467–75.

Liu, D. (1981). Biodegradation of Aroclor 1221 type PCBs in sewage wastewater. *Bulletin of Environmental Contamination and Toxicology*, **27**, 695–703.

Liu, Z., Laha, S. & Luthy, R. G. (1990). Surfactant solubilization of polycyclic aromatic hydrocarbons in soil-water suspensions. Presented at the 15th Biennial International Conference sponsored by the International Association on Water Pollution Research and Control, Kyoto, Japan.

Lunt, D. & Evans, W. C. (1970). The microbial metabolism of biphenyl. *Biochemical Journal*, **118**, 54–5.

Maeda, M., Chung, S-Y., Song, E. & Kudo, T. (1995). Multiple genes encoding 2,3-dihydroxybiphenyl 1,2-dioxygenase in the gram-positive polychlorinated biphenyl-degrading bacterium *Rhodococcus erythropolis* TA421, isolated from a termite ecosystem. *Applied and Environmental Microbiology*, **61**, 549–55.

Masse, R., Messier, F., Peloquin, L., Ayotte, C. & Sylvestre, M. (1984). Microbial biodegradation of 4 chlorobiphenyl, a model compound of chlorinated biphenyls. *Applied and Environmental Microbiology*, **47**, 947–51.

May, H. D., Boyle, A. W., Price, W. A. & Balke, C. K. (1992). Subculturing of a polychlorinated biphenyl-dechlorinating anaerobic enrichment on solid media. *Applied and Environmental Microbiology*, **58**, 4051–4.

McCullar, M. V., Brenner, V., Adams, R. H. & Focht, D. D. (1994). Construction of a novel polychlorinated biphenyl-degrading bacterium: utilization of 3,4′-dichlorobiphenyl by *Pseudomonas acidovorans* M3GY. *Applied and Environmental Microbiology*, 60, 3833–9.

McDermott, J. B., Unterman, R., Brennan, M. J., Brooks, R. E., Mobley, D. P., Schwartz, C. C. & Dietrich, D. K. (1989). Two strategies for PCB soil remediation: biodegradation and surfactant extraction. *Environmental Progress*, 8, 46–51.

Mihelcic, J. R. & Luthy, R. G. (1991). Sorption and microbial degradation of naphthalene in soil-water suspensions under denitrification conditions. *Environmental Science & Technology*, 25, 169–77.

Mokross, H., Schmidt, E. & Reineke, W. (1990). Degradation of 3-chlorobiphenyl by in vivo constructed hybrid pseudomonads. *FEMS Microbiology Letters*, 71, 179–86.

Mondello, F. J. (1989). Cloning and expression in *Escherichia coli* of *Pseudomonas* strain LB400 genes encoding polychlorinated biphenyl degradation. *Journal of Bacteriology*, 171, 1725–32.

Morris, P. J. & Pritchard, P. H. (1994). Concepts in improving polychlorinated biphenyl bioavailability to bioremediation strategies. In *Bioremediation of Chlorinated and*

Polycyclic Aromatic Hydrocarbon Compounds, ed. R. E. Hinchee, A. Leeson, L. Semprini & S. K. Ong, pp. 359–67. New York: Lewis Publishers.

Morris, P. J., Mohn, W. W., Quensen, J. F., Tiedje, J. M. & Boyd, S. A. (1992). Establishment of a polychlorinated biphenyl-degrading enrichment culture with predominantly *meta* dechlorination. *Applied and Environmental Microbiology*, 58, 3088–94.

Murphy, E. M., Zachara, J. M. & Smith, S. C. (1990). Influence of mineral-bound humic substances on the sorption of hydrophobic organic compounds. *Environmental Science & Technology*, 24(10), 1507–15.

Nadim, L., Schocken, M. J., Higson, F. J., Gibson, D. T., Bedard, D. L., Bopp, L. H. & Mondello, F. J. (1987). Bacterial oxidation of polychlorinated biphenyls. In *Proceedings of the 13th Annual Research Symposium on Land Disposal, Remedial Action, Incineration, and Treatment of Hazardous Waste*, pp. 395–402. U.S. Environmental Protection Agency, Cincinnati, OH (EPA/600/9-87/015).

Neidle, E. L., Hartnett, C., Ornston, L. N., Bairoch, A., Rekik, M. & Harayama, S. (1991). Nucleotide sequences of the *Acinetobacter calcoaceticus benABC* genes for benzoate 1,2-dioxygenase reveal evolutionary relationships among multicomponent oxygenases. *Journal of Bacteriology*, 173, 5385–95.

Nies, L. & Vogel, T. M. (1990). Effects of organic substrates on dechlorination of Aroclor 1242 in anaerobic sediments. *Applied and Environmental Microbiology*, 56, 2612–17.

Nies, L. & Vogel, T. M. (1991). Identification of the proton source for the microbial reductive dechlorination of 2,3,4,5,6-pentachlorobiphenyl. *Applied and Environmental Microbiology*, 57, 2771–4.

Ofjord, G. D., Puhakka, J. A. & Ferguson, J. F. (1994). Reductive dechlorination of Aroclor 1254 by marine sediment cultures. *Environmental Science & Technology*, 28, 2286–94.

Omori, T., Sugimura, K., Ishigooka, H. & Minoda, Y. (1986a). Purification and some properties of a 2-hydroxy-6-oxo-6-phenylhefa-2,4-dienoic acid hydrolyzing enzyme from *Pseudomonas cruciviae* S93B1 involved in the degradation of biphenyl. *Agricultural and Biological Chemistry*, 50, 931–7.

Omori, T., Ishigooka, H. & Minoda, Y. (1986b). Purification and some properties of 2-hydroxy-6-oxo-6-phenylhexa-2,4-dienoic acid (HOPDA) reducing enzyme from *Pseudomonas cruciviae* S93B1 involved in the degradation of biphenyl. *Agricultural and Biological Chemistry*, 50, 1513–18.

Opperhulzen, A., Gobas, F. A. P. C., Van der Steen, J. M. & Hutzinger, O. (1988). Aqueous solubility of polychlorinated biphenyls related to molecular structure. *Environmental Science & Technology*, 22, 638–46.

Packard, J., Breen, A., Sayler, G. S. & Palumbo, A. V. (1989). Monitoring populations of 4-chlorobiphenyl-degrading bacteria in soil and lake water microcosms using colony hybridization. In *HMCRI's Second National Conference, Biotreatment – The Use of Microorganisms in the Treatment of Hazardous Materials and Hazardous Wastes*, Washington, DC, November 27–29, pp. 119–25. Silver Spring, MD: Hazardous Materials, Control Research Institute.

Parsons, J. R. & Sijm, D. T. H. M. (1988). Biodegradation kinetics of polychlorinated biphenyls in continuous cultures of a *Pseudomonas* strain. *Chemosphere*, 17, 1755–66.

Peloquin, L. & Greer, C. W. (1993). Cloning and expression of the polychlorinated biphenyl-degradation gene cluster from *Arthrobacter* M5 and comparison to analogous genes from gram-negative bacteria. *Gene*, 125, 35–40.

Pettigrew, C. A. & Sayler, G. S. (1986). The use of DNA : DNA colony hybridization in the rapid isolation of 4-chlorobiphenyl degradative bacterial phenotypes. *Journal of Microbiological Methods*, 5, 205–13.

Pettigew, C. A., Breen, A., Corcoran, C. & Sayler, G. S. (1990). Chlorinated biphenyl mineralization by individual populations and consortia of freshwater bacteria. *Applied and Environmental Microbiology*, 56, 2036–45.

Quensen, J. F. III, Tiedje, J. M. & Boyd, S. A. (1988). Reductive dechlorination of polychlorinated biphenyls by anaerobic microorganisms from sediments. *Science*, 242, 752–4.

Quensen, J. F. III, Boyd, S. A. & Tiedje, J. M. (1990). Dechlorination of four commercial polychlorinated biphenyl mixtures (Aroclors) by anaerobic microorganisms from sediments. *Applied and Environmental Microbiology*, 56, 2360–9.

Reichardt, P. B., Chadwick, B. L., Cole, M. A., Robertson, B. R. & Button, D. K. (1981). Kinetic study of the biodegradation of biphenyl and its monochlorinated analogues by a mixed marine microbial community. *Environmental Science & Technology*, 15, 75–9.

Rhee, G-Y., Bush, B., Brown, M. P., Kane, M. & Shane, L. (1989). Anaerobic biodegradation of polychlorinated biphenyls in Hudson River sediments and dredged sediments in clay encapsulation. *Water Research*, 23, 957–64.

Rhee, G-Y., Sokol, R. C., Bush, B. & Bethoney, C. M. (1993a). Long-term study of the anaerobic dechlorination of Aroclor 1254 with and without biphenyl enrichment. *Environmental Science & Technology*, 27, 714–19.

Rhee, G-Y., Sokol, R. C., Bethoney, C. M. & Bush, B. (1993b). Dechlorination of polychlorinated biphenyls by Hudson River sediment organisms: Specificity to the chlorination pattern of congeners. *Environmental Science & Technology*, 27, 1190–2.

Roberts, L. (1987). Discovering microbes with a taste for PCBs. *Science*, 237, 975–7.

Rothmel, R. K., Shannon, M. J. R. & Unterman, R. (1993). Isolation and characterization of a new surfactant-producing PCB-degrading bacterial strain. In *Abstracts of the 93rd General Meeting of the American Society for Microbiology*, p. 374. Atlanta, Georgia, #Q-153.

Rothmel, R. K., Gaudet, J. L., Schulz, W. H., Shannon, M. J. R., Krishnamoorthy, R., Smith, J. R. & Unterman, R. (1994). Biostimulation versus bioaugmentation: Two strategies for treating PCB-contaminated soils and sediments. In *Abstracts of the 94th General Meeting of the American Society for Microbiology*, p. 442. Las Vegas, Nevada, #Q-309.

Sayler, G. S. & Colwell, R. R. (1976). Partitioning of mercury and polychlorinated biphenyl by oil, water, and suspended sediment. *Environmental Science & Technology*, 10, 1142–5.

Sayler, G. S., Shon, M. & Colwell, R. R. (1977). Growth of an estuarine *Pseudomonas* sp. on polychlorinated biphenyl. *Microbial Ecology*, 3, 241–55.

Sayler, G. S., Thomas, R. & Colwell, R. R. (1978). Polychlorinated biphenyl (PCB) degrading bacteria and PCB in estuarine and marine environments. *Estuarine & Coastal Marine Science*, 6, 553–67.

Scow, K. M., Simkins, S. & Alexander, M. (1986). Kinetics of mineralization of organic compounds at low concentration in soil. *Applied and Environmental Microbiology*, 51(5), 1028–35.

Shannon, M. J. R. & Bartha, R. (1988). Immobilization of leachable toxic soil pollutants by using oxidative enzymes. *Applied and Environmental Microbiology*, 54, 1719–23.

Shannon, M. J. R. & Unterman, R. (1993). Evaluating bioremediation: distinguishing fact from fiction. *Annual Review of Microbiology*, 47, 715–38.

Shannon, M. J. R., Rothmel, R. K., Blanchard, S. & Unterman, R. (1992). Effect of environmental and matrix conditions on the biotreatability of PCB-contaminated soils and sludges. In *Abstracts of the 92nd General Meeting of the American Society for Microbiology*, p. 369. New Orleans, Louisiana, #Q-204.

Shannon, M. J. R., Rothmel, R., Chunn, C. D. & Unterman, R. (1994). Evaluating polychlorinated biphenyl bioremediation process: from laboratory feasibility testing to pilot demonstrations. In *Bioremediation of Chlorinated and PAH Compounds*, ed. R. H. Hinchee, A. Leeson, L. Semprini & S. K. Ong, pp. 354–8. Boca Raton, FL: Lewis Publishers.

Sharma, A., Chunn, C. D., Rothmel, R. K. & Unterman, R. (1991). Studies on bacterial degradation of polychlorinated biphenyls: optimization of parameters for *in vivo* enzyme activity. In *Abstracts of the 91st General Meeting of the American Society for Microbiology*, p. 284. Washington, DC, #Q-48.

Shiaris, M. P. & Sayler, G. S. (1982). Biotransformation of PCB by natural assemblages of freshwater microorganisms. *Environmental Science & Technology*, 16, 367–9.

Shields, M. S., Hooper, S. W. & Sayler, G. S. (1985). Plasmid-mediated mineralization of 4-chlorobiphenyl. *Journal of Bacteriology*, 163, 882–9.

Smith, J. R., Tomicek, R. M., Swallow, P. V., Weightman, R. L., Nakles, D. V. & Helbling, M. (1995). Definition of biodegradation endpoints for PAH contaminated soils using a risk-based approach. In *Hydrocarbon Contaminated Soils*, vol. 5, ed. P. T. Kostecki, E. J. Calabese & M. Bonazountas, pp. 531–72. Amherst, MA: Amherst Scientific Publishers.

Sokol, R. C., Kwon, O-S., Bethoney, C. M. & Rhee, G-Y. (1994). Reductive dechlorination of polychlorinated biphenyls in St. Lawrence River sediments and variations in dechlorination characteristics. *Environmental Science & Technology*, 28, 2054–64.

Sondossi, M., Sylvestre, M., Ahmad, D. & Masse, R. (1991). Metabolism of hydroxybiphenyl and chloro-hydroxybiphenyl by biphenyl/chlorobiphenyl degrading *Pseudomonas testosteroni*, strain B-356. *Journal of Industrial Microbiology*, 7, 77–88.

Stokes, R. W., Bedard, D. L. & Deweerd, K. A. (1994). The use of slurry inoculum and bromobenzoic acid to stimulate PCB dechlorination in Woods Pond sediment. In *Abstracts of the 94th General Meeting of the American Society for Microbiology*, p. 444. Las Vegas, NV.

Stott, D. E., Martin, J. P., Focht, D. D. & Haider, K. (1983). Biodegradation, stabilization in humus, and incorporation into soil biomass of 2,4-D and chlorocatechol carbons. *Soil Science Society of American Journal*, 47, 66–70.

Stucki, G. & Alexander, M. (1987). Role of dissolution rate and solubility in biodegration of aromatic compounds. *Applied and Environmental Microbiology*, 53(2), 292–7.

Subba-Rao, R. V. & Alexander, M. (1982). Effect of sorption on mineralization of low concentrations of aromatic compounds in lake water samples. *Applied and Environmental Microbiology*, 44, 659–68.

Sun, S. & Boyd, S. A. (1991). Sorption of polychlorobiphenyl (PCB) congeners by residual PCB-oil phases in soils. *Journal of Environmental Quality*, 20, 557–61.

Sylvestre, M. & Fauteux, J. (1982). A new facultative anaerobe capable of growth on chlorobiphenyls. *Journal of General Applied Microbiology*, 28, 61–72.

Sylvestre, M., Masse, R., Messier, F., Fauteux, J., Bisaillon, J-G & Beaudet, R. (1982). Bacterial nitration of 4-chlorobiphenyl. *Applied and Environmental Microbiology*, 44, 871–7.

Sylvestre, M., Masse, R., Ayotte, C., Messier, F. & Fauteux, J. (1985). Total biodegradation of 4-chlorobiphenyl (4 CB) by a two-membered bacterial culture. *Applied Microbiology and Biotechnology*, 21, 192–5.

Taira, K., Hayase, N., Arimura, N., Yamashita, S., Miyazaki, T. & Furukawa, K. (1988). Cloning and nucleotide sequence of the 2,3-dihydroxybiphenyl dioxygenase gene from the PCB-degrading strain of *Pseudomonas paucimobilis* Q1. *Biochemistry*, 27, 3990–6.

Takase, I., Omori, T. & Minoda, Y. (1986). Microbial degradation products from biphenyl-related compounds. *Agricultural and Biological Chemistry*, 50, 681–6.

Thomas, J. M., Yordy, J. R., Amador, J. A. & Alexander, M. (1986). Rates of dissolution and biodegradation of water-insoluble organic compounds. *Applied and Environmental Microbiology*, 52(2), 290–6.

Timmis, K. N., Steffan, R. J. & Unterman, R. (1994). Designing microorganisms for the treatment of toxic wastes. *Annual Review of Microbiology*, 48, 525–57.

Tucker, E. S., Saeger, V. W. & Hicks, O. (1975). Activated sludge primary degradation of polychlorinated biphenyls. *Bulletin of Environmental Contamination and Toxicology*, 14, 705–12.

Tulp, M. Th. M., Schmitz, R. & Hutzinger, O. (1978). The bacterial metabolism of 4,4'-dichlorobiphenyl, and its suppression by alternative carbon sources. *Chemosphere*, 1, 103–8.

Umbreit, T. H., Hesse, E. J. & Gallo, M. A. (1986). Bioavailability of dioxin in soil from a 2,4,5-T manufacturing site. *Science*, 232, 497–9.

Unterman, R. (1991). What is the K_m of disappearase? In *Environmental Biotechnology for Waste Treatment*, ed. G. S. Sayler, R. Fox & J. W. Blackburn, pp. 159–62. New York: Plenum Press.

Unterman, R., Bedard, D. L., Bopp, L. H., Brennan, M. J., Johnson, C. & Haberl, M. L. (1985). Microbial degradation of polychlorinated biphenyls. In *Proceedings of the International Conference on New Frontiers for Hazardous Waste Management*, pp. 481–8. U.S. Environmental Protection Agency (EPA/600/9-85/025).

Unterman, R., Brennan, M. J., Brooks, R. E. & Johnson, C. (1987a). Biological degradation of polychlorinated biphenyls. In *Proceedings of the International Conference on Innovative Biological Treatment of Toxic Wastewaters*, Ed. R. J. Scholze *et al.*, pp. 379–89. Champaign, IL: U.S. Army, CERL, N-87/12.

Unterman, R., Mondello, F. J., Brennan, M. J., Brooks, R. E., Mobley, D. P., McDermott, J. B. & Schwartz, C. C. (1987b). Bacterial treatment of PCB contaminated soils: Prospects for the application of recombinant DNA technology. In *Proceedings of the Second International Conference on New Frontiers for Hazardous Waste Management*, pp. 259–64. U.S. Environmental Protection Agency (EPA6/600/9-87/018F).

Unterman, R., Bedard, D. L., Brennan, M. J., Bopp, L. H., Mondello, F. J., Brooks, R. E., Mobley, D. P., McDermott, J. B., Schwartz, C. C. & Dietrich, D. K. (1988). Biological approaches for PCB degradation. In *Reducing Risks From Environmental Chemicals Through Biotechnology*, ed. G. S. Omenn *et al.*, pp. 253–69. New York: Plenum Press.

Unterman, R., Chunn, C. D. & Shannon, M. J. R. (1991). Isolation and characterization of a PCB-degrading bacterial strain exhibiting novel aerobic congener specificity. In *Abstracts of the 91st General Meeting of the American Society for Microbiology*, p. 284. Dallas, Texas, #Q-49.

Uzgiris, E. E., Edelstein, W. A., Philipp, H. R. & Iben, I. E. T. (1995). Complex thermal desorption of PCBs from soil. *Chemosphere*, 30, 377–87.

Van Dort, H. & Bedard, D. L. (1991). Reductive *ortho* and *meta* dechlorination of polychlorinated biphenyl congeners by anaerobic microorganisms. *Applied and Environmental Microbiology*, 57, 1576–8.

Walia, A., Khan, A. & Rosenthal, N. (1990). Construction and applications of DNA probes for detection of polychlorinated biphenyl-degrading genotypes in toxic organic-contaminated soil environments. *Applied and Environmental Microbiology*, 56, 254–9.

Weissenfels, W. D., Klewer, H-J. & Langhoff, J. (1992). Adsorption of polycyclic aromatic hydrocarbons (PAHs) by soil particles: influence on biodegradability and biotoxicity. *Applied Microbiology and Biotechnology*, 36, 689–96.

Williams, W. A. (1994). Microbial reductive dechlorination of trichlorobiphenyls in anaerobic sediment slurries. *Environmental Science & Technology*, 28, 630–5.

Wszolek, P. C. & Alexander, M. (1979). Effect of desorption rate on the biodegradation of n-alkylamines bound to clay. *Journal of Agricultural and Food Chemistry*, 27, 410–14.

Yagi, O. & Sudo, R. (1980). Degradation of polychlorinated biphenyls by microorganisms. *Journal of the Water Pollution Control Federation*, 52, 1035–43.

Yates, J. R. & Mondello, F. J. (1989). Sequence similarities in the genes encoding polychlorinated biphenyl degradation by *Pseudomonas* strain LB400 and *Alcaligenes eutrophus* H850. *Journal of Bacteriology*, 171, 1733–5.

Zeddel, A., Majcherczyk, A. & Huttermann, A. (1994). Degradation and mineralization of polychlorinated biphenyls by white-rot fungi in solid-phase and soil incubation experiments. In *Bioremediation of Chlorinated and Polycyclic Aromatic Hydrocarbon Compounds*, ed. R. E. Hinchee, A. Leeson, L. Semprini & S. K. Ong, pp. 436–40. New York: Lewis Publishers.

Zhang, Y. & Miller, R. M. (1992). Enhanced octadecane dispersion and biodegradation by a *Pseudomonas rhamnolipid* surfactant (biosurfactant). *Applied and Environmental Microbiology*, 58, 3276–82.

Zylstra, G. J., Chauhan, S. & Gibson, D. T. (1990). Degradation of chlorinated biphenyls by *Escherichia coli* containing cloned genes of the *Pseudomonas putida* F1 toluene catabolic pathways. In *Proceedings of the 16th Annual Hazardous Waste Research Symposium: Remedial Action, Treatment, and Disposal of Hazardous Waste*, pp. 290–302. EPA/600/9-90/037. Cincinnati, OH: U.S. Environmental Protection Agency.

8

Bioremediation of chlorinated phenols

Jaakko A. Puhakka and Esa S. Melin

Chlorophenols are common environmental contaminants originating mainly from their use as wide-spectrum biocides in industry and agriculture, their formation during pulp bleaching and the incineration of organic material in the presence of chloride. Chlorophenol use as biocides has resulted in soil and groundwater contamination; pulp bleacheries discharge wastewaters into aquatic environments; and combustion results in the contamination of all environmental compartments. Chlorophenols have the potential of being biotransformed and/or mineralized in both aerobic and anaerobic systems and degradation pathways have been carefully delineated. In aquatic and terrestrial environments chlorophenols become bound to humic materials, affecting their bioavailability, and possibly offering a useful method of detoxification. Although chlorophenols persist in many environments, because of inappropriate conditions for biodegradation, biological methods have been utilized in groundwater and soil remediation of chlorophenols. Composting and landfarming are the most common soil remediation methods, while on-site bioreactors are used for groundwater remediation. *In situ* bioremediation for restoration of contaminated subsurface environments has gained significant interest and the first full-scale operations are in progress. The fate of chlorophenol contaminants in aquatic sediments has been widely studied and the results serve as a basis of future efforts in remediation of sediments.

List of abbreviations: BOD, biological oxygen demand; CA, chloroanisol; CCA, copper-chromate-arsenate; CP, chlorophenol; 2,4-D, dichlorophenoxyacetic acid; DCP, dichlorophenol; CFSTR, continuous-flow stirred tank reactor; FBBR, fluidized-bed biofilm reactor; MCP, monochlorophenol; NAPL, non-aqueous phase liquid; PAH, polycyclic aromatic hydrocarbon; PCPP, polychlorinated phenoxyphenol; PCDF, polychlorinated dibenzofuran; PCDD, polychlorinated dibenzodioxin; PCR, polymerase chain reaction; PCP, pentachlorophenol; PCA, pentachloroanisole; TeCP, tetrachlorophenol; TeCA, tetrachloroanisole; TCC, trichlorocatechol; TCP, trichlorophenol; TOC, total organic carbon; 2,4,5-T, trichlorophenoxyacetic acid; UASB, upflow anaerobic sludge blanket reactor; VSS, volatile suspended solids.

8.1 Sources of contamination

The forest industry has used extensive amounts of chloropheols (CPs) for the preservation of timber against blue sapstain fungi. In the U.S., pentachorophenol (PCP) ia mainly used (Cirelli, 1978), while in Europe and Japan, various CP congener mixtures are typical. Preservative solutions are prepared either by dissolution of CPs in sodium hydroxide to produce concentrated chlorophenate solutions, or dissolution in fuel oil or kerosene. If these solvents are not available, liquid petroleum gas, methylene chloride, isopropyl alcohol, or methanol are used.

Both dip treatment with chlorophenate solutions in Europe and high or low pressure spraying of PCPs in the U.S. and Canada have contaminated soil around wood-preserving facilities. For example, soil contaminated with 3500 mg/kg of CPs has been observed in Finland (Kitunen *et al.*, 1987), and 5700 mg PCP/kg in the U.S. (Jackson & Bisson, 1990). Wood dust and other organic substances mixed with soil have been found to contain up to 21 000 mg/kg of CPs. In a saw mill that had been abandoned for 28 years, the soil around one dipping basin contained 340 mg/kg of 2,3,4,6-tetrachlorophenol (TeCP) and 140 mg/kg of PCP (Knuutinen *et al.*, 1990). While soils around dipping basins are the most heavily contaminated, CPs have also been found from storage areas (Valo *et al.*, 1984).

Up to 47 mg/kg of CP-transformation products such as *o*- and *p*-hydroxylated phenols (Knuutinen *et al.*, 1990), as well as chloroanisoles such as 2,3,4,6-TeCA and PCA (Palm *et al.*, 1991), have often been found in CP-contaminated soils. An additional route for contamination is the CP-treated utility poles and railway ties that contaminate nearby environments via runoff and leaching (Wan, 1992).

The use of chlorophenate solutions has resulted in high levels of groundwater CP contamination. At some wood-treatment facilities, the CP from soil migrated into groundwater; up to 190 mg/l of CPs have been found (Valo *et al.*, 1984; Lampi *et al.*, 1990). The lake sediments adjacent to the treatment plant contained as much as 10 mg/kg of CPs due to the groundwater discharge into the lake (Lampi *et al.*, 1992b). An elevated risk of non-Hodgkins lymphoma has been observed in people exposed to this CP-contaminated drinking water and to CP-contaminated fish (Lampi *et al.*, 1992a).

Technical CP formulations contain impurities such as polychlorinated phenoxyphenols (PCPPs), polychlorinated dibenzodioxins (PCDDs), and poly-chlorinated dibenzofurans (PCDFs) (Humppi *et al.*, 1984; Kitunen *et al.*, 1985; Jackson & Bisson, 1990). Therefore, PCPPs, PCDFs, and PCDDs are often found in CP-contaminated wood treatment sites (Kitunen *et al.*, 1985, 1987; Jackson & Bisson, 1990; Kitunen & Salkinoja-Salonen, 1990; Trudell *et al.*, 1994). PCPP concentrations up to 78 mg/kg, PCDF up to 3.8 mg/kg, and PCDDs at 13 mg/kg have been detected in CP-contaminated sites (Kitunen *et al.*, 1987; Jackson & Bisson, 1990).

Other important sources of CP contamination include the synthesis and use (Table 8.1) of 2,4-dichlorophenoxyacetic acid (2,4-D) and 2,4,5-trich-

Table 8.1. *Principal uses of selected chlorophenols*

Principal use	Compounds
Wood preservation	Pentachlorophenol
	2,3,4,6-Tetrachlorophenol
	2,4,6-Trichlorophenol
Intermediates in industrial synthesis of:	
herbicides	2,4-Dichlorophenol
	2,3,5-Trichlorophenol
higher chlorophenols	2-Chlorophenol
	4-Chlorophenol
Additives[a]	Pentachlorophenol
	2,4,5-Trichlorophenol

[a]Inhibition of microbial growth in a diverse array of products.

lorophenoxyacetic acid (2,4,5-T) herbicides, pulp bleaching, chlorination of water, and combustion of municipal solid waste. Improper disposal of distillation wastes from 2,4-D synthesis is a typical source of CP contamination (Johnson *et al.*, 1984).

In 1984, the total worldwide production of PCP was 35 400 to 40 000 tons, 80% of which was used for wood preservation (Korte, 1987). CPs are widely distributed in the environment and many contaminated sites have been selected for cleanup in Europe and are on the National Priorities List of the Superfund Program in the U.S. (U.S.EPA, 1993). Some CPs are listed as priority pollutants by the U.S. EPA (Keith & Telliard, 1979).

8.2 Chemical and physical properties of chlorophenols

The mobility and bioavailability of CPs, and thus their bioremediation, are affected by their chemical and physical properties. The solubility of CPs in water decreases as the number of chlorine substituents increases. In addition, the increase in the number of chlorosubstituents in the phenol ring increases the lipophilicity of CPs and thus their tendency to bioaccumulate. The water solubilities, pK_a (acidity constant) and pK_{ow} (octanol–water partition constant) values for environmentally important CPs are summarized in Figure 8.1.

Environmental pH is the most important factor affecting CP adsorption and mobility (Choi & Aomine, 1972, 1974a,b; Christodoulatos *et al.*, 1994; Stapleton *et al.*, 1994). Since the dissociation constants (pK_a) of CPs are in the same range as the pH in groundwater, both protonated and deprotonated CPs may exist under natural conditions. Lower chlorinated phenols are more protonated in neutral environments than their polychlorinated congeners. With PCP, for example, the sorption to clay decreases threefold between pH 4 and 8.5 (Stapleton *et al.*, 1994). Low soil pH might also cause CP precipitation, especially from alkaline solution.

Soil organic content is another important factor affecting CP mobility (Choi & Aomine, 1972, 1974a,b; Schellenberg *et al.*, 1984; Christodoulatos *et al.*, 1994). The

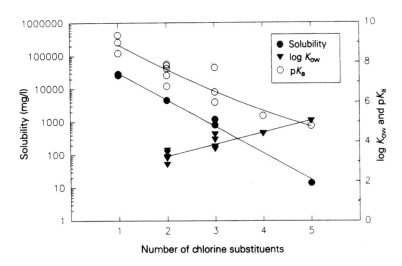

Figure 8.1. Chemical and physical characteristics of selected chlorophenols (Davis & Huang, 1990; Paukku, 1989; Lagas, 1988; Schellenberg et al., 1984; Westall et al., 1985).

natural dissolved organic matter content of groundwater may affect the bioavailability of PCP and thus its transport (Lafrance et al., 1994). In sediments, CP sorption mechanisms involved hydrogen-bond interactions between sorbate and sorbent in addition to purely hydrophobic interactions (Isaacson & Fink, 1984). Desorption is slower than sorption, and a significant fraction of CPs can be irreversibly held by sediments and organic soil (Isaacson & Fink, 1984; Warith et al., 1993). Other factors affecting CP adsorption are the presence of silt and clay and the clay type (Choi & Aomine, 1974a,b; Kitunen & Salkinoja-Salonen, 1990; Christodoulatos et al., 1994). Different retardation factors for different CPs cause different plume migration for different CPs (Johnson et al., 1984). PCP was estimated to have a retardation factor of 51 in soil with a high organic content (Warith et al., 1993). The presence of oil resulted in a substantial drop in the adsorption capacity of the soil and increased the mobility of PCP (Jackson & Bisson, 1990; Christodoulatos et al., 1994).

Dimeric impurities of technical-grade CP solutions are usually retained in the top 5 to 20 cm of surface soil (Kitunen et al., 1987). When CPs have been used with organic solvents, the migration of low solubility impurities, such as chlorinated dioxins, to considerable distances is possible (Pereira et al., 1985; Jackson & Bisson, 1990).

8.3 Biodegradation

8.3.1 Biodegradation mechanisms

CPs can be degraded by both aerobic and anaerobic microorganisms. The pathways of CP degradation have been delineated and several careful reviews have been published (Rochkind-Dubinsky et al., 1987; Tiedje et al., 1987; Kuhn & Suflita,

Table 8.2. *Aerobic chlorophenol mineralization stoichiometries*

PCP	$C_6Cl_5OH + 4.5O_2 + 2H_2O \rightarrow 6CO_2 + 5HCl$
TeCP	$C_6Cl_4HOH + 5O_2 + H_2O \rightarrow 6CO_2 + 4HCl$
TCP	$C_6Cl_3H_2OH + 5.5O_2 \rightarrow 6CO_2 + 3HCl$
DCP	$C_6Cl_2H_3OH + 6O_2 \rightarrow 6CO_2 + 2HCl + H_2O$
MCP	$C_6ClH_4OH + 6.5O_2 \rightarrow 6CO_2 + HCl + 2H_2O$

1989; Neilson *et al.*, 1991; Häggblom, 1992; Häggblom & Valo, 1995). The pathways of CP degradation are summarized in Figure 8.2.

Theoretically, aerobic mineralization of CPs follows the reaction stoichiometries presented in Table 8.2. The average oxidation state of carbon in CPs increases with the number of chlorine substituents and the energy available from biochemical oxidation thereby decreases. Reductive dechlorination of CPs involves large free energy changes (Dolfing & Harrison, 1992; Dolfing, 1995), but energy conservation from these reactions has not, so far, been demonstrated, with the exception of one bacterial isolate that reductively dechlorinates 2-CP (Cole *et al.*, 1994).

Aerobic CP degradation is initiated by the action of oxygenase enzymes that insert hydroxysubstituents on the aromatic ring (hydrolytic dechlorination), thereby making the structure more labile for further transformations. Two main pathways exist, both of which involve initial oxygenase action (Figure 8.2). MCPs and DCPs are *o*-hydroxylated to corresponding catechols, which is followed by ring cleavage and dechlorination. The resulting aliphatic compounds are subject to degradation via common catabolic pathways. The initial attack is catalyzed by a monooxygenase or a dioxygenase system. With polychlorophenols, the initial attack involves the substitution of chlorine from the *p*-position with a hydroxy group. Several reductive dechlorination steps follow, the ring is then cleaved and the resulting aliphatic structure is mineralized. A variety of aerobic bacteria is known to mediate these reactions (for review, see Häggblom, 1992; Häggblom & Valo, 1995). The occurrence of reductive dechlorinations by aerobic bacteria may be favored by lower redox potentials in the cytoplasm than in the growth medium itself.

Incomplete aerobic transformations may involve cometabolic transformations and reactions resulting in recalcitrant dead-end metabolites. Cometabolic *o*-hydroxylation of MCPs and DCPs by a phenol monooxygenase has been shown, for example, in a *Pseudomonas* sp. (Knackmuss & Hellwig, 1978) and by the toluene dioxygenase reaction in *Pseudomonas putida* (Spain & Gibson, 1988; Spain *et al.*, 1989). Cometabolic transformation of CPs is also possible in aerobic mixed culture systems. Phenol- and toluene-enriched cultures completely removed 2,4-DCP, and the toluene enrichment also removed 2,4,6-TCP and PCP (Ryding *et al.*, 1994). This PCP attack by the toluene enrichment involved an *o*-hydroxylation.

Insertion of methyl group(s) to the hydroxyl group(s) of CPs is a common

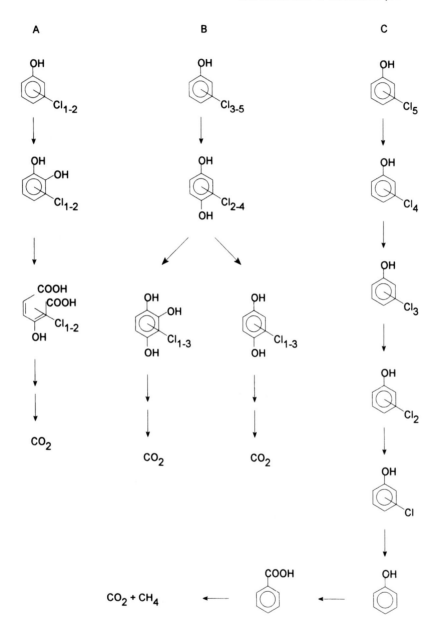

Figure 8.2. Pathways for chlorophenol degradation under aerobic (A,B) and anaerobic (C) conditions.

259

environmental phenomenon. *o*-Methylation of CPs is carried out by several species of bacteria and fungi (for review, see Häggblom, 1992). These transformations, which do not attack the carbon skeleton, change CPs into an environmentally more harmful form (Neilson *et al.*, 1984). The likelihood of these reactions increases in the presence of supplementary organic compounds (Suzuki, 1983; Allard *et al.*, 1987; Häggblom *et al.*, 1989).

The fungal degradation of CPs involves a nonspecific initial attack similar to that in the degradation of polymeric lignin or lignin-model compounds (Buswell & Odier, 1987; Kirk & Farrell, 1987). *Phaenerochaete chrysosporium*, a white-rot fungus, has an extracellular lignin degradative system which consists of heme peroxidases, lignin peroxidase, and manganese peroxidase, as well as an H_2O_2-generating system (Kirk & Farrell, 1987; Gold *et al.*, 1989; Hammel & Moen, 1991; Wariishi *et al.*, 1991). This system is expressed by white-rot fungi under nutrient-limiting conditions. Mineralization of several CPs by *P. chrysosporium* has been reported, including 2,4-DCP (Valli & Gold, 1991), 2,4,5-TCP (Joshi & Gold, 1993), and PCP (Mileski *et al.*, 1988; Lin *et al.*, 1990). In these studies, fungal mineralization remained partial. Besides mineralization, oxidative coupling of CPs with naturally occurring humic acid precursors is often biologically mediated via extracellular oxidoreductases (Bollag, 1992). In these reactions, phenolic reactants are linked through C–C and C–O bonds. The binding of CPs may decrease their availability for interaction with the biota and inhibit their movement via leaching. The first mechanism decreases toxicity and the latter helps in pollutant containment. Thus, complexation of CPs into humus may be an environmentally beneficial phenomenon. The use of oxidative coupling for soil decontamination raises concerns about the ultimate fate of CPs and the potential for forming dimers, such as PCDDs, PCDFs, and diphenyl ethers, which are more toxic than the parent compounds (Minard *et al.*, 1981; Svenson *et al.*, 1989; Öberg *et al.*, 1990).

Anaerobic CP degradation involves sequential reductive dehalogenations with MCPs or DCPs as final metabolites, or degradation may proceed to complete dechlorination to phenol, further transformation to benzoate and, ultimately, conversion to methane and carbon dioxide. Reductive dechlorination of PCP, for example, results in the formation of *meta-* or *para-*CPs as the end-products (e.g., Woods *et al.*, 1989; Madsen & Aamand, 1992), or anaerobic degradation may continue to complete mineralization (Boyd & Shelton, 1984; Mohn & Kennedy, 1992; Wu *et al.*, 1993). In reductive dechlorinations, CPs serve as electron acceptors and need a suitable electron donor.

Although a number of enrichment cultures are known to reductively dehalogenate CPs, only a few bacterial isolates are capable of mediating these reactions (Table 8.3). *Desulfomonile tiedjei* dechlorinates *m*-substituted CPs (Mohn & Kennedy, 1992), while a gram-positive, endospore-forming, nitrate-reducing bacterium dechlorinates CPs from *o*- and *m*-positions (Madsen & Licht, 1992). These bacteria do not seem to conserve energy from these transformations, although *D. tiedjei* is known to use 3-chlorobenzoate as a respiratory electron acceptor in oxidative

Table 8.3. *Bacteria that reductively dechlorinate chlorophenols*

Species	Dechlorination(s)/electron donor	Reference
Desulfomonile tiedjei (DCB-1)	*Meta*/formate	Mohn & Kennedy (1992)
Desulfitobacterium dehalogens	*Ortho*/pyruvate + yeast extract	Utkin *et al.* (1994)
Gram-positive sporeformer, DCB-2	*Ortho* and *meta*/pyruvate or H_2 from pyruvate fermentation	Madsen & Licht (1992)
2-MCP Dehalogenator	*Ortho*/acetate	Cole *et al.* (1994)

phosphorylation (Mohn & Tiedje, 1991). With *D. tiedjei*, CP dechlorination occurs only after the cells have been induced with 3-chlorobenzoate. A gram-positive isolate, *Desulfitobacterium dehalogenans*, *o*-dechlorinates several CPs (Utkin *et al.*, 1994) and a gram-negative, facultative anaerobe reduces 2-MCP to phenol (Cole *et al.*, 1994). Both of these isolates benefit from the dechlorination reaction. With the 2-MCP-dehalogenating isolate, the growth yield was about 3 g of protein per mole of 2-MCP dechlorinated (Cole *et al.*, 1994). The difficulty of isolating dehalogenating organisms may reflect the complexity of anaerobic communities and the requirement for interdependent microbial actions in reductive dehalogenations.

In the presence of a strong reductant, abiotic reductive dechlorination of CPs is catalyzed by the reduced form of Vitamin B_{12} (Gantzer & Wackett, 1991; Smith & Woods, 1994). These abiotic dechlorinations favor removal of *m*- and *p*-chlorines, which differs substantially from reductive dechlorinations by anaerobic microbial consortia or by the CP-dehalogenating bacteria isolated to date. This indicates that abiotic dechlorinations of CPs are not central reactions in environmental or engineered anaerobic systems.

8.3.2 Factors affecting biodegradation

The main factors influencing CP remediation include temperature, the properties of the environmental matrix, the toxicity of CPs or other compounds, and the composition of indigenous or added microbial cultures (Crawford & Mohn, 1985). Unlike many other xenobiotics, CPs undergo insignificant volatilization (Valo & Salkinoja-Salonen, 1986; Lamar *et al.*, 1990b; Mueller *et al.*, 1991a). The temperature, the contaminant bioavailability, the possible process amendments and the effects of additional contaminants are discussed below.

Temperature

Temperature is one of the main factors affecting the feasibility and design of bioremediation processes. Bioprocesses slow down when the temperature decreases.

Table 8.4. *Chlorophenol biodegradation at different temperatures*

Temperature optimum (°C)	Temperature range studied (°C)	Contaminants	Microbes	Reference
25–30	5–56	2,4,6-TCP	*Azotobacter*	Li *et al.* (1991)
37	30–43	2-MCP	Immobilized activated sludge	Sofer *et al.* (1990)
38	8–58	PCP	Aerobic mixed culture	Valo *et al.* (1985)
30	12–40	PCP	*Flavobacterium*	Crawford & Mohn (1985)
19–30	4–30	PCP	Aerobic sediment slurry	Pignatello *et al.* (1986)
20	0–20	PCP	*Pseudomonas* sp.	Trevors (1982)
25	5–25	PCP	Aerobic soil culture	Trudel *et al.* (1994)
31	4–62	4-MCP, 2,4-DCP	Anaerobic sediment	Zhang & Wiegel (1990)
20–30	7–72	2,4-DCP	Anaerobic sediment	Kohring *et al.* (1989)
50	10–90	2,3,6-TCP	Anaerobic sludge	Mohn & Kennedy (1992)

Groundwater, soil, and sediment temperatures are usually low. For environmental remedial applications, the range of operating temperatures, and the minimum temperatures in particular, are important. The optimum temperature for CP degradation is usually that at which mesophilic bacteria are active. These temperatures are, in many cases, higher than those possible for practical applications. Contamination often involves large quantities of groundwater, soil, or sediment and extensive heating prior to bioremediation is therefore impractical. During soil composting, the temperature generally rises by 5 to 15 °C above the ambient temperature (Valo & Salkinoja-Salonen, 1986; Holroyd & Caunt, 1995).

Biodegradation of CPs at low temperatures has been studied in some pure and mixed cultures. These results are summarized in Table 8.4. Slow and partial mineralization of CPs including PCP (Pignatello et al., 1986) and removal of 2-MCP, 2,4-DCP, 2,4,6-TCP and PCP (Smith & Novak, 1987) have been shown at 10 °C in long-term batch incubations. Trevors (1982) used three *Pseudomonad* isolates which removed 51–56% of PCP in 12 days at 20 °C and 11–51% in 80 days at 4 °C, but at 0 °C no degradation occurred in 100 days. Aerobic sediment cultures removed 24% of PCP in 30 days even at 0 °C (Baker et al., 1980); 45 to 78% of 2-MCP, 4-MCP and 2,4-DCP were removed in the same time. At 20 °C, 73 to 100% of CPs were removed by the same sediments. *Azotobacter* sp. degraded 2,4,6-TCP at 25 to 30 °C; at 20 °C the degradation rate was 40% of maximum and at 4 °C no degradation was observed (Li et al., 1991). Similarly, in a mixed culture, PCP mineralized at 28 °C but not at 8 °C (Valo et al., 1985); 50% of PCP was mineralized at 28 °C, while at 8 °C no removal was observed in 700 hours.

Temperature effects on CP bioremediation have been studied in a few cases. Activated municipal sludge removed 60–90% of 2,4,6-TCP, 2,3,4,6-TeCP, and PCP between 4 and 15 °C when fed at very low concentrations; the fate of CPs was not determined (Ettala et al., 1992). Trudell et al. (1994) observed slow PCP remediation at 5 °C by soil microbes; the half-life of PCP was increased from 60 days at 25 °C to 179 days at 5 °C. Crawford & Mohn (1985) inoculated soil with *Flavobacteria* and observed similar mineralization of PCP at temperatures between 24 and 35 °C; no mineralization occurred at 12 °C. *Rhodococcus chlorophenolicus* immobilized in a biofilter degraded 2,4,6-TCP, 2,3,4,6-TeCP and PCP from simulated groundwater at temperatures above 20 °C but not at 4 °C (Valo et al., 1990). In a pilot-scale fill-and-draw system, the temperature had to be raised to 21–25 °C from ambient groundwater temperature for biodegradation to occur (Valo & Hakulinen, 1990). An aerobic fluidized-bed enrichment mineralized 99% of 2,4,6-TCP, 2,3,4,6-TeCP and PCP at 10 °C (Järvinen & Puhakka, 1994), and 99.9% at 4 °C (Järvinen et al., 1994); at 7 °C, a CP loading rate of 740 mg/l/d was achieved with this mineralization efficiency.

Anaerobic CP transformation and treatment studies are usually performed at elevated temperatures close to those optimal for methanogenesis (e.g., Krumme & Boyd, 1988; Woods et al., 1989; Wu et al., 1993). Reductive dechlorination of CPs, however, seems to occur over a wide range of temperatures (Table 8.4).

In summary, most of the CP biodegradation and bioremediation studies have been conducted using mesophilic microorganisms. The effects of temperature on CP bioremediation have not been systematically studied, even though environmental temperatures are generally well below those suitable for mesophilic organisms. Therefore, future studies should more carefully consider the temperature constraints of bioremediation.

Bioavailability

A residual CP concentration is often observed after soil bioremediation. The 'leveling-off' of degradation is not due to decreased microbial activity, since freshly added CPs are rapidly degraded (Harmsen, 1993; Salkinoja-Salonen et al., 1989). The residual concentrations are explained by the gradual diffusion of pollutants deep into micropores, as well as by their adsorption onto soil organic matter (Harmsen, 1993). Lagas (1988) observed that the nonextractable fraction of CPs in sterile soil increased according to the square root of time as a consequence of diffusion into humic material.

In old contaminated soil sites, CPs are less accessible to biodegradation than CPs in more recently contaminated sites (Salkinoja-Salonen et al., 1989); inaccessibility correlates with unavailability of CPs to extraction by nonpolar solvents (diethyl ether, hexane). Remediation studies with artificially contaminated soil are likely to give lower residual CP concentrations than obtained in actual remediation. Decreased CP concentrations due to analytical limitations should not be interpreted as biodegradation.

Attempts to stimulate the biodegradation of residual (nonbioavailable) PCP by controlling soil moisture were unsuccessful (Harmsen, 1993). Extended time periods are needed for CPs to desorb and diffuse out from the micropores. On the other hand, low residual CP concentrations (0.3–0.5 mg/kg) have also been achieved by soil composting (Salkinoja-Salonen et al., 1989; Karlson et al., 1995).

In summary, bioavailability strongly affects CP biodegradation in soil and sediments. The partitioning of CPs to water-saturated subsurface solids also affects groundwater bioremediation. The organic content of the solid matrix increases the unavailability of CPs to biodegradation.

Effect of additional contaminants

In several wood-preserving facilities, other wood preservatives such as creosote and chromate–copper–arsenate (CCA) have been used in addition to PCP (e.g., Lamar & Glaser, 1994; Mueller et al., 1991a; Mahaffey et al., 1991). Environmental contamination by chemical mixtures is likely in these sites. When PCP has been dissolved in an organic carrier such as oil, soil is also contaminated with the solvent (Trudell et al., 1994; Lamar & Dietrich, 1990). Chlorinated dimeric impurities in technical CP formulations are also found in contaminated soil. Design of successful bioremediation must address the effects of other chemicals on CP biodegradation.

In a few cases, the effect of additional contaminants on CP degradation has been studied. *o*- and *m*-Cresol at 83 and 91 mg/l, respectively, and phenanthrene at 2.2 mg/l did not affect the PCP removal by *Arthrobacter* sp. On the other hand, naphthalene, 1-methylnapthalene, and 2-methylnaphthalene strongly inhibited PCP degradation by the same *Arthrobacter* strain (Edgehill, 1994). PCP degradation by *Flavobacterium* sp. was not prevented by 50 mg/l creosote (Topp & Hanson, 1990b). In this study, copper/chromate/arsenate completely inhibited PCP removal at concentrations of 2/2/10 mg/l.

In laboratory landfarming experiments with creosote- and PCP-contaminated soil, PCP decreased from 88 mg/kg to 25–40 mg/kg in 12 weeks during extensive biodegradation of low-molecular-weight creosote constituents (Mueller *et al.*, 1991a). During slurry-phase bioremediation of creosote-contaminated sediment and surface soil, 22% and 10% biodegradation of PCP was observed, respectively (Mueller *et al.*, 1991b). The presence of phenolic compounds can slow biodegradation of CPs (Namkoong *et al.*, 1989). At a site in Libby, Montana, PCP degraded faster than pyrene and was therefore not rate limiting in bioremediation (Piotrowski *et al.*, 1994). The lignin-degrading fungus *P. sordida* was inoculated into soil containing 673 mg/kg PCP and 4000 mg/kg polycyclic aromatic hydrocarbons (PAHs) and resulted in an 89% decrease in PCP as well as a decrease in low-molecular-weight PAHs (Lamar & Glaser, 1994).

Trudell *et al.* (1994) obtained over 99% and 83% PCP removal in 35 weeks in soil contaminated by 500 and 900 mg/kg PCP, respectively. This soil also contained 1.6% oil and 1.8% grease. A high concentration of mineral spirits in soil inhibited the growth of *P. chrysosporium* (Lamar & Dietrich, 1990). The mineral spirits had to be volatilized by soil mixing before the fungal inoculation. A high non-aqueous phase liquid (NAPL) content in groundwater may require oil/water separation before on-site treatment to prevent excessive organic loadings and operational problems (Piotrowsky *et al.*, 1994). The presence of 50 mg/kg of oil did not adversely affect biofilm treatment of PCP (Stinson *et al.*, 1991).

The effect of PCDDs, PCDFs, and PCDDs on CP-degrading microorganisms in soil has not been systematically studied. These compounds do not biodegrade and do not seem to inhibit CP bioremediation (Valo & Salkinoja-Salonen, 1986; McBain *et al.*, 1995). The treatment of soils contaminated simultaneously with PAHs and chlorinated dimeric compounds by white-rot fungi could be potentially advantageous since the peroxidase enzymes oxidize all these chemicals (e.g. Valli *et al.*, 1992; Gold *et al.*, 1994; Lamar & Glaser, 1994).

In anaerobic treatment, the presence of inorganic ions (sulfate, nitrate or reduced iron) may affect reductive dechlorination of CPs by serving as alternative electron acceptors for dehalogenating organisms. Sulfate in marine environments and metal ions in subsurface environments are examples of factors to be taken into account in bioremediation plans. The mechanisms of alternative electron acceptors may include direct inhibition of reductive dehalogenation or inhibition of enrichment of dehalogenating activity (for review, see Mohn & Tiedje, 1992). High inorganic ion concentrations may also cause toxicity and affect aqueous solubility of CPs.

In summary, the effects of additional contaminants seem to be site specific and should be estimated in preliminary laboratory biodegradation tests.

Remedial amendments

Biodegradation and bioremediation may be enhanced by amendments of nutrients, pH buffers, electron donors/acceptors, moisture and microorganisms. The availability of molecular oxygen is usually the limiting factor in contaminant biodegradation and needs to be supplied in both on-site and *in situ* bioremediation systems. Nitrogen and phosphorus are essential macronutrients for growth. For aerobic heterotrophic growth, a ratio $BOD_5 : N : P = 100 : 5 : 1$ has been used in the design of biological wastewater treatment systems (Metcalf & Eddy, 1991). For anaerobic growth, the nutrient requirements are lower. The contaminant concentrations in groundwater are usually significantly lower than the concentration of biodegradable organics in wastewater. Therefore, the nutrient supplements and their related costs are also much smaller than in wastewater treatment.

Electron donor amendments may be needed in anaerobic bioremediation to enhance reductive dechlorination of CPs. Hydrogen and several organic compounds can be used for this purpose (e.g., Perkins *et al.*, 1994). In anaerobic environments, the availability of electron acceptors is often the limiting resource in the turnover of organic matter. Inorganic electron acceptors such as sulfur oxyanions, carbon dioxide, nitrate and reduced metal ions affect the carbon and electron flow. Degradation of MCPs has been coupled to the reduction of sulfate (Häggblom & Young, 1990), nitrate (Häggblom *et al.*, 1993; Melin *et al.*, 1993), and iron (Häggblom *et al.*, 1993). Electron acceptors may be available or they may, in some cases, be amended to the treatment system.

The amendments may vary for soil and groundwater treatments and in on-site and *in situ* systems and are, therefore, discussed in detail in sections 8.4 and 8.5.

8.4 Groundwater remediation

8.4.1 Pump and treat

Pump-and-treat systems for groundwater bioremediation serve two purposes. They aim at containment of the contaminant plume and removal of contaminants via biodegradation.

The source of microorganisms and CP-degradation kinetics are key factors in on-site systems. Development of specific microbial cultures for bioremediation of CPs may involve pure and enrichment culture approaches. The approach of isolation and mass cultivation, followed by inoculation of the treatment system, has been applied to systems using, for example, *Flavobacteria* (O'Reilly & Crawford, 1989) and *Rhodococci* (Valo *et al.*, 1989). The *Flavobacterium*-based processes sometimes involve addition of supplemental carbon and energy sources during CP treatment.

Selective enrichment of desired CP-degraders from commonly available microbial sources such as activated sludge is the other approach for obtaining stable microbial cultures for CP remediation (Puhakka & Järvinen, 1992; Mäkinen *et al.*, 1993; Järvinen *et al.*, 1994; Puhakka *et al.*, 1995b). The general availability of microbial inoculum is an advantage over the use of pure culture processes.

The design, modeling, and operation of biological treatment systems requires an understanding of microbial degradation kinetics. Degradation of and growth on CPs have been modeled using Monod growth kinetics and the Haldane inhibition model. Relatively few detailed studies of CP-degradation kinetics have been reported (Table 8.5). A PCP-degrading mixed culture (Klecka & Maier, 1985) and *Flavobacterium* sp. (Hu *et al.*, 1994) both had low growth yields, 0.14 and 0.12 g cells/g of PCP, respectively. The K_i for these cultures were 1.38 and 1.5 mg/l respectively. The low K_i for the *Flavobacterium* is surprising because it degrades PCP at concentrations up to 150–200 mg/l. A fluidized-bed enrichment grown on a mixture of 2,3,4,6-TeCP and PCP had a net growth yield of 0.03 mg of VSS/mg of CP removed or 0.09 mg of VSS/mg of TOC removed (Mäkinen *et al.*, 1993).

CP removal has been reported in municipal (Melcer & Bedford, 1988; Ettala *et al.*, 1992) and industrial activated sludge (Salkinoja-Salonen *et al.*, 1984). In these uncharacterized mixed cultures, CPs have been trace constituents of the organic feed. The removal of CPs has been at least partially by way of partitioning in the wastewater solids and by incomplete transformations (Leuenberger *et al.*, 1985; Parker *et al.*, 1992). Similarly, anaerobic treatment of industrial effluents may involve CP removal and transformations (for review, see Rintala & Puhakka, 1994) and strong partitioning in the biosolids (Ballapragada *et al.*, 1994).

Several on-site bioremediation systems have been developed for CP-contaminated groundwater (Table 8.6). An on-site PCP-bioremediation system (BioTrol) based on the use of a *Flavobacterium* sp. as inoculum was demonstrated in pilot-scale experiments (Stinson *et al.*, 1991). The startup of the three-stage biofilm system involved liquid recycling after inoculation. Addition of nitrogen and phosphorus and heating of the influent to 21 °C were required. The heating costs were estimated for full-scale operation and accounted for 43 to 60% of total operating costs. A 96% removal efficiency with the effluent quality of 1 mg PCP/l was achieved at a CP loading rate of 365 mg/l/d. Chloride releases confirmed PCP mineralization in the process and toxicity removal was complete as shown by tests with *Daphnia magna* and freshwater minnows.

Mueller *et al.* (1993) treated PCP/PAH-contaminated groundwater in a two-stage pilot-reactor at 23 to 25 °C. The feed PCP and PAH concentrations were 25 and 1044 mg/l, respectively. The easily degradable PAHs were removed in a CSTR (continuously stirred tank reactor), which served as the first stage of treatment. The effluent from the first stage was pumped into three 227 l fill-and-draw batch reactors where the groundwater was held for 4 days. These reactors were inoculated daily by a mixture of PCP-degrading *Pseudomonas* sp. and PAH-degrading *P. paucimobilis*. During 14 days of field operation, the treatment system removed 78% of PCP; 24% was biodegraded and the rest was adsorbed. In a

Table 8.5. Examples of kinetic parameters for chlorophenol degradation modeled by Monod or Haldane (K_i involved) equations

Feed	q_{max} (day^{-1})	K_s (mg/l)	μ_{max} (day^{-1})	K_i (mg/l)	Y	b (day^{-1})	Reference
2-CP	1.88	0.025	—	—	—	—	Philbrook & Grady (1985)
2-CP	2.07	0.034	—	—	—	—	Philbrook & Grady (1985)
2-CP + lysine	1.43	0.011	—	—	—	—	Philbrook & Grady (1985)
2-CP + lysine	2.44	0.029	—	—	—	—	Philbrook & Grady (1985)
2-CP	—	17.0	0.60	—	0.49	—	Dang et al. (1989)
2-CP	—	16.0	0.48	—	0.35	—	Dang et al. (1989)
2-CP	—	12.6	7.0	1.1	0.67	—	Gaudy et al. (1988)
2-CP	—	26.6	18.6	1.7	0.67	—	Gaudy et al. (1988)
2-CP	—	—	—	80	0.236	—	Templeton & Grady (1988)
4-CP	—	2.66	6.0	114	0.41	—	Brown et al. (1990)
4-CP	—	7.2	7.0	48	0.31	—	Brown et al. (1990)
4-CP	—	126	32.59	4	0.392	—	Kennedy et al. (1990)
4-CP	0.57	0.59	—	—	—	—	Pitter & Chudoba (1990)
2,4-DCP	—	1.8	7.4	3.2	—	—	Bae et al. (1995)
2,4-DCP	0.18	1.3	—	—	3.1	0.065	Watkin & Eckenfelder (1984)
2,4-DCP + phenol	1.08	63	—	—	0.39	0.014	Beltrame et al. (1982)
2,4-DCP + methanol	5.59	0.6	2.1	—	0.37	0.86	Chudoba et al. (1989)
2,4-DCP	0.52	0.6	—	—	—	—	Pitter & Chudoba (1990)
2,4-DCP	22	11.5	—	12.5	—	—	Gu & Korus (1995)
2,4,5-TCP	—	0.16	0.91	5.68	0.19	—	Klecka & Maier (1988)
PCP	—	2.8	1.7	1.1	0.14	—	Bae et al. (1995)
PCP	—	0.06	1.78	1.38	—	—	Klecka & Maier (1985, 1988)
PCP + sodium glutamate	237	38	—	81	0.12	—	Gu & Korus (1995)
PCP + sodium glutamate	—	25	7.2	1.5	0.12	0.05	Hu et al. (1994)

bench-scale, two-stage continuous-flow experiment (hydraulic retention time 24 h in each stage) with similar groundwater, over 97% PCP removal was achieved; 72% was biodegraded. The continuous operation for 32 days required continuous inoculation with *Pseudomonas* sp. at 6×10^8 cells/h. The lower efficiency of PCP degradation in the pilot experimentation was attributed to poor-quality inoculum (low cell viability and incomplete induction).

Valo *et al.* (1990) used polyurethane-immobilized *R. chlorophenolicus* and another species of *Rhodococcus* in a biofilter treating a mixture of 2,4,6-TCP, 2,3,4,6-TeCP, and PCP. The process was operated in two stages. The first stage involved continuous flow at 4 °C for CP adsorption onto the carrier. Then, the reactor was heated to 25 °C and the groundwater was recirculated in the reactor to obtain biodegradation. In the first phase, CP adsorption occurred at a rate of 74 to 426 mg/l/d while microbial degradation took 5 to 16 days of recycling. In pilot-scale experiments, this process reduced CPs from 20–40 mg/l to 0.002–0.075 mg/l (Valo & Hakulinen, 1990) at a removal rate of 25–55 mg/l/d (Häggblom & Valo, 1995). Inorganic chloride measurements confirmed mineralization in these experiments.

An aerobic fluidized-bed treatment system has been developed for bioremediation of CP-contaminated groundwater. At room temperature 99% degradation of a mixture of 2,4,6-TCP, 2,3,4,6-TeCP and PCP was achieved at a loading rate of 1500 mg CP/l/d (Järvinen & Puhakka, 1994). At 5–7 °C, treatment of groundwater containing 7–11 mg/l of 2,4,6-TCP, 32–36 mg/l of 2,3,4,6-TeCP and 1.8–2.3 mg/l of PCP resulted in over 99.9% CP degradation at a loading rate of 740 mg CP/l/d (Järvinen *et al.*, 1994). The effluent concentrations were below 0.003 mg/l for each CP. Inorganic chloride releases were in conformity with the CP removals, indicating mineralization. This system used higher loading rates than previously reported for bioremediation and the effluent quality was close to drinking water standards. To our knowledge, this is the first report on high-rate bioremediation at ambient groundwater temperatures or lower. The operation of this fluidized-bed system was demonstrated in on-site pilot tests (Puhakka *et al.*, 1995a; Rintala *et al.*, 1995) and led to the construction of a full-scale process in early 1995. Full-scale bioremediation was started at ambient temperature (7.5 °C), and after startup, removed 98–99% of CPs from groundwater with a total feed CP concentration of 15 to 20 mg/l.

At the Libby, Montana, Superfund site, two 38 m^3 upflow, aerobic fixed-film reactors have been used to treat PCP/PAH-contaminated groundwater using an average combined hydraulic retention time of 42 h (Piotrowski *et al.*, 1994). The PAH-degrading and PCP-degrading microbes from contaminated groundwater were enriched in the first and second reactor, respectively. The influent groundwater was heated to 22 °C, since the PCP-degrading culture showed poor biodegradation at lower temperatures. During the second year of operation, PCP removal was 60% at PCP influent concentrations of 5 to 18 mg/l. During the fourth year of operation, the average PCP removal was 87% at 1.7 to 748 mg PCP/l feed concentration

Table 8.6. *Examples of chlorophenol-contaminated water treatment by continuous flow bioreactors*

Contaminant	Type	Size (l)	Culture	Mechanism	Temp. (°C)	Feed (mg/l)	CP load (mg/l/d)	Effluent (mg/l)	Removal (%)	HRT (h)	Reference
PCP	Fixed-film 3-stage	Pilot (2000)	Indigenous + *Flavobacterium* sp.	AE	25	42.0	112	0.13	99.8	9	Stinson et al. (1991)
PCP	Fixed-film 3-stage	Pilot (2000)	Indigenous + *Flavobacterium* sp.	AE	21	27.5	365	0.99	96.4	1.8	Stinson et al. (1991)
PCP, PAHs	Fixed-film 2-stage	Full 75700	Indigenous	AE	22	66	11	8.6	87	142	Piotrowsky et al. (1994) Woodward-Clyde & Champion International (1995)
2,4,6-TCP, 2,3,4,6-TeCP, PCP	FBBR	Lab 1.0	Enrichment from activated sludge	AE	24–29	46.9	1500	0.4	99	0.75	Järvinen & Puhakka (1994)
TCP, TeCP, PCP	FBBR	Lab 0.66	Enrichment from activated sludge	AE	5–7	43.7	740	<0.003	>99.9	1.4	Järvinen et al. (1994)
TCP, TeCP, PCP	FBBR	Pilot	Same as above	AE	16	22.6	217	0.125	99	2.5	Puhakka et al. (1995a) Rintala et al. (1995)
PCP	PUR in chemostat	Lab	*Flavobacterium* sp.	AE	NS	55	159	3.9	93	8.3	O'Reilly & Crawford (1989)
PCP, PAHs	CFSTR 2 stage	Lab 2	*Pseudomonas* sp.	AE	28.5	16.9	8.5	0.44	71.9	48	Mueller et al. (1993)

Compound	Reactor	Scale/Volume	Organism	Type							Reference
PCP, PAHs	CFSTR + 3 Fed-batch	Pilot 1135	*Pseudomonas* spp.	AE	23–25	25	2.5	5.6	77.6	239	Mueller *et al.* (1993)
2,4,6-TCP, 2,3,4,6-TeCP, PCP	Biofilter	Lab	*R. chlorophenolicus, Rhodococcus* sp.	AE	25	3–130	6–260	NS	>99.9	12	Valo *et al.* (1990)
TCP, TeCP, PCP	PUR Biofilter PUR	1.5 Pilot 2000	*R. chlorophenolicus, Rhodococcus* sp.	AE	20	25–40	10–70	0.002–NS 0.075	80	Valo & Hakulinen (1990)
PCP	UASB	Lab 0.325	Anaerobic granules and sludge	AN	28	60	95	<0.5	98–99	15	Wu *et al.* (1993)
PCP	UASB	Lab 2	Anaerobic granules	AN	37	3.0	1.5	0.006	99.8	48	Hendriksen *et al.* (1992)

AE, Aerobic; AN, anaerobic.
FBBR, fluidized-bed biofilm reactor; CFSTR, completely mixed stirred tank reactor; PUR, polyurethane immobilized cells; UASB, upflow anaerobic sludge blanket reactor; NS, not specified.

(Woodward-Clyde Consultants & Champion International, 1995). During this period, most of the PAHs and PCP were degraded in the first reactor. A typical cause of operational problems was excessive shock loadings caused by NAPLs in the feed.

CP removal from water by white-rot fungi has been studied in bioreactors. Lignin peroxidase activity and CP degradation were higher with immobilized than with suspended *P. chrysosporium* (Lewandowski *et al.*, 1990; Wang & Ruckstein, 1994). The following reactor types have been studied with *P. chrysosporium*: a batch-operated rotating biological contactor for 2,4-DCP (Tabak *et al.*, 1991), a batch silicone membrane biofilm reactor for PCP (Venkatadri *et al.*, 1992), a continuous-flow packed-bed reactor for 2-MCP (Lewandowski *et al.*, 1990; Armenante *et al.*, 1992), a continuous-flow air-lift column with alginate beads for 2-MCP, an upflow fixed-film bioreactor with nylon web media, and a fluidized-bed reactor with sand for PCP (Kang & Stevens, 1994). These processes require an additional carbon and energy source and a low pH optimum, and extracellular enzymes may be lost with the effluent. Simultaneous use of wood chips as a carrier and cosubstrate has been suggested for enhanced fungal water treatment (Lewandowski *et al.*, 1990).

Anaerobic treatment of CP-containing liquids has been reported in fixed-film reactors (Hendriksen *et al.*, 1991) and upflow anaerobic sludge bed reactors (UASB) with anaerobic granular sludge, resulting in less-chlorinated phenols (Hendriksen *et al.*, 1992; Mohn & Kennedy, 1992; Woods *et al.*, 1989) or mineralization (Wu *et al.*, 1993). With MCPs, the UASB treatment resulted in mineralization (Krumme & Boyd, 1988). UASB had better PCP removal efficiency and stability than anaerobic fixed-film reactors (Hendriksen & Ahring, 1993). Anaerobic remediation always requires the addition of an external electron donor.

In summary, several on-site biological methods for CP-contaminated groundwater have been developed. In both pure and mixed culture processes, biofilm-based systems are most commonly used. This is logical, because biofilm systems are known to tolerate higher toxicant concentrations and retain cells longer than suspended growth processes. The use of immobilized cell systems allows the operation of bioreactors at high flow rates. Internal recycling is used in many bioreactors, which efficiently dilutes the feed CP concentrations and thus avoids toxic effects.

8.4.2 In situ remediation

In situ treatment may involve the addition of air and nutrients to an aquifer to enhance indigenous microbial catabolism of contaminants. Commonly, on-site treatment involves pumping of groundwater, treatment, amendment of treated water, and reinjection to the aquifer in a controlled manner to achieve a groundwater flow through designated zones. In these zones, microbes form biofilms on the solid surfaces, where they come into contact with the passing contaminants.

Microbes can also be introduced into the aquifer, but they may not survive if they are first grown under laboratory conditions and then injected into the environmental conditions.

Subsurface environmental conditions are suboptimal with low temperatures and low concentrations of growth nutrients. The decline of bacterial inoculae by protozoan predation is of major concern in soil (Acea *et al.*, 1988; Acea & Alexander, 1988; Casida, 1989) but may not be a factor in saturated subsurface environments. Immobilization of cells to carrier material may enhance microbial survival in the environment through control of predation and supply of nutrients and moisture. Stormo & Crawford (1992) developed a cell immobilization technique for production of small beads (2–50 μm) consisting of agarose and cells of PCP-degrading *Flavobacterium* sp. Microencapsulated *Flavobacteria* efficiently degraded PCP and survived for two years in soil columns at environmental conditions (Stormo & Crawford, 1994). These results show that microencapsulation may be a very useful tool in *in situ* bioremediation.

Full-scale *in situ* bioremediation of PCP/PAH-contaminated groundwater is in operation at the Libby Superfund site in Montana (Piotrowski *et al.*, 1994; Woodward-Clyde Consultants & Champion International, 1995). In 1994, a two-injection well system was used to pump oxygenated water amended with nitrogen and phosphorus at a total injection rate of 300 to 8740 l/min. The groundwater temperatures varied from 2 to 19 °C. This water injection resulted in the reduction of PCP concentration from 686 μg/l to near or below detection level (3 μg/l) in monitoring wells at a distance of 30 to 60 m from the injection wells. Decreased contaminant concentrations were estimated to be partially due to dilution. On the other hand, the oxygen and nutrient consumptions indicate active subsurface biodegradation. After 6 years of operation, significant contaminant removal has been achieved. The limitations of the system seem to be related to the presence of contaminants as NAPLs and the complex hydrogeology of the aquifer.

The recent development of molecular methods for detection and identification of microorganisms in the environment (Ford & Olsen, 1988; Jain *et al.*, 1988; Ogram & Sayler, 1988; Pickup, 1991; Otte *et al.*, 1995) offers attractive tools for monitoring of *in situ* bioremediation. Direct molecular analyses based on 16S rRNA with a specific probe and polymerase chain reaction (PCR) primers have been developed for the polychlorophenol degrader *Mycobacterium chlorophenolicum* (Briglia, 1995) and genes involved in the first steps of PCP degradation have been characterized (Orser *et al.*, 1993a,b).

Examples of *in situ* bioremediation with CP-contaminated aquifers are limited. Good results of *in situ* bioremediation with hydrocarbon contaminants indicate the potential for low-cost bioremediation of other biodegradable compounds such as CPs. *In situ* bioremediation is very much affected by nonbiological factors. Therefore, further development of *in situ* methods requires close cooperation between hydrogeologists, engineers, and microbiologists.

8.5 Soil decontamination

Soil bioremediation techniques involve landfarming, composting, slurry reactors, and *in situ* treatment (Table 8.7 and 8.8). Decontamination may be enhanced by inoculation in all these processes.

In landfarming, a layer of contaminated soil is spread out on a specifically constructed site. Microbiological degradation is stimulated by soil tilling, irrigation, and addition of nutrients (nitrogen, phosphorus) and a specialized inoculum, if necessary. When the pollutant concentration is reduced to the target level, a new layer of polluted soil is spread and treated.

Full-scale landfarming has been successfully used for PCP/PAH-contaminated soil at the Libby Superfund site in Montana (Piotrowski *et al.*, 1994; Huling *et al.*, 1995). In 15 to 25 cm deep soil layers with weekly tilling and periodic irrigation, PCP concentrations ranging from 60 to 150 mg/kg were reduced to below the target level (37 mg/kg) within 21–75 days without added microbes. Nutrients (N and P) were added as needed. Ames and Microtox™ assays from water extracts showed decreased toxicity. Cleaned soil layers below the newly applied contaminated soil layers showed no toxicity in Microtox™ assays, indicating no vertical migration of pollutants during landfarming (Huling *et al.*, 1995). Another full-scale landfarming application was demonstrated in Trenton, Ontario, where 60 cm deep soil layers were remediated *in situ* or on-site with solid-phase nutrient amendments (Grace Dearborn, 1994).

In composting, bark chips or wood dust are frequently mixed in with soil to promote aeration and stimulate microbial activity. Bark chips protect microbes against toxicity by reversible adsorption of PCP (Apajalahti & Salkinoja-Salonen, 1984). Seech *et al.* (1991) reported that PCP degradation was stimulated in *Flavobacterium*-inoculated soil by adding finely ground red clover leaf and stem material to increase the organic matter content from 2.1 to 4.8%. Fungal inoculae are usually pregrown and added on wood chips (Lamar & Glaser, 1994). One strategy is to amend the soil with solid-phase, biodegradable organic soil which has a soil-specific particle size distribution, nutrient profile, and nutrient release kinetics. Use of this type of soil amendment significantly stimulated indigenous microorganisms in treatment of PCP/PAH-contaminated soil (Seech *et al.*, 1994; Grace Dearborn, 1994).

Windrow composting has been used successfully in Scandinavia to bioremediate CP-contaminated soils. In Denmark, 3500 metric tons of soil with an average PCP concentration of 30 mg/kg treated on site in 2 m high piles (Karlson *et al.*, 1995). Although the soil also had an active PCP-mineralizing microbial population the degradation rate was stimulated by adding *M. chlorophenolicum*. The inoculum was immobilized on 1 mm pieces of polyurethane at a rate of 10^6 to 10^7 cells per gram of soil. PCP degradation resulted in about 5 mg/kg concentration during the first two weeks. A residual concentration of 0.5 mg/kg was reached after 30 weeks. Total CAs were below 0.05 mg/kg during the treatment. In another detailed report,

Table 8.7. *Examples of chlorophenol-contaminated soil treatment*

Contaminant (mg/kg)	Scale	Culture (inoculum and size)	Temperature (°C)	Half-life (days)	Treatment time (days)	Residual concentration (mg/kg)	Other remarks	Reference
PCP (500)[a]	Laboratory 250 g	Indigenous	5	179	245	130	Oil and grease content 16 000 mg/kg Sand and silt Moisture 70% of field capacity	Trudell et al. (1994)
PCP (500)[a]	Laboratory 250 g	Indigenous	15	113	245	6	As above	Trudell et al. (1994)
PCP (500)[a]	Laboratory 250 g	Indigenous	25	60	245	4	As above	Trudell et al. (1994)
PCP (900)[a]	Laboratory 250 g	Indigenous	15	165	245	160	As above, except oil and grease content 18 000 mg/kg	Trudell et al. (1994)
PCP (15.4)[b]	Laboratory 5 g	Arthrobacter sp. 6.6×10^6 cells/g	30	0.125	ND	ND	Commercial quartz sand	Edgehill (1994)
PCP (15.4)[b]	Laboratory 50 g	Arthrobacter sp. 6.6×10^6 cells/g	30	NA	2	0.6	Soil contained 60% clay	Edgehill (1994)
PCP (185)[a]	Laboratory 3 kg; depth 10 cm	Indigenous	ND	ND	56	51	Sand and gravel Initial measured PAHs 6.0 g/kg + oil + petroleum hydrocarbons Soil moisture 50–70% of WHC	Mahaffey et al. (1991)

Table 8.7. *continued*

Contaminant (mg/kg)	Scale	Culture (inoculum and size)	Temperature (°C)	Half-life (days)	Treatment time (days)	Residual concentration (mg/kg)	Other remarks	Reference
PCP (680)[a]	Pilot 12 tonnes depth 0.5 m	*Pseudomonas resinovorans* 7×10^5 cells/g (isolated from treatment site)	11–34	21	207	6	Initial total PAH concentration 1.4 g/kg Total petroleum hydrocarbons 5.7–7.0 g/kg Sandy loam Organic carbon content 1.8% Soil moisture 80% of WHC Solid phase organic amendment	Seech *et al.* (1994)
PCP (20)[b]	Laboratory 50 g	*Arthrobacter* sp. 10^5 cells/g	30	1	2	2	Moist garden soil organic matter 4% Soil moisture 15–20%	Edgehill & Finn (1983)
PCP (182 mg/l)[b]	Outdoor $61 \times 61 \times 10$ cm (37 l)	*Arthrobacter* sp. 10^9/l	8–16	7.5[c]	12	28	As above	Edgehill & Finn (1983)
TCP, TeCP, PCP (total 212 mg/kg)[a]	Full scale Windrow compost 50 m^3	Indigenous	<0–32	75[c]	500	15	35 m^3 softwood bark 3 m^3 ash from fiberboard factory boiler gravel and sand, organic matter <2%	Valo & Salkinoja-Salonen (1986)
PCP (175)[b]	Laboratory 20 g	*Flavobacterium* sp. 10^6 cells/g (dry wt)	25	20[c] based on CO_2 recovery	170	5 mg/kg 68% ^{14}C-PCP recovered as CO_2	Silt loam soil, 2.1% organic carbon, soil moisture 60%	Seech *et al.* (1991)
PCP (100)[b]	Laboratory 100 g	*Flavobacterium* sp. 10^7 cells/g	30	6[c] based on CO_2 recovery	32	13 mg/kg	Sand; 2.5% combustibles	Crawford & Mohn (1985)

PCP (298)[a]	Laboratory 100 g	*Flavobacterium* sp. 20–40 d intervals 0.3–7 × 10⁷ cells/g	30	42[c]	100	58	Sand; 5.4% combustibles	Crawford & Mohn (1985)
PCP (30)[a]	Full scale 3500 tons in 2 m high piles	*Mycobacterium chlorophenolicum* 10⁶ to 10⁷ cells/g in PUR	Ambient	8[c]	210	0.5	Sandy soil, TOC 3.4 g/kg, pH 6.5	Karlson *et al.* (1995)
PCP (115)[a]	Full scale 4000 m² × 15 cm	Indigenous	Ambient	36	54	43	Soil moisture 40–70% field capacity Initial PAHs 280 mg/kg	Huling *et al.* (1995)
PCP (6–28)[a]	Full scale *in situ* 4800 m² × 0.6 m	Indigenous	Ambient	NA	305	<5	Sandy loam soil Organic matter 3.4% Initial PAHs 69–662 mg/kg addition of organic soil amendments Soil moisture 50–80% WHC	Grace Dearborn (1994)
PCP (102)[a]	Full scale on site 940 m² × 0.6 m	Indigenous	Ambient	27[c]	2	175	Initial PAHs 619 mg/kg Organic matter 4.6% Other same as above	Grace Dearborn (1994)
PCP (30)[b]	Laboratory 50 g	Anaerobic sewage sludge 5 g/kg	ND	13[c]	28	<0.5	Loamy sand, 1.1% organic carbon, >60% PCP recovered as dechlorination products	Mikesell & Boyd (1988)

[a]actual contaminated soil, [b]artificially contaminated soil, [c]estimated from data, WHC, water holding capacity; PUR, polyurethane foam. NA – not available, ND – no data.

Table 8.8. *Degradation of chlorophenols in soil by white rot fungi*

Inoculum	Initial PCP concentration (mg/kg)	PCP removal (GC) (%)	PCA formation (%)[b]	Mineralization to CO_2 (%)	Recovery of radioactivity in organic solvent (%)	Soil bound radioactivity (%)	Abiotic losses[a] (%)	Time (days)	Scale	Reference
P. chrysosporium sterile corncobs	100	~93	ND	6.8	31.2	12.6	3.8	56	Lab	Liang & McFarland (1994)
P. chrysosporium nonsterile corncobs	100	~93	ND	17.3	37.3	19.0	2.2	56	Lab	Liang & McFarland (1994)
P. chrysosporium[c] sterilized aspen pulpwood chips	50	~98		2.3	55	ND	0.8	56	Lab	Lamar et al. (1990b)
Same as above[c]	50	~98	ND	2.0	6.4	ND	1.8	56	Lab	Lamar et al. (1990b)
Same as above [c]	50	~98	ND	2.3	34.6	ND	0.9	56	Lab	Lamar et al. (1990b)
P. chrysosporium sterilized aspen pulpwood chips	80	96	46	ND	ND	ND	ND	64	Lab	Lamar et al. (1990a)
P. sordida sterilized aspen pulpwood chips	80	82	48	ND	ND	ND	ND	64	Lab	Lamar et al. (1990a)
P. chrysosporium sterilized aspen chips 3.35%, peat 1.9%	~250–300	82	14	ND	ND	ND	ND	46	Field	Lamar & Dietrich (1990)

P. sordida Other as above	~280–400	86	8	ND	ND	ND	ND	46	Field	Lamar & Dietrich (1990)
P. chrysosporium nutrient fortified grain–sawdust mixture 10%, wood chips 2.5%	705	72	15	ND	ND	ND	ND	56	Field	Lamar et al. (1993)
P. sordida Other as above	672	89	3	ND	ND	ND	ND	56	Field	Lamar et al. (1993)
Trametes hirsuta Other as above	399	55	Negligible	ND	ND	ND	ND	56	Field	Lamar et al. (1993)
P. chrysosporium straw, woodchips, sawdust, pine bark 5%	173–203[d]	85–94	BDL	ND	ND	ND	ND	730	Full	Holroyd & Caunt (1995)

[a] Volatilization and/or adsorption on flasks.
[b] Percentage conversion of removed chlorophenol to anisole at the end of experiment.
[c] Otherwise same conditions but soil was different.
[d] Mixture of TCP, TeCP, PCP.
ND, not determined; GC, gas chromatographic analysis; BDL, below detection limits

70 m^3 of CP (TCP, TeCP, and PCP) contaminated soil was mixed with 35 m^3 softwood bark and 3 m^3 ash and composted in two 50 m^3 compo ˙ piles and two pilot compost piles (Valo & Salkinoja-Salonen, 1986). During the four summer months, the CP concentration decreased to 30 mg/kg and after the second summer to 15 mg/kg. An active CP-mineralizing culture was present and no significant increase of CAs was observed. After two more years of composting a contaminant concentration of 0.3 mg/kg was achieved (Salkinoja-Salonen et al., 1989).

Slurry-phase bioremediation of contaminated soil in reactors maximizes biological degradation while minimizing abiotic losses (Mueller et al., 1991b). The potential benefits of slurry-phase treatment include better manageability and predictability of biodegradation than in in situ or solid-phase bioremediation. Mixing and intimate contact of microorganisms with pollutants and maintenance of optimum conditions (pH, dissolved oxygen, nutrients, substrate bioavailability, etc.) enhance biodegradation rates. The instances of slurry-phase CP remediation are limited (Mahaffey & Sanford, 1990; Compeau et al., 1991).

Compeau et al. (1990) reported a full-scale slurry-phase PCP remediation. The system consisted of soil washing and screening and resulted in clean soil and wash solution. The wash solution was a slurry containing PCP and <60-mesh-size soil particles at approximately 20% solids concentration. Slurry was treated subsequently in on-site slurry-phase bioreactors. A 50 m^3 slurry reactor was operated in batch mode and inoculated by an uncharacterized PCP-mineralizing culture (10^7 cells/ml of slurry). After 14 days, 370 mg PCP/kg slurry had been degraded to below 0.5 mg/kg. For effective biogradation to occur, inoculation was required.

In summary, landfarming is popular in North America. Another alternative for on-site soil remediation is windrow composting. In Finland, for example, it is almost solely used for soil CP treatment (Mikkola & Viitasaari, 1995). Bulking agents are used to improve aeration. Sometimes, the compost piles are equipped with drain pipes. In cold climates, the indigenous heat generation during composting gives it an advantage over other techniques.

8.5.1 Soil characteristics

Soil characteristics affect the rates and extent of CP degradation. Briglia et al. (1994) observed a higher degree of PCP mineralization by inoculated R. chlorophenolicus in sandy soil than in peaty soil. Soil type affected the initial PCP mineralization rate but not the total PCP removal by inoculated Flavobacterium sp. (Crawford & Mohn, 1985). Soil type also affects the nature of fungal transformation products. In different soils, PCP was converted to different extents into nonextractable soil-bound products by P. chrysosporium (Lamar et al., 1990b). In rich organic soil, 2,4-D degraded more slowly due to the stronger adsorption, lower pore water (soluble fraction) concentration, and lower rate of desorption (Greer & Shelton, 1992). In conclusion, soil properties affect bioremediation mostly through CP adsorption and reduced bioavailability.

8.5.2 Inhibition/toxicity

Soil CP concentration can be too high for a natural CP-degrading population to develop or for inoculated strains to function. With a *Flavobacterium* sp., an inverse relationship between PCP concentration and PCP mineralization rates was observed in soil. The bacterium did not degrade PCP at 500 mg/kg (Crawford & Mohn, 1985). Therefore, a waste dump soil containing 553 mg/kg PCP could not be decontaminated by *Flavobacterium* inoculation (Crawford & Mohn, 1985). On the other hand, sawdust-rich soil with 9000 mg/kg total CPs has been treated by indigenous microorganisms (Valo, 1990); the CP content decreased to 900 mg CP/kg in four years. An alternative treatment for highly contaminated soils could be soil washing at pH 8–9, followed by leachate treatment in separate bioreactors (Crawford & Mohn, 1985).

Studies have shown increasing lag phases in PCP degradation with increasing PCP concentration by mixed (Otte *et al.*, 1994) and pure cultures (Stanlake & Finn, 1982; Topp *et al.*, 1988; Gonzalez & Hu, 1991, 1995) and decreasing specific degradation rates with time (Topp *et al.*, 1988; O'Reilly & Crawford, 1989; Gonzalez & Hu, 1991). With a *Flavobacterium* sp., the lag was due to the initial loss of viable cells due to PCP toxicity (Topp *et al.*, 1988; Topp & Hanson, 1990a; Gonzalez & Hu, 1991, 1995). Gonzalez & Hu (1995) and Gu & Korus (1995) developed degradation models that accounted for the viable cell losses. The sensitivity of *Flavobacterium* sp. cells grown under carbon- or phosphate-limited conditions was higher than that of cells grown under ammonium or sulfate limitation (Topp & Hanson, 1990a). Therefore, in nutrient-deficient soil the numbers of indigenous CP degraders may remain low. The lag phase and decreased viable cell concentration were overcome by addition of glutamate or other readily metabolizable substrates (Topp *et al.*, 1988; Topp & Hanson, 1990a; Gonzalez & Hu, 1991).

Formation of CAs and soil-bound residues can be a major pathway for CPs in aerobic soil; higher levels of bound residues and PCA were observed in aerobic than in anaerobic soils (Murthy *et al.*, 1979; Schmitzer *et al.*, 1989). During controlled remediation, methylation of CP is insignificant. In *R. chlorophenolicus* inoculated soils, production of PCA was <3 mg/kg, which was less than 1% of PCP input (Briglia *et al.*, 1990, 1994). In *M. chlorophenolicum* amended compost ppb amounts of CPs were observed (Karlson *et al.*, 1995). Similarly, in soils amended with 120 mg/kg of total CPs, only 1 to 5 mg/kg CAs were detected, while 82 to 92% of CPs were removed (Valo & Salkinoja-Salonen, 1986). *o*-Methylating bacteria do not seem to compete effectively with PCP-mineralizing bacteria since the mineralization of PCP in soil by *R. chlorophenolicus* was similar both in the presence of equal amounts of *o*-methylating *R. rhodochrous* and in its absence, and the CA formation was low in both cases (Middeldorp *et al.*, 1990). During CP remediation by white-rot fungi, the formation of methylated phenols is of concern (Table 8.8).

8.5.3 Soil moisture

Soil moisture affects CP decontamination, although the direction is not always clear (Briglia *et al.*, 1994). In *Flavobacterium*-inoculated soil, PCP was removed most rapidly at soil water contents of 15 to 20% (Crawford & Mohn, 1985). Soil containing 50% water showed no PCP mineralization for the first 10 days but after that became very active; over an extended period the PCP mineralization was similar to that observed with drier soils. On the other hand, increasing soil water content from 30 to 60% enhanced PCP degradation in *Flavobacterium*-inoculated soil (Seech *et al.*, 1991). Soil saturation by water is not desirable and inhibits CP degradation by limiting oxygen diffusion (Harmsen, 1991; Briglia *et al.*, 1994). High moisture content increases the toxicity of CPs to degraders (Salkinoja-Salonen *et al.*, 1989).

8.5.4 Effect of inoculum

An indigenous CP-degrading microbial population may develop under suitable environmental conditions. For example, mixing with bark and nutrients helped to degrade the total CP concentration in an old sawmill soil from 200–300 mg/kg to 20–30 mg/kg during four summer months (Valo & Salkinoja-Salonen, 1986). In this case, the indigenous population was capable of more efficient mineralization than inoculated laboratory strains. In laboratory-scale experiments, the mineralization rate in this compost soil was higher than in sterilized soil amended with 10^9 rhodococci per gram of soil.

Without inoculation, the PCP (100 mg/kg) mineralization in soil was slower than it was with *Flavobacterium* inoculation (Crawford & Mohn, 1985). After ten days, PCP mineralization in uninoculated soil became significant; the same amount of PCP was eventually removed in both soils (Crawford & Mohn, 1985). In a waste dump soil, 320 mg/kg PCP was reduced to 41 mg/kg with and without inoculation by *Flavobacterium* (Crawford & Mohn, 1985).

Using multiple *Flavobacterium* inoculations the concentration of PCP decreased in landfill material from 298 mg/kg to 58 mg/kg in 100 days (Crawford & Mohn, 1985); no degradation occurred in uninoculated soils. Indigenous soil bacteria from the contaminated site did not degrade 100 μg/l PCP from PCP/PAH-contaminated groundwater during 14 days in shake-flask experiments, although extensive PAH degradation occurred (Mueller *et al.*, 1991c).

Increasing inoculum size initially increases CP degradation rates (Edgehill & Finn, 1983; Crawford & Mohn, 1985; Balfanz & Rehm, 1991; Edgehill, 1994). Crawford & Mohn (1985) observed increased PCP mineralization rates in soils receiving 3×10^3 *Flavobacterium* cells per gram of soil. Middeldorp *et al.* (1990) and Briglia *et al.* (1994) reported increased PCP mineralization with increasing inoculum size. Increasing *P. chrysosporium* inoculation from 5 to 10% increased PCP removal from 15 to 67% (Lamar *et al.*, 1993). The effective size for the inoculum depends on the soil type. In sandy soils an inoculum of 500 cells/g soil of *R.*

chlorophenolicus enhanced mineralization of PCP (Middeldorp *et al.*, 1990; Briglia *et al.*, 1994). In peaty soil, $8 \times 10^4 – 10^8$ cells/g were required to enhance PCP mineralization.

Calcium-alginate-immobilized *Flavobacterium* mineralized PCP in soil, whereas a suspended inoculum did not (Crawford *et al.*, 1989). Inoculating soil with a PCP-mineralizing culture was more successful with bark chips than inoculation with liquid culture (Salkinoja-Salonen *et al.*, 1989). Inoculation with polyurethane-immobilized cells resulted in improved mineralization (Karlson *et al.*, 1995). In full-scale composting, inoculation by 10^6 to 10^7 cells per gram of soil of polyurethane-foam-immobilized *M. chlorophenolicum* resulted in rapid CP removal and a decrease of 30 mg PCP/kg to 0.5 mg/kg within 30 weeks. In an uninoculated compost, a similar result was obtained within 50 weeks (Karlson *et al.*, 1995). *M. chlorophenolicum* was not detected by immunofluorescence microscopy after 2 to 3 years of full-scale composting when inoculated with suspension culture ($10^6 – 10^7$ colony forming units per gram) (Karlson *et al.*, 1995). These results indicate that immoblization enhances the efficiency of inoculation.

Long-term survival of inoculated microorganisms in soil is important since soil treatment may last for months or years. Good performance of isolates in laboratory liquid cultures does not necessarily indicate good survival and performance in soil (Briglia *et al.*, 1990, 1994; Lamar *et al.*, 1990a). Experiments with sterile soil may give an overly optimistic view of the survival of microbial inoculations (Topp & Hanson, 1990b; Jacobsen & Pedersen, 1992a). Inoculum size should be large enough to survive protozoan grazing or competition by indigenous bacteria for nutrients (Ramadan *et al.*, 1990). Both inoculum size and 2,4-D concentration increased the recovery of *P. cepacia* from soil (Jacobsen & Pedersen, 1992a). Optimal conditions for an introduced strain to utilize a given toxicant occurs when low numbers of indigenous degraders are present and high concentrations of toxicant substrate is available. When the toxicant concentration is low, mineralization with a small inoculum may be negligible (Jacobsen & Pedersen, 1992b). Enrichment of indigenous CP-degrading soil bacteria using a slurry reactor may enhance the survival of the inoculum after bioaugmentation (Otte *et al.*, 1994).

In summary, the primary advantage of inoculating CP-contaminated soil with high numbers of CP-degrading cells is usually an increased rate of decontamination as compared with uninoculated soils. The inoculum should be large enough to enhance biodegradation. Manipulation of natural CP-degrading microflora may be preferable to pure culture inoculations. In cold climate regions, due to the short summer it is preferable to obtain rapid CP removal by adding specialized microorganisms. Immobilization greatly improves inoculum performance.

Table 8.9. *Examples of chlorophenol degradation by sediment cultures*

Compound (μM)	Sediment origin	Terminal carbon and electron flow	Temperature (°C)	Metabolite(s)	Other remarks	References
2,4-DCP, (10–1000)	Marine	Sulfate reduction		4-CP, phenol		King (1988)
2,4-DCP (130), 4-CP (220–280)	Freshwater, Sandy Creek, GA	Sulfate reduction	19–40	4-CP	Sulfate amended	Kohring et al. (1989)
2-CP, 3-CP, 4-CP (500–900)	Freshwater, Escambia River, FL	Methanogenesis	30	Removal	Batch enrichment	Genthner et al. (1989)
2,4-DCP (100)	Intertidal estuarine, East River, NY	Sulfate reduction	30	4-CP	Sediment slurry	Häggblom & Young (1990)
2,6-DCP (100)	As above	As above	30	Phenol	Sediment slurry	Häggblom & Young (1990)
2-CP, 3-CP, 4-CP (100)	As above	As above	30	Removal	Sediment slurry	Häggblom & Young (1990)
2,4-DCP, 2,6-DCP (75)	Freshwater pond, Cherokee Trailer Park, GA	Methanogenesis	30	4-CP, 2-CP	p-Cresol and propionate enhanced	Häggblom et al. (1993)
2,4,6-TCP	Marine, Puget Sound, WA	Sulfate reduction	20	2,4-DCP	Semicontinuous feed	Puhakka et al. (submitted)
PCP (15)	Cherokee Trailer Pond, Clarke County			2,4-DCP, 3,4-DCP	Ethanol	Bryant et al. (1991)
PCP (7.5 and 37.5)	Lake sediment (Bagsvard), stream sediment (Molleaen)		50	3,4-DCP + higher chlorinated compounds	Ethanol	Larsen et al. (1991)

8.6 Potential for sediment remediation

The discharge of pulp bleaching effluents and contamination through use of wood preservatives, herbicides, and pesticides has polluted nearshore freshwater and marine environments with a variety of chlorophenolic compounds (Ahlborg & Thurnberg, 1980; Xie et al., 1986; Abrahamsson & Klick, 1989). The distribution, fate and persistence of chlorinated organic contaminants from pulp bleaching has been carefully reviewed by Neilson et al. (1991). In aquatic environments, CPs partition in the sediment (Neilson et al., 1991) and, therefore, may be mainly transformed by anaerobic biodegradative processes. Examples of reductive dehalogenations of CPs by sediment cultures are listed in Table 8.9. The potential of these transformations has been well characterized for freshwater sediment cultures, whereas much less is known about the CP transformations by marine sediment cultures. The extrapolation of laboratory results to in situ situations, however, is difficult and requires further research. Similarly, the methods for enhancement of desired reactions under sediment conditions may not be similar to those successfully used with laboratory cultures and enrichments.

In freshwater systems, methanogenesis is the main terminal path for carbon and electron flow in the degradation of organic matter. Other potential terminal electron-accepting processes include reduction of nitrate, sulfate and oxidized metal ions. These processes may compete with dehalogenations from reducing equivalents or some of these microorganisms equipped with electron transport phosphorylation mechanisms may be capable of catalyzing CP dehalogenations.

In marine systems, the presence of sulfate (20 to 30 mM) may regulate reductive dechlorination by two mechanisms. First, active sulfate reduction, as the main electron-accepting process in marine sediment (Capone & Kiene, 1988; Jorgensen & Fencher, 1974; Skyring, 1987), may maintain very low electron donor levels and prevent dechlorination (Genthner et al., 1989; Madsen & Aamand, 1992). Second, competition for reduction equivalents may be intracellular, as shown for the sulfate-reducing bacterium D. tiedjei (DeWeerd et al., 1991). Anaerobic freshwater sediment studies indicate that sulfate inhibits the development of cultures capable of aryl dechlorination (Gibson & Suflita, 1986, 1990; Genthner et al., 1989; Allard et al., 1992). On the other hand, coupling of sulfate, sulfite, and thiosulfate reduction with MCP degradation has been demonstrated (Häggblom & Young, 1990, 1995). Our recent results showed similar PCP, 2,4,6-TCP, and 3,4,5-TCC dechlorination by a marine sediment enrichment in the presence and absence of seawater concentrations of sulfate (27 mM) (Puhakka et al., submitted).

The studies with sediment cultures indicate natural degradation potential for aquatic sediments exposed to anthropogenic CP pollution. However, in situ remediation rates for CP-contaminated sediments may be difficult to enhance. Possibilities involve nutrient and electron donor/acceptor amendments. Ex situ remediation could involve sediment dredging and application of methods developed for soil decontamination, such as slurry reactors and composting.

8.7 Conclusion

Bioremediation aims at innocuous end-products. In aerobic systems, CP mineralization is frequently achieved. Anaerobic processes often produce less-chlorinated end-products, which are environmentally less harmful and easier to degrade by aerobic microorganisms than the starting compounds. The pathways of CP biodegradation are well characterized.

Temperature is a factor that often limits bioremediation. The first cold-temperature groundwater bioremediations are in progress. In soil treatment, heat generation during composting may overcome the temperature limitations. Bioavailability is another factor controlling bioremediation and strongly affects the residual concentrations achieved.

Chlorinated dimeric impurities such as PCDDs and PCDFs are frequently present in contaminated sites. These impurities are very toxic and recalcitrant towards biodegradation and, thus, make the completion of site decontamination very difficult.

On-site bioremediation has been successfully used for CP-contaminated groundwater and soil but not, so far, for sediments. On-site groundwater treatment helps in plume containment and reduces the amount of contaminants. The treatment requires long-term remediation due to the partitioning of contaminants to aquifer solids. *In situ* groundwater bioremediation of CPs is a developing technology with a lot of promise. Amendment of microencapsulated microorganisms into the aquifer is a promising method for enhancing *in situ* remediation.

In the remediation of CPs from soil, composting and landfarming are the most common technologies and may be accomplished with or without an external microbial inoculum. Unlike groundwater treatment, *in situ* remediation of unsaturated zones has not been carefully studied.

For control of bioremediation processes, reliable monitoring methods are needed. Molecular techniques offer useful tools for monitoring of on-site and *in situ* bioremediation.

References

Abrahamsson, K. & Klick, S. (1989). Distribution and fate of halogenated organic substances in an anoxic marine environment. *Chemosphere*, 18, 2247–56.

Acea, M. J. & Alexander, M. (1988). Growth and survival of bacteria introduced into carbon-amended soil. *Soil Biology and Biochemistry*, 20, 703–9.

Acea, M. J., Moore, C. R. & Alexander, M. (1988). Survival and growth of bacteria introduced into soil. *Soil Biology and Biochemistry*, 20, 509–15.

Ahlborg, U. G. & Thurnberg, T. M. (1980). Chlorinated phenols: occurrence, toxicity, metabolism, and environmental impact. *Critical Reviews in Toxicology*, 7, 1–35.

Allard, A-S., Remberger, M. & Neilson, A. H. (1987). Bacterial O-methylation of halogen substituted phenols. *Applied and Environmental Microbiology*, 53, 839–45.

Allard, A-S., Hynning, P-A., Remberger, M. & Neilson, A. H. (1992). Role of sulfate concentration in dechlorination of 3,4,5-trichlorocatechol by stable enrichment cultures grown with coumarin and flavanone glycones and aglycones. *Applied and Environmental Microbiology*, 58, 961–8.

Apajalahti, J. H. & Salkinoja-Salonen, M. S. (1984). Absorption of pentachlorophenol (PCP) by bark chips and its role in microbial PCP degradation. *Microbial Ecology*, 10, 359–67.

Armenante, P. M., Lewandowski, G. & Haq, I. U. (1992). Mineralization of 2-chlorophenol by *P. chrysosporium* using different reactor designs. *Hazardous Waste & Hazardous Materials*, 9, 213–29.

Bae, B., Autenrieth, R. L. & Bonner, J. S. (1995). Kinetics of multiple phenolic compounds degradation with a mixed culture in a continuous-flow reactor. *Water Environment Research*, 67, 215–23.

Baker, M. D., Mayfield, C. I. & Inniss, W. E. (1980). Degradation of chlorophenols in soil, sediment and water at low temperature. *Water Research*, 14, 1765–71.

Balfanz, J. & Rehm, H-J. (1991). Biodegradation of 4-chlorophenol by adsorptive immobilized *Alcaligenes* sp. A 7-2 in soil. *Applied Microbiology and Biotechnology*, 35, 662–8.

Ballapragada, B. S., Magar, V. S., Puhakka, J. A., Stensel, H. D. & Ferguson, J. F. (1994). Fate and biodegradation of tetrachloroethene and pentachlorophenol in anaerobic digesters. In *Proceedings of the Water Environment Federation 67th Annual Conference & Exposition*. Chicago, IL, October 16–20, pp. 265–76.

Beltrame, P., Beltrame, P. L., Carniti, P. & Pitea, D. (1982). Kinetics of biodegradation of mixtures containing 2,4-dichlorophenol in a continuous stirred reactor. *Water Research*, 16, 429–33.

Bollag, J. M. (1992). Decontaminating soil with enzymes. *Environmental Science and Technology*, 26, 1876–81.

Boyd, S. A. & Shelton, D. R. (1984). Anaerobic biodegradation of chlorophenols in fresh and acclimated sludge. *Applied and Environmental Microbiology*, 47, 272–7.

Briglia, M. (1995). Chlorophenol-degrading actinomycetes: molecular ecology and bioremediation properties. Ph.D. Thesis, Department of Applied Chemistry and Microbiology, University of Helsinki, Helsinki, Finland.

Briglia, M., Nurmiaho-Lassila, E-L., Vallini, G. & Salkinoja-Salonen, M. (1990). The survival of the pentachlorophenol-degrading *Rhodococcus chlorophenolicus* PCP-1 and *Flavobacterium* sp. in natural soil. *Biodegradation*, 1, 273–81.

Briglia, M., Middeldorp, P. J. M. & Salkinoja-Salonen, M. (1994). Mineralization performance of *Rhodococcus chlorophenolicus* strain PCP-1 in contaminated soil simulating on site conditions. *Soil Biology and Biochemistry*, 26, 377–85.

Brown, S. C., Grady, C. P. L., Jr & Tabak, H. H. (1990). Biodegradation kinetics of substituted phenolics: demonstration of a protocol based on electrolytic respirometry. *Water Research*, 24, 853–61.

Bryant, F. O., Hale, D. D. & Rogers, J. E. (1991). Regiospecific dechlorination of pentachlorophenol by dichlorophenol-adapted microorganisms in freshwater, anaerobic sediment slurries. *Applied and Environmental Microbiology*, 57, 2293–301.

Buswell, J. A. & Odier, E. (1987). Lignin degradation. *Critical Review in Biotechnology*, 6, 1–60.

Capone, D. G. & Kiene, R. P. (1988). Comparison of microbial dynamics in marine and freshwater sediments: contrasts in anaerobic carbon metabolism. *Limnology and Oceanography*, 33, 725–49.

Casida, L. E. Jr (1989). Protozoan response to the addition of bacterial predators and other bacteria to soil. *Applied and Environmental Microbiology*, 55, 1857–9.

Choi, J. & Aomine, S. (1972). Effects of the soil on the activity of pentachlorophenol. *Soil Science and Plant Nutrition*, 18, 255–60.

Choi, J. & Aomine, S. (1974a). Mechanisms of pentachlorophenol adsorption by soils. *Soil Science and Plant Nutrition*, 20, 371–9.

Choi, J. & Aomine, S. (1974b). Adsorption of pentachlorophenol by soils. *Soil Science and Plant Nutrition*, 20, 135–44.

Christodoulatos, C., Korfiatis, G. P., Talimcioglu, N. M. & Mohiuddin, M. (1994). Adsorption of pentachlorophenol by natural soils. *Journal of Environmental Science and Health*, A29, 883–98.

Chudoba, J., Albokova, J., Lentge, B. & Kümmel, R. (1989). Biodegradation of 2,4-dichlorophenol by activated sludge microorganisms. *Water Research*, 23, 1439–42.

Cirelli, D. P. (1978). Patterns of pentachlorophenol usage in the United States of America: an overview. In *Pentachlorophenol: Chemistry, Pharmacology and Environmental Toxicology*, ed. K. R. Rao, pp. 13–17. New York: Plenum Press.

Cole, J. R., Cascarelli, A. L., Mohn, W. W. & Tiedje, J. M. (1994). Isolation and characterization of a novel bacterium growing via reductive dehalogenation of 2-chlorophenol. *Applied and Environmental Microbiology*, 60, 3536–42.

Compeau, G. C., Mahaffey, W. D. & Patras, L. (1991). Full-scale bioremediation of contaminated soil and water. In *Environmental Biotechnology for Waste Treatment*, ed. G. S. Sayler, R. Fox & J. W. Blackburn, pp. 91–109. New York: Plenum Press.

Crawford, R. L. & Mohn, W. W. (1985). Microbial removal of pentachlorophenol from soil using a *Flavobacterium*. *Enzyme and Microbial Technology*, 7, 617–20.

Crawford, R. L., O'Reilly, K. T. & Tao, H-L. (1989). Microorganism stabilization for *in situ* degradation of toxic chemicals. In *Advances in Applied Biotechnology*, Vol. 4, ed. D. Kamely, A. Chakrabarty & G. S. Omenn, pp. 203–11. Houston: Gulf Publishing Company.

Dang, J. S., Harvey, M., Jobbagy, A. & Grady, C. P. L., Jr (1989). Evaluation of biodegradation kinetics with respirometric data. *Research Journal Water Pollution Control Federation*, 61, 1711–21.

Davis, A. P. & Huang, C. P. (1990). The removal of substituted phenols by a photocatalytic oxidation process with cadmium sulfide. *Water Research*, 24, 534–50.

DeWeerd, K. A., Concannon, F. & Suflita, J. M. (1991). Relationship between hydrogen consumption, dehalogenation, and the reduction of sulfur oxyanions by *D. tiedjei*. *Applied and Environmental Microbiology*, 57, 1929–34.

Dolfing, J. (1995). Letter to the editor: Regiospecificity of chlorophenol reductive dechlorination by Vitamin B_{12s}. *Applied and Environmental Microbiology*, 61, 2450–1.

Dolfing, J. & Harrison, B. K. (1992). Gibbs free energy of formation of halogenated aromatic compounds and their potential role as electron acceptors in anaerobic environments. *Environmental Science and Technology*, 26, 2213–18.

Edgehill, R. U. (1994). Pentachlorophenol removal from slightly acidic mineral salts, commercial sand, and clay soil by recovered *Arthrobacter* strain ATCC 33790. *Applied Microbiology and Biotechnology*, 41, 142–8.

Edgehill, R. U. & Finn, R. K. (1983). Microbial treatment of soil to remove pentachlorophenol. *Applied and Environmental Microbiology*, 45, 1122–5.

Ettala, M., Koskela, J. & Kiestilä, A. (1992). Removal of chlorophenols in a municipal treatment plant using activated sludge. *Water Research*, 26, 797–804.

Ford, S. F. & Olsen, B. (1988). Methods for detecting genetically engineered microorganisms in the environment. *Advances in Microbial Ecology*, 10, 45–79.

Gantzer, C. J. & Wackett, L. P. (1991). Reductive dechlorination catalyzed by bacterial transition-metal coenzymes. *Environmental Science and Technology*, 25, 715–22.

Gaudy, A. F., Jr, Lowe, W., Rozich, A. & Colvin, R. (1988). Practical methodology for predicting critical operating range of biological systems treating inhibitory substrates. *Journal Water Pollution Control Federation*, 60, 77–85.

Genthner, B. R. S., Price, W. A. & Pritchard, P. H. (1989). Anaerobic degradation of chloroaromatic compounds in aquatic sediments under a variety of enrichment conditions. *Applied and Environmental Microbiology*, 55, 1466–71.

Gibson, S. A. & Suflita, J. M. (1986). Extrapolation of biodegradation results to groundwater aquifers: reductive dehalogenation of aromatic compounds. *Applied and Environmental Microbiology*, 52, 681–8.

Gibson, S. A. & Suflita, J. M. (1990). Anaerobic biodegradation of 2,4,5-trichlorophenoxyacetic acid in samples from a methanogenic aquifer. Stimulation by short-chain organic acids and alcohols. *Applied and Environmental Microbiology*, 56, 1825–32.

Gold, M. H., Wariishi, H. & Valli, K. (1989). Extracellular peroxidases involved in lignin degradation by the white rot basidiomycete *Phaenerochaete chrysosporium*. *ACS Symposium Series*, 389, 127–40.

Gold, M. H., Joshi, D. K., Valli, K. & Wariishi, H. (1994). Degradation of chlorinated phenols and chlorinated dibenzo-*p*-dioxins by *Phanerochaete chrysosporium*. In *Bioremediation of Chlorinated and Polycyclic Aromatic Hydrocarbon Compounds*, ed. R. E. Hinchee, A. Leeson, L. Semprini & S. K. Ong, pp. 231–8. Boca Raton, FL: Lewis Publishers.

Gonzalez, J. F. & Hu, W-S. (1991). Effect of glutamate on the degradation of pentachlorophenol by *Flavobacterium* sp. *Applied Microbiology and Biotechnology*, 35, 100–4.

Gonzalez, J. F. & Hu, W-S. (1995). Pentachlorophenol biodegradation: simple models. *Environmental Technology*, 16, 287–93.

Grace Dearborn, Inc. (1994). *Daramend^{TM} Bioremediation of Soils Containing Chlorophenols and Polynuclear Aromatic Hydrocarbons (Full-scale Demonstration): Final Report*. Environmental/Engineering Group, Grace Dearborn Inc., Missisauga, Ontario.

Greer, L. E. & Shelton, D. R. (1992). Effect of inoculant strain and organic matter content on kinetics of 2,4-dichlorophenoxyacetic acid degradation in soil. *Applied and Environmental Microbiology*, 58, 1459–65.

Gu, Y. & Korus, R. A. (1995). Kinetics of pentachlorophenol degradation by a *Flavobacterium* species. *Applied Microbiology and Biotechnology*, 43, 374–8.

Häggblom, M. M. (1992). Microbial breakdown of halogenated aromatic pesticides and related compounds. *FEMS Microbiology Reviews*, 103, 29–72.

Häggblom, M. M. & Valo, R. J. (1995). Bioremediation of chlorophenol wastes. In *Microbiological Transformation of Toxic Organic Chemicals*, ed. L. Y. Young & C. Cerniglia, pp. 389–434. Wiley–Liss.

Häggblom, M. M. & Young, L. Y. (1990). Chlorophenol degradation coupled to sulfate reduction. *Applied and Environmental Microbiology*, 56, 3255–60.

Häggblom, M. M. & Young, L. Y. (1995). Anaerobic degradation of halogenated phenols by sulfate-reducing consortia. *Applied and Environmental Microbiology*, 61, 1546–50.

Häggblom, M. M., Janke, D., Middeldorp, P. J. M. & Salkinoja-Salonen, M. S. (1989). O-methylation of chlorinated phenols in the genus *Rhodococcus*. *Archives of Microbiology*, 152, 6–9.

Häggblom, M. M., Rivera, M. D. & Young, L. Y. (1993). Influence of alternative electron acceptors on the anaerobic biodegradability of chlorinated phenols and benzoic acids. *Applied and Environmental Microbiology*, 59, 1162–7.

Hammel, K. E. & Moen, M. A. (1991). Depolymerization of a synthetic lignin *in vitro* by lignin peroxidase. *Enzyme Microbial Technology*, 13, 15–18.

Harmsen, J. (1991). Possibilities and limitations of landfarming for cleaning contaminated soils. In *On-site Bioreclamation: Processes for Xenobiotic Hydrocarbon Treatment*, ed. R. E. Hinchee & R. F. Olfenbuttel, pp. 255–72. Stoneham, MA: Butterworth-Heinemann.

Harmsen, J. (1993). Managing bio-availability: an effective element in the improvement of biological soil-cleaning? In *Integrated Soil and Sediment Research: A Basis for Proper Protection*, ed. H. Eijsacker & T. Hamers, pp. 235–9. Dordrecht: Kluwer Academic Publishers.

Hendriksen, H. V. &·Ahring, B. K. (1993). Anaerobic dechlorination of pentachlorophenol in fixed-film and upflow anaerobic sludge blanket reactors using different inocula. *Biodegradation*, 3, 399–408.

Hendriksen, H. V., Larsen, S. & Ahring, B. K. (1991). Anaerobic degradation of PCP and phenol in fixed-film reactors: the influence of an additional substrate. *Water Science and Technology*, 24, 431–6.

Hendriksen, H. V., Larsen, S. & Ahring, B. K. (1992). Influence of a supplemental carbon source on anaerobic dechlorination of pentachlorophenol in granular sludge. *Applied and Environmental Microbiology*, 58, 365–70.

Holroyd, M. L. & Caunt, P. (1995). Large scale soil bioremediation using white rot fungi. In *Bioaugmentation for Site Remediation*, ed. R. E. Hinchee, J. Fredrickson & B. C. Alleman, pp. 181–7. Columbus, OH: Battelle Press.

Hu, Z-C., Korus, R. A., Levinson, W. E. & Crawford, R. L. (1994). Adsorption and biodegradation of pentachlorophenol by polyurethane-immobilized *Flavobacterium*. *Environmental Science and Technology*, 28, 491–6.

Huling, S. G., Pope, D. F., Matthews, J. E., Sims, J. L., Sims, R. L. & Sorenson, D. L. (1995). Wood preserving waste-contaminated soil: treatment and toxicity response. In *Bioremediation of Recalcitrant Organics*, ed. R. E. Hinchee, D. E. Hoeppel & D. B. Anderson, pp. 101–9. Columbus, OH: Battelle Press.

Humppi, T., Knuutinen, J. & Paasivirta, J. (1984). Analysis of polychlorinated phenoxyphenols in technical chlorophenol formulations and in sawmill environment. *Chemosphere*, 11, 1235–41.

Hwang, H-M., Hodson, R. E. & Lee, R. F. (1986). Degradation of phenol and chlorophenols by sunlight and microbes in estuarine water. *Environmental Science and Technology*, 20, 1002–7.

Isaacson, P. J. & Frink, C. R. (1984). Nonreversible sorption of phenolic compounds by sediment fractions: the role of sediment organic matter. *Environmental Science and Technology*, 18, 43–8.

Jackson, D. R. & Bisson, D. L. (1990). Mobility of polychlorinated aromatic compounds in soils contaminated with wood-preserving oil. *Journal of Air & Waste Management Association*, 40, 1129–33.

Jacobsen, C. S. & Pedersen, J. C. (1992a). Growth and survival of *Pseudomonas cepacia* DB01(pRO101) in soil amended with 2,4-dichlorophenoxyacetic acid. *Biodegradation*, 2, 245–52.

Jacobsen, C. S. & Pedersen, J. C. (1992b). Mineralization of 2,4-dichlorophenoxyacetic acid (2,4-D) in soil inoculated with *Pseudomonas cepacia* DB01(pRO101), *Alcaligenes eutrophus* AEO106(pRO101) and *Alcaligenes eutrophus* JMP134(pJP4): effects of inoculation level and substrate concentration. *Biodegradation*, 2, 253–63.

Jain, R. K., Burlage, R. S. & Sayler, G. S. (1988). Methods for detecting recombinant DNA in the environment. *CRC Critical Reviews in Biotechnology*, 8, 33–84.

Järvinen, K. T. & Puhakka, J. A. (1994). Bioremediation of chlorophenol contaminated groundwater. *Environmental Technology*, 15, 823–32.

Järvinen, K. T., Melin, E. S. & Puhakka, J. A. (1994). High-rate bioremediation of chlorophenol contaminated groundwater at low temperatures. *Environmental Science and Technology*, 28, 2387–92.

Johnson, R. L., Brillante, S. M., Isabelle, L. M., Houck, J. E. & Pankow, J. F. (1984). Migration of chlorophenolic compounds at the chemical waste disposal site at Alkali Lake, Oregon: 2. Contaminant distributions, transport, and retardation. *Ground Water*, 23, 652–66.

Jorgensen, B. B. & Fencher, T. (1974). The sulfur cycle of a marine sediment model system. *Marine Biology*, 24, 189–201.

Joshi, D. K. & Gold, M. H. (1993). Degradation of 2,4,5-trichlorophenol by the lignin-degrading basidiomycete *Phanerochaete chrysosporium*. *Applied and Environmental Microbiology*, 59, 1779–85.

Kang, G. & Stevens, D. K. (1994). Degradation of pentachlorophenol in bench scale bioreactors using the white rot fungus *Phanerochaete chrysosporium*. *Hazardous Waste & Hazardous Materials*, 11, 397–410.

Karlson, U., Miethling, R., Schu, K., Hansen, S. S. & Uotila, J. (1995). Biodegradation of PCP in soil. In *Bioremediation of Recalcitrant Organics*, ed. R. E. Hinchee, R. E. Hoeppel & D. B. Anderson, pp. 83–92. Columbus, OH: Battelle Press.

Keith, L. H. & Telliard, W. A. (1979). Priority pollutants I: a perspective view. *Environmental Science and Technology*, 13, 416–23.

Kennedy, M. S., Grammas, J. & Arbuckle, W. B. (1990). Parachlorophenol degradation using bioaugmentation. *Research Journal Water Pollution Control Federation*, 62, 227–33.

King, G. M. (1988). Dehalogenation in marine sediments containing natural sources of halophenols. *Applied and Environmental Microbiology*, 54, 3097–185.

Kirk, T. K. & Farrell, R. L. (1987). Enzymatic 'combustion': the microbial degradation of lignin. *Annual Reviews in Microbiology*, 41, 465–505.

Kitunen, V. H. & Salkinoja-Salonen, M. S. (1990). Soil contamination at abandoned sawmill areas. *Chemosphere*, 20, 1671–7.

Kitunen, V. H., Valo, R. J. & Salkinoja-Salonen, M. S. (1985). Analysis of chlorinated phenols, phenoxyphenols and dibenzofurans around wood preserving facilities. *International Journal of Environmental and Analytical Chemistry*, 20, 13–20.

Kitunen, V. H., Valo, R. J. & Salkinoja-Salonen, M. S. (1987). Contamination of soil around wood-preserving facilities by polychlorinated aromatic compounds. *Environmental Science and Technology*, 21, 96–101.

Klecka, G. M. & Maier, W. J. (1985). Kinetics of microbial growth on pentachlorophenol. *Applied and Environmental Microbiology*, 49, 46–53.

Klecka, G. M. & Maier, W. J. (1988). Kinetics of microbial growth on mixtures of pentachlorophenol and chlorinated aromatic compounds. *Biotechnology and Bioengineering*, 31, 328–35.

Knackmuss, H-J. & Hellwig, M. (1978). Utilization and cooxidation of chlorinated phenols by *Pseudomonas* sp. B13. *Archives of Microbiology*, 117, 1–7.

Knuutinen, J., Palm, H., Hakala, H., Haimi, J., Huhta, V. & Salminen, J. (1990). Polychlorinated phenols and their metabolites in soil and earthworms of sawmill environment. *Chemosphere*, 20, 609–23.

Kohring, G.-W., Zhang, X. & Wiegel, J. (1989). Anaerobic dechlorination of 2,4-dichlorophenol in freshwater lake sediments at different temperatures. *Applied and Environmental Microbiology*, 55, 348–53.

Korte, F. (1987). *Lehrbuch der Ökologishen Chemie*, 2nd edn. Stuttgart: George Thieme Verlag.

Krumme, M. L. & Boyd, S. A. (1988). Reductive dechlorination of chlorinated phenols in anaerobic upflow bioreactors. *Water Research*, 22, 171–7.

Kuhn, E. P. & Suflita, J. M. (1989). Dehalogenation of pesticides by anaerobic microorganisms in soils and groundwater: a review. In *Reactions and Movements of Organic Chemicals in Soils*, ed. B. L. Sawhney & K. Brown, pp. 111–80. Madison, WI: Soil Science Society of America and American Society of Agronomy.

Lafrance, P., Martineau, L., Perreault, L. & Villenueve, J-P. (1994). Effect of natural dissolved organic matter found in groundwater on soil adsorption and transport of pentachlorophenol. *Environmental Science and Technology*, 28, 2314–20.

Lagas, P. (1988). Sorption of chlorophenols in soil. *Chemosphere*, 17, 205–16.

Lamar, R. T. & Dietrich, D. M. (1990). In situ depletion of pentachlorophenol from contaminated soil by *Phanerochaete* spp. *Applied and Environmental Microbiology*, 56, 3093–100.

Lamar, R. T. & Glaser, J. A. (1994). Field evaluations of the remediation of soils contaminated with wood-preserving chemicals using lignin-degrading fungi. In *Bioremediation of Chlorinated and Polycyclic Aromatic Hydrocarbon Compounds*, ed. R. E. Hinchee, A. Leeson, L. Semprini & S. K. Ong, pp. 239–47. Boca Raton, FL: Lewis Publishers.

Lamar, R. T., Larsen, M. J. & Kirk, T. K. (1990a). Sensitivity to and degradation of pentachlorophenol by *Phanerochaete* spp. *Applied and Environmental Microbiology*, 56, 3519–26.

Lamar, R. T., Glaser, J. A. & Kirk, T. K. (1990b). Fate of pentachlorophenol (PCP) in sterile soils inoculated with the white-rot basidiomycete *Phanerochaete chrysosporium*: mineralization, volatilization and depletion of PCP. *Soil Biology and Biochemistry*, 4, 433–40.

Lamar, R. T., Evans, J. W. & Glaser, J. A. (1993). Solid-phase treatment of a pentachlorophenol-contaminated soil using lignin-degrading fungi. *Environmental Science and Technology*, 27, 2566–71.

Lampi, P., Vartiainen, T., Tuomisto, J. & Hesso, A. (1990). Population exposure to chlorophenols, dibenzo-*p*-dioxins and dibenzofurans after a prolonged ground water pollution by chlorophenols. *Chemosphere*, 20, 625–34.

Lampi, P., Hakulinen, T., Luostarinen, T., Pukkala, E. & Teppo, L. (1992a). Cancer incidence following chlorophenol exposure in a community in southern Finland. *Archives of Environmental Health*, 47, 167–75.

Lampi, P., Tolonen, K., Vartiainen, T. & Tuomisto, J. (1992b). Chlorophenols in lake bottom sediments: a retrospective study of drinking water contamination. *Chemosphere*, 24, 1805–24.

Larsen, S., Hendriksen, H. V. & Ahring, B. K. (1991). Potential for thermophilic (50 °C) anaerobic dechlorination of pentachlorophenol in different ecosystems. *Applied and Environmental Microbiology*, 57, 2085–90.

Leuenberger, C., Giger, W., Coney, R., Greydon, J. W. & Molnar-Kubica, E. (1985). Persistent chemicals in pulp mill effluents. Occurrence and behavior in an activated sludge plant. *Water Research*, 19, 885–94.

Lewandowski, G. A., Armenante, P. M. & Pak, D. (1990). Reactor design for hazardous waste treatment using a white rot fungus. *Water Research*, 24, 75–82.

Li, D-Y., Eberspächer, J., Wagner, B., Kuntzer, J. & Lingens, F. (1991). Degradation of 2,4,6-trichlorophenol by *Azotobacter* sp. strain GP1. *Applied and Environmental Microbiology*, 57, 1920–8.

Liang, R. & McFarland, M. J. (1994). Biodegradation of pentachlorophenol in soil amended with the white rot fungus *Phanerochaete chrysosporium*. *Hazardous Waste & Hazardous Materials*, 11, 411–21.

Lin, J-E., Wang, H. Y. & Hickey, R. F. (1990). Degradation kinetics of pentachlorophenol by *Phanerochaete chrysosporium*. *Biotechnology and Bioengineering*, 35, 1125–34.

Madsen, T. & Aamand, J. (1992). Anaerobic transformation and toxicity of trichlorophenols in a stable enrichment culture. *Applied and Environmental Microbiology*, 58, 557–61.

Madsen, T. & Licht, D. (1992). Isolation and characterization of an anaerobic chlorophenol transforming bacterium. *Applied and Environmental Microbiology*, 58, 2874–8.

Mahaffey, W. R. & Sanford, R. A. (1990). Bioremediation of pentachlorophenol contaminated soil: bench scale to full scale implementation. In *Gas, Oil, Coal, and Environmental Biotechnology II*, ed. C. Akin & J. Smith, pp. 117–43, Chicago: Institute of Gas Technology.

Mahaffey, W. R., Compeau, G., Nelson, M. & Kinsella, J. (1991). Developing strategies for PAH and TCE bioremediation. *Water, Environment and Technology*, 3(10), 83–8.

Mäkinen, P. M., Theno, T. J., Ferguson, J. F., Ongerth, J. E. & Puhakka, J. A. (1993). Chlorophenol toxicity removal and monitoring in aerobic treatment: recovery from process upsets. *Environmental Science and Technology*, 27, 1434–9.

McBain, A., Cui, F., Herbert, L. & Ruddick, J. N. R. (1995). The microbial degradation of chlorophenolic preservatives in spent, pressure-treated timber. *Biodegradation*, 6, 47–55.

Melcer, H. & Bedford, W. K. (1988). Removal of pentachlorophenol in municipal activated sludge process. *Journal Water Pollution Control Federation*, 60, 622–6.

Melin, E. S., Puhakka, J. A. & Shieh, W. K. (1993). Degradation of 4-chlorophenol in denitrifying fluidized-bed process. *Journal of Environmental Science and Health*, A28, 1801–11.

Metcalf & Eddy, Inc. (1991). *Wastewater Engineering: Treatment, Disposal and Reuse*. New York: McGraw-Hill. 1334 pp.

Middeldorp, P. J. M., Briglia, M. & Salkinoja-Salonen, M. (1990). Biodegradation of pentachlorophenol in natural soil by inoculated *Rhodococcus chlorophenolicus*. *Microbial Ecology*, 20, 123–39.

Mikesell, M. D. & Boyd, S. A. (1988). Enhancement of pentachlorophenol degradation in soil through induced anaerobiosis and bioaugmentation with anaerobic sewage sludge. *Environmental Science and Technology*, 22, 1411–14.

Mikkola, M. & Viitasaari, S. (1995). *Kloorifenolien Saastuttamien Saha-alueiden Kunnostuksen Nykytila*. Mimeograph Series of the National Board of Waters and the Environment no. 636, 63 pp., Helsinki: National Board of Waters and Environment. (*The Present Situation of Remediation at Sawmill Sites Contaminated by Chlorophenols*; in Finnish, abstract in English.)

Mileski, G. J., Bumpus, J. A., Jurek, M. A. & Aust, S. D. (1988). Biodegradation of pentachlorophenol by the white rot fungus *Phanerochaete chrysosporium*. *Applied and Environmental Microbiology*, 54, 2885–9.

Minard, R. D., Liu, S-Y. & Bollag, J-M. (1981). Oligomers and guinones from 2,4-dichlorophenol. *Journal of Agricultural and Food Chemistry*, 29, 250–2.

Mohn, W. W. & Kennedy, K. J. (1992). Reductive dehalogenation of chlorophenols by *Desulfomonile tiedjei* DCB-1. *Applied and Environmental Microbiology*, 58, 1367–70.

Mohn, W. W. & Tiedje, J. M. (1991). Evidence for chemiosmotic coupling of reductive dechlorination and ATP synthesis in *Desulfomonile tiedjei* DCB-1. *Archives of Microbiology*, 157, 1–6.

Mohn, W. W. & Tiedje, J. M. (1992). Microbial reductive dehalogenation. *Microbiology Reviews*, 56, 482–507.

Mueller, J. G., Lantz, S. E., Blattmann, B. O. & Chapman, P. J. (1991a). Bench-scale evaluation of alternative biological treatment processes for the remediation of pentachlorophenol- and creosote-contaminated materials: solid-phase bioremediation. *Environmental Science and Technology*, 25, 1045–55.

Mueller, J. G., Lantz, S. E., Blattmann, B. O. & Chapman, P. J. (1991b). Bench-scale evaluation of alternative biological treatment processes for the remediation of pentachlorophenol- and creosote-contaminated materials: slurry-phase bioremediation. *Environmental Science and Technology*, 25, 1055–61.

Mueller, J. G., Middaugh, D. P., Lantz, S. E. & Chapman, P. J. (1991c). Biodegradation of creosote and pentachlorophenol in contaminated groundwater: chemical and biological assessment. *Applied and Environmental Microbiology*, 57, 1277–85.

Mueller, J. G., Lantz, S. E., Ross, D., Colvin, R. J., Middaugh, D. P. & Pritchard, P. H. (1993). Strategy using bioreactors and specially selected microorganisms for bioremediation of groundwater contaminated with creosote and pentachlorophenol. *Environmental Science and Technology*, 27, 691–8.

Murthy, N. B. K., Kaufman, D. D. & Fries, G. F. (1979). Degradation of pentachlorophenol (PCP) in aerobic and anaerobic soil. *Journal of Environmental Science and Health*, B14, 1–14.

Namkoong, W., Loehr, R. C. & Malina, J. F., Jr (1989). Effects of mixture and acclimation on removal of phenolic compounds in soil. *Journal Water Pollution Control Federation*, 61, 242–50.

Neilson, A. H., Allard, A-S., Reiland, S., Remberger, M., Tärnholm, A., Victor, T. & Lendner, L. (1984). Tri- and tetrachloroveratrole, metabolites produced by bacterial O-methylation of tri- and tetrachloroguaiacol: an assessment of their bioconcentration potential and their effects on fish reproduction. *Canadian Journal of Fisheries and Aquatic Science*, 41, 1502–12.

Neilson, A. H., Allard, A-S., Hynning, P-A. & Remberger, M. (1991). Distribution, fate and persistence of organochlorine compounds formed during production of bleached pulp. *Toxicological and Environmental Chemistry*, 30, 3–41.

O'Reilly, K. T. & Crawford, R. L. (1989). Degradation of pentachlorophenol by polyurethane-immobilized *Flavobacterium* cells. *Applied and Environmental Microbiology*, 55, 2113–18.

Öberg, L. G., Glas, B., Swanson, S. E. & Rappe, K. G. (1990). Peroxidase-catalyzed oxidation of chlorophenols to polychlorinated dibenzo-*p*-dioxins and dibenzofurans. *Archives of Environmental Contamination and Toxicology*, 19, 930–8.

Ogram, A. V. & Sayler, G. S. (1988). The use of gene probes in the rapid analysis of natural microbial communities. *International Journal of Microbiology*, 3, 281–92.

Orser, C. S., Dutton, J., Lange, C., Jablonski, P., Xun, L. & Hargis, M. (1993a). Characterization of a *Flavobacterium* glutathione S-transferase gene involved in reductive dechlorination. *Journal of Bacteriology*, 175, 2640–4.

Orser, C. S., Lange, C. C., Xun, L., Zahrt, T. C. & Schneider, B. J. (1993b). Cloning, sequence analysis, and expression of the *Flavobacterium* pentachlorophenol 4-monooxygenase gene in *Escherichia coli*. *Journal of Bacteriology*, 175, 411–16.

Otte, M-P., Gagnon, J., Comeau, Y., Matte, N., Greer, C. W. & Samson, R. (1994). Activation of an indigenous microbial consortium for bioaugmentation of pentachlorophenol/creosote contaminated soils. *Applied Microbiology and Biotechnology*, 40, 926–32.

Otte, M-P., Greer, C. W., Comeau, Y. & Samson, R. (1995). Effect of soil on the stimulation of pentachlorophenol biodegradation. In *Bioremediation of Recalcitrant Organics*, ed. R. E. Hinchee, D. E. Hoeppel & D. B. Anderson, pp. 111–16. Columbus, OH: Battelle Press.

Palm, H., Knuutinen, J., Haimi, J., Salminen, J. & Huhta, V. (1991). Methylation products of chlorophenols, catechols and hydroquinones in soil and earthworms of sawmill environments. *Chemosphere*, 23, 263–7.

Parker, W. J., Bell, J. P. & Melcer, H. (1992). Modelling the fate of chlorinated phenols in wastewater treatment plants. In *Proceedings of the Water Environment Federation 65th Annual Conference and Exposition, New Orleans, Louisiana, September 20–24*. Alexandria, VA: Water Environment Federation.

Paukku, R. (1989). Kloorifenolit ja kloorianisolit ympäristössä. M.S. Thesis, University of Jyväskylä, Jyväskylä, Finland. (Chlorophenols and chloroanisols in the environment; in Finnish.)

Pereira, W. E., Rostad, C. E. & Sisak, M. E. (1985). Geochemical investigations of polychlorinated dibenzo-*p*-dioxins in the subsurface environment at an abandoned wood-treatment facility. *Environmental Toxicology and Chemistry*, 4, 629–39.

Perkins, P. S., Komisar, S. J., Puhakka, J. A. & Ferguson, J. F. (1994). Effects of electron donors and inhibitors on reductive dechlorination of 2,4,6-trichlorophenol. *Water Research*, 28, 2101–7.

Philbrook, D. M. & Grady, C. P. L., Jr (1985). Evaluation of biodegradation kinetics for priority pollutants. In *The Proceedings of 40th Industrial Waste Conference, Purdue University, West Lafayette, IN, May 14–16*, ed. J. M. Bell, pp. 795–804. Chelsea, MI: Lewis Publishers.

Pickup, R. W. (1991). Development of molecular methods for the detection of specific bacteria in the environment. *Journal of General Microbiology*, 137, 1009–19.

Pignatello, J. J., Johnson, L. K., Martinson, M. M., Carlson, R. E. & Crawford, R. L. (1986). Response of the microflora in outdoor experimental streams to pentachlorophenol: environmental factors. *Canadian Journal of Microbiology*, 32, 38–46.

Piotrowski, M. R., Doyle, J. R., Cosgriff, D. & Parsons, M. C. (1994). Bioremedial progress at the Libby, Montana, Superfund site. In *Applied Biotechnology for Site Remediation*, ed. R. E. Hinchee, D. B. Anderson, F. B. Metting, Jr & G. D. Sayles, pp. 240–55. Boca Raton, FL: Lewis Publishers.

Pitter, P. & Chudoba, J. (1990). *Biodegradability of Organic Substances in the Aquatic Environment*. Boston: CRC Press. 306 pp.

Puhakka, J. A. & Järvinen, K. T. (1992). Aerobic fluidized-bed treatment of polychlorinated phenolic wood preservative constituents. *Water Research*, 26, 765–70.

Puhakka, J. A., Melin, E. S., Järvinen, K. T., Koro, P. M., Rintala, J. A., Hartikainen, P., Nevalainen, I., Shieh, W. K. & Ferguson, J. F. (1995a). Fluidized-bed biofilms for chlorophenol mineralization. *Water Science and Technology*, 31(1), 227–35.

Puhakka, J. A., Herwig, R. P., Koro, P. M., Wolfe, G. V. & Ferguson, J. F. (1995b). Biodegradation of chlorophenols by mixed and pure cultures from a fluidized-bed reactor. *Applied Microbiology and Biotechnology*, 42, 951–7.

Puhakka, J. A., Lee, J. H., Herwig, R. P. & Ferguson, J. F. Reductive transformations of chlorophenolic compounds by anaerobic marine sediment cultures. *Systematic and Applied Microbiology* (submitted).

Ramadan, M. A., El-Tayeb, O. M. & Alexander, M. (1990). Inoculum size as a factor limiting success of inoculation for biodegradation. *Applied and Environmental Microbiology*, 56, 1392–6.

Rintala, J. A. & Puhakka, J. A. (1994). Anaerobic treatment in pulp- and paper-mill waste management: a review. *Bioresource Technology*, 47, 1–18.

Rintala, J., Järvinen, K., Hartikainen, P., Melin, E. & Puhakka, J. A. (1995). Kloorifenoleilla saastuneen pohjaveden biotekninen puhdistus: pilot-demonstraatio. *Vesitalous*, 36, 27–9. (Biochemical treatment of groundwater polluted by chlorophenols; in Finnish, abstract in English.)

Rochkind-Dubinsky, M. L., Sayler, G. S. & Blackburn, J. W. (1987). *Microbiological Decomposition of Chlorinated Aromatic Compounds*. New York: Marcel Dekker.

Ryding, J. M., Puhakka, J. A., Strand, S. E. & Ferguson, J. F. (1994). Degradation of chlorinated phenols by a toluene enriched microbial culture. *Water Research*, 28, 1897–906.

Salkinoja-Salonen, M., Valo, R., Apajalahti, J., Hakulinen, R., Silakoski, L. & Jaakkola, T. (1984). Biodegradation of chlorophenolic compounds in wastes from wood processing industry. In *Current Perspectives in Microbial Ecology*, ed. M. J. Klug & C. A. Reddy, pp. 668–76. Washington, D.C.: American Society for Microbiology.

Salkinoja-Salonen, M., Middeldorp, B., Briglia, M., Valo, R., Häggblom, M. & McBain, A. (1989). Cleanup of old industrial sites. In *Advances in Applied Biotechnology*, Vol. 4, ed. D. Kamely, A. Chakrabarty & G. S. Omenn, pp. 347–67. Houston: Gulf Publishing Company.

Schellenberg, K., Leuenberger, C. & Schwarzenbach, R. P. (1984). Sorption of chlorinated phenols by natural sediments and aquifer materials. *Environmental Science and Technology*, 18, 652–7.

Schmitzer, J., Bin, C., Scheunert, I. & Korte, F. (1989). Residues and metabolism of 2,4,6-trichlorophenol-^{14}C in soil. *Chemosphere*, 18, 1383–8.

Seech, A. G., Trevors, J. T. & Bulman, T. L. (1991). Biodegradation of pentachlorophenol in soil: the response to physical, chemical and biological treatments. *Canadian Journal of Microbiology*, 37, 440–4.

Seech, A. G., Marvan, I. J. & Trevors, J. T. (1994). On-site/ex situ bioremediation of industrial soils containing chlorinated phenols and polycyclic aromatic hydrocarbons. In *Bioremediation of Chlorinated and Polycyclic Aromatic Hydrocarbon Compounds*, ed. R. E. Hinchee, A. Leeson, L. Semprini & S. K. Ong, pp. 451–5. Boca Raton, FL: Lewis Publishers.

Skyring, G. W. (1987). Sulfate reduction in coastal ecosystems. *Geomicrobiology Journal*, 5, 295–374.

Smith, J. A. & Novak, J. T. (1987). Biodegradation of chlorinated phenols in subsurface soil. *Water, Air and Soil Pollution*, 33, 29–42.

Smith, M. H. & Woods, S. L. (1994). Regiospecificity of chlorophenol reductive dechlorination by Vitamin B_{12s}. *Applied and Environmental Microbiology*, 60, 4111–15.

Sofer, S. S., Lewandowski, G. A., Lodaya, M. P., Lakhwala, W. F. S., Yang, K. C. & Singh, M. (1990). Biodegradation of 2-chlorophenol using immobilized activated sludge. *Research Journal Water Pollution Control Federation*, 62, 73–80.

Spain, J. C. & Gibson, D. T. (1988). Oxidation of substituted phenols by *Pseudomonas putida* F1 and *Pseudomonas* sp. strain JS6. *Applied and Environmental Microbiology*, 54, 1399–404.

Spain, J. C., Zylstra, G. J., Blake, C. K. & Gibson, D. T. (1989). Monohydroxylation of phenol and 2,5-dichlorophenol by toluene dioxygenase in *Pseudomonas putida* P1. *Applied and Environmental Microbiology*, 55, 2648–52.

Stanlake, G. J. & Finn, R. K. (1982). Isolation and characterization of a pentachlorophenol-degrading bacterium. *Applied and Environmental Microbiology*, 44, 1421–7.

Stapleton, M. G., Sparks, D. L. & Dentel, S. K. (1994). Sorption of pentachlorophenol to HDTMA-clay as a function of ionic strength and pH. *Environmental Science and Technology*, 28, 2330–5.

Stinson, M. K., Skovronek, H. S. & Chresand, T. J. (1991). EPA site demonstration of BioTrol aqueous treatment system. *Journal of Air and Waste Management Association*, 41, 228–33.

Stormo, K. E. & Crawford, R. L. (1992). Preparation of encapsulated microbial cells for environmental applications. *Applied Environmental Microbiology*, 58, 727–30.

Stormo, K. E. & Crawford, R. L. (1994). Pentachlorophenol degradation by microencapsulated flavobacteria and their enhanced survival for in situ aquifer bioremediation. In *Applied Biotechnology for Site Remediation*, ed. R. E. Hinchee, D. B. Anderson, F. B. Metting & G. D. Sayles, pp. 422–7. Boca Raton, FL: Lewis Publishers.

Suzuki, T. (1983). Methylation and hydroxylation of pentachlorophenol by *Mycobacterium* sp. isolated from soil. *Journal of Pesticide Science*, 8, 419–28.

Svenson, A., Kjeller, L-O. & Rappe, C. (1989). Enzyme-mediated formation of 2,3,7,8-tetrasubstituted chlorinated dibenzodioxins and dibenzofurans. *Environmental Science and Technology*, 23, 900–2.

Tabak, H. H., Glaser, J. A., Strohofer, S., Kupferle, M. J., Scarpino, P. & Tabor, M. W. (1991). Characterization and optimization of treatment of organic wastes and toxic organic compounds by a lignolytic white rot fungus in bench-scale bioreactors. In *On-Site Bioreclamation: Processes for Xenobiotic Hydrocarbon Treatment*, ed. R. E. Hinchee & R. F. Olfenbuttel, pp. 341–65. Stoneham, MA: Butterworth-Heinemann.

Templeton, L. L. & Grady, C. P. L., Jr (1988). Effect of culture history on the determination of biodegradation kinetics by batch and fed-batch techniques. *Journal Water Pollution Control Federation*, 60, 651–8.

Tiedje, J. M., Boyd, S. A. & Fathepure, B. Z. (1987). Anaerobic degradation of chlorinated aromatic hydrocarbons. *Developments in Industrial Microbiology*, 27, 117–27.

Topp, E. & Hanson, R. S. (1990a). Degradation of pentachlorophenol by a *Flavobacterium* grown in continuous culture under various nutrient limitations. *Applied and Environmental Microbiology*, 56, 541–4.

Topp, E. & Hanson, R. S. (1990b). Factors influencing the survival and activity of a pentachlorophenol-degrading *Flavobacterium* sp. in soil slurries. *Canadian Journal of Soil Science*, 70, 83–91.

Topp, E., Crawford, R. D. & Hanson, R. S. (1988). Influence of readily metabolizable carbon on pentachlorophenol metabolism by a pentachlorophenol-degrading *Flavobacterium* sp. *Applied and Environmental Microbiology*, 54, 2452–9.

Trevors, J. T. (1982). Effect of temperature on the degradation of pentachlorophenol by *Pseudomonas* species. *Chemosphere*, 11, 471–5.

Trudell, M. R., Marowitch, J. M., Thomson, D. G., Fulton, C. W. & Hoffmann, R. E. (1994). In situ bioremediation at a wood-preserving site in cold, semi-arid climate: feasibility and field pilot design. In *Bioremediation of Chlorinated and Polycyclic Aromatic Hydrocarbon Compounds*, ed. R. E. Hinchee, A. Leeson, L. Semprini & S. K. Ong, pp. 99–116. Boca Raton, FL: Lewis Publishers.

U.S. Environmental Protection Agency (1993). EPA Updates: CERCLA priority list of hazardous substances. *Hazardous Waste Consulting*. May/June 2.26–2.30.

Utkin, I., Woese, C. & Wiegel, J. (1994). Isolation and characterization of *Desulfitobacterium dehalogenans* gen. nov., sp. nov., an anaerobic bacterium which reductively dechlorinates chlorophenolic compounds. *International Journal of Systematic Bacteriology*, 44, 612–19.

Valli, K. & Gold, M. H. (1991). Degradation of 2,4-dichlorophenol by the lignin-degrading fungus *Phanerochaete chrysosporium*. *Journal of Bacteriology*, 173, 345–52.

Valli, K., Wariishi, H. & Gold, M. H. (1992). Degradation of 2,7-dichlorodibenzo-*p*-dioxin by the lignin-degrading basidiomycete *Phanerochaete chrysosporium*. *Journal of Bacteriology*, 174, 2131–7.

Valo, R. (1990). Occurrence and metabolism of chlorophenolic wood preservative in the environment. Ph.D. thesis, University of Helsinki, Helsinki, Finland.

Valo, R. & Hakulinen, R. (1990). Bakteerit puhdistamaan kemiallisesti likaantunutta pohjavettä (Treatment of chlorophenol contaminated groundwater by bacteria). *Kemia-Kemi*, 17, 811–13 (in Finnish).

Valo, R. & Salkinoja-Salonen, M. (1986). Bioreclamation of chlorophenol-contaminated soil by composting. *Applied Microbiology and Biotechnology*, 25, 68–75.

Valo, R., Kitunen, V., Salkinoja-Salonen, M. S. & Räisänen, S. (1984). Chlorinated phenols as contaminants of soil and water in the vicinity of two Finnish sawmills. *Chemosphere*, 13, 835–44.

Valo, R., Apajalahti, J. & Salkinoja-Salonen. M. S. (1985). Studies on the physiology of microbial degradation of pentachlorophenol. *Applied Microbiology and Biotechnology*, 21, 313–9.

Valo, R. J., Häggblom, M. M. & Salkinoja-Salonen, M. S. (1990). Bioremediation of chlorophenol containing simulated groundwater by immobilized bacteria. *Water Research*, 24, 253–8.

Venkatadri, R., Tsai, S-P., Vukanic, N. & Hein, L. B. (1992). Use of biofilm membrane reactor for the production of lignin peroxidase and treatment of pentachlorophenol by fungus *Phanerochaete chrysosporium*. *Hazardous Waste & Hazardous Materials*, 9, 231–43.

Wan, M. T. (1992). Utility and railway right-of-way contaminants in British Columbia: chlorophenols. *Journal of Environmental Quality*, 21, 225–31.

Wang, X. & Ruckenstein, E. (1994). Immobilization of *Phanerochaete chrysosporium* on porous polyurethane particles with application to biodegradation of 2-chlorophenol. *Biotechnology Techniques*, 8, 339–44.

Wariishi, H., Valli, K. & Gold, M. H. (1991). In vitro depolymerization of lignin by manganese peroxidase of *Phanerochaete chrysosporium*. *Biochemical Biophysical Research Communications*, 176, 269–75.

Warith, M. A., Fernandes, L. & La Forge, F. (1993). Adsorption of pentachlorophenol on organic soil. *Hazardous Waste & Hazardous Materials*, 10, 13–25.

Watkin, A. T. & Eckenfelder, W. W., Jr (1984). Development of pollutant specific models for toxic organic compounds in the activated sludge process. *Water Science and Technology*, 17, 279–89.

Westall, J. C., Leuenberger, C. & Schwarzenbach, R. P. (1985). Influence of pH and ionic strength on the aqueous-nonaqueous distribution of chlorinated phenols. *Environmental Science and Technology*, 19, 193–8.

Woods, S. L., Ferguson, J. F. & Benjamin, M. M. (1989). Characterization of chlorophenol and chloromethoxybenzene biodegradation during anaerobic treatment. *Environmental Science and Technology*, 23, 62–8.

Woodward-Clyde Consultants & Champion International (1995). 1994 *Annual Operations Report for the Upper Aquifer*. Submitted to the U.S. Environmental Protection Agency Region VIII, Helena, Montana Office. February 28, 1995.

Wu, W-M., Bhatnagar, L. & Zeikus, G. (1993). Performance of anaerobic granules for degradation of pentachlorophenol. *Applied and Environmental Microbiology*, 59, 389–97.

Xie, T-M., Abrahamsson, K., Fogelqvist, G. & Josefsson, B. (1986). Distribution of chlorophenolics in a marine environment. *Environmental Science and Technology*, 20, 457–63.

Zhang, X. & Wiegel, J. (1990). Sequential anaerobic degradation of 2,4-dichlorophenol in freshwater sediments. *Applied and Environmental Microbiology*, 56, 1119–27.

9

Biodegradation of chlorinated aliphatic compounds

Lawrence P. Wackett

9.1 Chlorinated aliphatic compounds in the environment

9.1.1 Natural products and synthetic compounds

Chlorinated aliphatic compounds are abundant in nature. Among the more than 1500 natural product organohalides that have been identified (Gribble, 1992) are a significant number of chlorinated aliphatic compounds. Quantitatively, chloromethane is the most significant. It is estimated that 5×10^9 kg are produced annually, principally by soil fungi (Rasmussen, Khalil & Dalluge, 1980). The biochemical reaction producing chloromethane has been investigated (Wuosman & Hager, 1990) and soil bacteria have been identified that grow on chloromethane as a sole carbon and energy source (Hartmans et al., 1986; Traunecker, Preub & Dieckert, 1991). Hence, chloromethane is one of the almost innumerable intermediates in the global carbon cycle.

Chlorinated aliphatic compounds of industrial origin are perhaps more widely known. They include chloroalkanes, chloroalkenes, and chlorinated cycloaliphatic compounds (Figure 9.1). The principal usages of these compounds are as solvents and synthetic intermediates. 1,2-Dichloroethane is one of the most heavily used commodity chemicals, with 17.95 billion pounds produced in the United States in 1993 (Reisch, 1994). Industrial solvents such as dichloromethane, trichloroethylene (TCE), and tetrachloroethylene (PCE) have been valuable because of their relative chemical inertness, ease of evaporative transfer, and relatively low mammalian toxicity. Other prominent solvents are 1,1,1-trichloroethane, chloroform, and carbon tetrachloride. The latter two have seen greatly decreased usage due to demonstrated toxicity and carcinogenicity (Anders & Pohl, 1985). Chlorofluorocarbons used as refrigerants are predominantly C_1 and C_2 alkanes. Although they show excellent heat transfer and chemical inertness characteristics, their well-documented ozone-depleting effects are leading to drastic constraints on their use (Molina & Rowland, 1974). Given their chemical inertness, these molecules are projected to

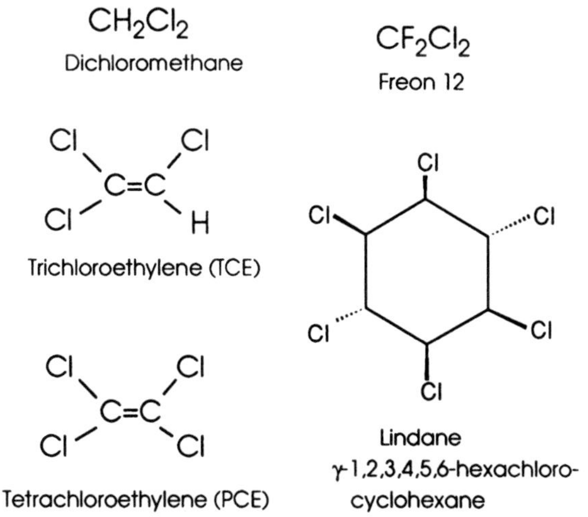

CH_2Cl_2
Dichloromethane

CF_2Cl_2
Freon 12

Trichloroethylene (TCE)

Lindane
γ-1,2,3,4,5,6-hexachloro-
cyclohexane

Tetrachloroethylene (PCE)

Figure 9.1. Chlorinated aliphatic compounds originating from chemical synthesis.

have long residence times in the environment. Cycloaliphatic compounds have historic interest as prominent pesticides. This class, including lindane and heptachlor, has also seen decreasing usage.

9.1.2 Distribution

Chlorinated aliphatic compounds are globally distributed. Natural product organohalides such as chloromethane and chloramphenicol are probably produced in diverse soil environments. Many halogenated organic compounds are biosynthesized by marine organisms (Neidleman & Geigert, 1986), so the oceans are a source as well.

The sites for production and usage of synthetic organohalides are more well known and can thus be determined more precisely. However, industrial organohalides are globally distributed as the result of atmospheric deposition. A 'cold-finger' effect serves to trap out environmentally persistent organohalides in cold regions of the earth. Thus, polar regions that have not been directly exposed to industry nevertheless contain chlorinated pesticides. In other cases, pollution from chlorinated aliphatic compounds has occurred directly from sources of usage or disposal. Chlorinated solvents such as trichloroethylene are the most common groundwater pollutants (Storck, 1987).

9.1.3 Toxicity

Chlorinated organic compounds are the largest single group of priority pollutants as classified by the United States Environmental Protection Agency (Leisinger, 1983). This designation emanates from the mammalian toxicity and carcinogenicity manifested by certain members of this group. It is impossible to generalize about the routes of toxicity for all chlorinated aliphatic compounds. However, toxicity and carcinogenicity are generally linked to metabolic activation of the compounds, as the example below illustrates.

Dichloromethane (CH_2Cl_2) is acutely toxic at high concentrations and is a carcinogen upon long-term exposure of mammals to lower concentrations (Anders & Pohl, 1985). Both effects are an outgrowth of metabolism that occurs principally in the livers of mammals (Figure 9.2). Cytochrome P450 monooxygenases hydroxylate CH_2Cl_2 to produce an unstable *gem*-chloroalcohol that spontaneously eliminates HCl to yield formyl chloride. Formyl chloride is a highly reactive intermediate that can alkylate cellular macromolecules including DNA. The resultant DNA lesions are thought to underlie the potential of CH_2Cl_2 to initiate tumor formation. Formyl chloride can undergo a second HCl elimination to yield carbon monoxide. Carbon monoxide binds to hemoglobin, the mammalian oxygen-carrier protein, and is only slowly displaced. Thus, rats exposed to over 400 ppm of CH_2Cl_2 in air have been reported to die of asphyxiation.

Chlorinated alkenes are also metabolized by liver cytochrome P450 monooxygenases, in this case yielding epoxides and chlorinated aldehydes (Henschler, 1985). Chlorinated epoxides are typically unstable and undergo isomerization and hydration. The resultant products, acyl chlorides and α-chloroketones, are potent alkylating agents. As an example, the carcinogenicity of vinyl chloride is thought to derive from these metabolic and spontaneous chemical reactions leading to DNA alkylation.

9.2 Challenges for microbial metabolism

9.2.1 Environmental persistence

The classification of many chlorinated organic compounds as EPA priority pollutants is based on their toxicity combined with environmental persistence.

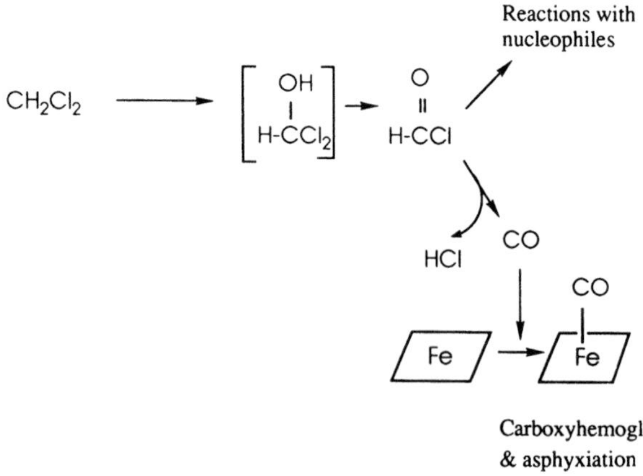

Figure 9.2. Metabolic activation of dichloromethane to bring about toxic and carcinogenic effects.

Longevity in the environment, in turn, is linked to chemical and metabolic inertness. Microorganisms are the primary agents for the cycling of most organic carbon on the earth. Many chlorinated compounds are poorly recycled, which was an important feature, in many cases, for their commercial application. For example, chlorinated pesticides and herbicides were made to persist in the environment so that repeated and costly applications would not be necessary. As a general rule, increasing the degree of chlorination increases environmental lifetime.

9.2.2 Chemistry of chlorinated aliphatic compounds

Chloroalkanes

Some general trends of chemical reactivity can be described, and these considerations are important in understanding the biodegradation of chloroalkanes. As an example, consider trends in reactivity for the chloromethane series: CH_3Cl, CH_2Cl_2, $CHCl_3$ and CCl_4. The reactivity of the series members is dependent on the specific mechanism of C–Cl bond cleavage. For nucleophilic substitution mechanisms, the order of reactivity decreases with increasing chlorine substitution: $CH_3Cl > CH_2Cl_2 > CHCl_3 > CCl_4$. Practically, CH_3Cl is reasonably reactive with nucleophiles; CH_2Cl_2 and more highly chlorinated methanes are not. Despite this, CH_2Cl_2 does undergo an enzyme-catalyzed nucleophilic displacement reaction in the environment, as discussed in a subsequent section. CH_3Cl may undergo significant spontaneous hydrolysis to yield CH_3OH, while CH_2Cl_2, $CHCl_3$, and CCl_4 are functionally stable in water.

In the presence of reactive reducing agents, chloromethanes can undergo a two-electron reduction reaction yielding chloride and a C–H bond in place of a C–Cl bond. In this type of reaction, the order of reactivity is: $CCl_4 > HCCl_3 > H_2CCl_2 > H_3CCl$. This shows up clearly in the reactions of reduced cobalamins with chloromethanes where reactivity decreases by approximately one order of magnitude with each decrease in chlorine substituent. In the context of biodegradation, CCl_4 is more likely to undergo biotransformation in reduced anaerobic ecosystems than in aerobic environments.

Chloromethanes may undergo C–Cl bond cleavage following oxygenation of a C–H bond to yield a *gem*-chloromethanol which undergoes spontaneous dechlorination:

$$R_2ClCOH \rightarrow HCl + R_2C=O.$$

The biological reactivity of chloromethane C–H bonds to oxygenation is more likely governed by steric factors in enzyme active sites than by large differences in C–H bond reactivity for the series.

Elimination reactions of chloroaliphatic compounds are important only for chloroform, $CHCl_3$, in the chloromethane series (Hine, Dowell & Singley, 1956). Chloroform undergoes deprotonation in base and the resultant anion undergoes *gem*-elimination, yielding a carbene. Dichlorocarbene is very reactive; for example, it rapidly hydrolyzes in water as shown (Figure 9.3A). With chloroalkanes containing two or more carbon atoms, β-elimination reactions figure prominently in both biological and abiotic transformation routes. *Beta*-elimination of halide substituents occurs during reductive dechlorination (as shown in Figure 9.3B) or by a deprotonation and elimination sequence, known as an E1cB mechanism in organic chemistry. This latter mechanism occurs abiotically at an appreciable rate with pentachloroethane (Roberts *et al.*, 1992) and enzymatically with hexa-chlorocyclohexane (Imai *et al.*, 1991).

Lastly, chloroalkanes undergo photochemical homolytic C–Cl bond cleavage to generate radical products. Carbon-centered radicals are generally stabilized by chlorine substituents. For example, $\cdot CCl_3$ is more stable than $\cdot CH_3$. These reactions are most important in the context of atmospheric chemistry. Volatile chlorinated methanes, if not degraded on land or in water, may undergo radical-based dechlorination in the lower atmosphere. Chlorofluorocarbons are sufficiently stable to reach the stratosphere, where ultraviolet light induces C–Cl bond cleavage generating $Cl\cdot$, which initiates an ozone-depleting chain reaction (Molina & Rowland, 1974). Elucidation of this fundamental chemistry has led to worldwide accords to phase out the commercial usage of chlorofluorocarbons.

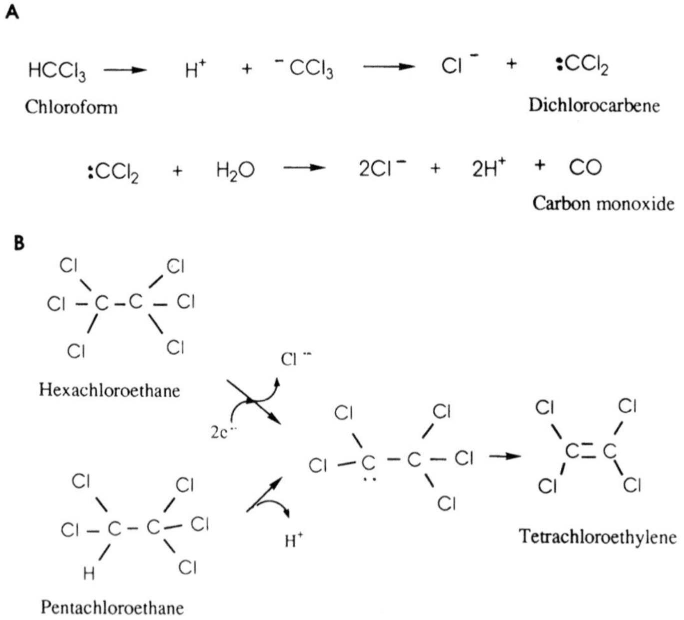

Figure 9.3. Elimination reactions of importance in the transformation of chlorinated aliphatic compounds. (A) *gem*-elimination; (B) β-elimination.

Chloroalkenes

The chemical reactivity, and hence the biological fate, of chloroalkenes differs markedly from that of chloroalkanes. In general, they are more stable than corresponding chloroalkanes. Chloroalkenes undergo direct vinylic nucleophilic substitution only under stringent conditions. They undergo reductive dehalogenation more sluggishly than do corresponding chloroalkanes (Gantzer & Wackett, 1991). Chloroalkenes do not readily eliminate halide substituents to yield alkynes. This unreactivity of chloroalkenes such as trichloroethylene and tetrachloroethylene underlies their usage as industrial solvents and degreasers.

Chloroalkenes, however, are potentially reactive to addition reactions and oxygenation. Some of these reactions are not particularly facile; addition of water to the double bond of trichloroethylene is extremely slow, for example. Enzymes have been identified that are suggested to add water to 3-chloroacrylic acid (van Hylckama Vliey & Janssen, 1992), indicating that biological systems use this chemistry with at least some chlorinated olefins. In the last seven years, numerous reports have described the biological oxygenation of chloroalkene solvents and vinyl chloride (Nelson, Montgomery & Pritchard, 1988; Wackett & Gibson, 1988; Oldenhuis *et al.*, 1989; Tsien *et al.*, 1989; Ewers, Freier-Schröder & Knackmuss, 1990; Harker & Kim, 1990; Winter, Yen & Ensley, 1989). Previously, it had been known that bacterial oxygenases, such as soluble methane monooxygenase from methanotrophs, oxidize ethylene to ethylene oxide (Colby, Stirling & Dalton, 1977). Ethylenes substituted with increasing numbers of chlorines will theoretically show decreasing reactivity with reactive electrophilic oxygen species. As an example, increasing chlorination of chloroalkenes leads to a decrease in the rate of reaction with ozone (Nikl *et al.*, 1982).

Do bacterial oxygenases oxidize chloroalkanes at rates consistent with the general chemical reactivity as determined by the number of chlorine substituents? For those cases in which data are available, the answer is no. An excellent example is the oxidation of a series of chloroethylenes by purified soluble methane monooxygenase (Fox *et al.*, 1990). Ethylene, vinyl chloride, the isomeric dichloroethylenes, and trichloroethylene are oxidized at rates (both V_{max} and V_{max}/K_M) that are independent of the degree of chlorination. Tetrachloroethylene is the most unreactive compound in this series and it is not oxidized by soluble methane monooxygenase. However, this is probably due to steric and not electronic effects of the chlorine substituents, since chlorotrifluoroethylene is oxidized by soluble methane monooxygenase (Fox *et al.*, 1990). It should not be surprising that an enzyme capable of oxidizing inert substances like methane has sufficient catalytic potential to attack the highly electron-deficient pi-electron cloud of perhalogenated alkenes.

9.3 Bioremediation

Chlorinated aliphatic compounds are prevalent soil and groundwater contaminants and, thus, are increasingly becoming targets for bioremediation. Traditionally, waters have been cleansed of volatile chlorinated aliphatic compounds by air stripping. This approach, which transfers the contaminant(s) from water to air, is increasingly being constrained by regulatory agencies. Adsorption onto activated carbon has been used, but this process is costly. Against this backdrop, bioremediation can provide an alternative treatment regime.

9.3.1 General principles

For bioremediation of chlorinated aliphatic compounds to succeed, microorganisms must be identified that are capable of metabolizing these compounds. Twenty years ago, most chlorinated compounds were considered to be nonbiodegradable. Recently, many microbial successes in metabolizing chlorinated compounds have been uncovered. Four general enzymatic mechanisms are enlisted to cleave carbon–chloride bonds of organohalides (Wackett *et al.*, 1992) and these mechanisms can serve to funnel the carbon atoms into central metabolic pathways. The biological mechanism(s) operative with a specific chlorinated compound is predicated on the chemical reactivity of the substance and the ability of evolutionary forces to produce enzyme catalysts capable of exploiting that unique reactivity.

An understanding of the mechanisms involved in organohalide metabolism is directly relevant to the design of bioremediation systems. For example, one of the first design decisions is to choose between aerobic or anaerobic biotreatment systems. This decision is dependent on the chlorinated compound(s) to be treated. A partially chlorinated alkene would likely be treated best aerobically, while perchlorinated alkanes and alkenes would be treated anaerobically. If the chlorinated compound to be degraded is not a carbon and energy source for the metabolizing organism, alternative growth-sustaining compounds must be present at the site or supplemented as part of the bioremediation protocol. Chlorinated aliphatic compounds can be toxic to bacteria, either by their solvent effect, which disrupts biological membranes, or by metabolic activation that generates toxic intermediates which react with cellular macromolecules, or by both effects. A knowledge of the organism(s) and the reaction mechanisms involved in halocarbon metabolism is crucial to maximize the bioremediation potential of any system.

9.3.2 Commercial developments

Interest in the bioremediation of chlorinated solvents is widespread. It is impractical to consider all the excellent efforts underway. An example will be given for both a chlorinated alkane and a chlorinated alkene. Both compounds, dichloromethane and trichloroethylene, are commonly used industrial solvents and degreasers that are important soil and water pollutants.

Dichloromethane or methylene chloride is an Environmental Protection Agency priority pollutant. Its mode of human toxicity and carcinogenicity is described above. Dichloromethane has a high vapor pressure and a relatively high aqueous solubility of 150 mM (Dean, 1985), properties that contribute to its distribution in the environment. Biological treatment schemes have been devised to address air pollution from factory off-gases and aqueous industrial waste streams. These are discussed separately below.

Aerobic and anaerobic CH_2Cl_2-metabolizing bacteria have been identified and the aerobic organisms have been most widely exploited for bioremediation. The aerobic bacteria useful for both air and water treatment are facultative methylotrophic bacteria that can metabolize dichloromethane as the sole source of carbon and energy. Several reviews describe the relevant microbial taxonomy and physiology of this interesting group of bacteria (Gälli, 1986; Leisinger et al., 1994). Fundamentally, CH_2Cl_2 is dechlorinated by dichloromethane dehalogenase, yielding formaldehyde (Stucki et al., 1981; Kohler-Staub & Leisinger, 1985). This is thought to be the rate-determining reaction in CH_2Cl_2 metabolism (Wackett et al., 1992). Formaldehyde is oxidized to CO_2, generating electrons to drive ATP synthesis, and assimilated into cell carbon. In batch culture, dichloromethane concentrations above 10 mM are toxic to the bacteria, likely due to solvent-mediated membrane disruption.

Water treatment systems use fluidized-bed reactors and a pump-and-treat design. A description of such reactors has been published (Gälli & Leisinger, 1985) and used commercially by Celgene Corporation (Summit, New Jersey, USA). The former system was a 4.3 liter fluidized-bed reactor with methylotrophic bacteria immobilized on sand particles of 0.25–0.40 mm. This pilot-scale system was run unsterilized, simulating actual field conditions. At time zero, the entire bacterial population in the reactor consisted of methylotrophic bacteria, which in laboratory fermenters had exhibited a maximum CH_2Cl_2 degradation rate of 0.86 g/l/h. Over time, two of the original three methylotrophic bacteria were shed from the reactor and the nonmethylotrophic bacteria, which were presumably unreactive with CH_2Cl_2, accounted for 80% of the biomass. Yet, the CH_2Cl_2 degradation rate went up to 1.6 g/l/h by day 35 and remained constant thereafter. The inlet CH_2Cl_2 concentration into the reactor was 120 ± 10 mM and the effluent concentration was <0.01 mM. These data showed the potential efficacy of CH_2Cl_2-degrading methylotrophs for treating aqueous CH_2Cl_2-containing waste streams.

Dichloromethane is vented from factories to protect workers, necessitating treatment schemes for gas-phase wastes. A bioremediation scheme based on methylotrophic bacteria has been described (Diks & Ottengraf, 1991) and implemented in the Netherlands. The design resembles a trickling-filter bed used in conventional wastewater treatment. The bed contains CH_2Cl_2-metabolizing bacteria. Dichloromethane-saturated air is forced through the bed with a compressor. Desiccation of the bed is prevented by a sprinkler system that moistens the bed. The continued presence of dichloromethane maintains selective pressure to keep the active bacterial population in the system.

Trichloroethylene (TCE) is the most prevalent organic contaminant of drinking water supplies in the United States. Air stripping, the once conventional treatment method, is now largely disallowed. Extensive efforts have been made to develop bacterial systems to remove TCE from water. Efforts to obtain bacteria that metabolize TCE as a carbon and energy source have been unsuccessful. One report of bacterial growth on TCE is not well documented and the report has not been confirmed five years after the publication date (Vandenbergh & Kunka, 1988). In this context, studies have focused on using bacteria that express broad-specificity oxygenases that fortuitously oxidize TCE. This necessitates differences from treatment strategies in which the contaminant is used by the bacteria for carbon and energy.

Different bacterial metabolic groups, and thus different oxygenases, have been discovered to oxidize TCE. The oxygenases are biosynthesized for the purpose of oxidizing aromatic hydrocarbons, gaseous alkanes, or ammonia. When the oxygenase is expressed, the respective organisms oxidize TCE fortuitously. In general, hydrocarbon-oxidizing oxygenases have broad specificity that extends to non-growth substrates (Leadbetter & Foster, 1960). While not all bacterial catabolic oxygenases oxidize TCE (Wackett *et al.*, 1989), a surprising number show some activity with this substrate. Different organisms, and their oxygenases, oxidize TCE at widely different rates, from 0.1 nmol/min. to 100 nmol/min. per mg of protein (Fox *et al.*, 1990). The highest rates, under conditions of substrate saturation (V_{max}), are exhibited by methanotrophic bacteria that express soluble methane monooxygenase (sMMO). This has focused attention on the use of methanotrophic bacteria for treatment of TCE-contaminated water. Other TCE-oxidizing bacteria have also been considered. As discussed below, the oxidation rate at V_{max} conditions is only one of the considerations for implementing a particular organism in biotreatment.

A comparison of desirable characteristics for a TCE-oxidizing bacteria can be illustrated using *Methylosinus trichosporium* OB3b containing sMMO and *Pseudomonas cepacia* G4 containing toluene 2-monooxygenase. Toluene 2-monooxygenase is thought to oxidize toluene, phenol and TCE. *M. trichosporium* OB3b and other methanotrophs expressing sMMO oxidize TCE more rapidly than other bacteria, under idealized conditions in the laboratory. Another kinetic parameter of some importance is the K_s, the *in vivo* equivalent of the K_M, the Michaelis constant, used in a steady-state kinetic treatment of enzymes. The K_s for TCE with *P. cepacia* G4 is 3 μM (Folsom, Chapman & Pritchard, 1990), 50-fold lower than the K_s with *M. trichosporium* (Brusseau *et al.*, 1990). In a biological treatment, with an effective available TCE concentration of 2 μM (0.25 ppm), both systems would be operating at similar rates despite the much higher V_{max} for the methanotroph. Another important factor is the hardiness of the bacteria. TCE oxidation produces reactive intermediates that can harm the organism and diminish the activity of the oxygenase that metabolizes TCE. This has been demonstrated *in vivo* for *P. putida* F1 containing toluene dioxygenase (Wackett & Householder, 1989) and both *in vivo* (Alvarez-Cohen & McCarty, 1991) and *in vitro* (Fox *et al.*, 1990) for methanotrophs and sMMO, respectively. With *P. cepacia* G4 in the laboratory, it has been observed that TCE oxidation rates do not diminish significantly over time. This needs further investigation but current data suggest that the mechanism of

TCE oxidation by *P. cepacia* G4 might generate a lesser amount of reactive products, thus preventing marked cytotoxicity. This phenomenon has important implications for TCE bioremediation.

In a shallow aquifer, the oxidation of TCE and other chloroalkenes was compared under conditions of phenol or methane amendments (Hopkins, Semprini & McCarty, 1993). The objective was to stimulate the growth and metabolism of bacteria comparable to *P. cepacia* G4 and *M. trichosporium* OB3b, respectively. It is possible, however, that the methane-stimulated bacteria expressed particulate (membrane-bound) methane monoxygenase, which is known to oxidize TCE much more slowly than sMMO (DiSpirito *et al.*, 1992). Under the conditions used, the phenol stimulation led to greater removal of TCE and dichloroethylenes. Ammonia addition, thought to stimulate ammonia-oxidizing bacteria that oxidize TCE (Vanelli *et al.*, 1990), showed the lowest extent of TCE oxidation in groundwater microcosm tests (Hopkins *et al.*, 1993).

Acknowledgements

Research on biodegradation of chlorinated aliphatic compounds has been supported in my laboratory by research grants from the National Institutes of Health (GM 41235) and the Environmental Protection Agency (CR820771-01).

References

Alvarez-Cohen, L. & McCarty, P. L. (1991). Product toxicity and cometabolic competitive inhibition modeling of chloroform and trichloroethylene transformation by methanotrophic resting cells. *Applied and Environmental Microbiology*, 57, 1031–7.

Anders, M. W. & Pohl, L. R. (1985). Halogenated alkanes. In *Bioactivation of Foreign Compounds*, ed. M. W. Anders, pp. 283–315. New York: Academic Press.

Brusseau, G. A., Tsien, H. C., Hanson, R. S. & Wackett, L. P. (1990). Optimization of trichloroethylene oxidation by methanotrophs and the use of a colorimetric assay to detect soluble methane monooxygenase activity. *Biodegradation*, 1, 19–29.

Colby, J., Stirling, D. I. & Dalton, H. (1977). The soluble methane monooxygenase of *Methylococcus capsulatus* (Bath): Its ability to oxygenate n-alkanes, n-alkenes, ethers and alicyclic, aromatic and heterocyclic compounds. *Biochemical Journal*, 165, 395–402.

Dean, J. A. (1985). Lange's Handbook of Chemistry, 13th edn. New York: McGraw-Hill.

Diks, R. M., Ottengraf, S. P. (1991). Verification studies of a simplified model for the removal of dichloromethane from waste gases using a biological trickling filter (Part I). *Bioprocess Engineering*, 6, 93–9.

DiSpirito, A. A., Gulledge, J., Schiemke, A. K., Murrell, J. C., Lidstrom, M. E. & Crema, C. L. (1992). Trichloroethylene oxidation by the membrane-associated methane monooxygenase in Type I, Type II, and Type X methanotrophs. *Biodegraduation*, 2, 151–64.

Ewers, J., Freier-Schröder, D. & Knackmuss, H-J. (1990). Selection of trichloroethene (TCE) degrading bacteria that resist inactivation by TCE. *Archives of Microbiology*, 154, 410–13.

Folsom, B. R., Chapman, P. J. & Pritchard, P. H. (1990). Phenol and trichloroethylene degradation by *Pseudomonas cepacia* G4: kinetics and interactions between substrates. *Applied and Environmental Microbiology*, 56, 1279–85.

Fox, B. G., Borneman, J. G., Wackett, L. P. & Lipscomb, J. D. (1990). Haloalkane oxidation by the soluble methane monooxygenase from *Methylosinus trichosporium* OB3b: mechanistic and environmental implications. *Biochemistry*, 29, 6419–27.

Gälli, R. (1986). Ph.D. dissertation, Swiss Federal Institute of Technology (ETH) Zürich.

Gälli, R. & Leisinger, T. (1985). Specialized bacterial strains for the removal of dichloromethane from industrial waste. *Conservation & Recycling*, 8, 91–100.

Gantzer, C. J. & Wackett, L. P. (1991). Reductive dehalogenation catalyzed by bacterial transition-metal coenzymes. *Environmental Science & Technology*, 25, 715–22.

Gribble, G. W. (1992). Naturally occurring organohalogen compounds: a survey. *Journal of Natural Products*, 55, 1353–95.

Harker, A. R. & Kim, Y. (1990). Trichloroethylene degradation by two independent aromatic-degrading pathways in *Alcaligenes eutrophus* JMP134. *Applied and Environmental Microbiology*, 56, 1179–81.

Hartmans, S., Schmuckle, A., Cook, A. M. & Leisinger, T. (1986). Methyl chloride: naturally occurring toxicant and C-1 growth substrate. *Journal of General Microbiology*, 132, 1139–42.

Henschler, D. (1985). Halogenated alkenes and alkynes. In *Bioactivation of Foreign Compounds*, ed. M. W. Anders, pp. 317–47. New York: Academic Press.

Hine, J., Dowell, A. M. & Singley, J. E. (1956). Carbon dihalides as intermediates in the basic hydrolysis of haloforms. IV Relative reactivities of haloforms. *Journal of the American Chemical Society*, 78, 479–82.

Hopkins, G. D., Semprini, L. & McCarty, P. L. (1993). Microcosm and in situ field studies of enhanced biotransformation of trichloroethylene by phenol-utilizing microorganisms. *Applied and Environmental Microbiology*, 59, 2277–85.

Imai, R. Y., Nagata, Y., Fukuda, M., Takagi, M. & Yano, K. (1991). Molecular cloning of a Pseudomonas paucimobilis gene encoding a 17-kilodalton polypeptide that eliminates HCl molecules from γ-hexachlorocyclohexane. *Journal of Bacteriology*, 173, 6811–19.

Kohler-Staub, D. & Leisinger, T. (1985). Dichloromethane dehalogenase of *Hyphomicrobium* sp. strain DM2. *Journal of Bacteriology*, 162, 676–81.

Leadbetter, E. R. & Foster, J. W. (1960). Bacterial oxidation of gaseous alkanes. *Archiv für Mikrobiologie*, 35, 92–104.

Leisinger, T. (1983). Microorganisms and xenobiotic compounds. *Experientia*, 29, 1183–91.

Leisinger, T., Bader, R., Hermann, R., Schmid-Appert, M. & Vuilleumier, S. (1994). Microbes, enzymes, and genes involved in dichloromethane utilization. *Biodegradation*, 5, 237–48.

Molina, M. J. & Rowland, F. S. (1974). Straptospheric sink for chlorofluoromethanes: chloride atom-catalyzed destruction of ozone. *Nature*, 249, 810–12.

Neidleman, S. L. & Geigert, J. (1986). *Biohalogenation: Principles, Basic Roles and Applications.* Chichester, England: Ellis Horwood.

Nelson, M. J., Montgomery, S. O. & Pritchard, P. H. (1988). Trichloroethylene metabolism by microorganisms that degrade aromatic compounds. *Applied and Environmental Microbiology*, 54, 604–6.

Nikl, H., Maker, P. D., Savage, C. M., Breitenbach, L. P., Martinez, R. I. & Herron, J. T. (1982). A Fourier transform infrared study of the gas-phase reactions of O_3 with chloroethylenes. *Journal of Physical Chemistry*, 86, 1858–61.

Oldenhuis, R., Vink, R. L., Vink, J. M., Janssen, D. B. & Witholt, B. (1989). Degradation of chlorinated aliphatic hydrocarbons by Methylosinus trichosporium OB3b expressing soluble methane monooxygenase. *Applied and Environmental Microbiology*, 55, 2819–26.

Rasmussen, R. A., Khalil, M. A. R. & Dalluge, R. W. (1980). Concentration and distribution of methyl chloride in the atmosphere. *Journal of Geophysical Research*, 85, 7350–6.

Reisch, M. S. (1994). Top 50 chemicals production rose modestly last year. *Chemical and Engineering News*, 72(15), 12–16.

Roberts, A. L., Sanborn, P. N. & Gschwend, P. M. (1992). Nucleophilic substitution reactions of dihalomethanes with the bisulfide ion HS⁻. *Environmental Science & Technology*, 26, 2263–74.

Storck, W. (1987). Chlorinated solvent use hurt by federal rules. *Chemical Engineering News*, 65, 11.

Stucki, G., Gälli, R., Ebersold, H. R. & Leisinger, T. (1981). Dehalogenation of dichloromethane by cell extracts of *Hyphomicrobium* DM2. *Archives of Microbiology*, 130, 366–71.

Traunecker, J., Preub, A. & Diekert, G. (1991). Isolation and characterization of a methyl chloride utilizing strictly anaerobic bacterium. *Archives of Microbiology*, 156, 416–21.

Tsien, H.-C., Brusseau, G. A., Hanson, R. S. & Wackett, L. P. (1989). Biodegradation of trichloroethylene by *Methylosinus trichosporium* OB3b. *Applied and Environmental Microbiology*, 55, 3155–61.

van Hylckama Vliey, J. E. T. & Janssen, D. B. (1992). Bacterial degradation of 3-chloroacrylic acid and the characterization of *cis*- and *trans*-specific dehalogenases. *Biodegradation*, 2, 139–50.

Vandenbergh, P. A. & Kunka, B. S. (1988). Metabolism of volatile chlorinated aliphatic hydrocarbons by *Pseudomonas fluorescens*. *Applied and Environmental Microbiology*, 54, 2578–9.

Vanelli, T., Logan, M., Arciero, D. M. & Hooper, A. B. (1990). Degradation of halogenated aliphatic compounds by the ammonia-oxidizing bacterium *Nitrosomonas europeae*. *Applied and Environmental Microbiology*, 56, 1169–71.

Wackett, L. P. & Gibson, D. T. (1988). Degradation of trichloroethylene by toluene dioxygenase in whole cell studies with *Pseudomonas putida* F1. *Applied and Environmental Microbiology*, 54, 1703–8.

Wackett, L. P. & Householder, S. R. (1989). Toxicity of trichloroethylene to *Pseudomonas putida* F1 is mediated by toluene dioxygenase. *Applied and Environmental Microbiology*, 55, 2723–5.

Wackett, L. P., Brusseau, G. A., Householder, S. R. & Hanson, R. S. (1989). Survey of microbial oxygenases: Trichloroethylene degradation by propane-oxidizing bacteria. *Applied and Environmental Microbiology*, 55, 2960–4.

Wackett, L. P., Logan, M. S. P., Blocki, F. A. & Bao-li, C. (1992). A mechanistic perspective on bacterial metabolism of chlorinated methanes. *Biodegradation*, 3, 19–36.

Winter, R. B., Yen, K-M. & Ensley, B. D. (1989). Efficient degradation of trichloroethylene by a recombinant *Escherichia coli*. *Biotechnology*, 7, 282–5.

Wuosman, A. M. & Hager, L. P. (1990). Methyl chloride transferase: a carbocation route for the synthesis of halometabolites. *Science*, 249, 160–2.

10

Microbial remediation of metals

T. M. Roane, I. L. Pepper and R. M. Miller

10.1 Metals in the environment

Metal pollution is a widespread problem; in fact, in industrially developed countries it is normal to find elevated levels of metal ions in the environment. In addition, it has been estimated that approximately 37% of sites in the U.S. contaminated with organic pollutants, such as pesticides, are additionally polluted with metals (Kovalick, 1991). Despite this, biological treatment or bioremediation of contaminated sites has largely focused on the removal of organic compounds, and only recently has attention turned to the treatment of metal-contaminated wastes (Brierley, 1990; Summers, 1992). Due to their toxic nature, the presence of metals in organic-contaminated sites often complicates and limits the bioremediation process. Such metals include the highly toxic cations of mercury and lead, but many other metals are also of concern, including arsenic, beryllium, boron, cadmium, chromium, copper, nickel, manganese, selenium, silver, tin and zinc.

Metals are ubiquitous in nature and even those metals generally considered as pollutants are found in trace concentrations in the environment (see Table 10.1). For the most part, metal pollution problems arise when human activity either disrupts normal biogeochemical cycles or concentrates metals (see Table 10.2). Examples of such activities include mining and ore refinement, nuclear processing, and industrial manufacture of a variety of products including batteries, metal alloys, electrical components, paints, preservatives, and insecticides (Gadd, 1986; Hughes & Poole, 1989; Suzuki, Fukagawa & Takama, 1992). Metals in these products or metal wastes from manufacturing processes can exist as individual metals or more often as metal mixtures. Unfortunately, past waste disposal practices associated with mining and manufacturing activities have been such that air, soil, and water contamination was common, and as a result there are many metal-contaminated sites that pose serious health risks.

A case in point is the Bunker Hill Mining Company located near Kellogg, Idaho.

Table 10.1. *Typical background levels of heavy metals in soil and aquatic environments*

Metal	Aquatic[a] concentration (μg/l)	Soil[b] concentration (μg/g)
Gold (Au)	ND[c]	0.50
Aluminium (Al)	Trace[d]	7.09×10^5
Arsenic (As)	Trace	0.49
Barium (Ba)	ND	4.34×10^3
Cadmium (Cd)	0.06	0.60
Cobalt (Co)	0.70	79.00
Chromium (Cr)	Trace	990.00
Cesium (Cs)	Trace	59.40
Copper (Cu)	0.63	296.00
Mercury (Hg)	Trace	0.29
Manganese (Mn)	10.00	6043.00
Nickel (Ni)	ND	397.00
Lead (Pb)	0.06	99.00
Tin (Sn)	Trace	101.00
Zinc (Zn)	19.62	496.00

[a]*Source:* Goldman & Horne (1983); Leppard (1981); Sigg (1985). [b]*Source:* Lindsay (1979).
[c]ND = no data reported. [d]trace = levels usually below detection.

In 1972, the Environmental Protection Agency (EPA) ordered the Company to limit dumping of its smelter wastes. At this time, typical smelter releases were up to approximately 27 215 kg lead per $1.6 \, km^2$ on surrounding communities within a six month period. As a result, there was little or no vegetation within 1.6 km of the smelter and, according to a 1974 study conducted by the Idaho Department of Health and Welfare, approximately 40% of the children tested living in the Kellogg area had abnormally high blood lead levels (Mink, Williams & Wallace, 1971; Keely *et al.*, 1976; von Lindern, 1981).

Elevated metal concentrations in the environment have wide-ranging impacts on animal, plant, and microbial species. For example, human exposure to a variety of metals causes symptoms such as hypophosphatemia, heart disease and liver damage, cancer, neurological disorders, central nervous system damage, encephalopathy and paresthesia (Carson, Ellis & McCann, 1986; Hammond & Foulkes, 1986). Exposure to metals is the cause of most morphological and mutational changes observed in plants (Brooks, 1983). These include shortening of roots, leaf scorch, chlorosis, nutrient deficiency and increased vulnerability to insect attack (Canney *et al.*, 1979; Carlisle *et al.*, 1986). Likewise, microbial growth is often slowed or inhibited completely in the presence of excessive amounts of metals (Duxbury, 1981; Baath, 1989).

Some plant and microbial species have developed unique and sometimes high tolerance for metals. Plant species of *Agrostis*, *Minuartia*, and *Silene* are known for their tolerance to heavy metals (Sieghardt, 1990; Verkleij *et al.*, 1991). In a study

Table 10.2. *Typical metal concentrations in contaminated environments*

Location	Source	Metal	Concentration	Reference
River Godavari, India	Paper plant	Zn Mn	233.41 µg/l 157.91 µg/l	Sudhakar, Jyothi & Venkateswarlu (1991)
Otago Harbor sediment, New Zealand	Tannery effluent	Cr	7000 µg/g	Johnson, Flower & Loutit (1981)
California coast sediment	Sewage, industrial effluents	Cr	1317 µg/g	Mearns & Young (1977)
Tennessee Valley (wastestream)	Leaking coal slurry dike	Fe Mn	6900 µg/l 9300 µg/l	Douglas (1992)
Clark Fork River sediment, Montana	Nonpoint sources	As	100 µg/g	Moore, Ficklin & Johns (1988)
Ninemile Creek sediment, NY	Chemical effluents	Hg	5.45 µg/g	Barkay & Olson (1986)
Ley Creek sediment, NY	Chemical effluents	Hg	4.25 µg/g	Barkay & Olson (1986)
Quebec, Canada (soil)	Sewage sludge	Cu Pb Zn	879.50 µg/g 193.5 µg/g 762.5 µg/g	Couillard & Zhu (1992)
Silver Valley, Idaho (soil)	Mining activity	Cu Mn Pb	197.64 µg/g 1.28×10^4 µg/g 2.2×10^4 µg/g	Roane (1994)

by Sieghardt (1990), leaves of *Minuartia verna* were able to accumulate, on average, 2900 mg zinc/kg. *Silene vulgaris*, on the other hand, accumulated up to 1000 mg lead/kg in root tissue. Similarly, microorganisms have developed a variety of strategies to deal with high metal concentrations in the environment. These include binding of metals to the cell surface or cell wall, translocation of the metal into the cell, and metal transformations including precipitation and volatilization (Hughes & Poole, 1989; Francis, 1990). It is the observation of these resistance phenomena in plants and microorganisms that has helped lead to the development of biologically based metal-remediation strategies.

Perhaps the most difficult inherent problem in metal remediation is that metals, although they may be released in the breakdown of a metal-containing compound, are not degradable in the same sense as carbon-based molecules. The metal atom is not the major building block for new cellular components, and while a significant amount of carbon is released to the atmosphere as CO_2, the metal atom is often not volatilized. Both incorporation into cell mass and volatilization facilitate carbon removal from environmental systems. In contrast, metals, unless removed completely from a system through intervention, will persist indefinitely.

10.2 Physical and chemical remediation of metal-contaminated sites

The following sections review the traditional physico-chemical approaches that have been used to remediate metal-contaminated sites and wastestreams. It is the complications and cost involved in use of such traditional remediation technologies that has led to interest in, and development of, alternative remediation approaches.

10.2.1 Soils

Traditional metal remediation technologies have typically involved physical removal by excavation and transport of contaminated soils to hazardous waste landfills (see Figure 10.2). This is thought by some to be the best method for metal reclamation (Gieger, Federer & Sticher, 1993). However, the increasing cost of excavation and transport, and shrinking available landfill space, make alternative options attractive. Two alternative strategies for remediation of metal-contaminated sites are immobilization and metal removal by soil washing, or pump and treat. Both of these strategies use or depend on pH to influence metal solubility.

Immobilization

Metal solubility in soil generally decreases with increasing pH. Thus, pH can be used to immobilize metals, effectively making them less toxic and preventing their movement into uncontaminated areas. As pH and cationic exchange capacity is increased, the electrostatic attraction between metals and soil constituents, including soil particles and organic matter, is enhanced (Sposito, 1989). As a result, basic soils are characterized by precipitated calcic and phosphoric metal-containing minerals. Similarly, in some situations, liming can be used to increase soil pH, causing precipitation of soluble contaminating metals. Addition of organic matter can be used to aid this process by acting as a sorbent for free metal in the soil matrix (see Section 10.3.2).

Metal removal

Metal removal can be achieved by *in situ* or *ex situ* soil washing, which is used primarily for surface soil contamination, or by pump-and-treat technology, which is used at sites with deep soil or aquifer contamination. Removal by soil washing or pump and treat is often difficult and time-consuming because soils and sediments sorb metals so strongly because of interactions with colloids such as humics and negatively charged clays. Therefore, approaches have been developed to enhance removal during the washing/flushing process (see Figure 10.2).

Soil washing with acidic solutions is one way to facilitate metal removal. Under acidic conditions, sorbed metals are released as the increase in hydrogen ions causes competition for available phosphate and results in formation of phosphoric acid. The released metals have increased solubility and thus are more easily removed from the system during treatment. For example, Tuin & Tels (1991) describe the use of concentrated acid or phosphate solutions to wash metal-contaminated soil.

The metals in wash effluents are then extracted by complexation with added resins. A second approach to recovery of metals is to use a flocculant to separate the metals from soil particles following an acid wash. The metals are then concentrated and recovered by sodium hydroxide precipitation.

As an alternative to acid washing, soils can also be flushed with chelating agents. Examples of effective chelating agents include ethylenediaminetetraacetic acid (EDTA) and nitrilotriacetic acid (NTA), both of which readily bind and solubilize metals. Using this approach, Peters & Shem (1992) have recently reported on the removal of lead from a contaminated soil. In this study, 0.1M EDTA removed 60% of the lead in a soil containing 10 000 mg lead/kg.

10.2.2 Sediments

Metal reclamation of sediments uses many of the same approaches as for soils, except that sediment access is often more difficult. Once removed from the bottom of a lake or river, sediments can be treated and replaced, or landfilled in a hazardous waste containment site. The actual removal of sediments involves dredging. This can pose serious problems since dredging includes the excavation of sediments from benthic anaerobic conditions to more atmospheric oxidizing conditions. This can result in increased solubilization of metals, along with increased bioavailability (see Section 10.3) and potential toxicity, and increased risk of contaminant spreading (Moore, Ficklin & Johns, 1988; Jorgensen, 1989; Moore, 1994). There are ongoing discussions as to whether it is more detrimental to remove sediments, whether for treatment or removal, or simply to leave them in place.

10.2.3 Aquatic systems

Metal removal from surface water, groundwater or wastewater streams is more straightforward than that from soils. Typically, removal is achieved by concentration of the metal within the wastestream using flocculation, complexation, and/or precipitation. For example, the use of lime or caustic soda will cause the precipitation and flocculation of metals as metal hydroxides. Alternatively, ion exchange, reverse osmosis, and electrochemical recovery of metals can be used for metal removal (Chalkley et al., 1989; Moore, 1994). Unfortunately, these techniques can be expensive, time-consuming and sometimes ineffective, depending on the metal contaminant present.

10.3 Metal speciation and bioavailability

An important aspect of metal–microbe interactions, but one that is rarely addressed, is metal speciation and metal bioavailability. It is the metal species present and their relative bioavailability rather than total metal concentration in the environment that determines the overall physiological and toxic effects on biological systems (Bernhard, Brinckman & Sadler, 1986; Hughes & Poole, 1989; Morrison, Batley & Florence, 1989).

10.3.1 Metal speciation

The speciation of a metal in any environment is a result of the combined effects of pH, redox potential, and, to a somewhat lesser extent, ionic strength (Sposito, 1989; Alloway, 1990; Brierley, 1990; Moore, 1994). At high pH, metals are predominantly found as insoluble metal mineral phosphates and carbonates while at low pH they are more commonly found as free ionic species or as soluble organometals.

Besides pH, redox potential also influences speciation. While the redox potential of an environment is determined primarily by environmental factors, microorganisms and their metabolic activities play essential roles in establishing redox potential as well. Redox potential is established by oxidation–reduction reactions in the environment, reactions that are relatively slow, particularly in soils. Reduced (anaerobic) conditions (negative E_h) found in saturated soils often result in metal precipitation due, in part, to the presence of carbonates, ferrous (Fe^{2+}) ions, and microbial reduction of sulfate to sulfides by sulfate-reducing bacteria such as *Desulfovibrio*. Under these conditions metals may combine with such sulfides (S^{2-}) to form nontoxic, insoluble sulfide deposits. Under oxidizing conditions (positive E_h), metals are more likely to exist in their free ionic form and exhibit increased water solubility. In addition, pH may decrease slightly or dramatically under oxidizing conditions, the classical example being acid mine drainage, where sulfur is oxidized by *Thiobacillus thiooxidans* to sulfuric acid, resulting in pH values ranging as low as 2 or less.

The susceptibility of metals to changes in pH and redox demonstrates how important it is to define clearly the chemical parameters of metal-contaminated systems. System pH, redox potential, and ionic strength will strongly influence the success of remediation. As an aid to prediction of metal speciation in the environment, there are several geochemical equilibrium speciation modeling programs which may be used in laboratory and natural settings, such as MINTEQA2 (EPA, Washington, DC) and PHREEQE (National Water Research Unit, Ontario, Canada). These models predict metal distribution between the dissolved, adsorbed and solid phases under a variety of environmental conditions.

10.3.2 Bioavailability

In contrast to metal speciation, metal bioavailability is determined by the solubility of metal species present and the sorption of metal species by solid surfaces including soil minerals, organic matter, and colloidal materials (Babich & Stotzky, 1980, 1985; Bernhard et al., 1986; Morrison et al., 1989; Alloway, 1990). For the purpose of this chapter, bioavailability is defined as metals in solution that are not bound to solid phase particles. Organic matter is a significant source of metal complexation, especially in soils (Stevenson, 1982; Alloway, 1990). Living organisms, organic debris and humus sorb metals, reducing metal solubility and bioavailability. Organic matter consists of humic and nonhumic material. Nonhumic substances include amino acids, carbohydrates, organic acids, fats and waxes which have not been chemically altered from their original biological form (Alloway, 1990). Humics comprise high molecular weight compounds altered from their original structure. Anionic functional groups on such compounds, including carboxyl, carbonyl, phenolic, hydroxyl and ester groups, bind cationic metals, sequestering metal activity. Some organic complexing agents form soluble complexes with metals while others form insoluble complexes. For instance, amino acids, simple aliphatic acids and other microbially produced agents can form soluble chelates. The formation of these soluble structures is of concern since such structures are prone to transport and may result in contaminant spreading. Insoluble complexes form when metals bind to high molecular weight humic materials. In this case, toxic metal concentrations may be reduced to nontoxic levels.

Metal bioavailability is generally increased with decreasing pH. This is due to the presence of phosphoric, sulfuric and carbonic acids, which increasingly solubilize organic- and particulate-bound metals. Particulate-bound metals are considered those bound to secondary minerals, for example, clays, iron and aluminum oxides, carbonates and sulfidic and phosphoric minerals. Due to the heterogeneous nature of soils and sediments, wide fluctuations in pH can exist in a given environment. For instance, metals may be more soluble in surface layers where plant exudates, microbial activity, moisture and leaching lower pH.

The toxicity of a metal to a biological system is dependent upon metal bioavailability. One problem with metal remediation is that metals in the environment are subject to chemically and biologically mediated oxidation–reduction reactions, which can alter both metal speciation and bioavailability. Generally, the most dangerous inorganic form of a metal is the metal cation (McLean & Beveridge, 1990), a metal species that has greater solubility with decreasing pH. As the solubility of the metal increases, metal toxicity increases due to enhanced mobility and bioavailability to biological systems. In general, metals are more soluble when oxidized than reduced and are thus more likely to exist as free ionic species. For example, trivalent chromium, Cr(III), is highly insoluble and poses little threat to the health of an environment. Environmental oxidation of Cr(III), however, results in the release of the highly toxic and soluble chromate iron, Cr(VI).

10.4 Metal toxicity to microorganisms and microbial resistance mechanisms

Metals are essential components of microbial cells; for example, sodium and potassium regulate gradients across the cell membrane, and copper, iron and manganese are required for activity of key metalloenzymes in photosynthesis and electron transport. However, metals can also be extremely toxic to microorganisms, impacting microbial growth, morphology and biochemical activities as a result of specific interactions with cellular components (Foster, 1983; Gadd, 1986; Beveridge & Doyle, 1989; Gadd, 1992; Freedman, 1995). Perhaps the most toxic metals are the non-essential metals such as cadmium, lead, and mercury.

Mechanisms of metal toxicity differ. Toxicity may occur as a result of binding of the metal to ligands containing reactive sulfhydryl, carboxyl or phosphate groups such as proteins or nucleic acids. The larger metal ions such as mercury or cadmium cations readily bind sulfhydryl groups, while smaller, more highly electropositive metal ions such as tin react with carboxyl and phosphate groups. Other interactions which cause inhibition of cell growth include metal-catalyzed decomposition of essential metabolites and analog replacement of structurally important cell components. A good example of analog replacement is arsenic, which is bactericidal because it acts as an analog of phosphate, disrupting nucleic acid structure and enzyme action. Photosynthetic and nitrogen-fixing organisms are particularly sensitive to metals, and high concentrations of lead and nickel retard cell division. Another sensitive cell component is the cell membrane, which is susceptible to disruption by copper and zinc (Baath, 1989).

As a consequence of, or perhaps in spite of, metal toxicity, some microorganisms have developed various resistance mechanisms to prevent metal toxicity. The strategies are either to prevent entry of the metal into the cell or actively to pump the metal out of the cell. This can be accomplished by either sequestration, active transport or chemical transformation through metal oxidation or reduction.

10.4.1 Sequestration

Sequestration involves metal complexation with microbial products such as extracellular polymeric substances (EPS) and metallothionein-like proteins. Sequestration may also involve the binding to electronegative components in cellular membranes. Regardless of the agent, the goal is to reduce or eliminate metal toxicity via complexation. This may be accomplished by binding the metal extracellularly to prevent metal entry into the cell, as is the case with exopolymers and, to some extent, cell surface binding (Rudd, Sterritt & Lester, 1984a,b; Stupakova, Demina & Dubinina, 1988; Beveridge & Doyle, 1989; Ehrlich, 1990; Gadd, 1992; Grappeli et al., 1992; Shuttleworth & Unz, 1993). Alternatively, metals may enter a cell and be concentrated there as a result of passive diffusion or by active transport. *Pseudomonas aeruginosa* has been shown to accumulate uranium by both passive diffusion and, in some instances, by a metabolism-dependent translocation process (Strandberg, Shumate & Parrott, 1981; Hughes & Poole, 1989). Cadmium uptake in *Staphylococcus aureus* occurs via the manganese transport system (Silver, Misra & Laddaga, 1989). Following metal uptake, some cells sequester metals intracellularly utilizing low molecular weight, cysteine-rich proteins called metallothioneins. Such proteins, initially discovered in fungi, rapidly bind metals as they enter the cell, effectively reducing their toxicity (Robinson & Tuovinen, 1984; Gadd, 1990b). The complexed metal may then either be transported back out of the cell or stored as intracellular granules.

Which mechanism predominates, intracellular or extracellular sequestration, is dependent on the organism involved. For example, uranium accumulated extracellularly as needle-like fibrils in a layer approximately 0.2 μm thick on the surface of a yeast, *Saccharomyces cerevisiae*, but formed dense intracellular deposits in *Pseudomonas aeruginosa* (Strandberg *et al.*, 1981).

10.4.2 Active transport

As already mentioned, active transport of metals out of the cell is one mechanism of microbial resistance to metal toxicity. Highly specific efflux systems can rapidly pump out toxic metal ions that have entered the cell. Such efflux pumps may derive their energy from membrane potential or from ATP, and in fact ATP-dependent efflux pumps have been identified that are specific for arsenic, cadmium and chromium (Horitsu *et al.*, 1986; Laddaga *et al.*, 1987; Hughes & Poole, 1989; Rani & Mahadevan, 1989; Nies, 1992; Silver, 1992). For cadmium, a plasmid gene, *cad*A, encodes a cadmium-specific ATPase that transports cadmium out of the cell. This gene has been found in several genera, including *Staphylococcus aureus*, *Pseudomonas putida*, and *Bacillus subtilis*. Similarly, *Escherichia coli* utilizes a plasmid-encoded arsenic ATPase for arsenate efflux (Silver & Misra, 1988).

10.4.3 Oxidation–reduction

Microbially facilitated oxidation–reduction (redox) reactions constitute a third mechanism for microbial metal resistance. Microorganisms may either oxidize to mobilize or reduce to immobilize a metal and both reactions are used to prevent metal entry into the cell. For example, the reduction of mercuric ion, Hg(II), to methyl mercury volatilizes the mercury, which can then diffuse away from the organism (Robinson & Tuovinen, 1984; Beveridge, 1989; Beveridge & Doyle, 1989; Silver *et al.*, 1989; Gadd, 1992). Other toxic metals undergo microbial reduction. Microbial hydrogen sulfide production is widespread and confers resistance through the precipitation of insoluble metal sulfides (Gadd & Griffiths, 1978; Ehrlich, 1990). Enzymatically produced phosphate, for example, by *Citrobacter* spp., has been shown to precipitate and detoxify lead and copper (Aiking *et al.*, 1984; Brierley, 1990). The ubiquitous nature of microorganisms and their ability to tolerate a variety of anthropogenic and environmental stresses, including heavy metals, have made them primary candidates for exploitation in the remediation of metal-contaminated systems. Some plants, especially of the genera *Silene* and *Agrostis*, are particularly metal resistant, and it should be noted that plants may be used in the bioremediation of metal-contaminated systems. However, the role of plants in metal reclamation will not be discussed in this chapter. Several excellent reviews on phytoremediation are available (Mitsch & Jorgensen, 1989; Otte, 1991; Verkleij *et al.*, 1991).

10.5 Origin of microbially based metal remediation

Microbial systems have long been used to treat efficiently domestic waste, and to recover precious metals from ores. It is therefore logical to extend the application of microbial systems for use in remediation of metal-contaminated soils and wastestreams. Perhaps the inception of microbially based metal remediation methods was in the eighteenth century in Rio Tinto, Spain, during the extraction of copper from copper ore. Although unknown at the time, copper solubilization during the extraction process was a result of microbial activity. The use of microorganisms in the recovery of precious metals, such as copper and gold, was later coined biohydrometallurgy. This is an efficient and cost-effective process, and by 1989, more than 30% of US copper production utilized biohydrometallurgy, the key component of which was the microorganism *Thiobacillus ferrooxidans* (Debus, 1990).

Bioleaching generally involves the oxidation of iron or sulfur-containing minerals. Three major organisms with unique capabilities are involved in the bioleaching process. *Thiobacillus ferrooxidans* is thought to be the dominant organism in acidic environments where it oxidizes iron sulfides such as bornite (Cu_5FeS_4) and pyrite (FeS_2). In the oxidation of ferrous iron (Fe^{2+}) to ferric iron (Fe^{3+}) by *Thiobacillus ferrooxidans*, the produced ferric ion is able to oxidize sulfide (S^{2-}) to sulfate (SO_4^{2-}), resulting in the production of sulfuric acid. The sulfuric acid lowers the pH of the environment and facilitates additional leaching. *Thiobacillus thiooxidans*, on the other hand, is unable to oxidize iron but can readily oxidize other minerals, such as zinc sulfides, where leaching is dependent on elemental sulfur oxidation. *Leptospirillum ferrooxidans* is also commonly associated with bioleaching. This organism oxidizes pyrite under acidic conditions; however, *Leptospirillum ferrooxidans* cannot oxidize sulfur, limiting its ability to solubilize certain mineral sulfides such as chalcopyrite (Cu_2S). The discovery of these microorganisms and others has led to increased interest in metal–microbe relations (Lodi, Del Borghi & Ferraiolo, 1989; Ehrlich & Brierley, 1990).

10.6 Microbial interactions with metals

Francis (1990) has summarized the numerous possible microbially mediated reactions resulting in the mobilization or immobilization of metals and found that major interactions include oxidation–reduction processes, biosorption and immobilization by cell biomass and exudates, and mobilization by microbial metabolites. A profound issue in metal remediation is that through microbial action, metals can readily be re-mobilized, creating toxicity issues in sites where metals are not completely removed.

10.6.1 Oxidation–reduction reactions

As briefly discussed, microorganisms can decrease metal toxicity by the oxidation or reduction of metals. Some microorganisms actively reduce metals to decrease bioavailability while others may oxidize metals to facilitate their removal from the environment. Early laboratory studies by Konetzka (1977) found *Pseudomonas* spp.

capable of the aerobic oxidation of arsenite, As(OH)$_3$ to the less toxic arsenate, As(O)(OH)$_3$. Selinite, Se^{4+}, and selenate, Se^{6+}, reduction under aerobic conditions to elemental selenium, Se^0, has been observed in certain fungi as a mechanism for tolerance (Konetzka, 1977; Gharieb, Wilkinson & Gadd, 1995). Tomei *et al.* (1995) found that *Desulfovibrio desulfuricans* anaerobically reduced selenate and selenite to elemental selenium as a means of selenium detoxification. Bacterial reduction of chromates (CrVI) to less toxic and less soluble Cr(III) under aerobic conditions by a chromium-resistant *Pseudomonas fluorescens*, isolated from chromium-contaminated sediments from the Hudson River, New York, has been reported by Bopp, Chakrabarty & Ehrlich (1983). Similarly, Ishibashi, Cervantes & Silver (1990) observed chromium reduction in *Pseudomonas putida* using a soluble chromate reductase.

In some cases, metals are solubilized microbiologically through oxidation to facilitate metal removal from a contaminated system. With some metals, toxicity increases with oxidation since bioavailability is also increased. The greater mobility of metals in these circumstances and the likelihood of removal from a microorganism's habitat acts as a mechanism of metal tolerance. *Thiobacillus* spp. and *Leptothrix* spp. readily solubilize a variety of metals via oxidation, including manganese, uranium and copper (Ehrlich & Brierley, 1990). *Bacillus megaterium* has been shown to oxidize elemental selenium to selenite, increasing selenium mobilization (Sarathchandra & Watkinson, 1981). Francis & Dodge (1990) have reported that a N$_2$-fixing *Clostridium* sp. is capable of anaerobic solubilization of cadmium, chromium, lead and zinc.

What is apparent in the examination of microbial metal redox reactions is that the metal one microorganism can immobilize, another is capable of solubilizing. This can lead to inadequate metal detoxification and complexation in systems where metal removal is not performed.

10.6.2 Complexation

Bacteria, algae, fungi and yeasts have all been found to complex or sorb metals (Gadd, 1990b); however, bacteria have been most extensively studied. Complexation of metals by microorganisms occurs in two ways: (1) the metals may be involved in nonspecific binding to cell wall surfaces, the slime layer, or the extracellular matrix, or (2) they may be taken up intracellularly. Studies have shown that both types of metal complexation are used to reduce metal toxicity and mobility (Bitton & Freihofer, 1978; Scott & Palmer, 1988; Gadd, 1990b; Marques *et al.*, 1990; Roane, 1994; Roane & Kellogg, 1994).

There are numerous studies of metal complexation by whole cells indicating that complexation depends on the bacterium, the metal, and pH. It has been found that at low pH, cationic metal complexation is reduced; however, binding of anionic metals such as chromate (CrO$_4^{2-}$) and selenate (SeO$_4^{2-}$) is increased. Bacteria seem to show selective affinities for different metals. This was demonstrated by a study of the complexation of four metals by four different bacterial genera that showed that the affinity series for bacterial removal of the metals studied decreased in the order Ag > La > Cu > Cd (Mullen *et al.*, 1989). Of the bacteria studied, *Pseudomonas aeruginosa* was the most efficient at metal complexation while *Bacillus cereus* was least. Complexation capacity is reported to be species specific as well as genus specific. A study of metal complexation between different *Pseudomonas* spp. showed up to one order of magnitude difference in complexation of several metals including lead, iron, and zinc (Corpe, 1975). The efficiency of metal complexation by microorganisms has resulted in development and use of several bioremediation processes (Brierley, 1990).

The presence of soil complicates metal removal because soils sorb metals strongly and can also affect microbial–metal complexation. Walker *et al.* (1989) showed that purified preparations of cell walls from *Bacillus subtilis* and *Escherichia coli* (423 to 973 mmol metal/g cell wall) were more effective than either of two clays, kaolinite (0.46 to 37 mmol metal/g clay), or smectite (1 to 197 nmol metal/g clay), in the binding of seven different metals. However, in the presence of cell-wall/clay mixtures, binding was reduced. In summary, there are several parameters that affect metal complexation. These include specific surface properties of the organism, cell metabolism, metal type, and the physicochemical parameters of the environment.

10.6.3 Methylation

Methylation of metals generally results in the volatilization and increased toxicity of a metal. The addition of methyl or alkyl groups on metals increases their lipophilicity and their permeability across biological membranes (Hughes & Poole, 1989). Such conversions from inorganic to organic metallic forms are of considerable interest from an ecological standpoint since toxicity and mobilization of the metal is modified. Methylation of the mercuric ion, Hg^{2+}, results in formation of monomethyl, CH_3Hg^+, or dimethylmercury, $(CH_3)_2Hg$, which can be 10 to 100 times more toxic than inorganic mercury. As seen with mercury, however, although volatilization increases metal toxicity, microorganisms use volatilization to facilitate metal diffusion away from their surroundings, thus avoiding the toxic effects. In addition to mercury, others including tin, lead, arsenic and selenium are among the metals commonly methylated by microorganisms (Reamer & Zoller, 1980; Thayer & Brinckman, 1982; Barkay & Olson, 1986; Gilmour, Tuttle & Means, 1987; Barkay *et al.*, 1992). The volatilization of metals can be a significant source of metal loss in estuarine and freshwater sediments, as well as heavily contaminated soils and sewage (Robinson & Tuovinen, 1984; Compeau & Bartha, 1985; Lester, 1987; Barkay, Liebert & Gillman, 1989). Studies conducted by Reamer & Zoller (1980) and Frankenberger & Karlson (1995) have observed selenium methylation in soils, sediments and sewage sludge. Saouter, Turner & Barkay (1994) found that microbially induced mercury volatilization can be a significant mechanism of removal from a mercury-contaminated pond. Both studies found that biomethylation was part of the detoxification process for microorganisms.

10.6.4 Biosurfactants and siderophores

Although their role in nature is still not clear, there are extracellular microbial products, such as biosurfactants, bioemulsifiers and siderophores, that complex or chelate metals quite efficiently. Examples of these compounds are shown in Figure 10.1.

Biosurfactants/bioemulsifiers

There are several reviews that discuss biosurfactants and their properties and applications (Zajic & Seffens, 1984; Rosenberg, 1986; Lang & Wagner, 1987; Miller, 1995b). It has been suggested that biosurfactants may play a role in the adhesion and desorption of microbial cells to surfaces (Rosenberg, 1986). In addition, biosurfactants have been shown to enhance the bioavailability of hydrocarbons with low water solubilities, thereby stimulating growth and biodegradation of such hydrocarbons (Zhang & Miller, 1994, 1995). Recent work has suggested that in addition to these roles, biosurfactants complex metals very

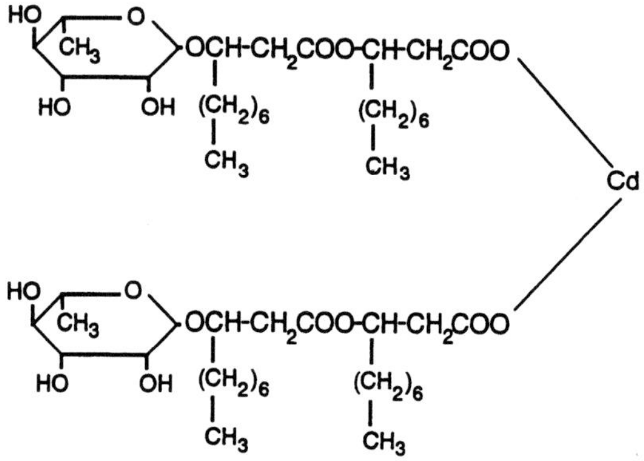

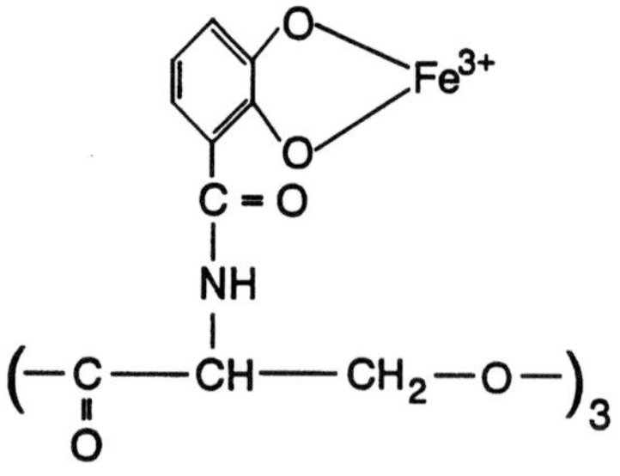

Figure 10.1. (a) Rhamnolipid from *P. aeruginosa* ATCC 9027 showing cadmium binding. (b) Structure of the iron–siderophore complex of enterobactin.

efficiently. A rhamnolipid biosurfactant (see Figure 10.1a) has been shown to complex metals such as cadmium, lead, and zinc with a complexation capacity comparable to those reported for exopolysaccharides (Miller, 1995a). The stability constant ($\log K = -2.47$) was higher than those reported by cadmium–sediment (-6.08 to -5.03) and cadmium–humic acid systems (-6.02 to -4.92), indicating that the cadmium–rhamnolipid complex was much stronger than the cadmium–sediment or cadmium–humic acid complex (Tan et al., 1994).

Similarly, Zosim, Gutnik & Rosenberg (1983) have reported uranium binding by emulsan (up to 240 mg uranium (UO_2^{+2})/mg emulsan), a bioemulsifier produced by Acinetobacter calcoaceticus RAG-1. The intriguing aspect of this system is that in a hexadecane–water solution, the emulsan preferentially binds to the hexadecane–water interface, which effectively concentrates the complexed uranium for easy recovery.

Siderophores

Siderophores are chelating agents often produced under iron-limiting conditions. They contain reactive groups such as dicarboxylic acids, polyhydroxy acids, and phenolic compounds. As shown in Figure 10.2, siderophores complex metals, and they have been reported to facilitate movement of metals in soil (Duff, Webley & Scott, 1963; Bolter, Butz & Arseneau, 1975; Cole, 1979). Siderophores can complex gallium, chromium, nickel, uranium, and thorium, in addition to iron (Hausinger, 1987; Macaskie & Dean, 1990). However, limited progress has been made regarding the role of siderophores in metal recovery and/or removal.

10.7 Innovative approaches to microbial remediation of metal-contaminated environments

Current approaches to metal bioremediation are based upon the complexation, oxidation–reduction, and methylation reactions just discussed. Until recently, interest was focused on technologies that could be applied to achieve in situ immobilization of metals. However, within the last few years, the focus has begun to shift toward actual metal removal, because it is difficult to guarantee that metals will remain immobilized indefinitely.

Numerous bioremediation strategies have potential applicability in the removal of metals from contaminated environments. Unfortunately, the number of field-based studies has been few for several reasons. Often the institutions that perform the basic research and feasibility studies are not equipped to gear up to a field application, making this a difficult and expensive transition. Second, it is sometimes difficult to change entrenched attitudes which in the past have been favorable toward excavation and removal of contamination and unfavorable toward bioremediation processes. Finally, bioremediation has not always been a predictable technology; however, as the field record of success grows, it is expected that acceptance of bioremediation as a viable technology will also grow.

10.7.1 Soils and sediments

Microbial leaching

Microbial leaching of metals from metal-containing soils is an extension of the practice of bioleaching of metals from ores (Lodi *et al.*, 1989; see Figure 10.2). In bioleaching, copper, lead, and zinc have successfully been recovered through metal solubilization by *Thiobacillus ferrooxidans* and *Thiobacillus thiooxidans*. This process has been used to leach uranium from nuclear-waste-contaminated soils (Hutchins *et al.*, 1986; McCready & Gould, 1990; Rossi & Ehrlich, 1990; Macaskie, 1991), and to remove copper in the bioremediation of copper tailings (Gokcay & Onerci, 1994). In a recent study by Phillips, Landa & Lovley (1995), uranium was removed from a contaminated soil using bicarbonate. The uranium was then precipitated out of the soil effluent using *Desulfovibrio desulfuricans* to reduce U(VI) to U(IV) with removal efficiencies ranging from 20–100%.

While used extensively in mining and metal recovery, microbial leaching of

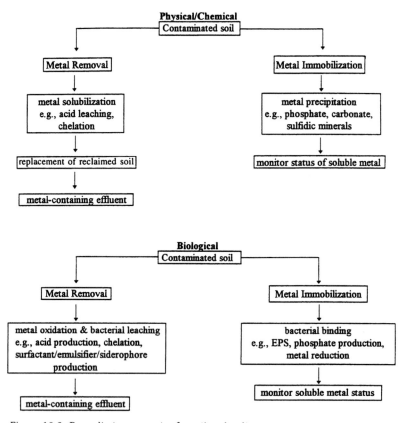

Figure 10.2. Remediation strategies for soil and sediments.

metals has, unfortunately, received little attention as a microbial-based technology for metal remediation. One potential application is the treatment of sewage sludge earmarked for disposal in soil. Sludge-amended soils have increased plant-available nutrients and improved productivity, but, unfortunately, sludge addition to soils also increases metal content (Reimers, Akers & White, 1989; Sauerbeck, 1991). Concern over the metallic and organic contaminants in sludge has limited its application despite the benefits of nutrients and organic matter. This concern has, in turn, stimulated interest in microbial removal of heavy metals from sewage sludge-amended soils. *Thiobacillus ferrooxidans* and *Thiobacillus thiooxidans* have been used to leach metals from contaminated sludge before soil application. This process was successful in making the sludge suitable for agricultural application (Couillard & Zhu, 1992).

Biosurfactants/bioemulsifiers

The use of bacterial surfactants for metal remediation of soils is gaining attention. These molecules (see Figure 10.1a) are water soluble and have low molecular weights ($\gg 500$ to 1500). Above a critical micelle concentration (CMC), biosurfactant monomers aggregate to form structures such as micelles, vesicles and bilayers. Because of their small size, biosurfactant monomers as well as smaller aggregate structures, as in micelles and small vesicles, may freely move through soil pores. It was estimated in one study of a rhamnolipid biosurfactant and cadmium that the biosurfactant–metal complexes ranged up to 50 nm in diameter (Champion *et al.*, 1995). Particles of that size should not be removed by physical straining in most soils (Gerba, Yates & Yates, 1991).

Bacterially produced surfactants consist of a variety of molecular structures which may show metal specificity and thus may be optimized for that metal's removal (Miller, 1995a). Herman, Artiola & Miller (1995) demonstrated removal of cadmium (56%) and lead (42%) from soil by a rhamnolipid biosurfactant produced by *Pseudomonas aeruginosa*. These studies indicate the potential for metal-containing soils to be flushed with a surfactant-containing solution, thereby causing metal desorption and removal. The use of biosurfactants is a promising treatment of soil systems because it offers the possibility of using an *in situ* treatment, eliminating the need for excavation and external washing (see Figure 10.2).

Similarly, bioemulsifiers, such as emulsan produced by *Acinetobacter calcoaceticus*, have been shown to aid in removal of metals. Potential for remediation of soils using bacterial exopolymers is indicated by a study which showed that purified exopolymers from 13 bacterial isolates removed cadmium and lead from an aquifer sand with efficiencies ranging from 12 to 91% (Chen *et al.*, 1995). Although such molecules have much larger molecular weights ($\sim 10^6$) than biosurfactants, this study showed that sorption by the aquifer sand was low, suggesting that in a porous medium with a sufficiently large mean pore size, use of exopolymers may be feasible.

Volatilization

Although methylation is not a desirable reaction for most metals, particularly arsenic or mercury, methylation and volatilization have been proposed as techniques for remediation of selenium-contaminated soils and sediments (Huysmans & Frankenberger, 1990; Karlson & Frankenberger, 1990; Frankenberger & Karlson, 1995). Selenium-contaminated soils from San Joaquin, California, were remediated using selenium volatilization stimulated by the addition of pectin in the form of orange peel. In their study, Karlson & Frankenberger (1989) found that the addition of pectin enhanced the rate of selenium alkylation, ranging from 11.3 to 51.4% of added selenium, and that this was a feasible approach to treat soils contaminated with selenium.

10.7.2 Aquatic systems

Metal reclamation from acid mine drainage and contaminated surface- and groundwater and wastewaters has been extensively studied. Technologies for metal removal from solution are based on the microbial–metal interactions discussed earlier: the binding of metal ions to microbial cell surfaces; the intracellular uptake of metals; the volatilization of metals; and the precipitation of metals via complexation with microbially produced ligands.

Acid mine drainage

Acid mine drainage and mining effluent waters containing high amounts of zinc, copper, iron, manganese, as well as lead, cadmium and arsenic, are commonly remediated using wetlands. Wetland treatment of acid mine drainage has proved to be cost effective and less labor intensive than traditional chemical treatments. The Tennessee Valley Authority reports an 80% success rate in meeting discharge standards using wetlands alone to remediate acid drainage (Douglas, 1992). Wildeman *et al.* (1994) reported the following metal reductions in contaminated water in response to wetland treatment: zinc, 150 to 0.2 mg/l; copper, 55 to <0.05 mg/l; iron, 700 to 1 mg/l; and manganese, 80 to 1 mg/l.

Wetland remediation involves a combination of interactions including microbial adsorption of metals, metal bioaccumulation, bacterial oxidation of metals, and sulfate reduction (Fennessy & Mitsch, 1989; Kleinmann & Hedin, 1989). Sulfate reduction produces sulfides which in turn precipitate metals and reduce aqueous metal concentrations. The high organic matter content in wetland sediments provides the ideal environment for sulfate-reducing populations and for the precipitation of metal complexes. Some metal precipitation may also occur in response to the formation of carbonate minerals (Kleinmann & Hedin, 1989). In addition to the aforementioned microbial activities, plants, including cattails, grasses, and mosses, serve as biofilters for metals (Brierley, Brierley & Davidson, 1989).

While proving to be an effective method of treatment for a number of metals, wetlands may need additional inputs of organic matter which can be added in the form of mushroom compost, as an example. Wetland technology is limited by the problems of build-up and disposal of contaminated biomass, plant and microbial, and by the possible toxic effects of the metal-containing influent upon wetland plant and microbial communities. In response, some researchers are investigating the possibility of treating acid mine drainage biologically without wetlands.

Surface- and groundwater

Microbial biofilms are a common treatment technology for metal-contaminated surface waters and groundwater. Immobilized biofilms, viable and nonviable, essentially trap metals as the contaminated water is pumped through (Macaskie & Dean, 1989; Summers, 1992). A variety of microorganisms has been used to create such biofilms (Brierley et al., 1989; Gadd, 1992). For example, live immobilized biofilms of Citrobacter spp. have been used to remove uranium from contaminated wastewater in both laboratory and field studies (Macaskie, 1991). Biofilms of Streptomyces viridochromogens have similarly been used in uranium removal. Arthrobacter sp. biomass and exopolymers have been used in the capture of cadmium, chromium, lead, copper, and zinc from fluid wastes in the laboratory (Grappelli et al., 1992). Nonliving films of Bacillus spp. have been used to remove cadmium, chromium, copper, mercury, nickel, uranium, and zinc from wastewater (Brierley et al., 1989). Fungi and yeasts have also been shown to be effective in metal removal from aquatic systems (Gadd, 1990a). In response to silver concentrations ranging from 0 to 2 mM and copper concentrations from 0 to 3 mM, strains of Saccharomyces cerevisiae and Candida sp. were able to accumulate from 0.03 to 0.19 mmol silver/g dry weight and 0.05 to 0.18 mmol copper/g dry weight, respectively (Simmons, Tobin & Singleton, 1995).

In response to the success of microbially mediated cleanup of metal-contaminated waters, some commercial bioremediation products such as BIOCLAIM, AlgaSORB, and BIO-FIX are available. More detailed descriptions are provided by Brierley (1990). BIOCLAIM and BIO-FIX use immobilized bacterial preparations while AlgaSORB utilizes a nonviable algae matrix for metal removal. In addition to these products, there are several proposed proprietary processes including the use of immobilized Rhizopus arrhizus biomass for uranium recovery (Tsezos, McCready & Bell, 1989). A froth flotation method for enhanced contact between biomass and contaminated water has been proposed by Smith, Yang & Wharton (1988).

Similar to the microbial biofilm preparations described above, free-floating, viable microbial mats are also successful in removal of metals from solution (Bender & Phillips, 1994; Vatcharapijarn, Graves & Bender, 1994). Consisting primarily of algae, cyanobacteria and bacteria, microbial mats perform a number of activities which promote metal complexation and subsequent removal. The mat contains oxidizing and reducing zones that aid in the immobilization and precipitation of metals. There is an increased pH which promotes metal precipitation as metal hydroxides, and the mat releases negatively charged extracellular substances to promote flocculation.

Microbial mat formation may also stimulate metal removal through sulfate reduction. Barnes, Scheeren & Buisman (1994) have developed a process that specifically uses sulfate-reducing bacteria to treat metal-contaminated groundwater. In this process, as groundwater is pumped through the water treatment plant, sulfide produced by sulfate-reducing bacteria precipitates the metals in the water. Metal concentrations in the treated water were reportedly reduced to $\mu g/l$ quantities and the water was suitable for release into the environment.

Marine water

Few studies have evaluated the potential for use of microorganisms in the remediation of sea water; however, the problems encountered are similar to those of other aquatic systems. Stupakova *et al.* (1988) proposed the use of the marine bacteria *Deleya venustus* and *Moraxella* sp. for copper uptake from sea water. Additionally, Corpe (1975) performed metal-binding studies with copper using exopolymer from film-producing marine bacteria and found that insoluble copper precipitates formed, effectively decreasing copper toxicity.

Wastewater

Metal removal from domestic waste is a better-studied system. The efficiencies of metal removal vary with respect to both waste and metal type (Lester, 1987). However, the removal of metals from wastewater is generally an efficient process. Cheng, Patterson & Minear (1975) studied heavy metal uptake by activated sludge and found that in an activated sludge effluent containing 2100–25 200 μg copper/l, 89% of the copper was removed during treatment. Furthermore, 98% of the lead was removed from effluent containing 2100–25 500 μg lead/l. In a similar study, in effluent with 9000 μg zinc/l, 89% of the zinc was removed during sewage treatment (Barth *et al.*, 1965).

Metal removal in wastewater treatment can also be done efficiently on a smaller scale, as reported by Whitlock (1990). The Homestake Goldmine in Lead, South Dakota, utilizes a biological treatment plant to detoxify four million gallons of wastewater that are discharged daily to Whitewood Creek. The wastewater matrix mainly contains elevated levels of cyanide, ammonia and metals. The plant consists of a train of five rotating biological contactors (RBCs) which sequentially treat the wastewater. The plant had been in successful operation for 9 years at the time of the report with the following results: treatment capacity and resistance to upset improved with time, copper and iron were removed at 95 to 98% efficiency while nickel, chromium and zinc removal were inconsistent, ammonia and cyanide were removed to acceptable levels, and finally, Whitewood Creek was successfully re-established as a trout fishery.

The removal of metals from wastewater streams and sewage relies primarily on the immobilization and complexation of metals by extracellular compounds, e.g. EPS (Brown & Lester, 1982a,b; Rudd *et al.*, 1984a). The genus *Zoogloea*, an important organism in sewage treatment, readily forms an anionic slime matrix. Toxic metals complex with the matrix and precipitate out of solution. *Klebsiella aerogenes* is another common sewage bacterium which binds metal ions with extracellular polymers. Complexed metals are then removed from the wastewater via sedimentation during the treatment process.

Other proposed mechanisms of metal removal from sewage include physical capture by microbial flocs, cellular accumulation and volatilization by such organisms as *Klebsiella, Pseudomonas, Zoogloea* and *Penicillium* spp. (Brown & Lester, 1979). In laboratory and pilot-scale studies, up to 98% metal removal by these mechanisms combined has been documented (Lester, 1987).

10.7.3 Nuclear waste

More recent concern about metal release into the environment comes from the nuclear industry. Nuclear waste can contain a combination of nonradioactive metals, such as lead and copper, and radioactive metals including uranium (^{235}U), thorium (^{230}Th) and radium (^{226}R). Nuclear waste disposal, perhaps better termed as storage, has the potential to introduce both radioactive and nonradioactive metals into surrounding aquifers, soils and surface waters as a result of poor construction or placement of storage facilities. The risk of environmental contamination via nuclear waste is not yet well understood and has received relatively little attention (Macaskie, 1991; Haggin, 1992).

Radioactive metal wastes from the nuclear industry are of increasing concern as the amount of waste to be disposed of increases. Current treatment of nuclear wastewater involves the addition of lime, which is effective in precipitating most metals out of solution with the exception of radium (Tsezos & Keller, 1983). Barium chloride ($BaCl_2$) is used to precipitate radium from sulfur-rich effluents as barium–radium sulfate. Other treatment methods include incineration for some solid wastes, and filtration, adsorption and crystallization for liquid wastes (Godbee & Kibbey, 1981).

In the removal of nuclides from contaminated systems, biological adsorbents are superior to conventional adsorbents, such as zeolite and activated carbon. For example, *Penicillium chrysogenum* was found to adsorb up to 5×10^4 nCi radium/g biomass from an initial radium concentration of 1000 pCi/l, compared with adsorption of only 3600 nCi/g by activated carbon under the same conditions (Tsezos & Keller, 1983; Tsezos, Baird & Shemilt, 1987). There are a variety of biological products including tannins, melanins, bacterial chelating agents, microbial polysaccharides and metallothioneins, whole microbial cells, including yeasts and fungi, and chitin that have been investigated for application in the bioremediation of nuclear wastes. An extensive review of biological treatment of nuclear wastes is provided by Macaskie (1991).

Due to the danger uranium poses to human health and to its ubiquity in nuclear by-products, uranium is the most extensively studied radionuclide. Several microbiological systems have been proposed for the removal of uranium. Phosphate-producing *Citrobacter* spp. have been studied for the precipitation of uranium and lanthanum metals (Plummer & Macaskie, 1990; Macaskie *et al.*, 1992). Microbially released products, such as emulsan from *Acinetobacter* sp. and an exopolysaccharide from a *Pseudomonas* sp., have also been suggested for use in the recovery of uranium (Zosim, Gutnick & Rosenberg, 1983; Marques *et al.*, 1990). Tsezos, McCready & Bell (1989) reported the use of immobilized biomass of *Rhizopus arrhizus* in uranium recovery from contaminated water. As more studies are performed, microbial reclamation of these wastes will be better understood.

10.8 Concluding remarks

Bioremediation of metal-contaminated environments is a rapidly advancing field. The use of microbial biomass, exopolysaccharides, biosurfactants, bioemulsifiers, siderophores, and microbially induced oxidation–reduction and methylation reactions, offer promising alternatives to traditional technologies in the treatment of heavy metals. Future research focuses on *in situ* mining to help reduce contaminated waste and development of microbial applications in metal removal from industrial and mining wastes before release into the environment will provide still more treatment options. In the attempt to develop new microbially mediated remediation technologies, more research needs to be conducted on the significance of microbial reduction of metals in environmental systems which can increase metal mobility, on the effects of co-contaminants, including the presence of other metals, in metal reclamation, and on the speciation and bioavailability of metals in the environment, an area of utmost importance in biological remediation. Finally, more field-based studies need to be performed since, while laboratory studies are necessary, *in situ* studies can provide essential information about environmental interactions. While there is much to be done, the future promises more innovative applications in the field of metal bioremediation.

References

Aiking, H., Stijnman, A., van Garderen, C., van Heerikhuizen, H. & van't Riet, J. (1984). Inorganic phosphate accumulation and cadmium detoxification in *Klebsiella aerogenes* NCTC 418 growing in continuous culture. *Applied and Environmental Microbiology*, 47, 374–7.

Alloway, B. J. (1990). *Heavy Metals in Soils.* New York: John Wiley.

Baath, E. (1989). Effects of heavy metals in soil on microbial processes and populations (a review). *Water, Air, and Soil Pollution*, 47, 335–79.

Babich, H. & Stotzky, G. (1980). Environmental factors that influence the toxicity of heavy metal and gaseous pollutants to microorganisms. *Critical Reviews in Microbiology*, 8, 99–145.

Babich, H. & Stotzky, G. (1985). Heavy metal toxicity to microbe–mediated ecologic processes: a review and potential application to regulatory policies. *Environmental Research*, 36, 111–37.

Barkay, T. & Olson, B. H. (1986). Phenotypic and genotypic adaptation of aerobic heterotrophic sediment bacterial communities to mercury stress. *Applied and Environmental Microbiology*, 52, 403–6.

Barkay, T., Liebert, C. & Gillman, M. (1989). Environmental significance of the potential for *mer*(Tn21)-mediated reduction of Hg^{2+} to Hg^0 in natural waters. *Applied and Environmental Microbiology*, 55, 1196–202.

Barkay, T., Turner, R., Saouter, E. & Horn, J. (1992). Mercury biotransformations and their potential for remediation of mercury contamination. *Biodegradation*, 3, 147–59.

Barnes, L. J., Scheeren, P. J. M. & Buisman, C. J. N. (1994). Microbial removal of heavy metals and sulfate from contaminated groundwaters. In *Emerging Technology for Bioremediation of Metals*, ed. J. L. Means & R. E. Hinchee, pp. 38–49. London: Lewis Publishers.

Barth, E. F., Ettinger, M. B., Salotto, B. V. & McDermott, G. N. (1965). Summary report on the effects of heavy metals on the biological treatment processes. *Journal Water Pollution Control Federation*, 37, 86–96.

Bender, J. & Phillips, P. (1994). Implementation of microbial mats for bioremediation. In *Emerging Technology for Bioremediation of Metals*, ed. J. L. Means & R. E. Hinchee, pp. 85–98. London: Lewis Publishers.

Bernhard, M., Brinckman, F. E. & Sadler, P. J. (1986). *The Importance of Chemical Speciation in Environmental Processes*. Berlin: Springer-Verlag.

Beveridge, T. J. (1989). Metal ions and bacteria. In *Metal Ions and Bacteria*, ed. T. J. Beveridge & R. J. Doyle, pp. 1–29. New York: John Wiley.

Beveridge, T. J. & Doyle, R. J. (1989). *Metal Ions and Bacteria*. New York: John Wiley.

Bitton, G. & Freihofer, V. (1978). Influence of extracellular polysaccharides on the toxicity of copper and cadmium toward *Klebsiella aerogenes*. *Microbial Ecology*, 4, 119–25.

Bolter, E., Butz, T. & Arseneau, J. F. (1975). Mobilization of heavy metals by organic acids in the soils of a lead mining and smelting district. In *Trace Substances in Environmental Health–IX*, ed. D. D. Hemphill, pp. 107–12. Missouri: University of Missouri.

Bopp, L. H., Chakrabarty, A. M. & Ehrlich, H. L. (1983). Chromate resistance plasmid in *Pseudomonas fluorescens*. *Journal of Bacteriology*, 155, 1105–9.

Brierley, C. L. (1990). Bioremediation of metal-contaminated surface and groundwaters. *Geomicrobiology Journal*, 8, 201–23.

Brierley, C. L., Brierley, J. A. & Davidson, M. S. (1989). Applied microbial processes for metals recovery and removal from wastewater. In *Metal Ions and Bacteria*, ed. T. J. Beveridge & R. J. Doyle, pp. 359–82. New York: John Wiley.

Brooks, R. R. (1983). *Biological Methods for Prospecting Minerals*. New York: John Wiley.

Brown, M. J. & Lester, J. N. (1979). Metal removal in activated sludge: the role of bacterial extracellular polymers. *Water Research*, 13, 817–37.

Brown, M. J. & Lester, J. N. (1982a). Role of bacterial extracellular polymers in metal uptake in pure bacterial culture and activated sludge–I: effects of metal concentration. *Water Research*, 16, 1539–48.

Brown, M. J. & Lester, J. N. (1982b). Role of bacterial extracellular polymers in metal uptake in pure bacterial culture and activated sludge–II: effects of mean cell retention time. *Water Research*, 16, 1549–60.

Canney, F. C., Cannon, H. L., Cathrall, J. B. & Robinson, K. (1979). Autumn colors, insects, plant disease, and prospecting. *Economic Geology*, 74, 1673–76.

Carlisle, D., Berry, W. L., Kaplan, I. R. & Watterson, J. R. (1986). *Mineral Exploration: Biological Systems and Organic Matter*. New Jersey: Prentice-Hall.

Carson, B. L., Ellis, H. V. & McCann, J. L. (1986). *Toxicology and Biological Monitoring of Metals in Humans*. Michigan: Lewis Publishers, Inc.

Chalkley, M. E., Conrad, B. R., Lakshmanan, V. I. & Wheeland, K. G. (1989). *Tailings and Effluent Management*. New York: Pergamon Press.

Champion, J. T., Gilkey, J. C., Lamparski, H., Rettner, J. & Miller, R. M. (1995). Electron microscopy of rhamnolipid (biosurfactant) morphology: effects of pH, cadmium and octadecane. *Journal of Colloid and Interface Science*, 170, 569–74.

Chen, J-H., Lion, L. W., Ghiorse, W. C. & Shuler. M. L. (1995). Mobilization of adsorbed cadmium and lead in aquifer material by bacterial extracellular polymers. *Water Research*, 29, 421–30.

Cheng, M. H., Patterson, J. W. & Minear, R. A. (1975). Heavy metals uptake by activated sludge. *Journal Water Pollution Control Federation*, 47, 362–76.

Cole, M. (1979). Solubilization of heavy metal sulfides by heterotrophic soil bacteria. *Soil Science*, 127, 313–17.

Compeau, G. C. & Bartha, R. (1985). Sulfate-reducing bacteria: principal methylators of mercury in anoxic estuarine sediment. *Applied and Environmental Microbiology*, 50, 498–502.

Corpe, W. A. (1975). Metal-binding properties of surface material from marine bacteria. *Developments in Industrial Microbiology*, 16, 249–55.

Couillard, D. & Zhu, S. (1992). Bacterial leaching of heavy metals from sewage sludge for agricultural application. *Water, Air, and Soil Pollution*, 63, 67–80.

Debus, K. H. (1990). Mining with microbes. *Technology Review*, 93, 50–7.

Douglas, J. H. (1992). Constructed wetlands reduce cost of treating acid drainage. *Environmental Update*, April, 4–5, 15.

Duff, R. B., Webley, D. M. & Scott, R. O. (1963). Solubilization of minerals and related materials by 2-ketogluconic acid–producing bacteria. *Soil Science*, 95, 105–14.

Duxbury, T. (1981). Toxicity of heavy metals to soil bacteria. *FEMS Microbiological Letters*, 11, 217–20.

Ehrlich, H. L. (1990). *Geomicrobiology*. New York: Marcel Dekker.

Ehrlich, H. L. & Brierley, C. L. (1990). *Microbial Mineral Recovery*. New York: McGraw-Hill Publishing.

Fennessey, M. S. & Mitsch, W. J. (1989). Design and use of wetlands for renovation of drainage from coal mines. In *Ecological Engineering: An Introduction to Ecotechnology*, ed. W. J. Mitsch & S. E. Jorgensen, pp. 231–53. New York: John Wiley.

Foster, T. J. (1983). Plasmid-determined resistance to antimicrobial drugs and toxic metal ions in bacteria. *Microbiological Reviews*, 47, 361–409.

Francis, A. J. (1990). Microbial dissolution and stabilization of toxic metals and radionuclides in mixed wastes. *Experientia*, 46, 840–51.

Francis, A. J. & Dodge, C. J. (1990). Anaerobic microbial remobilization of toxic metals coprecipitated with iron oxide. *Environmental Science and Technology*, 24, 373–8.

Frankenberger, W. T. Jr, Karlson, U. (1995). Volatilization of selenium from a dewatered seleniferous sediment: a field study. *Journal of Industrial Microbiology*, 14, 226–32.

Freedman, B. (1995). *Environmental Ecology: The Ecological Effects of Pollution, Disturbance, and Other Stresses.* San Diego: Academic Press.

Gadd, G. M. (1986). Fungal responses towards heavy metals. In *Microbes in Extreme Environments*, ed. R. A. Herbert & G. A. Codd, pp. 83–110. New York: Academic Press.

Gadd, G. M. (1990a). Fungi and yeasts for metal accumulation. In *Microbial Mineral Recovery*, ed. H. L. Ehrlich & C. L. Brierley, pp. 249–75. New York: McGraw-Hill.

Gadd, G. M. (1990b). Heavy metal accumulation by bacteria and other microorganisms. *Experientia*, 46, 834–40.

Gadd, G. M. (1992). Microbial control of heavy metal pollution. In *Microbial Control of Pollution*, ed. J. C. Fry, G. M. Gadd, R. A. Herbert, C. W. Jones & I. A. Watson-Craik, pp. 59–88. Cambridge: Cambridge University Press.

Gadd, G. M. & Griffiths, A. J. (1978). Microorganisms and heavy metal toxicity. *Microbial Ecology*, 4, 303–17.

Gerba, C. P., Yates, M. V. & Yates, S. R. (1991). Quantitation of factors controlling viral and microbial transport in the subsurface. In *Modeling the Environmental Fate of Microorganisms*, ed. C. J. Hurst, pp. 77–88. Washington, D.C.: American Society for Microbiology.

Gharieb, M. M., Wilkinson, S. C. & Gadd, G. M. (1995). Reduction of selenium oxyanions by unicellular, polymorphic and filamentous fungi: cellular location of reduced selenium and implications for tolerance. *Journal of Industrial Microbiology*, 14, 300–11.

Gieger, G., Federer, P. & Sticher, H. (1993). Reclamation of heavy metal-contaminated soils: field studies and germination experiments. *Journal of Environmental Quality*, 22, 201–7.

Gilmour, C. C., Tuttle, J. H. & Means, J. C. (1987). Anaerobic microbial methylation of inorganic tin in estuarine sediment slurries. *Microbial Ecology*, 14, 233–42.

Godbee, H. W. & Kibbey, A. H. (1981). Unit operations used to treat process and/or waste streams at nuclear power plants. *Nuclear and Chemical Waste Management*, 2, 71–88.

Gokcay, C. F. & Onerci, S. (1994). Recycling of copper flotation tailings and bioremediation of copper-laden dump sites. In *Emerging Technology for Bioremediation of Metals*, ed. J. L. Means & R. E. Hinchee, pp. 61–73. London: Lewis Publishers.

Goldman, C. R. & Horne, A. J. (1983). *Limnology.* New York: McGraw-Hill.

Grappelli, A., Campanella, L., Cardarelli, E., Mazzei, F., Cordatore, M. & Pietrosanti, W. (1992). Metals removal and recovery by *Arthrobacter* sp. biomass. *Water Science Technology*, 26, 2149–52.

Haggin, J. (1992). Microbially based treatment process removes toxic metals, radionuclides. *Chemical and Engineering News*, 70, 35–6.

Hammond, P. B. & Foulkes, E. C. (1986). Metal ion toxicity in man and animals. In *Metal Ions in Biological Systems*, ed. H. Sigel, pp. 157–200. New York: Marcel Dekker.

Hausinger, R. P. (1987). Nickel utilization by microorganisms. *Microbiological Reviews*, 51, 22–42.

Herman, D. C., Artiola, J. F. & Miller, R. M. (1995). Removal of cadmium, lead, and zinc from soil by a rhamnolipid biosurfactant. *Environmental Science and Technology*, 29, 2280–5.

Horitsu, H., Yamamoto, K., Wachi, S., Kawai, K. & Fukuchi, A. (1986). Plasmid-determined cadmium resistance in *Pseudomonas putida* GAM-1 isolated from soil. *Journal of Bacteriology*, 165, 334–5.

Hughes, M. N. & Poole, R. K. (1989). *Metals and Micro-organisms.* New York: Chapman & Hall.

Hutchins, S. R., Davidson, M. S., Brierley, J. A. & Brierley, C. L. (1986). Microorganisms in reclamation of metals. *Annual Review of Microbiology*, 40, 311–36.

Huysmans, K. D. & Frankenberger, W. T. Jr, (1990). Arsenic resistant microorganisms isolated from agricultural drainage water and evaporation pond sediments. *Water, Air, and Soil Pollution*, 53, 159–68.

Ishibashi, Y., Cervantes, C. & Silver, S. (1990). Chromium reduction in *Pseudomonas putida*. *Applied and Environmental Microbiology*, 56, 2268–70.

Johnson, I., Flower, N. & Loutit, M. W. (1981). Contribution of periphytic bacteria to the concentration of chromium in the crab *Helice crassa*. *Microbial Ecology*, 7, 245–52.

Jorgensen, S. E. (1989). Changes in redox potential in aquatic ecosystems. In *Ecological Engineering: An Introduction to Ecotechnology*, ed. W. J. Mitsch & S. E. Jorgensen, pp. 341–55. New York: John Wiley.

Karlson, U. & Frankenberger, W. T. Jr, (1989). Accelerated rates of selenium volatilization from California soils. *Soil Science Society of America Journal*, 53, 749–53.

Karlson, U. & Frankenberger, W. T. Jr, (1990). Volatilization of selenium from agricultural evaporation pond sediments. *The Science of the Total Environment*, 92, 41–54.

Keely, J. F., Hutchinson, F. I., Sholley, M. G. & Wai, C. M. (1976). *Heavy Metal Pollution in the Coeur d'Alene Mining District*. Project Technical Report to National Science Foundation. Moscow: University of Idaho.

Kleinmann, R. L. P. & Hedin, R. (1989). Biological treatment of mine water: an update. In *Tailings and Effluent Management*, ed. M. E. Chalkley, B. R. Conrad, V. I. Lakshmanan & K. G. Wheeland, pp. 173–9. New York: Pergamon Press.

Konetzka, W. A. (1977). Microbiology of metal transformations. In *Microorganisms and Minerals*, ed. E. D. Weinberg, pp. 317–42. New York: Marcel Dekker.

Kovalick, W. (1991). Perspectives on health and environmental risks of soil pollution and experiences with innovative remediation technologies. In *Abstracts of the 4th World Congress of Chemical Engineering*, abstr. 3.3-1. Karlsruhe, Germany.

Laddaga, R. A., Chu, L., Mistra, T. K. & Silver, S. 1987. Nucleotide sequence and expression of the mercurial-resistance operon from *Staphylococcus aureus* plasmid pI258. *Proceedings of the National Academy of Science USA*, 84, 5106–10.

Lang, S. & Wagner, F. (1987). Structure and properties of biosurfactants. In *Biosurfactants and Biotechnology*, ed. N. Kosaric, W. L. Cairns & N. C. C. Gray, pp. 21–45. New York: Marcel Dekker.

Leppard, G. G. (1981). *Trace Element Speciation in Surface Waters*. New York: Plenum Press.

Lester, J. N. (1987). Biological treatment. In *Heavy Metals in Wastewater and Sludge Treatment Processes: Treatment and Disposal*, Vol. 2, ed. J. N. Lester, pp. 15–40. New York: CRC Press.

Lindsay, W. L. (1979). *Chemical Equilibria in Soils*. New York: John Wiley.

Lodi, A., Del Borghi, M. & Ferraiolo, G. (1989). Biological leaching of inorganic materials. In *Biological Waste Treatment*, ed. A. Mizrahi, pp. 133–58. New York: Alan R. Liss.

Macaskie, L. E. (1991). The application of biotechnology to the treatment of wastes produced from the nuclear fuel cycle: biodegradation and bioaccumulation as a means of treating radionuclide-containing streams. *Critical Reviews in Biotechnology*, 11, 41–112.

Macaskie, L. E. & Dean, A. C. R. (1989). Microbial metabolism, desolubilization, and deposition of heavy metals: metal uptake by immobilized cells and application to the detoxification of liquid wastes. In *Biological Waste Treatment*, ed. A. Mizrahi, pp. 160–201. New York: Alan R. Liss.

Macaskie, L. E. & Dean, A. C. R. (1990). Metal-sequestering biochemicals. In *Biosorption of Heavy Metals*, ed. B. Volesky, pp. 199–248. Boca Raton, FL: CRC Press.

Macaskie, L. E., Empson, R. M., Cheetham, A. K., Grey, C. P. & Skarnulis, A. J. (1992). Uranium bioaccumulation by a *Citrobacter* sp. as a result of enzymically mediated growth of polycrystalline HUO_2PO_4. *Science*, 257, 782–4.

Marques, A. M., Bonet, R., Simon-Pujol, M. D., Fuste, M. C. & Congregado, F. (1990). Removal of uranium by an exopolysaccharide from *Pseudomonas* sp. *Applied Microbiology and Biotechnology*, 34, 429–31.

McCready, R. G. L. & Gould, W. D. (1990). Bioleaching of uranium. In *Microbial Mineral Recovery* ed. H. L. Ehrlich & C. L. Brierley, pp. 107–25. New York: McGraw-Hill.

McLean, R. J. C. & Beveridge, T. J. (1990). Metal-binding capacity of bacterial surfaces and their ability to form mineralized aggregates. In *Microbial Mineral Recovery*, ed. H. L. Ehrlich & C. L. Brierley, pp. 185–222. New York: McGraw-Hill.

Mearns, A. J. & Young, D. R. (1977). Chromium in the southern California environment. In *Pollutant Effects on Marine Organisms*, ed. C. S. Giam, pp. 125–42. Lexington: Lexington Books.

Miller, R. M. (1995a). Biosurfactant-facilitated remediation of metal-contaminated soils. *Environmental Health Perspectives*, 103, 59–62.

Miller, R. M. (1995b). Surfactant-enhanced bioavailability of slightly soluble organic compounds. In *Bioremediation-Science & Applications*, ed. H. Skipper, pp. 35–54. Madison, WI: Soil Society of America.

Mink, L. L., Williams, R. E. & Wallace, A. (1971). *Effect of Industrial and Domestic Effluents on the Water Quality of the Coeur d'Alene River Basin*. Idaho Bureau of Mines and Geology, Pamphlet 149.

Mitsch, W. J. & Jorgensen, S. E. (1989). *Ecological Engineering: An Introduction to Ecotoxicity*. New York: John Wiley.

Moore, J. N. (1994). Contaminant mobilization resulting from redox pumping in a metal-contaminated river-reservoir system. In *Environmental Chemistry of Lakes and Reservoirs*, ed. L. A. Baker, pp. 451–71. Washington, D.C.: American Chemical Society.

Moore, J. N., Ficklin, W. H. & Johns, C. (1988). Partitioning of arsenic and metals in reducing sulfidic sediments. *Environmental Science and Technology*, 22, 432–7.

Morrison, G. M., Batley, G. E. & Florence, T. M. (1989). Metal speciation and toxicity. *Chemistry in Britain*, 8, 791–5.

Mullen, M. D., Wolf, D. C., Ferris, T. J., Beveridge, T. J., Flemming, C. A. & Bailey, G. W. (1989). Bacterial sorption of heavy metals. *Applied and Environmental Microbiology*, 55, 3143–9.

Nies, D. H. (1992). Resistance to cadmium, cobalt, zinc, and nickel in microbes. *Plasmid*, 27, 17–28.

Otte, M. L. (1991). Contamination of coastal wetlands with heavy metals: factors affecting uptake of heavy metals by salt marsh plants. In *Ecological Responses to Environmental Stresses*, ed. J. Rozema & J. A. C. Verkleij, pp. 126–33. London: Kluwer Academic.

Peters, R. W. & Shem, L. (1992). Use of chelating agents for remediation of heavy metal contaminated soil. In *Environmental Remediation*, ed. American Chemical Society, pp. 70–84. Washington, D.C.: American Chemical Society.

Phillips, E. J. P., Landa, E. R. & Lovley, D. R. (1995). Remediation of uranium contaminated soils with bicarbonate extraction and microbial U(VI) reduction. *Journal of Industrial Microbiology*, 14, 203–7.

Plummer, E. J. & Macaskie, L. E. (1990). Actinide and lanthanum toxicity towards a

Citrobacter sp.: uptake of lanthanum and a strategy for the biological treatment of liquid wastes containing plutonium. *Bulletin of Environmental Contamination and Toxicology*, 44, 173–80.

Rani, D. B. R. & Mahadevan, A. (1989). Plasmid-encoded metal resistance in bacteria. *Journal of Scientific Industrial Research*, 48, 338–45.

Reamer, D. C. & Zoller, W. H. (1980). Selenium biomethylation products from soil and sewage sludge. *Science*, 208, 500–2.

Reimers, R. S., Akers, T. G. & White, L. (1989). Use of applied fields in biological treatment of toxic substances, wastewater, and sludges. In *Biological Waste Treatment*, ed. A. Mizrahi, pp. 236–72. New York: Alan R. Liss.

Roane, T. M. (1994). Microbial community analyses in heavy metal contaminated soil (thesis). Moscow, Idaho: University of Idaho.

Roane, T. M. & Kellogg, S. T. (1994). A study of microbial communities in lead-contaminated soils. In *Proceedings of 9th Annual Conference of Hazardous Waste Remediation*, ed. L. E. Erickson, D. L. Tellinson, S. C. Grant & J. P. McDonald, pp. 247–59. Manhattan, Kansas: Kansas State University.

Robinson, J. B & Tuovinen, O. H. (1984). Mechanisms of microbial resistance and detoxification of mercury and organomercury compounds: physiological, biochemical, and genetic analyses. *Microbiological Reviews*, 48, 95–124.

Rosenberg, E. (1986). Microbial surfactants. *Critical Reviews in Biotechnology*, 3, 109–32.

Rossi, G. & Ehrlich, H. L. (1990). Other bioleaching processes. In *Microbial Mineral Recovery*, ed. H. L. Ehrlich & C. L. Brierley, pp. 149–70. New York: McGraw-Hill.

Rudd, T., Sterritt, R. M. & Lester, J. N. (1984a). Complexation of heavy metals by extracellular polymers in the activated sludge process. *Journal Water Pollution Control Federation*, 56, 1260–8.

Rudd, T., Sterritt, R. M. & Lester, J. N. (1984b). Formation and conditional stability constants of complexes formed between heavy metals and bacterial extracellular polymers. *Water Research*, 18, 379–84.

Saouter, E., Turner, R. & Barkay, T. (1994). Mercury microbial transformations and their potential for the remediation of a mercury-contaminated site. In *Emerging Technology for Bioremediation of Metals*, ed. J. L. Means & R. E. Hinchee, pp. 99–104. London: Lewis Publishers.

Sarathchandra, S. U. & Watkinson, J. H. (1981). Oxidation of elemental selenium to selenite by *Bacillus megaterium*. *Science*, 211, 600–1.

Sauerbeck, D. R. (1991). Plant, element and soil properties governing uptake and availability of heavy metals derived from sewage sludge. *Water, Air, and Soil Pollution*, 57–58, 227–37.

Scott, J. A. & Palmer, S. J. (1988). Cadmium bio-sorption by bacterial exopolysaccharide. *Biotechnology Letters*, 10, 21–4.

Shuttleworth, K. L., & Unz, R. F. (1993). Sorption of heavy metals to the filamentous bacterium *Thiothrix* strain A1. *Applied and Environmental Microbiology*, 59, 1274–82.

Sieghardt, H. (1990). Heavy-metal uptake and distribution in *Silene vulgaris* and *Minuartia verna* growing on mining-dump material containing lead and zinc. *Plant and Soil*, 123, 107–11.

Sigg, L. (1985). Metal transfer mechanisms in lakes; the role of settling particles. In *Chemical Processes in Lakes*, ed. W. Stumm, pp. 283–310. New York: John Wiley.

Silver, S. (1992). Plasmid-encoded metal resistance mechanisms: range and overview. *Plasmid*, 27, 1–3.

Silver, S. & Misra, T. K. (1988). Plasmid-mediated heavy metal resistances. *Annual Review of Microbiology*, 42, 717–43.

Silver, S., Misra, T. K. & Laddaga, R. A. (1989). Bacterial resistance to toxic heavy metals. In *Metal Ions and Bacteria*, ed. T. J. Beveridge & R. J. Doyle, pp. 121–39. New York: John Wiley.

Simmons, P., Tobin, J. M. & Singleton, I. (1995). Considerations on the use of commercially available yeast biomass for the treatment of metal-containing effluents. *Journal of Industrial Microbiology*, 14, 240–6.

Smith, R. W., Yang, Z. & Wharton, R.A. (1988). Flotation of microorganisms – implications in the removal of metal ions from aqueous streams. In *Metallurgical Processes for the Year 2000 and Beyond*, University of Nevada, pp. 395–409. Nevada: University of Nevada.

Sposito, G. (1989). *The Chemistry of Soils*. New York: Oxford University Press.

Stevenson, F. J. (1982). *Humus Chemistry: Genesis, Composition, Reactions*. New York: John Wiley.

Strandberg, G. W., Shumate, S. E. & Parrott, J. R. Jr, (1981). Microbial cells as biosorbents for heavy metals: accumulation of uranium by *Saccharomyces cerevisiae* and *Pseudomonas aeruginosa*. *Applied and Environmental Microbiology*, 41, 237–45.

Stupakova, T. P., Demina, L. L. & Dubinina, G. A. (1988). Copper accumulation by bacteria in sea water. *Geokhimiya*, 10, 1492–502.

Sudhakar, G., Jyothi, B. & Venkateswarlu, V. (1991). Metal pollution and its impact on algae in flowing waters in India. *Archives of Environmental Contamination and Toxicology*, 21, 556–66.

Summers, A. O. (1992). The hard stuff: metals in bioremediation. *Current Opinion in Biotechnology*, 3, 271–6.

Suzuki, S., Fukagawa, T. & Takama, K. (1992). Occurrence of tributyltin-tolerant bacteria in tributyltin- or cadmium-containing seawater. *Applied and Environmental Microbiology*, 58, 3410–12.

Tan, H., Champion, J. T., Artiola, J. F., Brusseau, M. L & Miller, R. M. (1994). Complexation of cadmium by a rhamnolipid biosurfactant. *Environmental Science and Technology*, 28, 2402–6.

Thayer, J. S. & Brinkman, F. E. (1982). The biological methylation of metals and metalloids. *Advances in Organometallic Chemistry*, 20, 313–57.

Tomei, F. A., Barton, L. L., Lemanski, C. L., Zocco, T. G., Fink, N. H. & Sillerud, L. O. (1995). Transformation of selenate and selenite to elemental selenium by *Desulfovibrio desulfuricans*. *Journal of Industrial Microbiology*, 14, 329–36.

Tsezos, M. & Keller, D. M. (1983). Adsorption of radium-226 by biological origin adsorbents. *Biotechnology and Bioengineering*, 25, 201–15.

Tsezos, M., Baird, M. H. I. & Shemilt, L. W. (1987). The elution of radium adsorbed by microbial biomass. *The Chemical Engineering Journal*, 34, B57–B64.

Tsezos, M., McCready, R. G. L. & Bell, J. P. (1989). The continuous recovery of uranium from biologically leached solutions using immobilized biomass. *Biotechnology and Bioengineering*, 34, 10–17.

Tuin, B. J. W. & Tels, M. (1991). Continuous treatment of heavy metal contaminated clay soils by extraction in stirred tanks and in a countercurrent column. *Environmental Technology*, 12, 178–90.

Vatcharapijarn, Y., Graves, B. & Bender, J. (1994). Remediation of mining water with microbial mats. In *Emerging Technology for Bioremediation of Metals*, ed. J. L. Means & R. E. Hinchee, pp. 124–9. London: Lewis Publishers.

Verkleij, J. A. C., Lolkema, P. C., de Neeling, A. L. & Harmens, H. (1991). Heavy metal resistance in higher plants: biochemical and genetic aspects. In *Ecological Responses to Environmental Stresses*, ed. J. Rozema & J. A. C. Verkleij, pp. 8–19. London: Kluwer Academic.

von Lindern, I. H. (1981). *Ambient Air Analyses of Bunker Hill Lead Emissions*. Washington: Environmental Protection Agency, region X, no. IV0117NASX.

Walker, S. G., Flemming, C. A., Ferris, F. G., Beveridge, T. J. & Bailey, G. W. (1989). Physiochemical interaction of *Escherichia coli* cell envelopes and *Bacillus subtilis* cell walls with two clays and ability of the composite to immobilize heavy metals from solution. *Applied and Environmental Microbiology*, 55, 2976–84.

Whitlock, J. L. (1990). Biological detoxification of precious metal processing wastewaters. *Geomicrobiology Journal*, 8, 241–9.

Wildeman, T. R., Updegraff, D. M., Reynolds, J. S. & Bolis, J. L. (1994). Passive bioremediation of metals from water using reactors or constructed wetlands. In *Emerging Technology for Bioremediation of Metals*, ed J. L. Means & R. E. Hinchee, pp. 13–25. London: Lewis Publishers.

Zajic, J. E. & Seffens, W. (1984). Biosurfactants. *Critical Reviews in Biotechnology*, 1, 87–109.

Zhang, Y. & Miller, R. M. (1994). Effect of a *Pseudomonas* rhamnolipid biosurfactant on cell hydrophobilicity and biodegradation of octadecane. *Applied and Environmental Microbiology*, 60, 2101–6.

Zhang, Y. & Miller, R. M. (1995). Effect of rhamnolipid (biosurfactant) structure on solubilization and biodegradation of n-alkanes. *Applied and Environmental Microbiology*, 61, 2247–51.

Zosim, Z., Gutnik, D. & Rosenberg, E. (1983). Uranium binding by emulsan and emulsanosols. *Biotechnology and Bioengineering*, 25, 1725–35.

11

Molecular techniques in bioremediation

Malcolm S. Shields and Stephen C. Francesconi

11.1 Introduction

Hydrocarbon degradation by microorganisms is a natural manifestation of the carbon cycle. Only when hydrocarbon pollutants occur in excessively high concentrations, or prove highly toxic or are recalcitrant to existing bacterial degradative processes do they become the focus of remediation efforts. Intrinsic (or natural) bioremediation therefore defines the fate of most environmental chemicals, whether natural or anthropogenic. Indeed, through chemical modifications such as halogenation, a great deal of effort has gone into creating classes of hydrocarbons, such as the polychlorinated biphenyls (PCBs), that are more environmentally stable than their unmodified analogs. The recalcitrance of these compounds has led us to attempt to determine why the organisms normally responsible for carbon cycling fail when challenged by certain pollutants. Through the process of genetic engineering, strain development, enzyme redesign and pathway construction we can then undertake the molecular manipulation of biodegradative systems to improve their utility in bioremediation.

Biodegradation depends on the microbial production of enzymes capable of catalyzing chemical reactions that will transform (or, ideally, mineralize) pollutants. Bioremediation technologies can be enhanced in several ways, but the existing biochemical capabilities of the organisms should be reviewed before any anticipated genetic alterations are considered.

The most frequent means of stimulating bioremediation include:

(1)   Chemical modifications of the environment (Nelson et al., 1986; Mueller et al., 1991a,b, 1992; Pritchard & Costa, 1991; McCarty & Semprini, 1994) can include the addition of chemicals which act as electron acceptors or supplemental electron donors, or which enhance bioavailability (Rosenberg et al., 1979; Georgiou, Lin & Sharma, 1992).

(2)   Physical modifications of polluted sites may relate to *ex situ* or *in situ* treatments. They generally address limitations of electron acceptor supply or delivery (Semprini *et al.*, 1990; Huling, Bledsoe & White, 1992; Hopkins, Semprini & McCarty, 1993; Hinchee, 1994; McCarty & Semprini, 1994).

Chemical and physical modifications are sufficient to achieve bioremediation if native bacterial populations can destroy the pollutants. Such modifications fall into the area of bioremediation known as biostimulation. Such measures do nothing to expand the range of biological routes available at a given site, but rather augment the natural processes as they affect the rate of pollutant degradation. To change fundamentally the biology on site, a third approach must be considered.

(3)   Biological modification, or bioaugmentation, is the addition of microbes to the polluted site to accelerate degradation (Jobson *et al.*, 1974; Ismailov, 1985; Mueller *et al.*, 1989, 1990, 1991b, 1992, 1993a,b; Kelly & Cerniglia, 1991; Lin, Wang & Hickey, 1991; Lin, Mueller & Pritchard, 1992; Stormo & Crawford, 1992; Thomas & Ward, 1994). Bioaugmentation is also the only way to use genetically altered microbes (Timmis, Steffan & Unterman, 1994).

The purpose of this chapter is to review the applications of molecular biology to the genetic manipulation of organisms to be used in bioremediation. Molecular biological techniques will be considered in light of certain practical limitations inherent to bioremediation.

11.2 Pathway construction

11.2.1 Biochemical background

For several reasons, bacterial aromatic catabolism has received the most attention in terms of genetic manipulations relevant to bioremediation. A great deal of research has already gone into describing the enzymology, genetic organization and control mechanisms of these pathways. Aromatics also represent the largest single structural category of recalcitrant environmental chemicals with well defined toxicities and chemical behavior. Additionally, they are frequently attractive in terms of genetic manipulation for bioremediation because of the tendency of bacteria to sequester many of the genes responsible for the degradation of aromatics on self-transmissible plasmids. There is much information on degradative plasmids specifying the degradation of aromatics, including toluene (TOL) (Worsey & Williams, 1975), toluene and methylphenols (TOM) (Shields *et al.*, 1995), methylphenols (pSVI150) (Shingler *et al.*, 1989), naphthalene (NAH) (Dunn & Gunsalus, 1973), and 2,4-dichlorophenoxyacetic acid (pJP4 pRC10 and pEST4011) (Don & Pemberton, 1985; Chaudry & Huang, 1988; Mäe *et al.*, 1993). Aromatic substrates are therefore easily justified targets both in terms of need and pre-existing data. An analogous situation exists for the degradation of nonaromatics, such as the turpenes and linear alkanes, by members of the genus *Pseudomonas*, where this degradative ability is encoded by plasmids CAM (Chakrabarty & Gunsalus, 1971) and OCT (Eggink *et al.*, 1987b), respectively.

The aromatic degradative pathways receiving the most attention include those for toluene, benzoate, PCBs, and naphthalene. Five completely independent routes of toluene catabolism have been well characterized for aerobic bacteria (Figure 11.1). To date, all five pathways have been described in *Pseudomonas* and *Burkholderia*, despite the fact that at least two have been shown to be encoded by broad-host-range degradative plasmids.

Aerobic bacteria usually metabolize aromatics through one of four metabolic intermediates (substituted to various degrees depending on the starting compound): gentisate, homogentisate, catechol, and protocatechuic acid. These di-hydroxy-substituted intermediates are in turn metabolized through oxidative routes that result in cleavage of the aromatic ring. Routes producing catechol or protocatechuate intermediates (the most prevalent for aromatic utilization among bacteria) undergo oxygen addition in either *ortho* or *meta* position, relative to the adjacent paired hydroxyls. A summary of these two pathways in bacteria is given in Figure 11.2.

The NAH and TOL plasmids which encode naphthalene and toluene degradation represent discrete, mobile genetic systems that are easily transferred and selected. This tendency to focus on systems according to the ease of genetic manipulation led

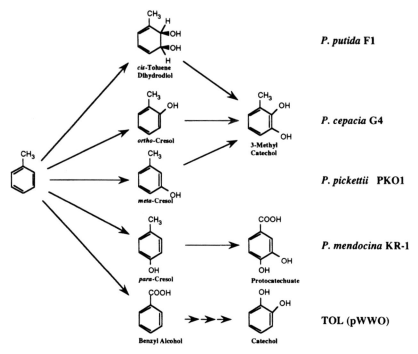

Figure 11.1. The five known aerobic pathways for toluene catabolism.

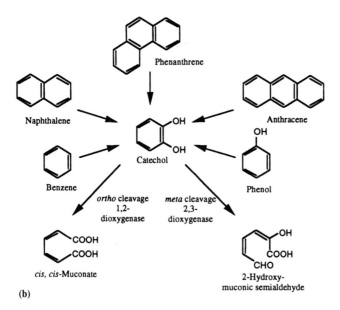

Figure 11.2. Examples of aromatics metabolized through (a) protocatechuate, and (b) catechol, from Stanier & Ornston (1973); after Gottschalk (1979).

to the first experiments in pathway manipulation. The best described of these involved the conjugal transfer of TOL, NAH, or pJP4 to recipient bacteria, thereby adding fundamental oxidative catabolic capabilities by virtue of the xylene, benzoate, or catechol oxygenases (Reineke & Knackmuss, 1979; Kellogg, Chatterjee & Chakrabarty, 1981; Harayama & Don, 1985; Gerger et al., 1991; Timmis et al., 1986). Consequently, there has been a general expectation that genetic manipulation of bacteria would allow us to overcome many currently perceived biochemical limitations through the introduction of missing genes or the modification of existing ones.

Investigations, beginning about 1974, by K. N. Timmis, H-J. Knackmuss, and colleagues have contributed substantially to our understanding of the degradation of substituted benzoates. Their early experiments on benzoate and chlorobenzoate catabolism by a chemostat selected bacterium, *Pseudomonas* sp. B13, were particularly informative on the consequences of catabolism blockages (Dorn et al., 1974; Knackmuss & Hellwig, 1978; Reineke & Knackmuss, 1980; Wigmore & Ribbons, 1981; Weightman et al., 1984; Timmis, Rojo & Ramos, 1987), and the effects of added pathway junctures.

The information base in aromatic catabolism allowed halogenated aromatics to emerge as the logical models for the biodegradation of recalcitrants, including the chlorobenzoates (Walker & Harris, 1970; Horvath & Alexander, 1970; Dorn et al., 1974; Reineke & Knackmuss, 1980), 2,4-dichlorophenoxyacetic acid (2,4-D) (Evans et al., 1971), chlorobenzenes (Spain & Nishino, 1987; Gerger et al., 1991), pentachlorophenol (PCP) (Steiert & Crawford, 1986), chlorosalicylates (Rubio, Engesser & Knackmuss, 1986), and PCBs (Furakawa et al., 1989, 1993; Yates & Mondello, 1989; Ahmad, Masse & Sylvestre, 1990; Taira et al., 1992; Erickson & Mondello, 1992, 1993; Asturias & Timmis, 1993; Hofer et al., 1993; Peloquin & Greer, 1993; Hirose et al., 1994). For a concise synopsis of bacterial pathways involved in haloaromatic-degrading bacteria and selected mutants, the reader is referred to an excellent 1987 review of the subject by P. J. Chapman (Chapman, 1987).

11.2.2 Operon deregulation

Trichloroethylene cometabolism

In terms of absolute scale of contamination, the largest single category of groundwater pollutants on earth is the volatile organics, which include the chloroaliphatics. The members of this group found most abundantly by far in groundwater are the chloroethylenes: trichloroethylene (TCE); tetrachloroethylene; *trans*-1,2-dichloroethylene (DCE); 1,1-DCE; and vinyl chloride (Rajagopal, 1986). This class of organic pollutant has also received a great deal of interest in terms of pathway construction for several reasons.

Despite an exceptionally low rate of biological transformation of TCE in the environment, there have been numerous reports of bacteria capable of metabolizing TCE and related isomers. Anaerobes seem to utilize TCE as a terminal electron acceptor. Aerobes, however, co-oxidize TCE through the coincidental expression of a wide variety of oxygenases that are not produced in response to TCE, but accept it as a co-substrate. The more diverse heterotrophic bacteria that co-oxidize TCE are found primarily among the genus *Pseudomonas*, but also include representatives of *Burkholdia, Mycobacterium, Xanthobacter, Alcaligenes, Rhodococcus, Nocardia, Methylosinus*, and *Streptomyces*.

These bacteria are induced for TCE oxidation by a wide variety of organic compounds including methane (Little *et al.*, 1988; Oldenhuis *et al.*, 1989; Moore, Vira & Fogel, 1989; Uchiyama *et al.*, 1992), ammonia (Arciero *et al.*, 1989; Vannelli *et al.*, 1990), isoprene (Ewers, Freier-Schroder & Knackmuss, 1990), propane (Wackett *et al.*, 1989), propylene (Ensign, Hyman & Arp, 1992), phenol (Nelson *et al.*, 1986; Montgomery *et al.*, 1989; Harker & Kim, 1990), 2,4-dichlorophenoxyacetic acid (Harker & Kim, 1990), toluene (Nelson *et al.*, 1986, 1987, 1988; Wackett & Gibson, 1988; Winter, Yen & Ensley, 1989; Kaphammer, Kukor & Olson, 1990) and isopropyl benzene (Dabrock *et al.*, 1992).

When considered for use in TCE bioremediation, such bacteria are presented with two fundamental difficulties unrelated to the biochemistry of substrate utilization. TCE must compete for the same active site on the degradative enzyme as the inducer substrate molecule. In addition, inducer requirements mean that the organisms are active only within an environmental zone where an effective concentration of an inducer can be maintained. As a result, such TCE-degrading bacteria remain 'tethered' to the inducing substrate. These TCE-degradative bacteria are targets for molecular manipulation because of such limitations that hamper the design of both *in situ* and bioreactor applications. Since some of their pathways and enzymes are reasonably well understood, effective manipulation is possible. Regulatory complications arise when environmental application of inducing substrates such as toluene or phenol is contemplated. Although these problems may be overcome with genetically altered variants that do not require inducer substrates, the related environmental and regulatory complications may be no less formidable.

Molecular strategies to overcome these limitations have been applied by a number of labs. Several TCE co-oxidative enzymes have been cloned and expressed in *E. coli* (Winter, Yen & Ensley, 1989; Kaphammer, Kukor & Olsen, 1990; Kukor & Olsen, 1990; Shields & Reagin, 1992). These constructs all demonstrate the ability to degrade TCE when induced through the regulatory system of the cloning vector. While this route has contributed significantly to our understanding of these genes and their regulation, such clones remain unusable for bioremediation, primarily because of regulatory restrictions. Also, the questionable utility of *E. coli* under environmental conditions can better be addressed by constructions in environmental strains instead. Other methods to improve natural strains include the use of transposon and chemical mutants in which the native inducible regulatory mechanism is inactivated, and from which subsequent constitutive mutants may be isolated. *Methylosinus trichosporium, B. cepacia*, and *P. mendocina* strains have been constructed that constitutively express their TCE cooxidative enzymes, and that are therefore free of inducer requirements (and consequently, potential competitive inhibition by them) (Phelps *et al.*, 1992; Shields & Reagin, 1992; Steffan & Tugusheva, 1993; Tugusheva & Steffan, 1993). Recently, chemically induced copper-tolerant mutants of methanotrophs have been constructed that produce the more biodegradatively active soluble form of the methane monooxygenase in the presence of environmentally significant levels of copper (G. S. Sayler, pers. commun.), whereas the wild type cannot.

11.2.3 Vectors

The ability to clone, express, delete, fuse, and deliver various genes in organisms used for bioremediation is now possible using a variety of vectors. Plasmid systems developed for cloning and expression in *E. coli* are not particularly well suited to bacteria other than the enterobacteriaceae, in part because of inability to replicate (Schmidhauser, Ditta & Helsinki, 1988), unrecognized promoter consensus sequences (Rothmel *et al.*, 1991), and use of uncommon codons because of different G + C content (West & Iglewski, 1988). Plasmids able to replicate in many different bacterial hosts are collectively termed broad-host-range vectors. Their development has allowed genes to be cloned and stably maintained by various bacterial strains with relative ease. More recent reports have described the inclusion of inducible promoters upstream of multiple cloning sites, providing high levels of expression.

A second class of vectors has been described that are transposon based, relying on insertion into the host genome rather than extra-chromosomal maintenance for stability. Genes of interest can be cloned into multiple cloning regions downstream of an inducible promoter, and inserted into the genome as an appropriately mutated copy. Some of the transposable elements can be used for insertion mutagenesis, and in conjunction with a reporter gene, promoter probing. Because many of the transposon-based vectors were designed for bioremediation purposes, they offer alternative selectable markers, such as resistance to mercury salts, organomercuric compounds, herbicides or arsenite. Although an extensive description of these plasmids and their use is not included here, representative examples are listed in Table 11.1.

11.2.4 Hybrid pathways and enzymes

Once a metabolic limitation is identified, and if analogous pathways are known, an isofunctional enzyme with broader specificity can be introduced (Reineke & Knackmuss, 1980; Gerger *et al.*, 1991). New pathways, once conceived, can be logically built from components of well-described ones. This concept serves as a persuasive rebuttal to the view that such manipulations are not warranted because all enzymes necessary for biodegradation already exist. Even if such enzymes do exist, there is certainly no guarantee they will be suitably induced or will even occur at the place and time required for the requisite interaction with site and pollutant.

Table 11.1. *Vectors*

Plasmid	Based on	Regulator	Inducer	Mode of delivery	Selection	Use	Reference
pNM185	TOL	XylS/P_m	Benzoate *m*-toluate	Self mobilizable	Str^r	Regulated high expression of foreign genes in *Pseudomonas*, up to 5% total cellular protein. Expression in *P. putida* was up to 600 fold inducible.	Mermod *et al.* (1986)
pVDtac39/24	pMMB22/24	lacI^q	IPTG	Triparental with HB 101 and helper plasmid pRK2013	Ap^r	Subcloning and expression of *Pseudomonas* genes, protein fusions	Deretic *et al.* (1987)
pMON7197 pMON7190	Tn7				Gm^r	Allowed stable insertion of *lacZY* into the chromosomes of soil bacteria	Barry (1988)
pAV10	IS50 promoters		Constitutive	Conjugation	Tc^r	Uses promoters normally producing transposase in Tn5 to transcribe genes in a variety of gram–bacteria	Kozlowski *et al.* (1988)
pQF40	*Pseudomonas*				Cb^r, Tc^g	Promoterless Tet gene	Farinha & Kropinski (1989)
pQF26	Phage ϕPLS27				Cb^r, Cm^g	Promoterless Cat gene. Helps find promoters by cloning genomic DNA and selecting for Cm^r	

Vector / construct	Based on	Regulator / promoter	Inducer	Delivery	Resistance	Application	Reference
Mini-Tn5 lacZ Mini-Tn5phoA Mini-Tn5 xylE Mini-Tn5 lux Mini-Tn5 Tc, Km, Cm, etc.	Tn5	N/A	N/A	Transposons on R6K suicide delivery plasmid, that provides *tnp* in *cis*.	Km^r Sp^r Tc^r Cm^r	Insertional mutation Gene fusion Promoter probes Cloning into chromosome of many gram – bacteria	de Lorenzo *et al.* (1990)
pLOF/pUTKm pLOF/pUTPt pLOF/pUTHg pLOF/pUTArs	Tn10, Tn5			Filter mating with pGP704	Km^r ptt^r Hg^r As^r	Cloning, insertional mutagenesis, chromosomal insertion of foreign genes.	Herrero, de Lorenzo & Timmis (1990)
pKMY299	NAH7	nahR/P_G	salicylate		Tc^r	Allows high levels of expression, up to 10% of cellular protein in *P. putida*	Yen (1991a)
pPZ10 pPZ20 pPZ30	Broad-host-range vector pNM480	N/A	N/A	Transformation of *P. aeruginosa*	Ap^r	Allow isolation of protein fusion to βgal in all three reading frames.	Schweizer (1991)
pEHK455	pWW0	XylR	Toluene, m-xylene	Conjugative, triparentals	Sm^r	Regulated high level expression, <600 fold induction in Enterobacteriaceae and *Pseudomonas*. No induction in *Agrobacterium* and *Rhizobium*.	Keil & Keil (1992)
MiniD-181 pEB8,11,12,14	T7	lacuv5/$lacI^q$ T7 Promoter	IPTG	Mobilize with bhr helper plasmid pRK2013	Tc^r Ap^r	Overexpression of genes in *P. aeruginosa*. Up to 20% total cellular protein.	Brunschwig & Darzins (1992)
P4::Tn5AP-2 pKG2/4	P4 (phage) P4, pKT231			Infection, transduction onto any P4 sensitive strain	Km^r Km^r, Sm^r	Suicide transposon vector: mutagenesis, phagmid vector, cloning.	Polissi *et al.* (1992)
pVLT31/33/35	pMMB207	P_{tac}/$lacI^q$	IPTG	Conjugative, triparentals	Tc^r, KM^r, Sm/Sp^r	Plasmid-based high level expression in *Pseudomonas*.	de Lorenzo *et al.* (1993b)

Table 11.1. *continued*

Plasmid	Based on	Regulator	Inducer	Mode of delivery	Selection	Use	Reference
pUT/mini-Tn5-lacIq/P_{trc} Mini Tn5 xylS/Pm::T7 Pol.	TOL, T7	P_{trc}/lacIq Xyl S/Pm T7/P_{T7}	IPTG Benzoate	Conjugative, triparentals Mini-Tn5 transposition	Kmr Tcr	Chromosomal expression, mutagenesis, etc. System allows cloning of gene to be expressed into mini-Tn5 transposon using lacZ blue/white screen, then delivering the transposon to stable insertion in host chromosome. Another mini-Tn5 provides T7 polymerase inducible by benzoate. Seven different antibiotic selections available to allow multiple genes expressed in same host.	Herrero et al. (1993)
pVNB# (1–4)	P_m(TOL) P_u(TOL) = P_{sac}(NAH)	Xyl S Xyl R Nah R	Alkyl, halo benzoates halotoluenes salicylates	Tn5 suicide	Varies	Outward facing promoters allow controlled induction from chromosomal locations. Conditional phenotypes dependent on specific effectors.	de Lorenzo et al. (1993a)
pUCP20/21 pUCP22/23 pUCP24/25 pUCP26/27	pUC18/19			Blue/white cloning in *Pseudomonas* and related organisms	Apr Apr, Gmr Gmr Tcr	Extraneous DNA removed, gentamicin selection added, amp selection deleted, tetracyclin selection added.	West et al. (1994)

r, resistance to gentamicin (Gm), ampicillin (Ap), herbicide bialaphos (ptc), inorganic and organomercurials (Hg), arsenite (As), kanamycin (Km), streptomycin/spectinomycin (Sm/Sp), Tetracycline (Tc), Carbenicillin (Cb), Chloramphenicol (Cm); IGP, utilization of the detergent IGEPAL; and not applicable (N/A).

Pseudomonas sp. B13

Chlorobenzoates

A general approach for the molecular redesign of a blocked pathway involves the joining of two or more central pathways before the point of dysfunctionality in a process sometimes referred to as enzyme recruitment. The two most common junctures of aromatic catabolism by bacteria are those leading to protocatechuic acid and catechol (Figure 11.2). One of the first bacteria modified in these central catabolic functions was *Pseudomonas* sp. B13 (Knackmuss & Hellwig, 1978). Extensive modifications of B13 have been reported in the last 15 years. In 1980, Reineke and Knackmuss demonstrated that *Pseudomonas* sp. B13 carrying the TOL plasmid could grow with 4-chlorobenzoate or 3,5-dichlorobenzoate, whereas B13 alone could not (Reineke & Knackmuss, 1980). Shortly thereafter, Lehrbach *et al.* (1984) demonstrated that the TOL-encoded toluate dioxygenase (*xyl*XYZ) and carboxylate dehydrogenase (*xyl*L), along with their native promoter (Pm) and positive regulatory gene (*xyl*S), caused the same effect when cloned and introduced to B13. *Pseudomonas* sp. B13 containing *xyl*XYZ, *xyl*L, Pm, and *xyl*S grew with both the B13 native substrate 3-chlorobenzoate and 4-chlorobenzoate. However, this recombinant did *not* use 3,5-dichlorobenzoate (Timmis, Rojo & Ramos, 1987).

The rationale for expanding the repertoire of chloroaromatics degraded by *Pseudomonas* sp. B13 was as follows (and is presented diagrammatically in Figure 11.3). While B13 was able to oxidize 3-chlorobenzoate through 3-chlorocatechol and subsequent *ortho*-cleavage to readily metabolizable aliphatic acids, it was unable to utilize 4-methyl phenol, 4-methylbenzoate, or 4-chlorobenzoate. It was theorized that through a combination of introduced genes from other organisms and mutational alteration of native genes these compounds could be rendered utilizable.

The TOL-encoded enzymes demonstrate a far more relaxed substrate specificity than the benzoate oxygenase/dehydrogenase system native to B13. *P. putida* (TOL) can degrade mono- and di-substituted alkyl benzoates, such as 3- and 4-methylbenzoate, 3,4-dimethylbenzoate and 3-ethylbenzoate, to their corresponding alkyl-substituted catechols by virtue of these two enzymes. These alkyl-catechols are subject to *meta*-cleavage in TOL-containing cells through dioxygenation by the TOL-encoded catechol 2,3-dioxygenase (*xyl*E). The utilization of chlorocatechols produced by these relaxed-specificity oxygenases is, however, not productive through the TOL-encoded *meta*-fission catechol 2,3-dioxygenase, which works well for the alkyl-substituted catechols, but is rapidly inactivated by chlorocatechols. B13, on the other hand, does not degrade alkyl benzoates, but does oxidize 3-chlorobenzoate to 3-chlorocatechol. This occurs through a similar route, utilizing the benzoate 1,2-dioxygenase, which accepts substrates chlorinated at the *meta*-position. The chromosomally encoded catechol 1,2-dioxygenase of *Pseudomonas* sp. B13 allows subsequent catabolism via the *ortho*-cleavage of 3-chlorocatechol. It has even been suggested that some catechol 1,2-dioxygenases may have evolved

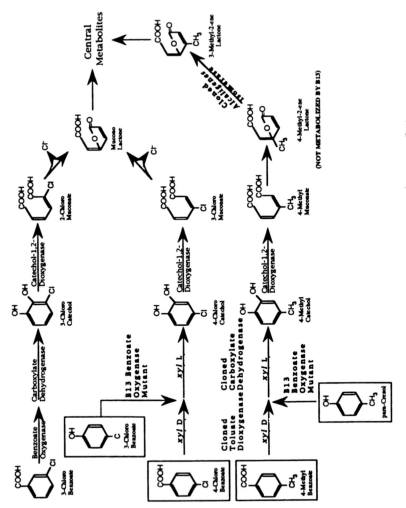

Figure 11.3 Pathway construction in *Pseudomonas* sp. B13.

with a high degree of specificity for certain halocatechols (van der Meer *et al.*, 1991a). The resulting chloromuconate is then productively metabolized by B13.

4-Chlorocatechol and 3,5-dichlorocatechol are similarly oxidized by the same catechol, 1,2-dioxygenase. However, neither of their potential precursors, 4-chlorobenzoate or 3,5-dichlorobenzoate, are substrates for benzoate 1,2-dioxygenase. Therefore, the *ortho*-fission pathway of the unmodified B13 is never presented with the catechol derivatives of 4-chlorobenzoate or 3,5-dichlorobenzoate. The toluate 1,2-dioxygenase of TOL was chosen as an isofunctional enzyme supplement to the benzoate 1,2-dioxygenase. Toluate-1,2-dioxygenase has the requisite relaxed substrate specificity necessary to accept 3- and 4-chlorobenzoate, 3,5-dichlorobenzoate, 4-methyl phenol, or 4-methylbenzoate.

These strain alterations were performed in three phases to address three classes of substrate (Timmis *et al.*, 1987: 61) resulting in a genetically altered bacterium with a much broader range of aromatic growth substrates.

4-Chlorobenzoate

Introduction of *xyl*XYZ and *xyl*L was via a recombinant Tn5::*xyl*XYZLS. The resultant strain FR1 (*Pseudomonas* sp. B13::Tn5 *xyl*XYZLS]) remained unable to use 4-methyl phenol and continued to utilize 3-chlorobenzoate via *ortho*-fission of the 3-chlorocatechol oxidation product as it did before the pathway was altered. However, unlike B13, FR1 was able to grow with 4-chlorobenzoate as the sole carbon source. The newly introduced toluate dioxygenase accepted 4-chlorobenzoate as a substrate, and the carboxylate dehydrogenase re-aromatized the resulting dihydrodiol to 4-chlorocatechol. 4-Chlorocatechol proved to be an acceptable substrate for the native catechol 1,2-dioxygenase. Its immediate product, 3-chloromuconate, was completely metabolized.

4-Chlorophenol

The next phase was to address the inability of FR1 to utilize *para*-cresol and 4-chlorophenol. This was accomplished for the second of these substrates not through introduction of a new gene but simply by mutation of the existing chromosomal benzoate oxygenase to accept *para*-cresol and 4-chlorophenol as substrates. The mutant could not grow with *para*-cresol as the sole carbon source because 4-methyl-2-ene lactone produced by the combined action of the chromosomally encoded catechol 1,2-dioxygenase and mucono-lactone isomerase accumulated as a dead-end metabolite. It could, however, grow with 4-chlorophenol as the sole source of carbon and energy because the mutant benzoate oxygenase produced 4-chlorocatechol. Like 4-chlorobenzoate, this intermediate is utilized by B13.

para-Cresol

The most recent alteration of B13 involved the introduction of another foreign gene to allow complete metabolism of *para*-cresol through transformation of the dead-end metabolite, 4-methyl-2-ene lactone, accumulated by the benzoate dioxygenase mutant FR1. A lactone isomerase (which converts 4-methyl-2-ene lactone to 3-methyl-2-ene lactone) gene was cloned from an *Alcaligenes*. 3-Methyl-2-ene lactone was readily utilized by FR1. The final biological product of these constructions is a genetic chimera composed of five independent pathways, including the gene resources of two bacteria and two degradative plasmids; i.e., the benzoate oxygenase of B13 is at least transiently associated with the degradative plasmid pWR1 (Weightman *et al.*, 1984).

Chlorosalicylates and salicylate

In a separate demonstration, *nah*G cloned from NAH7 expanded the catabolic range of B13 to include the utilization of salicylate as well as 3-, 4-, and 5-chlorosalicylate as carbon and energy sources (Lehrbach *et al.*, 1984).

These alterations stand as important examples of the flexibility of pathway construction when enough is known about the individual components of these pathways.

Burkholderia (Pseudomonas) cepacia G4

Chlorobenzene and chlorophenol

In a case analogous to the combination of the two plasmid encoded *ortho*- and *meta*-fission pathways in B13 for specific halo-aromatic utilization, another bacterium was modified with a hybrid pathway for this purpose. *B. cepacia* G4 degrades toluene, phenol, benzene, and *ortho*-cresol via a unique toluene *ortho*-monooxygenase (Tom) pathway (Shields *et al.*, 1989) encoded by the toluene catabolic plasmid TOM (Shields *et al.*, 1995).

However, like TOL, TOM encodes the *meta*-fission of the resulting catechol (Figure 11.4). Consequently, it is unproductive in the assimilation of chloroaromatics like chlorobenzene and 2-chlorophenol, which are also metabolized to 3-chlorocatechol because of catechol 2,3-dioxygenase inactivation. To avoid a build-up of this product an *ortho*-cleavage enzyme and downstream enzymes for the complete catabolism of the product were recruited through the introduction of the 2,4-dichlorophenoxyacetic-acid-degradative plasmid pRO101 (Kaphammer, Kukor & Olsen, 1990).

The two plasmids combine to encode parallel *ortho*- and *meta*-fission pathways for the complete catabolism of the chlorinated aromatics chlorobenzene and 2-chlorophenol, as well as for the non-chlorinated and methylated aromatics, where neither plasmid alone can encode these pathways.

A similar addition of a catabolic plasmid was used to construct a *Pseudomonas* strain capable of utilizing 2,4,6-trinitrotoluene (TNT) as a carbon and energy source (Duque *et al.*, 1993). *Pseudomonas* sp. clone A utilizes the nitro-groups of TNT as the sole source of nitrogen while growing with fructose as the carbon source. Toluene is produced as a consequence of this activity, which cannot be metabolized by clone A. Introduction of TOL allowed complete utilization of TNT without the need for an ancillary source of carbon and energy.

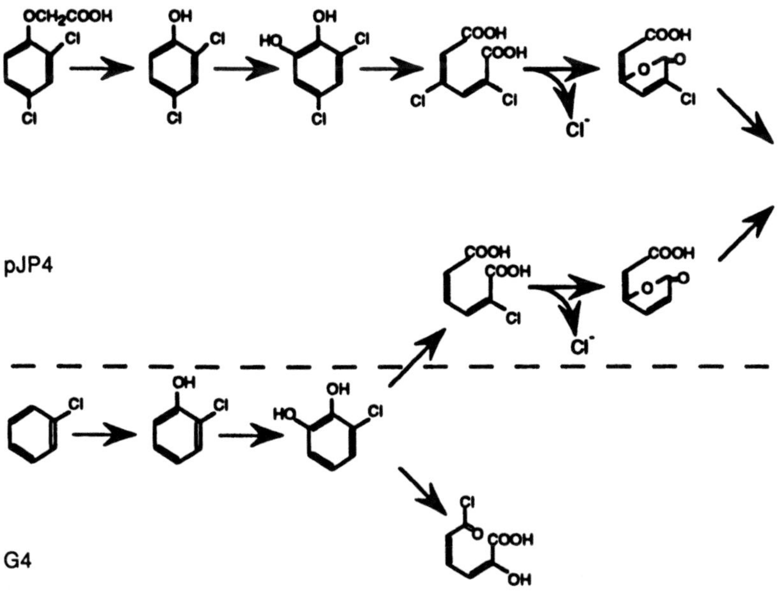

Figure 11.4. Chloroaromatic utilization by *P. cepacia* G4 (pJP4).

P. putida F1

Haloalkanes

P. putida F1 has the capacity to metabolize toluene via a chromosomally encoded toluene-2,3-dioxygenase (*tod*ABC1C2DE) to its corresponding dihydrodiol (Figure 11.1). This same enzyme has been shown to co-oxidize chlorinated ethenes (i.e., TCE) to oxidized organic acid products. This system, however, has an absolute requirement for unsaturated carbons, and will not add oxygen without a double bond present. As a result, it has no effect on haloalkanes. In another unique combination of enzymatic processes utilizing yet another catabolic plasmid Wackett *et al.* (1994) combined the oxidative capacity of Tod (cloned under the control of a constitutive promoter) with the anaerobic reductive activity of *cyt* P450*cam* (camphor inducible from the CAM plasmid) in a single cell. Under reduced oxygen conditions the *cyt*P450*cam* enzyme can reduce pentachloroethane to TCE (Figure 11.5).

In this recombinant pathway construct, under conditions of 50% atmospheric O_2, the pentachloroethane is reduced to TCE, which is subsequently oxidized by the constitutively expressed Tod.

Chloroaromatic catabolic transposons

The recent discovery of specific catabolic transposons that encode oxidative enzymes for chloroaromatic catabolism has created the possibility of some

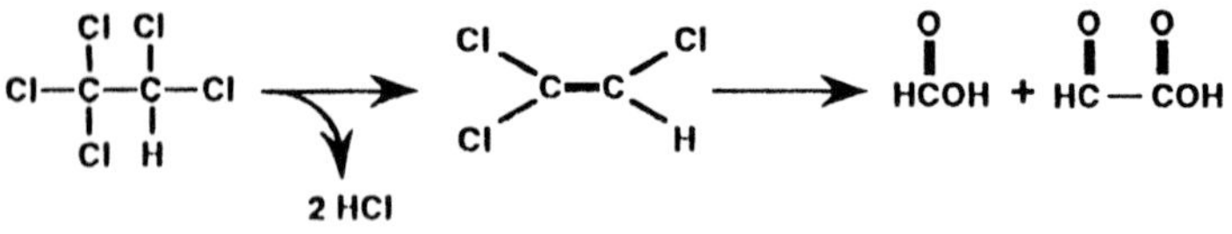

Figure 11.5. P-450 CAM and Tod construction.

interesting variations to the molecular modifications just discussed. Chlorobenzene catabolism by Tn*5280* (discovered in a *Pseudomonas* isolate) (van der Meer, Zehnder & de Vos, 1991b) would offer an interesting alternative for expansion of the catabolic potential of toluene degraders with only *meta*-fission pathways. Tn*5271*, a transposon from an *Alcaligenes* isolate which encodes chlorobenzoate catabolism, was found to move to native sediment bacterial populations in microcosms dosed with 3-chlorobenzoate, 4-chloroaniline, and 3-chlorobiphenyl, but not with 2,4-dichlorophenoxyacetic acid (Nakatsu *et al.*, 1991; Fulthorpe & Wyndham, 1992). Both represent interesting, 'nonrecombinant' mechanisms that may already exist in nature, in which gene clusters can be mobilized and recruited to forge novel pathways under appropriate selection conditions.

Non-catabolic genes for catabolic pathway constructions

Bacterial hemoglobin

Pathway alternations are made with the express purpose of improving a given metabolic process. We generally view these modifications of catabolic pathways as the addition or modification of existing catabolic enzymes. However, certain ancillary pathways and enzymes exist that are not necessary to the catabolism of a given substrate, yet nevertheless can influence the overall metabolic process.

Vitreoscilla are obligate aerobes capable of utilizing carbon sources under microaerophilic conditions. They produce a globin protein (encoded by the *vgb* gene) which, when combined with the proper heme-prosethic moiety, aids in cellular oxygen acquisition and transport by an as-yet-undescribed mechanism (Wakabayashi & Webster, 1986). Recombinant *Acremonium chrysogenum* or *E. coli* expressing *Vitreoscilla* hemoglobin have been shown to have improved fermentative yields of cephalosporin and α-amylase in batch fermentations, as well as enhanced *E. coli* growth characteristics (Khosla & Bailey, 1988; Khosravi, Webster & Stark, 1990; DeModena *et al.*, 1993). Presumably, these effects are due to an increased efficiency in the transport and, therefore, the utilization of oxygen.

Experiments have been undertaken in our laboratory to determine if the introduction of this oxygen transport protein to *B. cepacia* $PR1_{31}$ would accentuate its oxygen-scavenging capacity (Shields *et al.*, 1994). Subcloning of *vgb* was accomplished through PCR amplification using synthetic oligonucleotide primers (Figure 11.6). This approach allowed a specific construction containing *vgb* free of its native promoter while retaining the native ribosomal binding site and the introduction of unique cloning sites for *Eco*RI and *Bam*HI. These sites, in turn, allowed the cloning of this gene under an IPTG-controllable p*tac* promoter in the broad-host-range vector pMMB277 (an Inc Q replicon, RSF1010, constructed cloning vector).

Expression of *vgb* by this strain was found to accelerate the rate of oxygen uptake and the oxidative processes linked to the constitutive *ortho*-monooxygenase activity (i.e., oxidation of toluene, phenol, and TCE) by more than 400%. Its utility under environmental or bioreactor conditions remains to be determined.

Poly-beta-hydroxybutryate

A particular problem associated with bioaugmentation using constructed bacteria is competition with indigenous species. For the most part, molecular applications have little bearing because so little is known about what makes a particular bacterium a successful competitor. In a particular niche, such a technology might be of use when the target pollutant cannot serve as an adequate carbon and energy source because its concentration is too low (e.g., PCBs or dioxins), or it is the target of a cometabolic system (e.g., aerobic co-oxidation of chlorinated ethenes) that results in no benefit to the degrading cells. It is virtually impossible to specifically 'feed' a given bacterial strain, either *in situ* or in bioreactors, with wild type competitors present. One way to compensate for this is to engineer the degradative bacteria to store the reducing equivalent within themselves prior to environmental release. Once the engineered bacteria are released, such an alteration would necessarily be transient. Whether availability of such short-term energy sources would be beneficial to the treatment process would depend on the specific situation.

A possible solution to this internal energy source limitation would include the use of strains that accumulate carbon storage polymers under conditions of nitrogen-limited growth. An example of such a polymer, common among environmental bacteria, is poly-beta-hydroxybutryate (PHB). The gene system for PHB production has been cloned from *Alcaligenes eutrophus* as a three-gene anabolic pathway (Peoples & Sinskey, 1989). Inclusion of these genes as part of a cometabolic pathway construct in cells that can catabolize PHB is an intriguing alternative to the energy limitations of constructed strains in a highly competitive environment.

11.3 Rational enzyme redesign

Site-directed mutagenesis and molecular modeling approaches have contributed dramatically to the understanding of enzyme structure and function. For example, much work has been done in the modification of the glutathione reductase (GR) of *E. coli*, which has led to alterations in substrate binding (Perham, Scrutton & Berry, 1991) and coenzyme specificity (Scrutton, Berry & Perham, 1990; Perham *et al.*, 1991; Rescigno & Perham, 1994; Mittl *et al.*, 1993; Bocanegra, Scrutton & Perham, 1993). Phenotypic changes resulting from primary structure alteration are now more predictable. By comparing GR with trypanothione reductase (another

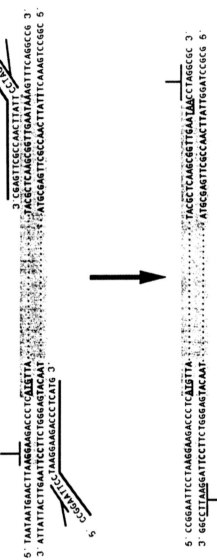

Figure 11.6. Cloning of bacterial hemoglobin.

flavoprotein disulfide oxidoreductase), primary sequence alterations of GR have been made through site-directed mutagenesis, creating recombinant *E. coli* (Henderson *et al.*, 1991) and human GRs (Bradley, Bucheler & Walsh, 1991) that accept trypanothione instead of glutathione. Other researchers are on the threshold of performing similar modifications with the related *E. coli* enzyme thioredoxin reductase (Kuriyan *et al.*, 1991; Waksman *et al.*, 1994).

Examples of other recombinant enzymes in which an alteration using site-directed mutagenesis resulted in altered substrate binding efficiencies, rates of catalysis, or stability include carbonic anhydrase (Alexander, Nair & Christianson, 1991), lactate dehydrogenase (Feeney, Clarke & Holbrook, 1990), and several industrially important proteases (Wells *et al.*, 1987; Siezen *et al.*, 1991; Teplyakov *et al.*, 1992; Aehle *et al.*, 1993; Rheinnecker *et al.*, 1994).

11.3.1 Trichloroethylene

Presumably, such methods could also be applied to enzymes useful in bioremediation, notably the aromatic oxygenases. One of the best-studied classes of oxygenases is the cytochrome P450s. Because they have been successfully crystallized and structurally characterized, it has been possible to create eukaryotic transgeneric hybrids with predictably altered substrate activities following specific structural changes (Murakami *et al.*, 1987; Yabusaki *et al.*, 1988; Shibata *et al.*, 1990; Sakaki *et al.*, 1994). The only bacterial P450 to have been studied in this depth is the P450$_{CAM}$ (Poulos, Finzel & Howard, 1986; Poulos & Howard, 1987; Poulos, Finzel & Howard, 1987; Raag & Poulos, 1991; Poulos & Raag, 1992), which catalyzes the oxidation of camphor and is encoded by the CAM catabolic plasmid (Eggink *et al.*, 1987a). P450$_{CAM}$ has therefore become a principal candidate for rational enzyme redesign. On the basis of computer simulations it has been proposed that P450$_{CAM}$ should be capable of TCE orientation, to result in either oxidative (cyt p450$_{CAM-ox}$) (hydrogen abstraction) or reductive (cyt p450$_{CAM-red}$) dehalogenations of TCE in the same active site normally bound by camphor (Ornstein, 1991). The oxidative dehalogenation described here would be analogous to the dehalogenation of TCE via microbial oxygenases, presumably via an oxo-intermediate (Hales, Ho & Thompson, 1987; Wackett & Gibson, 1988; Wackett *et al.*, 1989). Reductive dehalogenation of a chlorinated solvent by P450$_{CAM}$ has been demonstrated, not with TCE, but with the chlorinated alkane tetrachloroethane (described in this chapter – see section 11.2.4, Figure 11.5).

11.3.2 PCBs

Site-directed mutagenesis

The biphenyl dioxygenases (BPDO) (Mondello, 1989) of *Pseudomonas* sp. LB400 (Bopp, 1986) and *P. pseudoalcaligenes* KF707 (Furukawa & Arimura, 1987; Furukawa *et al.*, 1993; Taira *et al.*, 1992) have also become targets for enzyme bioengineering. The BPDO of LB400 has now been successfully crystallized (Eltis *et al.*, 1993) and is reported to exist as an octamer. Both of these enzymes have an unusual substrate range that includes many of the higher chlorinated PCB congeners, thus ensuring an interest in their structure/function relationships. The cloning of these oxygenases has led to attempts to engineer them with the goal of altering the substrate range and rates of congener oxidation (Erickson & Mondello, 1993). High similarities (95.6% amino acid sequence similarity) in the large subunits of the BPDOs of LB400 and *P. pseudoalcaligenes* KF707 (Yates & Mondello, 1989; Erickson & Mondello, 1992; Hofer *et al.*, 1993) (i.e., the ISP reductase components) fail to reflect relatively dramatic differences in PCB congener specificities. Site-directed mutagenesis of four nucleotides that cause a change in four amino acids in a stretch of seven has been performed in the *bph*A gene encoding the LB400 BPDO reductase component. These seven now match the same seven of the corresponding *bph*A1 subunit of KF707 (amino acid positions 335, 336, 338, and 341), resulting in a significant (150 to 670%) increase in the capability to degrade di-, tri- and tetra-*para*-substituted congeners (more representative of the KF707 congener range).

Subunit mixing

Such alterations illustrate the vast potential for the alteration of oxygenases and reductases useful in bioremediation provided enough molecular information is available. Where there is not enough information on structure or activity to infer active sites or probable effects of amino acid substitution, site-directed alterations of the gene would not be an effective way to approach enzyme modification. In these cases it may be feasible to alter enzyme specificity by substituting whole enzyme subunits. Candidates here include BPDO as well as the various toluene oxygenases. The toluene dioxygenase (TDO) of *P. putida* F1 has been shown to be a three-component oxygenase with recognized reductase, ferredoxin reductase, and iron sulfur oxidase (Zylstra & Gibson, 1989), and the toluene monooxygenase of *P. mendocina* (Richardson & Gibson, 1984) and the dimethylphenol hydroxylase of *Pseudomonas* sp. C600 are both five-subunit enzymes (Shingler *et al.*, 1989; Yen *et al.*, 1991). Extensive homology is evident between much of the latter two operon sequences, with some open reading frames exhibiting greater than 70% homology at the amino acid level. Determination of the role of these individual subunits would not only contribute to an understanding of the enzyme, but also aid in directing the engineering of hybrid enzymes. Rational substitution of those subunits responsible for substrate binding and coenzyme utilization could create enzyme activities and substrate ranges beyond the capacity of either of the parent enzymes.

Such multicomponent enzyme complexes have been created through incorporation of oxidase and electron transport protein subunits from both the TDO (*tod*C1C2BA) of F1 and the BPDO (*bph*A1A2A3A4) of KF707. The hybrid proteins exhibit substrate specificities similar to those of both parents, but some exhibit altered substrate oxidation rates (Hirose *et al.*, 1994). One hybrid, *tod*C1*bph*A2A3A4, demonstrated a substantially greater rate of TCE cometabolism in *E. coli* (Furukawa *et al.*, 1994), which points to an apparent freedom of substitution of the electron transfer reductase components (*bph*A3A4 and *tod*BA) without substantial effects on enzyme activity. However, hybrids of the large and small subunits of the terminal dioxygenase components (*bph*A1A2 and *tod*C1C2) create unpredictable changes in the substrate recognition capacity and turnover rates of the hybrid holoenzyme that can be addressed at present only through trial and error.

11.4 GEM survival

11.4.1 Promoting GEM survival: implications for bioremediation

The ability to manipulate heritable characteristics in microorganisms has been seen as a double-edged sword. By changing the substrate specificity of enzymes, recruiting novel pathways into a given organism, or changing the manner in which an operon is regulated, more useful biodegradative bacteria can be engineered. However, there is considerable concern that these GEMs would harm the environment by displacing native biota. In virtually all scenarios, the anticipated utility of GEMs in deliberate release hinges on their continued viability in order to provide a source of enzyme to transform the pollutant compound. Because the released cells are alive, they can grow, proliferate, and even spread their genetic material to native organisms. By the very nature of the genetic manipulations, the GEMs are phenotypically distinct, creating the possibility of altering native organisms as well. Many worry that these recombinants may change the microbial ecology where they are applied, and even spread outside the treatment area. While such strains may appear harmless in the laboratory, it is virtually impossible to be certain that they would remain so in the field.

More practical questions are whether introduced GEMs will survive in polluted matrices, or whether their continued viability will dramatically affect the success of remediation. A recent development in this area is the isolation of bacterial strains which can respond to selective agents under field conditions. These strains have been called 'field application vectors' or FAVs, since their intended purpose is to express foreign genes in environments not conducive to the use of the parent organism (Lajoie *et al.*, 1993). The proposed use of FAVs illustrates the need to control bacterial growth and survival under the real-world conditions where bacterial remediation has to occur.

As a management tool, molecular monitoring techniques could provide information on how often the cells must be reintroduced, and whether a particular chemical addition aids in their survival and activity. Questions about a particular organism's effectiveness, persistence, and deleterious effects can be answered by microcosm-scale investigations. Several such studies designed to measure the survival of individual GEMs in aquifer, soil, and lake microcosms and in activated sludge have been published in the last few years (Table 11.2).

Table 11.2. *GEM survival*

Microcosm type	Organism	Survival (days)	Conditions	Comments	Reference
Soil	P. fluorescens, and Tn5 marked derivatives	16	Loamy sand	In soil inoculated with both parent and Tn5 derivatives, parent showed increased survival.	van Elsas et al. (1994)
Soil	P. fluorescens, and Tn5 derivative	75	Clay soil	Declined 5 orders by 75 days.	Araujo et al. (1994)
			Sandy soil	Could not detect after 25 days. All varied by temp and pH.	
Activated sludge	E. coli (pBH500) P. putida (pBH500)	40	1. Fill and draw 2. Continuous flow	Declines quickly then stabilizes after decreasing ~4 orders.	Fujita et al. (1994)
Soil	P. fluorescens: wild type and regulatory gene gacA mutant		Various soils	Cells decreased in numbers by 5–6 orders over 3 months, but varied as to endpoint by soil type. Antibiotic producing wild type had competitive advantage over mutant. Effect more profound in unplanted soil than rhizosphere.	Natsch et al. (1994)
Soil	P. aeruginosa (pR68.45)	<70	Unsaturated flow; soil remained wet	GEM slowly declined in number from start until total disappearance by day 70. Transconjugants of indigenous population persisted and spread.	Lovins et al. (1994)
Soil	E. coli (pRP4)		Clay loam, sandy loam, sandy microcosms	Measured conjugation frequencies of broad host range plasmids to indigenous microflora.	Naik et al. (1994)
Aquifer	P. putida	30		Studies designed to test the mobilization and persistence of recombinant catabolic plasmid pD10.	Hill et al. (1994)

Environment	Organism	Duration	Conditions	Results	Reference
Soil	*E. coli*	160	Experiment performed at 4 temps	Estimated survival times of 23.3 months at 5 °C.	Sjorgren (1994)
Soil	*P. aeruginosa* UG2 from oil contaminated soil, also UG2 with chromosomal reporter genes	12 weeks	15 g soil in bottles, 10^8–10^9 CFU/g added. Also, oil spiked tests	UG2 and derivatives declined 2–3 orders throughout 12 week study in soil microcosm, while indigenous bacteria stabilized. However, in oil contaminated, it competed well with indigenous.	Flemming *et al.* (1994)
Aquifer	*P. putida* with TOL and RK2	56		Showed by gene probes that GEM persisted without selective pressure.	Jain *et al.* (1987)
Activated sludge	*P. putida*	>56	Lab. scale activated sludge unit	Yes, and was able to transfer plasmid (pD10); transconjugants had greater rate of 3-chlorobenzoate breakdown.	McClure *et al.* (1989)
Aquatic	Engineered and wild type *E. coli* and *P. putida*	20	Varied temperature	Recombinant DNA plasmids had no significant effect on survival of recombinant strains.	Awong *et al.* (1990)
Soil	*Erwinia carotovora* and Tn5 derivative	60	200 g soil in 1 l glass jars, with humidified sterile air circulation.	Compared wild type with Tn5 variant for survival, competitiveness, and effects upon other genera in soil microcosms, and found virtually no difference.	Orvos *et al.* (1990)
Soil	*P. cepacia* AC1100	42	Soil microcosms with radish seeds planted on top	Found persistence of *P. cepacia* AC1100 in microcosms with and without addition of 2,4,5-T.	Bej *et al.* (1991)
Soil	*P. putida* PPO301 (pRO103) degrades 2,4-dichlorophenoxyacetate	53 days		Measured ecological effects on soil by GEM versus nonrecombinant parent uninoculated control. Microcosms supplemented with 2,4-dichlorophenoxyacetate, glucose, or unamended.	Doyle *et al.* (1991)
Soil	*Enterobacter cloacae*, *Escherichia coli*, *P. putida*	43 days	Soil column	Measured ammonification, nitrification, and denitrification of the soil in GEM, and its nonrecombinant parent.	Jones *et al.* (1991)

Table 11.2. *continued*

Microcosm type	Organism	Survival (days)	Conditions	Comments	Reference
Aquatic	*Pseudomonas* sp. B13 (degrades 3-chlorobenzene)	28	Intact sediment cores with overlaid water column	GEM survived in sediment throughout the length of study. Microcosm was healthy, and presence of GEM did not affect total numbers of bacteria present. GEM enhanced the rate of 3-chlorobenzoate and 4-methylbenzoate added to sediment.	Pipke *et al.* (1992)
Aquatic	*P. putida*	60	Growth in lake water	Measured growth rate and plasmid loss in prototrophic and auxotrophic strains.	Sobecky *et al.* (1992)
Groundwater Aquifer-sediment	*P. cepacia* $PR1_{23}$ (constitutive degrader of TCE)	70	20 ml groundwater in 100 ml serum bottles	Groundwater microcosms: degraded 25 μM TCE within 24 h at high cell density, but not at all at lower densities. Sediment microcosms: high densities removed all TCE overnight, lower densities took much longer (weeks).	Krumme *et al.* (1993)
Aquifer	*Pseudomonas* sp. B13 (degrades 3-chlorobenzene)	>365		Compared survival of previously introduced nonrecombinant (B13) in containment plume with microcosm composed of the same aquifer material and introduced both B13 and the genetically modified FR120.	Krumme *et al.* (1994)

11.4.2 Preventing GEM survival: suicide containment systems

As early as 1975, participants in the Asilomar conference expressed their concerns over the safety and wisdom of releasing GEMs. At that time the primary concern was accidental release, but we are now pursuing intentional releases. The successful release of several recombinant bacteria has done little to alleviate concerns over perceived ecological dangers (Halvorson *et al.*, 1985).

In an attempt to address these concerns, some have suggested that genetic engineering itself can be the solution as well as the cause. The use of the naturally occurring lethal genes that are carried by several low-copy plasmids has been suggested for the control of released GEMs. When linked to a gene control network on a recombinant vector, the expression of such lethal genes could presumably be controlled.

One decade after the Asilomar conference, S. Molin and colleagues described a recombinant strain engineered in such a manner that it could be 'instructed' to kill itself (Molin *et al.*, 1987). Since then, many refinements have been added. The crux of the problem is to design a system that can be tightly controlled, will cost little or nothing to activate, is as simple as possible (for in complexity lies a whole subset of problems), and will ultimately be completely effective (reviewed by Molin *et al.*, 1993).

Molin's first strain utilized the *hok* gene (so named because it encodes the *h*ost *k*illing function of the plasmid R1), part of the *gef* family of genes of *E. coli*. The *gef* family (composed of *hok*, *relF*, and *gef*) encode small (50–52 amino acid) polypeptides that are probably implicated in low-copy number plasmid maintenance (Gerdes, 1988; Poulsen *et al.*, 1989). The *hok* gene was initially put under the control of the *E. coli trp* promoter/operator, such that when the strain is grown in the presence of tryptophan, the promoter is repressed by virtue of the *trp* repressor. Theoretically, without added tryptophan, the strong *trp* promoter is derepressed, allowing the transcription of the *hok* gene, and nearly instant death for cells harboring the plasmid. The actual results were not as encouraging. A decrease in strain survival of only about two orders of magnitude was reported. It was speculated that the high rate of survival was caused by insufficient promoter strength. A similar construct using the *E. coli lac* promoter to drive the *hok* gene was examined using a *Bacillus subtilis* shuttle vector. When induced with IPTG, only about 75% of cells in the culture died, presumably because of limited expression via suboptimal Shine–Dalgarno sequences present downstream from P_{lac}.

When *hok* was placed under the invertible *fimA* promoter (which randomly switches to constitutive expression by virtue of the *fimB* and *fimE* regulators) a stochastic killing was observed, and the population of viable cells in the culture slowly decreased. Unfortunately, the *fimA* promoter only works in *E. coli*, and therefore would be of little utility in actual bioremediation scenarios.

Shortly after this, other workers also described suicide vectors using the *hok* gene under the control of the *lac* promoter (Bej, Perlin & Atlas, 1988; Atlas *et al.*, 1989). They found that when cells were grown without selective pressure for the plasmid and induced, nearly all the cells died; survivors had lost the plasmid. When cells were grown under selective pressure (carbenicillin in the medium) and induced, viable cells initially declined, but quickly recovered and returned to log growth. Analysis of survivors demonstrated the presence of intact *hok* genes, so survivors presumably arose via second site mutations (affecting such functions as the *lac*I^q gene or *lac* promoter, IPTG membrane permeability, or membrane targets of *hok*). Applications of the cells to soil microcosms demonstrated that cell death could be induced, albeit to only 90–99%. Survival was due to emergence of *hok*-resistant bacteria.

In 1991, several factors were demonstrated to affect the efficiency of *rel*F controlled killing of *E. coli* when under *lac* promoter control (Knudsen & Karlström, 1991). Cells escaped suicide primarily because of the mutation rate and the leakiness of repression during normal growth. When *rel*F was under *lac*UV5 promoter/operator control, the inactivation of suicide function through spontaneous mutation occurred at a frequency of $<5 \times 10^{-9}$. Knudsen has further theorized that if the number of suicide minus (mutant) bacteria in a culture can be kept at zero before induction, all cells will die. This can be achieved in two ways: provide the suicide function in duplicate, and control the suicide gene expression so stringently, with a chromosomally located *lac*I^q gene, that a basal level of the suicide function to which cells can adapt will not be present (Knudsen *et al.*, 1995).

In an attempt to provide a strain that was more responsive to environmental conditions and constraints, a containment system was designed which can be used in conjunction with organisms that degrade aromatics via the TOL *meta*-cleavage pathway and in which the promoter (P_m) of the TOL plasmid is fused to the *lac*I gene (Contreras, Molin & Ramos, 1991). Another site on the plasmid contains the *gef* gene controlled by P_{tac} and a functional *xyl*S. When *xyl*S effectors such as benzoate and substituted benzoates are present, the XylS protein activates P_m, producing LacI. LacI in turn represses the killing function *gef*. However, in the absence of effectors, LacI is not synthesized. P_{tac} becomes derepressed, allowing expression of *gef*, causing death of the cells. There are two problems, however: the concentration of effector at which P_m is no longer activated is still relatively high (especially for environmental purposes), and the appearance of Gef-resistant cells (probably due to mutation).

Jensen *et al.* (1993) demonstrated that bacteria bearing two copies of the killing function have many times fewer survivors (i.e., $\sim 10^{-8}$ survivors per cell per generation). They did not address the usefulness of a control system requiring high concentrations of controlling substrates.

In 1993, suicide by an *E. coli* containing the *npt*II-*sac*R-B suicide cassette (*npt*II gene encodes kanamycin resistance) was reported (Recorbet *et al.*, 1993). In the absence of sucrose the *sac*R gene does not induce expression of the *sac*B gene. To induce cell death, sucrose was added to the soil in which the organism was released. This causes the cells to produce levansucrase (a *sac*B product), which in turn causes the periplasm of the cells to fill with levan. This leads to cell death by lysis. Sucrose addition to sterile soil caused a thousand-fold reduction in *E. coli* population density. Mutants arise at a frequency of 10^{-5}. Advantages of this system are its adaptability to a wide range of organisms and its use of an inducer as inexpensive as sucrose.

Recently, there has been speculation that causing the death of GEMs may not be sufficient to contain them (Klier, 1992), because the modified genotype may persist and in fact be taken up by natural biota (Kloos *et al.*, 1994). The killing genes discussed above may actually enhance the release of the recombinant DNA into the environment through the synchronized death of many of the introduced bacteria. Since bacteria have been long known to take up released DNA (Avery, MacLeod & McCarty, 1944, reviewed in Stotzky, 1989), several groups have experimented with nuclease genes as a mode of total GEM and recombinant DNA neutralization in one step.

The leader sequence of the *Serratia marcescens* extracellular nuclease has been removed and the gene for the resulting nontransportable protein cloned behind the leftward promoter (P_L) of lambda (Ahrenholtz, Lorenz & Wackernagel, 1994). In the presence of a temperature-sensitive repressor (cI_{857}), the cells can be induced to produce this altered enzyme by an increase in temperature. This produces the nuclease within the cell and, because it cannot be exported, its intracellular concentrations rise, rapidly degrading the cellular genome. Limitations to this system are that cell survival is reduced to only 2×10^{-5}.

The colicin E3 gene, which is a 16S RNA riboendonuclease, has been reported to be an effective suicide construct in several bacteria (Díaz *et al.*, 1994). Since E3 acts upon a highly conserved nucleotide sequence of the 16s RNA, it should be effective in most bacteria. However, some strains of *E. coli* carry a colicin immunity function, the *imm*E3 gene. The E3 protein binds to the RNAase domain of the colicin, blocking its nucleolytic activity. Very few strains have natural immunity to E3, and the gene can be placed on the chromosome of a desired GEM using mini-transposon vectors. The colicin E3 (killing) gene can be introduced on a plasmid encoding a desired trait. In such an engineered strain, the plasmid-encoded colicin E3 nuclease would be inactive by virtue of the chromosomally encoded immunity function and the cells would suffer no deleterious effects from promoter 'leakiness', thus creating a mutual dependency between the plasmid and chromosome. Where such cells are released, lateral gene spread of the plasmid by mating with indigenous bacteria or by their transformation with liberated DNA would be vastly reduced because of the unregulated production of E3 in the new host. When cloned to a promiscuous plasmid, E3 caused a decrease of four to five orders of magnitude in the frequency of successful conjugation.

Kloos *et al.* describe the induction of cellular autolysins by the ϕX174E gene, which causes cell lysis, and, apparently via a cellular nuclease, the degradation of cellular nucleic acids (Kloos *et al.*, 1994). Indeed, cellular DNA was shown to be degraded after induction of the plasmid carrying the E gene. These containment approaches are summarized in Table 11.3.

Table 11.3. *Suicide containment systems*

Gene	Product	Source	Reference
hok	52 aa, membrane associated	Unknown, plasmid encoded	Molin *et al.* (1987)
gef	50 aa, membrane associated	Works with *sok*, plasmid stability, chromosomally encoded	Contreras *et al.* (1991)
relF	51 aa, membrane associated	Likely plasmid stability, chromosomally encoded	Knudsen & Karlström (1991)
*sac*B	Levansucrase	Induction by sucrose causes levan to accumulate in periplasm, lysing cells	Recorbet *et al.* (1993)
nuc	Extracellular nuclease	Endonucleolytically cleaves RNA and DNA	Ahrenholtz *et al.* (1994)
E3	Colicin	Inhibits protein synthesis by destroying mRNA	Díaz *et al.* (1994)
SRRz	Bacteriophage λ lysis gene cluster	S: Creates lesions in cytoplasmic membrane R: Transglycosylase, degrades murein Rz: Cleaves peptidoglycan cross links?	Kloos *et al.* (1994)
E	Bacteriophage ϕX174 lysis gene	Induces and acts with cellular autolysins to produce a pore in the membrane	Kloos *et al.* (1994)

11.5 Molecular probes

The two most important qualities required of any method designed to detect and/or enumerate specific microorganisms in environmental samples are sensitivity and specificity. The ability to detect one organism in 1×10^9 cells relies not only on the ability to differentiate the lone cell from the milieu in which it is sampled, but also on the strength of the signal which allows its perception. If the method chosen is particularly time-consuming or expensive, or it if requires elaborate instrumentation, the turnaround time of sampling can become the limiting factor in accurately gathering the data. Therefore, the speed of assay, cost, and amount of technology necessary to perform a given test also become important factors. The assay must not only require inexpensive reagents, but, ideally, also require little training of the technicians. The ideal assay must be very reproducible and include controls appropriate to the samples.

Methods that require the re-isolation and growth of cells encounter a separate series of problems. Cells isolated from environmental sources do not always grow up on minimal or selective media as do lab-cultured cells. Environmental isolates often contain resting cells and cells that have become dormant and that may seem nonculturable under the given growth conditions. Switching to rich growth media can cause proper enumeration to suffer because of faster-growing natural flora that outgrow the cells in question. And of course, methods dependent on the regrowth of cells are unable to detect dead cells.

All detection methods have limitations. The most severe are sensitivity and reproducibility. Happily, the past few years have seen a flurry of technologies developed specifically to detect cells at rather low environmental concentrations. The most prominent of these methods are outlined here, along with notes of their merits and reported sensitivity. Several excellent reviews have recently dealt with this topic (Drahos, 1991b; Atlas *et al.*, 1992; Edwards, 1993; Prosser, 1994).

11.5.1 Bioluminescence

The discovery that certain bacteria are able to produce visible light has been turned into a tool to probe whether certain genes are active *in vivo* (Carmi *et al.*, 1987), and an *in situ* method for detecting GEMs in the environment. The best-studied bioluminescent pathway is that of the *lux* operon of *Vibrio* (Meighen, 1988). The luciferase genes are localized to two operons composed of seven open reading frames: *lux*A and *lux*B encode the structural components of the luciferase, *lux*C, *lux*D, and *lux*E are genes whose products function in the synthesis and recycling of the aldehyde substrate, and *lux*I and *lux*R are regulatory genes that encode for the transcriptional regulators of the operon. Light production is dependent on oxygen, a long-chain aldehyde substrate, and reducing equivalents ($FMNH_2$). Photons are produced as a result of luciferase-catalyzed oxidation of the aldehyde (Figure 11.7).

A limitation of this type of detection is that cell viability and growth is essential for a positive result. Resting, injured, or dead cells (which still contain genetically modified nucleic acids) would remain undetected.

Light produced from the cells can be quantified by luminometry (i.e., a light meter), charge-coupled microscopy (Jovin & Arndt-Jovin, 1989) and autoradiography. Charge-coupled microscopy is the most sensitive. Utilizing a liquid nitrogen cooled camera, light quanta produced by individual cells are detected. A charge-coupled device (CCD) has also been used to visualize a genetically modified

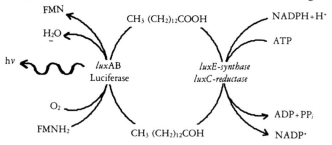

Figure 11.7. The *lux* bioluminescence system.

Xanthamonas campestris strain engineered to bioluminesce (Shaw *et al.*, 1992). The strain was applied to cabbage plants and surrounding soils in a limited field study conducted in 1990. The fate of the organisms was monitored by both CCD-microscopy and plate counts. Recombinant cells were found for up to six weeks, with the CCD and plate counts giving quantitatively similar results. Bioluminescence in this case was no more sensitive than plate counts, and required a precision instrument.

Bioluminescence was also demonstrated as a method of detecting *E. coli* made luminescent by incorporation of the *lux* operon (Rattray *et al.*, 1990). Roughly 100 cells were necessary to produce enough light to be detected by luminometry in liquid culture. Detection of cells inoculated into soil showed a reduced sensitivity of about one order of magnitude, where 2×10^2 to 6×10^3 cells/gram soil could be detected. This loss of sensitivity was probably a result of combined factors, including quenching of the emitted light by soil particles, attachment of cells to soil particles, and sensitivity of the luminometer. This method has the advantage of speed, requiring approximately 5 minutes per sample. Beauchamp *et al.* (Beauchamp, Kloepper & Lemke, 1993) used Tn5-*lux*AB mutants of *Pseudomonas* strains to study root colonization. CCD imagery allowed detection of bacterial cells colonizing roots of cucumbers, cotton, maize, and soybeans. However, cells were more sensitively enumerated by selective broth enrichment and dilution plating. Two areas in which bioluminescence may be eminently practical (though not as a probe) is as an on-line continuous assessment of bacterial activity in bioreactors (King *et al.*, 1990) or as reporter genes that respond to specific chemicals in environmental samples (Burlage, Sayler & Larimer, 1990; Tescione & Belfort, 1993).

11.5.2 Polymerase chain reaction

The polymerase chain reaction (PCR) has gained application throughout biology as an extremely sensitive and specific method. As an experimental tool, it allows researchers to amplify exact regions of the genome they wish to study. Such amplified fragments can be cloned, used as probes or directly sequenced. With the availability of inexpensive oligonucleotides, specific primers can be synthesized to precise regions surrounding or within genes, and the genomic DNA of a cell can be amplified, subcloned, or sequenced within a week.

In an applied role, PCR is invaluable as a 'fingerprinting' device in forensic applications (Budowle *et al.*, 1994), as a diagnostic tool of diseases (Cariolou *et al.*, 1993), and for identifying organisms from viruses (Ali & Jameel, 1993) to whales (Baker & Palumbi, 1994). Volumes have already been devoted to this powerful technique, despite its being only 10 years old.

The polymerase chain reaction utilizes a thermostable DNA polymerase to 'amplify' DNA through a series of temperature cycle steps. The key to the specificity of the reaction is the selection of oligonucleotide primers that hybridize to the opposite strands of the DNA being tested, about 400–2000 bp apart. If the sequence of the primers is unique within the genome, and the primers hybridize to the target DNA at a high enough temperature to avoid 'close matches' (various forms of computer software are readily available to assist the selection of efficient primers), a single amplified DNA fragment, visible as a single band on an agarose gel, can be produced. Template DNA having the 'target' sequence of interest is placed in a tube with the oligonucleotide primers, each of the deoxyribonucleotide triphosphates, buffer, magnesium, and the thermostable DNA polymerase (a number are currently available). A 'cycle' of *denature–anneal–extend* is repeated 25–30 times, the newly synthesized molecules now acting as templates in the next round, causing a geometric progression in the number of molecules. The amplified product is visualized on an agarose gel, either by staining or by transfer to a membrane and hybridizing with a probe of the target sequence.

There are, however, a number of stumbling blocks with the PCR assay. Very close attention must be paid to pipetting technique so that stocks (of nucleotide, primers, buffers) do not become contaminated with target DNA. Controls are *de rigueur*, and must be run at every amplification. Important, too, is the nature of the template. Because environmental samples usually contain soil, sediment, aquifer material, and/or plant parts, and because these materials also contain substances that interfere with and inactivate the PCR assay, such as humic acids, the bacterial DNA must be extracted free of these contaminating substances. Other compounds, such as metals, also inhibit the reaction, necessitating purification of the DNA prior to amplification.

Many methods used to extract total DNA from all the cells in a sediment and to amplify target sequences have been reported. Steffan and Atlas initially described a method of blending the cells and sediment, followed by cell lysis and clarifying through differential centrifugation (Steffan & Atlas, 1988). DNA was recovered and purified by CsCl isopycnic density gradient centrifugation before amplification. The resulting products were immobilized to a membrane by dot blot and probed with the target sequence. A sensitivity of 10 cells/g sediment was reported. Later experiments (Atlas *et al.*, 1989) compared two techniques for total DNA isolation from sediment and soil, and found that direct lysis of cells in soil followed by extraction of DNA yielded a sensitivity of about 1 cell/gram soil or sediment.

Extraction of DNA directly from soils and sediments improved DNA recovery 1000–5000% over methods requiring separation of cells from the environmental sample first. However, humic substances are often found in soils, sediments, and aquatic environments. Composed primarily of humic acid and fulvic acid, they strongly inhibit PCR. Tsai and Olson used gel filtration chromatography on an extracted sediment which had a high enough humic content to be unamplifiable to remove enough of the humics to allow amplification (Tsai & Olson, 1992b). Their reported sensitivity was less than 1 cell/g of sediment, since the sterile sediment they started with was seeded with 70 cells, and the eluate of the column was 100 μl, of which they used 1 μl in the PCR reaction to achieve a positive result.

So, sensitivity is obviously quite good despite the numerous manipulations required to produce an amplifiable sample, but what about the specificity? One of the problems in any PCR reaction is the appearance of extra bands caused by false priming of the substrate (genomic) DNA added to the reaction. The fault is commonly with the oligonucleotide primers, which are able to find homologous regions in the chromosomes where they can anneal and be extended. A number of strategies have been designed to deal with this problem in general (Sheffield *et al.*, 1989; Sarkar, Kapelner & Sommer, 1990), but a different approach has been taken by a few groups. Chaudhry *et al.* inserted a 300 bp region of napier grass genomic DNA into a 2,4-dichlorophenoxyacetic-acid-degrading plasmid, pRC10, and transferred it into *E. coli* (Chaudhry, Toranzos & Bhatti, 1989). Since the napier grass fragment does not share homology with prokaryotic sequences, the primers used to amplify the region proved quite specific. Furthermore, it easily differentiated the GEMs from other coliforms in samples from lake and sewage treatment water. A similar approach was used by Amici *et al.* who designed a synthetic 'signature' region (Amici *et al.*, 1991), which was inserted into the chromosome of *Pseudomonas putida* and released into freshwater microcosms. The results demonstrated that the PCR assay was able to detect cells one week after no cells could be recovered by culture techniques.

Detection by PCR can therefore be seen as having the advantages of high sensitivity and specificity, speed, and ability to assay many samples at the same time. Furthermore, the method does not require growth, nor an overly expensive apparatus. The requirement for competent technical assistance, the losses incurred from extraction and purification of samples, and per sample costs are among its disadvantages.

11.5.3 Immunological techniques

A paradigm for specificity is illustrated by antibody recognition of an epitope unique to a particular bacterial strain. Serotyping of medically important strains of *Salmonella, Streptococcus,* and *Escherichia*, to name just a few, underscores the sensitivity of this technique. The same specificity can be exploited in environmental samples as long as an antibody can be found with high avidity to a unique surface determinant of the GEM. Producing the antibody through monoclonal technology with high specificity is relatively easy, but finding the strain-specific determinant is more challenging, requiring screening against various organisms that would be encountered in nature. For a particular antigen to be of value in monitoring the fate of a GEM, the determinant must be expressed and subject to antibody binding at all times, and during all growth phases (including starvation and resting) (Raibaud & Shwartz, 1984).

Once such an antigen/antibody combination is identified, the detection of the organism can be greatly facilitated. Coupled with fluorescent secondary antibodies, the antigen/antibody complex can be observed directly by fluorescent microscopy, allowing the cells to be viewed nearly *in situ*. Cell cytometry can be automated, making the sampling and quantization of data much easier, and independent of extraction, purification, and/or growth losses inherent in other methods. This is especially true for aqueous samples, which can be filtered for concentration and analyzed in a matter of hours. Cells bearing the epitope of interest can also be detected by ELISA, most notably in dot-blot studies where the primary antibody bears a linked enzyme that produces a colorimetric response (horseradish peroxidase, alkaline phosphatase, luciferase, etc.), or by autoradiography.

Such an approach was reported by Ramos-Gonzales *et al.* working with TOL plasmid host *Pseudomonas putida* 2440 (Ramos-Gonzales *et al.*, 1992). A monoclonal antibody (Mab) which recognizes only *P. putida* via a lipopolysaccharide determinant of the outer membrane was developed. The strain was released into different growth conditions in nonsterile lake water. The investigators reported detecting as few as 100 cells/dot using a dot-blot immunoassay. A similar approach was used to detect *B. cepacia* $PR1_{23}$ by antibodies to an O-specific lipopolysaccharide in groundwater microcosms (Winkler, Timmis & Snyder, 1995), and a recombinant *P. putida* in a lake-water microcosm (Brettar *et al.*, 1994). A summary of the use of molecular methods to create organism- and gene-specific monitoring probes is given in Table 11.4.

11.5.4 Hybridization techniques

Nucleic acid hybridization is one of the most widely accepted molecular techniques used to detect the presence of a particular gene, or set of genes, in organisms. The method utilizes nucleic acid 'probes' which were historically radiolabeled, but can now be enzymaticaly tagged as well. There are two limitations: (1) getting enough of the target to be probed, and (2) determining the optimal stringency of hybridization.

One way to increase the signal strength of bacterial cells found in an environmental specimen is to grow them under nonselective conditions, and 'lift' the resulting colonies to membranes, where the cells are lysed, and the nucleic acids denatured and bound to the membrane for probe hybridization. Since a colony contains substantially more copies of the target DNA than an individual cell, it is much easier to detect cells bearing the target sequence. Environmental samples, however, commonly contain solids to which bacteria adhere, or in which they become entrapped, rendering their isolation difficult. Besides difficulties in physically separating the cells from solids without lysis, the physiological state and/or nutritional requirements of many bacteria will preclude their isolation on any particular medium (discussed in Jain, Burlage & Sayler, 1988). The requirement for growth, therefore, will probably lead to an underestimation of the number of cells present. Hybridization stringency may sometimes be defined rather easily, by inclusion of known target-containing and target-devoid colonies, co-processed with the test membranes. Difficulties are sometimes encountered when probes contain conserved sequences found in undesired targets. This then creates a range of hybridization responses beyond the desired positive/negative result.

Limitations notwithstanding, colony hybridization techniques have been used successfully for a number of years to monitor environmental samples. Representative examples include detection of naphthalene and toluene degraders in contaminated soils (Sayler *et al.*, 1985), recombinant organisms in groundwater (Jain *et al.*, 1987;

Table 11.4. *Molecular probes*

Method	Novelty	Detection limit	Reference
Bioluminescence	5 minute sampling	100 cells/ml (liquid) 1000 cells/ml (soil)	Rattray *et al.* (1990)
	Detects organisms in soil, leaves, etc.	As sensitive as plate counting	Shaw *et al.* (1992)
	Direct visualization of root colonization by *Pseudomonas* sp.	Less sensitive than dilution-plating and enrichment techniques	Beauchamp *et al.* (1993)
PCR	Combines amplification and hybridization for detection	10 cells/g sediment	Steffan & Atlas (1988)
	Napier grass genomic DNA acts as unique target	1 pg DNA acts as template	Chaudhry *et al.* (1989)
	Compared methods of DNA isolation prior to amplification	1 cell/gram	Atlas *et al.* (1989)
	72 base 'signature' region	20 cells	Amici *et al.* (1991)
	Amplify two 16S rRNA regions	<3 cells/g soil <10 cells/g sediment	Tsai & Olson (1992a)
	Removal of humics from sediments	<1 cell/gram	Tsai & Olson (1992b)
Antibodies	MAb to LPS determinant	100 cells/spot	Ramos-Gonzales *et al.* (1992)
	Highly specific	20 cells/ml	Brettar *et al.* (1994)
	MAb to LPS determinant	20 cells/ml	Winkler *et al.* (1995)
Enzymatically	Spectrophotometric assay based on cells enzymology	1000 cells/ml	King *et al.* (1991)
	Blue colonies on indicator media	<10 cfu/gram	Drahos (1991b)
Hybridization	RNA extraction, use of sense and anti sense probes	10^4 cells/100 ml seawater	Pichard & Paul (1991)
	Tn 5 *nptII* probe		Recorbet *et al.* (1992)
	Tn 5 *nptII* probe	4.3×10^4 cells/gram soil	Holben *et al.* (1988)
	P. cepacia repeat sequence RS-1100-1 probe	0.1 μg DNA from soil	Steffan *et al.* (1989)

Steffan *et al.*, 1989), PCB degraders in microbial consortia (Pettigrew & Sayler, 1986), and mercury resistance genes in gram-negative populations (Barkay, Fount & Olson, 1985). Assuming the probe works adequately, sensitivity is limited only by the number of colonies that can be processed.

A refinement of this approach occurred when it was found that nucleic acids could be isolated directly from environmental samples. In such cases, the total DNA from a particular sample is obtained through *in situ* lysis of cells, and purified through sequential steps aimed at removing contaminants. The purified DNA can then be immobilized to a membrane and subjected to hybridization analyses. Sensitivity is determined by the number of copies of the target nucleic acid immobilized to the membrane and the specific activity of the probe. This technique has been used to detect specific organisms in soil microcosms (Holben *et al.*, 1988; Jansson, Holben & Tiedje, 1989; Ka, Holben & Tiedje, 1994). Pichard & Paul (1991) isolated RNA from seawater filtrates, and, using specific gene probes, detected the *nptII* transcript in as few as 10^4 *Vibrio* cells/100 ml.

11.6 Summary and future directions

In our rush to produce bacteria better capable of degrading the vast array of pollutants in our environment, it is perhaps wise to contemplate previous experiences in the environmental release of recombinant bacteria. At least 29 approved field tests of recombinant bacteria were carried out between 1986 and 1990 worldwide. The first was the release of an engineered baculovirus as an insecticide in Wytham, Oxfordshire, U.K., in 1986 (Bishop, 1986), and again in 1987 by NERC Institute of Virology. In the U.S. the first two releases of recombinants both occurred in 1987. One was the highly publicized field applications of the ice-minus deletion constructs of *Pseudomonas* strains developed by S. Lindow for frost protection, which required more than five years of litigation and planning for a series of five field applications in Brentwood, California, and Tulelake, California, in 1987 (Baskin, 1987). The other was a *Pseudomonas* strain containing the *E. coli lacZY* genes as markers to assess the survival and spread of a recombinant in a test site at Clemson University, South Carolina (Drahos *et al.*, 1988). The intentional release of a recombinant *P. syringae* at Montana State University, Bozeman, Montana, without regulatory approval, certainly heightened public awareness of all such contemplated environmental releases for the next several years (Fox, 1987). There have also been a number of releases of recombinant nitrogen-fixing bacteria and various bacterial strains expressing cloned *Bacillus thuringiensis* forespore crystal endotoxin for insect control (primarily in corn) (Drahos, 1991a).

Despite these field trials, there has never been a release of a geneticaly engineered microorganism for bioremediation purposes. One should therefore bear in mind, when considering pathway construction as a route to bioremediation, that the constructed organism cannot carry out its programmed function without interacting with its environment. That we can carry out fairly complex alterations of the physiology of bacteria is without question. Our understanding of the consequences of these changes is barely in its infancy. Failure to account for the ecological realities in which the recombinants are to function tremendously lessens the likelihood of success, and makes it very difficult to interpret field results. This factor is exacerbated by the lack of knowledge of how to cause such bacteria to survive and function in nature. Perhaps the new approaches being undertaken in the field of molecular ecology will open a window on this dilemma.

References

Aehle, W., Sobek, H., Amory, A., Vetter, R., Wilke, D. & Schomburg, D. (1993). Rational protein engineering and industrial application: structure prediction by homology and rational design of protein-variants with improved 'washing performance': the alkaline protease from *Bacillus alcalophilus*. *Journal of Biotechnology*, 28, 31–40.

Ahmad, D., Masse, R. & Sylvestre, M. (1990). Cloning and expression of genes involved in 4-chlorobiphenyl transformation by *Pseudomonas testosteroni*: homology to poly-chlorobiphenyl-degrading genes in other bacteria. *Gene*, 86, 53–61.

Ahrenholtz, I., Lorenz, M. G. & Wackernagel, W. (1994). A conditional suicide system in *Escherichia coli* based on the intracellular degradation of DNA. *Applied and Environmental Microbiology*, 60, 3746–51.

Alexander, R. S., Nair, S. K. & Christianson, D. W. (1991). Engineering the hydrophobic pocket of carbonic anhydrase II. *Biochemistry*, 30, 11 064–72.

Ali, N. & Jameel, S. (1993). Direct detection of Hepatitis C virus RNA in serum by reverse transcription PCR. *Bio/Techniques*, 15, 40–2.

Amici, A., Bazzicalupo, M., Gallori, E. & Rollo, F. (1991). Monitoring a genetically engineered bacterium in a fresh water environment by rapid enzymatic amplification of a synthetic DNA 'number-plate'. *Applied Microbiooogy and Biotechnology*, 36, 222–7.

Araujo, M. A. V., Mendonça-Hagler, L. C., Hagler, A. N. & van Elsas, J. D. (1994). Survival of genetically modified *Pseudomonas fluorescens* introduced into subtropical soil microcosms. *FEMS Microbiology Ecology*, 13, 205–16.

Arciero, D., Vannelli, T., Logan, M. & Hooper, A. B. (1989). Degradation of trichloroethylene by the ammonia-oxidizing bacterium *Nitrosomonas europaea*. *Biochemical and Biophysical Communications*, 159, 640–3.

Asturias, J. A. & Timmis, K. N. (1993). Three different 2,3-dihydroxybiphenyl-1,2-dioxygenase genes (in the gram-positive polychlorobiphenyl-degrading bacterium Rhodococcus globerlus. *Journal of Bacteriology*, 175, 4631–40.

Atlas, R. M., Bej, A. K., Steffan, R. J. & Perlin, M. H. (1989). Approaches for monitoring and containing genetically engineered microorganisms released into the environment. *Hazardous Waste & Hazardous Materials*, 6, 135–44.

Atlas, R. M., Sayler, G., Burlage, R. S. & Bej, A. K. (1992). Molecular approaches for environmental monitoring of microorganisms. *Biotechniques*, 12, 706–17.

Avery, O. T., MacLeod, C. M. & McCarty, M. (1944). Studies on the chemical nature of the substance inducing transformation of *Pneumococcal* types. *Journal of Experimental Medicine*, 79, 137–58.

Awong, J., Bitton, G. & Chaudhry, G. R. (1990). Microcosm for assessing survival of genetically engineered microorganisms in aquatic environments. *Applied and Environmental Microbiology*, 56, 977–83.

Baker, C. S. & Palumbi, S. R. (1994). Which whales are hunted? A molecular genetic approach to monitoring whaling. *Science*, 265, 1538–9.

Barkay, T., Fount, D. L. & Olson, B. H. (1985). The preparation of a DNA probe for the detection of mercury resistance genes in Gram-negative bacterial communities. *Applied and Environmental Microbiology*, 49, 686–92.

Barry, G. F. (1988). A broad-host-range shuttle system for gene insertion into the chromosomes of gram-negative bacteria. *Gene*, 71, 75–84.

Baskin, Y. (1987). Testing engineered microbes in the field. *ASM News*, 53, 611–12.

Beauchamp, C. J., Kloepper, J. W. & Lemke, P. A. (1993). Luminometic analyses of plant root colonization by bioluminescent Pseudomonads. *Canadian Journal of Microbiology*, 39, 434–41.

Bej, A. K., Perlin, M. & Atlas, R. M. (1988). Model suicide vector for containment of genetically engineered microorganisms. *Applied and Environmental Microbiology*, 54, 2472–7.

Bej, A. K., Perlin, M. & Atlas, R. M. (1991). Effect of introducing genetically engineered microorganisms on soil microbial community diversity. *FEMS Microbiology Ecology*, 86, 169–76.

Bishop, D. H. L. (1986). UK release of genetically marked virus. *Nature*, 323, 496.

Bocanegra, J. H. A., Scrutton, N. S. & Perham, R. N. (1993). Creation of an NADP-dependent pyruvate dehydrogenase multienzyme complex by protein engineering. *Biochemistry*, 32, 2737–40.

Bopp, L. H. (1986). Degradation of highly chlorinated PCB's by *Pseudomonas* strain LB400. *Journal of Industrial Microbiology*, 1, 23–9.

Bradley, M., Bucheler, U. S. & Walsh, C. T. (1991). Redox enzyme engineering: conversion of human glutathione reductase into a trypanothione reductase. *Biochemistry*, 30, 6124–7.

Brettar, I., Ramos-Gonzales, M. I., Ramos, J. L. & Höfle, M. G. (1994). Fate of *Pseudomonas putida* after release into lake water mesocosms: different survival mechanisms in response to environmental conditions. *Microbial Ecology*, 27, 99–122.

Brunschwig, E. & Darzins, A. (1992). A two-component T7 system for the over expression of genes in *Pseudomonas aeruginosa*. *Gene*, 111, 35–41.

Budowle, B., Sajantila, A., Hochmeister, M. N. & Comey, C. T. (1994). The application of PCR to forensic science. In *The Polymerase Chain Reaction*, ed. K. B. Mullis, F. Ferré & R. A. Gibbs, pp. 243–56. Boston: Birkhauser.

Burlage, R. S., Sayler, G. S. & Larimer, F. (1990). Monitoring of naphthalene catabolism by bioluminescence with *nah-lux* transcriptional fusions. *Journal of Bacteriology*, 172, 4749–57.

Cariolou, M. A., Kakkafitou, A., Manoli, P. & Ioannou, P. (1993). Prenatal diagnosis for β-thalassemia by PCR from single chorionic villus. *BioTechniques*, 15, 32–4.

Carmi, O. A., Stewart, G. S. A. B., Ulitzur, S. & Kuhn, J. (1987). Use of bacterial luciferase to establish a promoter probe vehicle capable of nondestructive real-time analysis of gene expression in *Bacillus* sp. *Journal of Bacteriology*, 169, 2165–70.

Chakrabarty, A. M. & Gunsalus, I. C. (1971). Degradative pathways specified by extrachromosomal gene clusters in *Pseudomonas*. *Genetics*, 68, S10.

Chapman, P. J. (1987). Constructing microbial strains for degradation of halogenated aromatic hydrocarbons. In *Environmental Biotechnology, Reducing Risks from Environmental Chemicals Through Biotechnology*, ed. G. S. Omenn, pp. 81–95. New York: Plenum Press.

Chaudry, R. G. & Huang, G. H. (1988). Isolation and characterization of a new plasmid from a *Flavobacterium* sp. which carries the genes for the degradation of 2,4-dichlorophenoxyacetate. *Journal of Bacteriology*, 170, 3897–902.

Chaudhry, G. R., Toranzos, G. A. & Bhatti, A. R. (1989). Novel method for monitoring genetically engineered microorganisms in the environment. *Applied and Environmental Microbiology*, 55, 1301–4.

Contreras, A., Molin, S. & Ramos, J. L. (1991). Conditional-suicide containment system for bacteria which mineralize aromatics. *Applied and Environmental Microbiology*, 57, 1504–8.

Dabrock, B., Riedel, J., Bertram, J. & Gottschalk, G. (1992). Isopropylbenzene (cumene): a new substrate for the isolation of trichloroethene-degrading bacteria. *Archives of Microbiology*, 158, 9–13.

de Lorenzo, V., Herrero, M., Jakubzik, U. & Timmis, K. N. (1990). Mini-Tn5 transposon derivative for insertion mutagenesis, promoter probing, and chromosomal insertion of cloned DNA in gram-negative eubacteria. *Journal of Bacteriology*, 172, 6568–72.

de Lorenzo, V., Fernandez, S., Herrero, M., Jakubzik, U. & Timmis, K. N. (1993a). Engineering of alkyl- and haloaromatic-responsive gene expression with mini-transposons containing regulated promoters of biodegradative pathways of *Pseudomonas*. *Gene*, 130, 41–6.

de Lorenzo, V., Eltis, L., Kessler, B. & Timmis, K. N. (1993b). Analysis of *Pseudomonas* gene products using *lacI^q/Ptrp-lac* plasmids and transposons that confer conditional phenotypes. *Gene*, 123, 17–24.

DeModena, J., Gutiérrez, S., Velasco, J., Fernández, F., Fachini, R., Galazzo, J. & Hughes, D. (1993). The production of cephalosporin C by *Acremonium chrysogenum* is improved by the intracellular expression of a bacterial hemoglobin. *Biotechnology*, 11, 926–8.

Deretic, V., Chandrasekharappa, S., Gill, J. F., Chatterjee, D. K. & Chakrabarty, A. M. (1987). A set of cassettes and improved vectors for genetic and biochemical characterization of *Pseudomonas* genes. *Gene*, 57, 61–72.

Díaz, E., Munthali, M., deLorenzo, V. & Timmis, K. N. (1994). Universal barrier to lateral spread of specific genes among microorganisms. *Molecular Microbiology*, 13, 855–61.

Don, R. H. & Pemberton, J. M. (1985). Genetic and physical map of the 2,4-dichlorophenoxyacetic acid degradative plasmid pJP4. *Journal of Bacteriology*, 161, 466–8.

Dorn, E., Hellwig, M., Reineke, W. & Knackmuss, H-J. (1974). Isolation and characterization of a 3-chlorobenzoate degrading Pseudomonad. *Archives of Microbiology*, 99, 61–70.

Doyle, J. D., Short, K. A., Stotzky, G., King, R. J., Seidler, R. J. & Olsen, R. H. (1991). Ecologically significant effects of *Pseudomonas putida* PPO301 (pRO103), genetically engineered to degrade 2,4-dichlorophenoxyacetate, on microbial populations and processes in soil. *Canadian Journal of Microbiology*, 37, 682–91.

Drahos, D. J. (1991a). Field testing of genetically engineered microorganisms. *Biotechnology Advances*, 9, 157–71.

Drahos, D. J. (1991b). Current practices for monitoring genetically engineered microbes in the environment. *Agriculture Biotechnology News and Information*, 3, 39–48.

Drahos, D. J., Barry, G. F., Hemming, B. C., Brandt, E. J., Kline, E. L., Skipper, H. D., Kluepfel, D. A., Hughes, T. A. & Gooden, D. T. (1988). Pre-release testing procedures: U.S. field test of a *lacZY*-engineered soil bacterium. In *Release of Genetically Engineered Micro-organisms*, ed. M. Sussman, C. H. Collins, F. A. Skinner & D. E. Stewart-Tull, pp. 181–91. London: Academic Press.

Dunn, N. W. & Gunsalus, I. C. (1973). Transmissible plasmid coding early enzymes of naphthalene oxidation in *Pseudomonas putida*. *Journal of Bacteriology*, 114, 974–9.

Duque, E., Hidour, A., Godoy, F., Ramos, J. L. (1993). Construction of a Pseudomonas hybrid strain that mineralizes 2,4,6-trinitrotoluene. *Journal of Bacteriology*, 175, 2278–83.

Edwards, C. (1993). *Monitoring Genetically Manipulated Microorganisms in the Environment*. Chichester: John Wiley.

Eggink, G., Lageveen, R., Altenburg, B. & Witholt, B. (1987a). Controlled and functional expression of the *Pseudomonas oleovorans* alkane utilizing system in *Pseudomonas putida* and *Escherichia coli*. *Journal of Biological Chemistry*, 262, 17712–18.

Eggink, G., Van Lelyveld, P., Arnberg, A., Arfman, N., Witteveen, C. & Witholt, B. (1987b). Structure of the *Pseudomonas putida* alkBAC operon. *Journal of Biological Chemistry*, 262, 6400–6.

Eltis, L. D., Hofmann, B., Hecht, H. J., Lunsdorf, H. & Timmis, K. N. (1993). Purification and crystallization of 2,3-dihydroxybiphenyl-1,2-dioxygenase. *Journal of Biological Chemistry*, 268, 2727–32.

Ensign, S. A., Hyman, M. R. & Arp, D. J. (1992). Cometabolic degradation of chlorinated alkenes by alkene monooxygenase in a propylene-grown *Xanthobacter* strain. *Applied and Environmental Microbiology*, 58, 3038–46.

Erickson, B. D. & Mondello, F. J. (1992). Nucleotide sequencing and transcriptional mapping of the genes encoding biphenyl dioxygenase, a multicomponent polychlorinated-biphenyl-degrading enzyme in *Pseudomonas* strain LB400. *Journal of Bacteriology*, 174, 2903–12.

Erickson, B. D. & Mondello, F. J. (1993). Enhanced biodegradation of polychlorinated biphenyls after site-directed mutagenesis of a biphenyl dioxygenase gene. *Applied and Environmental Microbiology*, 59, 3858–62.

Evans, W. C., Smith, B. S. W., Fernley, H. N. & Davies, J. I. (1971). Bacterial metabolism of 2,4-dichlorophenoxyacetate. *Biochemical Journal*, 122, 543–51.

Ewers, J., Freier-Schroder, D. & Knackmuss, H-J. (1990). Selection of trichloroethene (TCE) degrading bacteria that resist inactivation by TCE. *Archives of Microbiology*, 154, 410–13.

Farinha, M. A. & Kropinski, A. M. (1989). Construction of broad-host-range vectors for general cloning and promoter selection in *Pseudomonas* and *Escherichia coli*. *Gene*, 77, 205–10.

Feeney, R., Clarke, A. R. & Holbrook, J. J. (1990). A single amino acid substitution in lactate dehydrogenase improves the catalytic efficiency with an alternative coenzyme. *Biochemical and Biophysical Research Communications*, 166, 667–72.

Flemming, C. A., Leung, K. T., Lee, H., Trevors, J. T. & Greer, C. W. (1994). Survival of lux-lac-marked biosurfactant-producing *Pseudomonas aeruginosa* UG2L in soil monitored by nonselective plating and PCR. *Applied and Environmental Microbiology*, 60, 1606–13.

Fox, J. L. (1987). Unauthorized Dutch Elm disease test in Montana. *ASM News*, 53, 615–17.

Fujita, M., Ike, M. & Uesugi, K. (1994). Operation parameters affecting the survival of genetically engineered microorganisms in activated sludge processes. *Water Research*, 28, 1667–72.

Fulthorpe, R. R. & Wyndham, R. C. (1992). Involvement of a chlorobenzoate catabolic transposon, Tn5271, in community adaptation to chlorobiphenyl, chloroaniline, and 2,4-dichlorophenoxyacetic acid in a freshwater ecosystem. *Applied and Environmental Microbiology*, 58, 314–25.

Furukawa, K. & Arimura, N. (1987). Purification and properties of 2,3-dihydroxybiphenyl dioxygenase from polychlorinated biphenyl-degrading *Pseudomonas pseudoalcaligenes* and *Pseudomonas aeruginosa* carrying the cloned *bphC* gene. *Journal of Bacteriology*, 169, 924–7.

Furukawa, K., Hayase, N., Taira, K. & Tomizuka, N. (1989). Molecular relationship of chromosomal genes encoding biphenyl/polychlorinated biphenyl catabolism: some soil bacteria possess a highly conserved *bph* operon. *Journal of Bacteriology*, 171, 5467–72.

Furukawa, K., Hirose, J., Suyama, A., Zaiki, T. & Hayashida, S. (1993). Gene components responsible for discrete substrate specificity in the metabolism of biphenyl (*bph* operon) and toluene (*tod* operon). *Journal of Bacteriology*, 175, 5224–32.

Furukawa, K., Hirose, J., Hayashida, S. & Nakamura, K. (1994). Efficient degradation of trichloroethylene by a hybrid aromatic ring dioxygenase. *Journal of Bacteriology*, 176, 2121–3.

Georgiou, G., Lin, S-C. & Sharma, M. M. (1992). Surface-active compounds from microorganisms. *Biotechnology*, 10, 60–5.

Gerdes, K. (1988). The *parB* (*hok/sok*) locus of plasmid R1: a general purpose plasmid stabilization system. *Bio/Technology*, 6, 1402–5.

Gerger, R. R., Winfrey, M. R., Reagin, M. & Shields, M. S. (1991). Introduction of chloroaromatic tolerance into a strain of *Pseudomonas cepacia* G4 constitutive for trichloroethylene degradation. In *Abstract, 91st General Meeting of the American Society for Microbiology*, abstr. K-2, p. 214. Washington, D.C.: The American Society for Microbiology.

Gottschalk, G. (1979). Growth with aromatic compounds. In *Bacterial Metabolism*, ed. M. P. Starr, pp. 126–31. New York: Springer Verlag.

Hales, D. B., Ho, B. & Thompson, J. A. (1987). Inter- and intramolecular deuterium isotope effects on the cytochrome P-450-catalyzed oxidative dehalogenation of 1,1,2,2-tetrachloroethane. *Biochemical and Biophysical Research Communications*, 149, 319–25.

Halvorson, H. O., Pramer, D. & Rogul, M. (1985). *Engineered Organisms in the Environment: Scientific Issues*. Washington, D.C.: American Society for Microbiology.

Harayama, S. & Don, R. H. (1985). Catabolic plasmids: their analysis and utilization in the manipulation of bacterial metabolic activities. In *Genetic Engineering: Principles and Methods*, Vol. 7, ed. J. K. Setlow and A. Hollaender, pp. 283–307. New York: Plenum Publishing Co.

Harker, A. R. & Kim, Y. (1990). Trichloroethylene degradation by two independent aromatic-degrading pathways in *Alcaligenes eutrophus* JMP134. *Applied and Environmental Microbiology*, 56, 1179–81.

Henderson, G. B., Murgolo, N. J., Kuriyan, J., Osapay, K., Kominos, D., Berry, A., Scrutton, N. S., Hinchliffe, N. W., Perham, R. N. & Cerami, A. (1991). Engineering the substrate specificity of glutathione reductase toward that of trypanothione reduction. *Proceedings of the National Academy of Sciences USA*, 88, 8769–73.

Herrero, M., de Lorenzo, V. & Timmis, K. N. (1990). Transposon vectors containing non-antibiotic resistance selection markers for cloning and stable chromosomal insertion of foreign genes in gram-negative bacteria. *Journal of Bacteriology*, 172, 6557–67.

Herrero, M., de Lorenzo, V., Ensley, B. & Timmis, K. N. (1993). A T7 RNA polymerase-based system for the construction of *Pseudomonas* strains with phenotypes dependent on TOL-*meta* pathway effectors. *Gene*, 134, 103–6.

Hill, K. E., Fry, J. C. & Weightman, A. J. (1994). Gene transfer in the aquatic environment: persistance and mobilization of the catabolic recombinant plasmid pD10 in the epilithon. *Microbiology*, 140, 1555–63.

Hinchee, R. E. (1994). Bioventing of petroleum hydrocarbons. In *Handbook of Bioremediation*, ed. R. D. Norris, R. E. Hinchee, R. Brown, P. L. McCarty, L. Semprini, J. T. Wilson, D. H. Kampbell, M. Reinhard, E. J. Bouwer, R. C. Borden, T. M. Vogel, J. M. Thomas & C. H. Ward, pp. 39–59. Boca Raton: Lewis Publishers.

Hirose, J., Suyama, A., Hayashida, S. & Furukawa, K. (1994). Construction of hybrid biphenyl (*bph*) and toluene (*tod*) genes for functional analysis of aromatic ring dioxygenases. *Gene*, 138, 27–33.

Hofer, B., Eltis, L. D., Dowling, D. N. & Timmis, K. N. (1993). Genetic analysis of a *Pseudomonas* locus encoding a pathway for biphenyl/polychlorinated biphenyl degradation. *Gene*, 130, 47–55.

Holben, W. E., Jansson, J. K., Chelm, B. K. & Tiedje, J. M. (1988). DNA probe method for the detection of specific microorganisms in the soil community. *Applied and Environmental Microbiology*, 54, 703–11.

Hopkins, G. D., Semprini, L. & McCarty, P. L. (1993). Microcosm and in situ field studies of enhanced biotransformation of trichloroethylene by phenol-utilizing microorganisms. *Applied and Environmental Microbiology*, 59, 2277–85.

Horvath, R. S. & Alexander, M. (1970). Cometabolism of *m*-chlorobenzoate by an *Arthrobacter*. *Applied and Environmental Microbiology*, 20, 254–8.

Huling, S. G., Bledsoe, B. E., White, M. V. (1992). The feasibility of utilizing hydrogen peroxide as a source of oxygen in bioremediation. In In situ *Bioreclamation*, ed. R. E. Hinchee & R. F. Olfenbuttel, pp. 83–102. Boston: Butterworth-Heinemann.

Ismailov, N. M. (1985). Biodegradation of oil hydrocarbons in soil inoculated with yeasts. *Microbiology*, 54, 670–5.

Jain, R. K., Sayler, G. S., Wilson, J. T., Houston, L. & Pacia, D. (1987). Maintenance and stability of introduced genotypes in groundwater aquifer material. *Applied and Environmental Microbiology*, 53, 996–1002.

Jain, R. K., Burlage, R. S. & Sayler, G. S. (1988). Methods for detecting recombinant DNA in the environment. *CRC Critical Reviews in Biotechnology*, 8, 33–84.

Jansson, J. K., Holben, W. E. & Teidje, J. M. (1989). Detection in soil of a deletion in an engineered DNA sequence by using DNA probes. *Applied and Environmental Microbiology*, 55, 3022–5.

Jensen, L. B., Ramos, J. L., Kaneva, Z. & Molin, S. (1993). A substrate-dependent biological containment system for *Pseudomonas putida* based on the *Escherichia coli gef* gene. *Applied and Environmental Microbiology*, 59, 3713–17.

Jobson, A., McLaughlin, M., Cook, F. D. & Westlake, D. W. S. (1974). Effect of amendments on the microbial utilization of oil applied to soil. *Applied and Environmental Microbiology*, 27, 166–71.

Jones, R. A., Broder, M. W. & Stotzky, G. (1991). Effects of genetically engineered microorganisms on nitrogen transformation and nitrogen-transforming microbial populations in soil. *Applied and Environmental Microbiology*, 57, 3212–19.

Jovin, T. M. & Arndt-Jovin, D. J. (1989). Luminescence digital imaging microscopy. *Annual Reviews of Biophysics and Biophysical Chemistry*, 18, 271–308.

Ka, J. O., Holben, W. E. & Tiedje, J. M. (1994). Use of gene probes to aid in recovery and identification of functionally dominant 2,4-dichlorophenoxyacetic acid-degrading populations in soil. *Applied and Environmental Microbiology*, 60, 1116–20.

Kaphammer, B. J., Kukor, J. J. & Olsen, R. H. (1990). Cloning and characterization of a novel toluene degradative pathway from *Pseudomonas pickettii* PKO1. *Abstract, 90th Annual Meeting of the American Society for Microbiology*, abstr K-145, p. 243. Washington, D.C.: The American Society for Microbiology.

Keil, S. & Keil, H. (1992). Construction of a cassette enabling regulated gene expression in the presence of aromatic hydrocarbons. *Plasmid*, 27, 191–9.

Kellogg, S. T., Chatterjee, D. K. & Chakrabarty, A. M. (1981). Plasmid-assisted molecular breeding: new technique for enhanced biodegradation of persistent toxic chemicals. *Science*, 214, 1133–5.

Kelly, I. & Cerniglia, C. E. (1991). The metabolism of fluoranthene by a species of *Mycobacterium*. *Journal of Industrial Microbiology*, 7, 19–26.

Khosla, C. & Bailey, J. (1988). Heterologous expression of bacterial hemoglobin improves the growth properties of recombinant *Escherichia coli*. *Nature*, 331, 633–5.

Khosravi, M., Webster, D. & Stark, B. (1990). Presence of the bacterial hemoglobin gene improves alpha-amylase production of a recombinant *Escherichia coli* strain. *Plasmid*, 24, 190–4.

King, J. M. H., DiGrazia, P. M., Applegate, B., Burlage, R., Sanseverino, J., Dunbar, P., Larimer, F. & Sayler, G. S. (1990). Rapid, sensitive bioluminescent reporter technology for naphthalene exposure and biodegradation. *Science*, 249, 778–81.

King, R. J., Short, K. A. & Seidler, R. J. (1991). Assay for detection and enumeration of genetically engineered microorganisms which is based on the activity of a deregulated 2,4-dichlorophenoxyacetate monooxygenase. *Applied and Environmental Microbiology*, 57, 1790–2.

Klier, A. (1992). Release of genetically modified microorganisms in natural environments: scientific and ethical problems. In *Gene Transfers and Environment*, ed. M. J. Gauthier, pp. 183–90. Berlin: Springer-Verlag.

Kloos, D-U., Strätz, M., Güttler, A., Steffan, R. J. & Timmis, K. N. (1994). Inducible cell lysis system for the study of natural transformation and environmental fate of DNA released by cell death. *Journal of Bacteriology*, 176, 7352–61.

Knackmuss, H-J. & Hellwig, M. (1978). Utilization and cooxidation of chlorinated phenols by *Pseudomonas* sp. B13. *Archives of Microbiology*, 117, 1–7.

Knudsen, S. & Karlström, O. H. (1991). Development of efficient suicide mechanisms for biological containment of bacteria. *Applied and Environmental Microbiology*, 57, 85–92.

Knudsen, S., Saadbye, P., Hansen, L. H., Collier, A., Jacobsen, B. L., Schlundt, J. & Karlström, O. H. (1995). Development and testing of improved suicide functions for biological containment of bacteria. *Applied and Environmental Microbiology*, 61, 985–91.

Kozlowski, M., Van Brunschot, A., Nash, N. & Davies, R. W. (1988). A novel vector allowing the expression of genes in a wide range of gram-negative bacteria. *Gene*, 70, 199–204.

Krumme, M. L., Timmis, K. N. & Dwyer, D. F. (1993). Degradation of trichloroethylene by *Pseudomonas cepacia* G4 and the constitutive mutant strain G4 5223 PR1 in aquifer microcosms. *Applied and Environmental Microbiology*, 59, 2746–9.

Krumme, M. L., Smith, R. L., Egestorff, J., Thiem, S. M., Tiedje, J. M., Timmis, K. N. &

Dwyer, D. F. (1994). Behavior of pollutant-degrading microorganisms in aquifers: predictions for genetically engineered organisms. *Environmental Science and Technology*, 28, 1134–8.

Kukor, J. J. & Olsen, R. H. (1990). Molecular cloning, characterization, and regulation of a *Pseudomonas pickettii* PK01 gene encoding phenol hydroxylase and expression of the gene in *Pseudomonas aeruginosa* PA01c. *Journal of Bacteriology*, 172, 4624–30.

Kuriyan, J., Krishna, T. S., Wong, L., Guenther, B., Pahler, A., Williams, C. H. & Model, P. (1991). Convergent evolution of similar function in two structurally divergent enzymes. *Nature*, 352, 172–4.

Lajoie, C. A., Zylstra, G. J., DeFlaun, M. F. & Strom, P. F. (1993). Development of field application vectors for bioremediation of soils contaminated with polychlorinated biphenyls. *Applied and Environmental Microbiology*, 59, 1735–41.

Lehrbach, P. R., Zeyer, J., Reineke, W., Knackmuss, H-J. & Timmis, K. N. (1984). Enzyme recruitment *in vitro*: use of cloned genes to extend the range of haloaromatics degraded by *Pseudomonas* sp. strain B13. *Journal of Bacteriology*, 158, 1025–32.

Lin, J-E., Wang, H. Y. & Hickey, R. F. (1991). Use of coimmobilized biological systems to degrade toxic organic compounds. *Biotechnology and Bioengineering*, 38, 273–9.

Lin, J-E., Mueller, J. G. & Pritchard, P. H. (1992). Use of encapsulated microorganisms as inoculants for bioremediation. *American Chemical Society's Special Session on Bioremediation of Soils and Sediments, September 21–23 (1992), Atlanta, Georgia.*

Little, C. D., Palumbo, A. V., Herbes, S. E., Lidstrom, M. E., Tyndall, R. L. & Gilmer, P. J. (1988). Trichloroethylene biodegradation by a methane-oxidizing bacterium. *Applied and Environmental Microbiology*, 54, 951–6.

Lovins, K. W., Angle, J. S., Wiebers, J. L. & Hill, R. L. (1994). Leaching of *Pseudomonas aeruginosa*, and transconjugants containing pR68.45 through unsaturated, intact soil columns. *FEMS Microbiology Ecology*, 13, 105–12.

Mäe, A. A., Martis, R. O., Ausmees, N. R., Koiv, V. M. & Heinaru, A. L. (1993). Characterization of a new 2,4-dichlorophenoxyacetic acid degrading plasmid pEST4011: Physical map and localization of the catabolic genes. *Journal of General Microbiology*, 139, 3165–70.

McCarty, P. L. & Semprini, L. (1994). Ground-water treatment for chlorinated solvents. In *Handbook of Bioremediation*, ed. R. D. Norris, R. E. Hinchee, R. Brown, P. L. McCarty, L. Semprini, J. T. Wilson, D. H. Kampbell, M. Reinhard, E. J. Bouwer, R. C. Borden, T. M. Vogel, J. M. Thomas & C. H. Ward, pp. 87–116. Boca Raton: Lewis Publishers.

McClure, N. C., Weightman, A. J. & Fry, J. C. (1989). Survival of *Pseudomonas putida* UWC1 containing cloned catabolic genes in a model activated-sludge unit. *Applied and Environmental Microbiology*, 55, 2627–34.

Meighen, E. A. (1988). Enzymes and genes from the lux operons of bioluminescent bacteria. *Annual Review of Microbiology*, 42, 151–76.

Mermod, N., Ramos, J. L., Lehrbach, P. R. & Timmis, K. N. (1986). Vector for regulated expression of cloned genes in a wide range of gram-negative bacteria. *Journal of Bacteriology*, 167, 447–54.

Mittl, P. R., Berry, A., Scrutton, N. S., Perham, R. N. & Schulz, G. E. (1993). Structural differences between wild-type NADP-dependent glurathione reductase from Escherichia coli and a redesigned NAD-dependent mutant. *Journal of Molecular Biology*, 231, 191–5.

Molin, S., Klemm, P., Poulsen, L. K., Biehl, H., Gerdes, K. & Anderson, P. (1987).

Conditional suicide system for containment of bacteria and plasmids. *Bio/Technology*, 5, 1315–18.

Molin, S., Boe, L., Jensen, L. B., Kristensen, C. S., Givskov, M., Ramos, J. L. & Bej, A. K. (1993). Suicidal genetic elements and their use in biological containment of bacteria. *Annual Review of Microbiology*, 47, 139–66.

Mondello, F. J. (1989). Cloning and expression in *Escherichia coli* of *Pseudomonas* strain LB400 genes encoding polychlorinated biphenyl degradation. *Journal of Bacteriology*, 171, 1725–32.

Montgomery, S. O., Shields, M. S., Chapman, P. J. & Pritchard, P. H. (1989). Identification and characterization of trichloroethylene-degrading bacteria. In *Abstract, 89th Annual Meeting of the American Society for Microbiology*, abstr. K-68, p. 256. Washington, DC: The American Society for Microbiology.

Moore, A. T., Vira, A. & Fogel, S. (1989). Biodegradation of *trans*-1,2-dichloroethylene by methane-utilizing bacteria in an aquifer simulator. *Environmental Science and Technology*, 23, 403–6.

Mueller, J. G., Chapman, P. J., Blattman, B. O. & Pritchard, P. H. (1989). Action of a fluoranthene-utilizing bacterial community on polycyclic aromatic hydrocarbon components of creosote. *Applied and Environmental Microbiology*, 55, 3085–90.

Mueller, J. G., Chapman, P. J., Blattman, B. O. & Pritchard, P. H. (1990). Isolation and characterization of a fluoranthene-utilizing strain of *Pseudomonas paucimobilis*. *Applied and Environmental Microbiology*, 56, 1079–86.

Mueller, J. G., Lantz, S. E., Blattmann, B. O. & Chapman, P. J. (1991a). Bench-scale evaluation of alternative biological treatment processes for the remediation of pentachlorophenol- and creosote-contaminated materials: slurry-phase bioremediation. *Environmental Science and Technology*, 25, 1055–61.

Mueller, J. G., Lantz, S. E., Blattmann, B. O. & Chapman, P. J. (1991b). Bench-scale evaluation of alternative biological treatment processes for the remediation of pentachlorophenol- and creosote-contaminated materials: solid-phase bioremediation. *Environmental Science and Technology*, 25, 1045–55.

Mueller, J. G., Resnick, S. M., Shelton, M. E. & Pritchard, P. H. (1992). Effect of inoculation on the biodegradation of weathered Prudhoe Bay crude oil. *Journal of Industrial Microbiology*, 10, 95–102.

Mueller, J. G., Lantz, S. E., Ross, D., Colvin, R. J., Middaugh, D. P. & Pritchard, P. H. (1993a). Strategy using bioreactors and specially-selected microorganisms for bioremediation of ground water contaminated with creosote and pentachlorophenol. *Environmental Science and Technology*, 27, 692–8.

Mueller, J. G., Lin, J-E., Lantz, S. E. & Pritchard, P. H. (1993b). Recent developments in cleanup technologies. *Remediation*, Summer, 369–81.

Murakami, H., Yabusaki, Y., Sakaki, T., Shibata, M. & Ohkawa, H. (1987). A genetically engineered P450 monooxygenase: construction of the functional fused enzyme between rat cytochrome P450c and NADPH-cytochrome P450 reductase. *DNA*, 6, 189–97.

Naik, G. A., Bhat, L. N., Chopade, B. A. & Lynch, J. M. (1994). Transfer of broad-host-range antibiotic resistance plasmids in soil microcosms. *Current Microbiology*, 28, 209–15.

Nakatsu, C., Ng, J., Singh, R., Straus, N. & Wyndham, C. (1991). Chlorobenzoate catabolic transposon Tn*5271* is a composite class I element with flanking class II insertion sequences. *Proceedings of the National Academy of Sciences USA*, 88, 8312–16.

Natsch, A., Keel, C., Pfirter, H. A., Haas, D. & Défago, G. (1994). Contribution of the global regulator gene gacA to persistence and dissemination of *Pseudomonas fluorescens* biocontrol strain CHA0 introduced into soil microcosms. *Applied and Environmental Microbiology*, 60, 2553–60.

Nelson, M. J. K., Montgomery, S. O., O'Neill, E. J. & Pritchard, P. H. (1986). Aerobic metabolism of trichloroethylene by a bacterial isolate. *Applied and Environmental Microbiology*, 52, 383–4.

Nelson, M. J. K., Montgomery, S. O., Mahaffey, W. R. & Pritchard, P. H. (1987). Biodegradation of trichloroethylene and involvement of an aromatic biodegradative pathway. *Applied and Environmental Microbiology*, 53, 949–54.

Nelson, M. J. K., Montgomery, S. O. & Pritchard, P. H. (1988). Trichloroethylene metabolism by microorganisms that degrade aromatic compounds. *Applied and Environmental Microbiology*, 54, 604–6.

Oldenhuis, R., Vink, R. L. J. M., Janssen, D. B. & Witholt, B. (1989). Degradation of chlorinated aliphatic hydrocarbons by *Methylosinus trichosporium* OB3b expressing soluble methane monooxygenase. *Applied and Environmental Microbiology*, 55, 2819–26.

Ornstein, R. L. (1991). Why timely bioremediation of synthetics may require rational enzyme redesign: preliminary report on redesigning cytochrome P450cam for trichloroethylene dehalogenation. In *On-Site Bioreclamation. Processes for Xenobiotic and Hydrocarbon Treatment*, ed. R. E. Hinchee and R. F. Olfenbuttel, pp. 509–14. Boston: Butterworth-Heinemann.

Orvos, D. R., Lacy, G. H. & Cairns, J., Jr (1990). Genetically engineered *Erwinia carotovora*: survival, intraspecific competition, and effects upon selected bacterial genera. *Applied and Environmental Microbiology*, 56, 1689–94.

Peloquin, L. & Greer, C. W. (1993). Cloning and expression of the polychlorinated biphenyl-degradation gene cluster from *Arthrobacter* M5 and comparison to analogous genes from gram-negative bacteria. *Gene*, 125, 35–40.

Peoples, O. P. & Sinskey, A. J. (1989). Poly-beta-hydroxybutryate (PHB) biosynthesis in Alcaligenes eutrophus H16. Identification and characterization of the PHB polymerase gene (*phbC*). *Journal of Biological Chemistry*, 264, 15 298–303.

Perham, R. N., Scrutton, N. S. & Berry, A. (1991). New enzymes for old: redesigning the coenzyme and substrate specificities of glutathione reductase. *Bioessays*, 13, 515–25.

Pettigrew, C. A. & Sayler, G. S. (1986). The use of DNA:DNA colony hybridization in the rapid isolation of 4-chlorobiphenyl degradative bacterial phenotypes. *Journal of Microbiological Methods*, 5, 205–13.

Phelps, P. A., Agarwal, S. K., Speitel, G. E., Jr & Georgiou, G. (1992). *Methylosinus trichosporium* OB3b mutants having constitutive expression of soluble methane monooxygenase in the presence of high levels of copper. *Applied and Environmental Microbiology*, 58, 3701–8.

Pichard, S. L. & Paul, J. H. (1991). Detection of gene expression in genetically engineered microorganisms and natural phytoplankton populations in the marine environment by mRNA analysis. *Applied and Environmental Microbiology*, 57, 1721–7.

Pipke, R., Wagner-Döbler, I., Timmis, K. N. & Dwyer, D. F. (1992). Survival and function of a genetically engineered Pseudomonad in aquatic sediment microcosms. *Applied and Environmental Microbiology*, 58, 1259–65.

Polissi, A., Bertoni, G., Acquati, F. & Deho, G. (1992). Cloning and transposon vectors derived from satellite bacteriophage P4 for genetic manipulation of *Pseudomonas* and other gram-negative bacteria. *Plasmid*, 28, 101–14.

Poulos, T. L. & Howard, A. J. (1987). Crystal structures of metyrapone- and phenylimidazole-inhibited complexes of cytochrome P-450cam. *Biochemistry*, 26, 8165–74.

Poulos, T. L. & Raag, R. (1992). Cytochrome P450cam: crystallography, oxygen activation, and electron transfer. *Federation of American Society Experimental Biology Journal*, 6, 674–9.

Poulos, T. L., Finzel, B. C. & Howard, A. J. (1986). Crystal structure of substrate-free *Pseudomonas putida* cytochrome P-450. *Biochemistry*, 25, 5314–22.

Poulos, T. L., Finzel, B. C. & Howard, A. J. (1987). High resolution crystal structure of cytochrome P-450cam. *Journal of Molecular Biology*, 195, 687–700.

Poulson, L. K., Larsen, N. W., Molin, S. & Andersson, P. (1989). A family of genes encoding a cell-killing function may be conserved in gram-negative bacteria. *Molecular Microbiology*, 3, 1463–72.

Pritchard, P. H. & Costa, C. F. (1991). EPA's Alaska oil spill bioremediation project. *Environmental Science and Technology*, 25, 372–9.

Prosser, J. I. (1994). Molecular marker system for detection of genetically engineered microorganisms in the environment. *Microbiology*, 140, 5–17.

Raag, R. & Poulos, T. L. (1991). Crystal structures of cytochrome P-450cam complexed with camphane, thiocamphor, and adamantane: factors controlling P-450 substrate hydroxylation. *Biochemistry*, 30, 2674–84.

Raibaud, O. & Schwartz, M. (1984). Positive control of transcription initiation in bacteria. *Annual Review of Genetics*, 18, 173–206.

Rajagopal, R. (1986). Conceptual design for a groundwater quality monitoring strategy. *The Environmental Professional*, 8, 244–64.

Ramos-Gonzales, M. I., Ruiz-Cabello, F., Brettar, I., Garrido, F. & Ramos, J. L. (1992). Tracking genetically engineered bacteria: monoclonal antibodies against surface determinants of the soil bacterium *Pseudomonas putida* 2440. *Journal of Bacteriology*, 174, 2978–85.

Rattray, E. A. S., Prosser, J. I., Killham, K. & Glover, L. A. (1990). Luminescence-based nonextractive technique for in situ detection of *Escherichia coli* in soil. *Applied and Environmental Microbiology*, 56, 3368–74.

Recorbet, G., Givaudan, A., Steinberg, C., Bally, R., Normand, P. & Faurie, G. (1992). Tn 5 to assess soil fate of genetically marked bacteria: screening for aminoglycoside-resistance advantage and labelling specificity. *FEMS Microbiology Ecology*, 86, 187–94.

Recorbet, G., Robert, C., Givaudan, A., Kudla, B., Normand, P. & Faurie, G. (1993). Conditional suicide system of *Escherichia coli* released into soil that uses the *Bacillus subtilus sac*B Gene. *Applied and Environmental Microbiology*, 59, 1361–6.

Reineke, W. & Knackmuss, H-J. (1979). Construction of haloaromatics utilising bacteria. *Nature (London)*, 277, 385–6.

Reineke, W. & Knackmuss, H-J. (1980). Hybrid pathway for chlorobenzoate metabolism in *Pseudomonas* sp. B13 derivatives. *Journal of Bacteriology*, 142, 467–73.

Rescigno, M. & Perham, R. N. (1994). Structure of the NADPH-binding motif of glutathione reductase: efficiency determined by evolution. *Biochemistry*, 33, 5721–7.

Rheinnecker, M., Eder, J., Pandey, P. S. & Fersht, A. R. (1994). Variants of subtilisin BPN' with altered specificity profiles. *Biochemistry*, 33, 221–5.

Richardson, K. L. & Gibson, D. T. (1984). A novel pathway for toluene oxidation in

Pseudomonas mendocina. In *Annual Meeting of the American Society for Microbiology*, abstr. K54, p. 156. St Louis, MI.

Rosenberg, E., Zuckerberg, A., Rubinowitz, H. & Gutnick, D. L. (1979). Emulsifier of Arthrobacter RAG-1: isolation and emulsifying properties. *Applied and Environmental Microbiology*, 37, 402–8.

Rothmel, R. K., Chakrabarty, A. M., Berry, A. & Darzins, A. (1991). Genetic systems in *Pseudomonas. Methods in Enzymology*, 204, 485–514.

Rubio, M. A., Engesser, K. H. & Knackmuss, H-J. (1986). Microbial metabolism of chlorosalicylates: Accelerated evolution by natural genetic exchange. *Archives of Microbiology*, 145, 123–5.

Sakaki, T., Kominami, S., Takemori, S., Ohkawa, H., Akiyoshi-Shibata, M. & Yabusaki, Y. (1994). Kinetic studies on a genetically engineered fused enzyme between rat cytochrome P4501A1 and yeast NADPH-P450 reductase. *Biochemistry*, 33, 4933–9.

Sarkar, G., Kapelner, S. & Sommer, S. S. (1990). Formamide can drastically improve the specificity of PCR. *Nucleic Acids Research*, 18, 7465.

Sayler, G. S., Shields, M. S., Breen, A., Tedford, E. T., Hooper, S., Sirotkin, K. M. & Davis, J. W. (1985). Application of DNA:DNA colony hybridization to the detection of catabolic genotypes in environmental samples. *Applied and Environmental Microbiology*, 49, 1295–1303.

Schmidhauser, T. J., Ditta, G. & Helinski, D. R. (1988). Broad-host-range plasmid cloning vectors for gram-negative bacteria. In *Vectors: A Survey of Molecular Cloning Vectors and their Uses*, ed. R. L. Rodriguez & D. T. Denhardt, pp. 287–332. Boston: Butterworth.

Schweizer, H. P. (1991). Improved broad-host-range *lac*-based vectors for the isolation and characterization of protein fusions in *Pseudomonas aeruginosa. Gene*, 103, 87–92.

Scrutton, N. S., Berry, A. & Perham, R. N. (1990). Redesign of the coenzyme specificity of a dehydrogenase by protein engineering. *Nature*, 343, 38–43.

Semprini, L., Roberts, P. V., Hopkins, G. D. & McCarty, P. L. (1990). A field evaluation of *in-situ* biodegradation of chlorinated ethenes: Part 2. Results of biostimulation and biotransformation experiments. *Ground Water*, 28, 715–27.

Shaw, J. J., Dane, F., Geiger, D. & Kloepper, J. W. (1992). Use of bioluminescence for detection of genetically engineered microorganisms released into the environment. *Applied and Environmental Microbiology*, 58, 267–73.

Sheffield, V. C., Cox, D. R., Lerman, L. S. & Myers, R. M. (1989). Attachment of a 40-base-pair G + C-rich sequence (GC-clamp) to genomic DNA fragments by the polymerase chain reaction results in improved detection of single-base changes. *Proceedings of the National Academy of Sciences USA*, 86, 232–6.

Shibata, M., Sakaki, T., Yabusaki, Y., Murakami, H. & Ohkawa, H. (1990). Genetically engineered P450 monooxygenases: construction of bovine P450c17/yeast reductase fused enzymes. *DNA Cell Biology*, 9, 27–36.

Shields, M. S. & Reagin, M. J. (1992). Selection of a *Pseudomonas cepacia* strain constitutive for the degradation of trichloroethylene. *Applied and Environmental Microbiology*, 58, 3977–83.

Shields, M. S., Montgomery, S. O., Chapman, P. J., Cuskey, S. M. & Pritchard, P. H. (1989). Novel pathway of toluene catabolism in the trichloroethylene-degrading bacterium G4. *Applied and Environmental Microbiology*, 55, 1624–9.

Shields, M. S., Blake, A., Reagin, M., Moody, T., Overstreet, K., Campbell, R., Francesconi, S. C. & Pritchard, P. H. (1994). Trichloroethylene (TCE) remediation using a plasmid specifying constitutive TCE degradation: alteration of bacterial strain designs based on field evaluations. Presented at *EPA Symposium on Bioremediation of Hazardous Wastes: Research, Development and Field Evaluations*. June 28–30. San Francisco. EPA/600/R-94/075.

Shields, M. S., Reagin, M. J., Gerger, R. R. & Somerville, C. (1995). TOM: A new aromatic degradative plasmid from *Burkholderia(Pseudomonas)cepacia* G4. *Applied and Environmental Microbiology* , 61, 1352–6.

Shingler, V., Franklin, F. C., Tsuda, M., Halroyd, D. & Bagdasarian, M. (1989). Molecular analysis of a plasmid-encoded phenol hydroxylase from *Pseudomonas* CF600. *Journal of General Microbiology*, 135, 1083–92.

Siezen, R. J., de Vos, W. M., Leunisen, J. A. & Dijkstra, B. W. (1991). Homology modelling and protein engineering strategy of subtilases, the family of subtilisin-like serine proteinases. *Protein Engineering*, 4, 719–37.

Sjogren, R. E. (1994). Prolonged survival of an environmental *Escherichia coli* in laboratory soil microcosms. *Water, Air, and Soil Pollution*, 75, 389–403.

Sobecky, P. A., Schell, M. A., Moran, M. A. & Hodson, R. E. (1992). Adaption of model genetically engineered microorganisms to lake water: growth rate enhancements and plasmid loss. *Applied and Environmental Microbiology*, 58, 3630–7.

Spain, J. C. & Nishino, S. F. (1987). Degradation of 1,4-dichlorobenzene by a *Pseudomonas* sp. *Applied and Environmental Microbiology*, 53, 1010–19.

Stanier, R. Y. & Ornston, L. N. (1973). The beta-ketoadipate pathway. *Advances in Microbial Physiology*, 9, 89–151.

Steffan, R. J. & Atlas, R. M. (1988). DNA amplification to enhance detection of genetically engineered bacteria in environmental samples. *Applied and Environmental Microbiology*, 54, 2185–91.

Steffan, R. J. & Tugusheva, M. (1993). Construction and selection of constitutive toluene monoxygenase (TMO) mutants of *Pseudomonas mendocina* KR1. In *Abstracts, 4th International Symposium on Pseudomonas*, abstr. p. 198. Vancouver, BC.

Steffan, R. J., Breen, A., Atlas, R. M. & Sayler, G. S. (1989). Application of gene probe methods for monitoring specific microbial populations in freshwater ecosystems. *Canacian Journal of Microbiology*, 35, 681–5.

Steiert, J. G. & Crawford, R. L. (1986). Catabolism of pentachlorophenol by a *Flavobacterium*. *Biochemical and Biophysical Research Communications*, 141, 825–30.

Stormo, K. E. & Crawford, R. L. (1992). Preparation of encapsulated microbial cells for environmental applications. *Applied and Environmental Microbiology*, 58, 727–30.

Stotzky, G. (1989). Gene transfer among bacteria in soil. In *Gene Transfer in the Environment*, ed. S. B. Levy & R. B. Miller, pp. 165–222. New York: McGraw-Hill.

Taira, K., Hirose, J., Hayashida, S. & Furukawa, K. (1992). Analysis of *bph* operon from the polychlorinated biphenyl-degrading strain of *Pseudomonas pseudoalcaligenes* KF707. *Journal of Biological Chemistry*, 267, 4844–53.

Teplyakov, A. V., van der Laan, J. M., Lammers, A. A., Kelders, H., Kalk, K. H., Misset, O., Mulleners, L. J. & Dijkstra, B. W. (1992). Protein engineering of the high-alkaline serine protease PB92 from *Bacillus alcalophilus*: functional and structural consequences of mutation at the S4 substrate binding pocket. *Protein Engineering*, 5, 413–20.

Tescione, L. & Belfort, G. (1993). Construction and evaluation of a metal ion biosensor. *Biotechnology and Bioengineering*, 42, 945–52.

Thomas, J. M. & Ward, C. H. (1994). Introduced organisms for substrate bioremediation. In *Handbook of Bioremediation*, ed. R. D. Norris, R. E. Hinchee, R. Brown, P. L. McCarty,

L. Semprini, J. T. Wilson, D. H. Kampbell, M. Reinhard, E. J. Bouwer, R. C. Borden, T. M. Vogel, J. M. Thomas & C. H. Ward, pp. 227–44. Boca Raton: Lewis Publishers.

Timmis, K. N., Gonzalez-Carrero, M. I., Sekizaki, T. & Rojo, F. (1986). Biological activities specified by antibiotic resistance plasmids. *Journal of Antimicrobial Chemotherapy*, 18 (Suppl. C), 1–12.

Timmis, K. N., Rojo, F. & Ramos, J. L. (1987). Prospects for laboratory engineering of bacteria to degrade pollutants. In *Environmental Biotechnology. Reducing Risks from Environmental Chemicals through Biotechnology*, ed. G. S. Omen, pp. 61–79. New York: Plenum Press.

Timmis, K. N., Steffan, R. J. & Unterman, R. (1994). Designing microorganisms for the treatment of toxic wastes. *Annual Reviews of Microbiology*, 48, 525–57.

Tsai, Y-L. & Olson, B. H. (1992a). Detection of low numbers of bacterial cells in soils and sediments by polymerase chain reaction. *Applied and Environmental Microbiology*, 58, 754–7.

Tsai, Y-L. & Olson, B. H. (1992b). Rapid method for separation of bacterial DNA from humic substances for polymerase chain reaction. *Applied and Environmental Microbiology*, 58, 2292–5.

Tugusheva, M. & Steffan, R. J. (1993). In vitro construction of constitutive toluene monooxygenase (TMO) mutants of *Pseudomonas mendocina* KR1. In *Abstract, 93rd Annual Meeting of the American Society for Microbiology*, abstr. K-89, p. 276. Washington, DC: The American Society for Microbiology.

Uchiyama, H., Nakajima, T., Yagi, O. & Tabuchi, T. (1992). Role of heterotrophic bacteria in complete mineralization of trichloroethylene by *Methylocystis* sp. strain M. *Applied and Environmental Microbiology*, 58, 3067–71.

van der Meer, J. R., Eggen, R. I., Zehnder, A. J. & de Vos, W. M. (1991a). Sequence analysis of the *Pseudomonas* sp. strain P51 *tcb* gene cluster, which encodes metabolism of chlorinated catechols: evidence for specialization of catechol 1,2-dioxygenases for chlorinated substrates. *Journal of Bacteriology*, 173, 2425–34.

van der Meer, J. R., Zehnder, A. J. & de Vos, W. M. (1991b). Identification of a novel composite transposable element, Tn5280, carrying chlorobenzene dioxygenase genes of *Pseudomonas* sp. strain P51. *Journal of Bacteriology*, 173, 7077–83.

van Elsas, J. D., Wolters, A. C., Clegg, C. D., Lappin-Scott, H. M. & Anderson, J. M. (1994). Fitness of genetically modified *Pseudomonas fluorescens* in competition for soil and root colonization. *FEMS Microbiology Ecology*, 13, 259–72.

Vannelli, T., Logan, M., Arciero, D. M. & Hooper, A. B. (1990). Degradation of halogenated aliphatic compounds by the ammonia-oxidizing bacterium *Nitrosomonas europaea*. *Applied and Environmental Microbiology*, 56, 1169–71.

Wackett, L. P. & Gibson, D. T. (1988). Degradation of trichloroethylene by toluene dioxygenase in whole-cell studies with *Pseudomonas putida* F1. *Applied and Environmental Microbiology*, 54, 1703–8.

Wackett, L. P., Brusseau, G. A., Householder, S. R. & Hanson, R. S. (1989). A survey of microbial oxygenases: trichloroethylene degradation by propane-oxidizing bacteria. *Applied and Environmental Microbiology*, 55, 2960–4.

Wackett, L. P., Sadowsky, M. J., Newman, L. M., Hur, H-G. & Li, S. (1994). Metabolism of polyhalogenated compounds by a genetically engineered bacterium. *Nature*, 368, 627–9.

Wakabayashi, S. & Webster, D. (1986). Primary sequence of a dimeric bacterial hemoglobin from *Vitreoscilla*. *Nature*, 322, 481–3.

Waksman, G., Krishna, T. S., Williams, C. H. & Kuriyan, J. (1994). Crystal structure of

Escherichia coli thioredoxin reductase refined at 2 A resolution. Implication of a large conformational change during catalysis. *Journal of Molecular Biology*, 236, 800–16.

Walker, N. & Harris, D. (1970). Metabolism of 3-chlorobenzoic acid by *Azotobacter* species. *Soil Biology and Biochemistry*, 2, 27–32.

Weightman, A. J., Don, R. H. J., Lehrbach, P. R. & Timmis, K. N. (1984). The identification and cloning of genes encoding haloaromatic catabolic enzymes and the construction of hybrid pathways for substrate mineralization. *Basic Life Sciences*, 28, 47–80.

Wells, J. A., Cunningham, B. C., Graycar, T. P. & Estell, D. A. (1987). Recruitment of substrate-specificity properties from one enzyme into a related one by protein engineering. *Proceedings of the National Academy of Sciences USA*, 84, 5167–71.

West, S. E. H & Iglewski, B. H. (1988). Codon usage in *Pseudomonas aeruginosa*. *Nucleic Acids Research*, 16, 9323–35.

West, S. E. H., Schweizer, H. P., Dall, C., Sample, A. K. & Runyen-Janecky, L. J. (1994). Construction of improved *Escherichia–Pseudomonas* shuttle vectors derived from pUC18/19 and sequence of the region required for their replication in *Pseudomonas aeruginosa*. *Gene*, 128, 81–6.

Wigmore, G. J. & Ribbons, D. W. (1981). Selective enrichment of *Pseudomonas* sp. defective in catabolism after exposure to halogenated substrates. *Journal of Bacteriology*, 146, 920–7.

Winkler, J., Timmis, K. N. & Snyder, R. A. (1995). Tracking the response of Burkholderia cepacia G4 5223-PR1 in aquifer microcosms. *Applied and Environmental Microbiology*, 61, 448–55.

Winter, R. B., Yen, K-M. & Ensley, B. D. (1989). Efficient degradation of trichloroethylene by a recombinant *Escherichia coli*. *Bio/Technology*, 7, 282–5.

Worsey, M. J. & Williams, P. A. (1975). Metabolism of toluene and xylenes by *Pseudomonas putida (arvilla)* mt-2: evidence for a new function of the TOL plasmid. *Journal of Bacteriology*, 124, 7–13.

Yabusaki, Y., Murakami, H., Sakaki, T., Shibata, M. & Ohkawa, H. (1988). Genetically engineered modification of P450 monooxygenases: functional analysis of the amino-terminal hydrophobic region and hinge region of the P450/reductase fused enzyme. *DNA*, 7, 701–11.

Yates, J. R. & Mondello, F. J. (1989). Sequence similarities in the genes encoding polychlorinated biphenyl degradation by *Pseudomonas* strain LB400 and *Alcaligenes eutrophus* H850. *Journal of Bacteriology*, 171, 1733–5.

Yen, K. M. (1991). Construction of cloning cartridges for development of expression vectors in gram-negative bacteria. *Journal of Bacteriology*, 173, 5328–35.

Yen, K. M., Karl, M. R., Blatt, L. M., Simon, M. J., Winter, R. B., Fausset, P. R., Lu, H. S., Harcourt, A. A. & Chen, K. K. (1991). Cloning and characterization of a *Pseudomonas mendocina* KR1 gene cluster encoding toluene-4-monooxygenase. *Journal of Bacteriology*, 173, 5315–27.

Zeyer, J., Lehrbach, P. R. & Timmis, K. N. (1985). Use of cloned genes of Pseudomonas TOL plasmid to effect biotransformation of benzoates to cis-dihydrodiols and catechols by *Escherichia coli* cells. *Applied and Environmental Microbiology*, 50, 1409–13.

Zylstra, G. J. & Gibson, D. T. (1989). Toluene degradation by *Pseudomonas putida* F1: nucleotide sequence of the *tod* C1C2BADE genes and their expression in *Escherichia coli*. *Journal of Biological Chemistry*, 264, 14 940–6.

Index

Index

Index

dredging, 316
Dutch PAH treatment standards, 147–8, 149

E gene, ϕX174, 367, 368
E3 gene, colicin, 367, 368
electron acceptors, 62, 158, 265, 266
electron donors, 265
emulsan, 110–12, 325, 327–8
emulsifiers, *see* bioemulsifiers
encapsulated microbes, 8, 163–5, 273, 283
engineered soil cell, 152, 153, 167
engineering, bioremediation processes, 13–31
Enterobacter cloacae, 201, 363
enterobactin, 324
enzymes
  hybrid, 347–58
  in microbial detection, 374
  rational redesign, 358–61
  recruitment, 351
  subunit mixing, 360–1
Erwinia carotovora, 363
Escherichia coli
  metal interactions, 320, 323
  recombinant, 9, 10, 346, 356, 362, 363, 370
ethylbenzene, 62, 79–83, 98
  see also BTEX hydrocarbons
ethylenediaminetetraacetic acid (EDTA), 316
Europe, PAH treatment standards, 147–8, 149
explosives, *see* munitions compounds
Exxon Valdez oil spill, 101, 114–16

F-1 fertilizer, 117–18
fertilizers, 113–16, 117–18, 162
field application vectors (FAV), 224–5, 361
fimbriae, hydrophobic, 109
fire-retardant fluid, 209
fixed film reactors, 22–3
  with plug flow, 22
  with upflow, 22–3, 269–72
Flavobacterium
  chlorinated phenol degradation, 262, 266, 267, 280, 281, 282
  microencapsulated, 8, 273, 283
  PCP pathway genes, 9, 10
flocculants, 316
fluidized-bed biofilm reactors (FBBR), 269, 270, 307
fluoranthene, 105, 126, 137, 139, 141, 163–5
formaldehyde, 307
formyl chloride, 302
fulvic acid, 37–8
fungi
  aromatic hydrocarbon degradation, 104, 105, 135–43
  chlorinated phenol degradation, 260, 272, 278–9
  inoculation, 167–8

ligninolytic, 135–43, 167–8
  metal interactions, 322, 329
  see also white-rot fungi

gas–water interface, 44
gef gene, 365, 366, 368
GEMs, *see* genetically engineered microorganisms
genetic transfer, 42–3, 226, 345, 367
genetically engineered microorganisms (GEMs), 9, 228, 342–68
  detection in environment, 369–70, 372–3
  environmental release, 375
  genetic transfer from, 42–3, 367
  pathway construction, 342–58
    biochemical background, 342–5
    enzyme subunit mixing, 360–1
    hybrid pathways and enzymes, 347–58
    operon deregulation, 345–7
    rational enzyme redesign, 358–61
    vectors, 347, 348–50
  PCB degradation, 216–17, 224–5, 359–61
  survival, 42, 361–7
    preventing, 365–7
    promoting, 361
Geobacter metallireducens, 74, 75, 76, 77
glutathione reductase (GR), 358–9
groundwater
  bioreactors, 21–5
  bioremediation methods, 6–7, 8
  chlorinated aliphatic compound-contaminated, 266–73, 309
  chlorinated phenol-contaminated, 255, 266–73
  GEM survival, 362–4
  metal removal, 329–30
  PAH bioremediation, 155–6, 157, 172–3
  PAH cleanup standards, 149
  PAH contamination, 133
group contribution approach, 19–20

Haifa beach, 117–18, 162
haloalkanes, 355
hemoglobin, bacterial, 356–8
Henry's Law constant, 15, 18, 19, 236
herbicides, 255, 256
hexachlorobenzene (HCB), 24
hexahydro-1,3,5-trinitro-1,3,5-triazine (RDX), 7, 195–6, 199, 200, 201
hok genes, 365–6, 368
humic acid, 37–8
humic materials, 37–8, 237
  chlorinated phenol complexation, 260
  metal complexation, 318
  PCR inhibition, 371
hybridization techniques, 373–5
hydrogen peroxide, 158, 172
hydrophilins, 109